Brownian Motion and Molecular Reality

OXFORD STUDIES IN PHILOSOPHY OF SCIENCE

Science, Truth, and Democracy
Philip Kitcher

The Devil in the Details: Asymptotic Reasoning in Explanation, Reduction, and Emergence
Robert W. Batterman

Science and Partial Truth: A Unitary Approach to Models and Scientific Reasoning
Newton C. A. da Costa and Steven French

The Book of Evidence
Peter Achinstein

Inventing Temperature: Measurement and Scientific Progress
Hasok Chang

The Reign of Relativity: Philosophy in Physics 1915–1925
Thomas Ryckman

Inconsistency, Asymmetery, and Non-Locality: A Philosophical Investigation of Classical Electrodynamics
Mathias Frisch

Making Things Happen: A Theory of Causal Explanation
James Woodward

Mathematics and Scientific Representation
Christopher Pincock

Simulation and Similarity: Using Models to Understand the World
Michael Weisberg

Systemacity: The Nature of Science
Paul Hoyningen-Huene

Causation and Its Basis in Fundamental Physics
Douglas Kutach

Reconstructing Reality: Models, Mathematics, and Simulations
Margaret Morrison

The Ant Trap: Rebuilding the Foundations of the Social Sciences
Brian Epstein

Understanding Scientific Understanding
Henk de Regt

The Philosophy of Science: A Companion
Anouk Barberousse, Denis Bonnay, and Mikael Cozic

Calculated Surprises: The Philosophy of Computer Simulation
Johannes Lenhard

Chance in the World: A Skeptic's Guide to Objective Chance
Carl Hoefer

Brownian Motion and Molecular Reality: A Study in Theory-Mediated Measurement
George E. Smith and Raghav Seth

Brownian Motion and Molecular Reality

A Study in Theory-Mediated Measurement

GEORGE E. SMITH AND RAGHAV SETH

OXFORD
UNIVERSITY PRESS

OXFORD
UNIVERSITY PRESS

Oxford University Press is a department of the University of Oxford. It furthers the University's objective of excellence in research, scholarship, and education by publishing worldwide. Oxford is a registered trade mark of Oxford University Press in the UK and certain other countries.

Published in the United States of America by Oxford University Press
198 Madison Avenue, New York, NY 10016, United States of America.

Library of Congress Cataloging-in-Publication Data
Names: Smith, George E. (George Edwin), 1938– author. | Seth, Raghav, author. |
Perrin, Jean, 1870–1942. Les atomes.
Title: Brownian motion and molecular reality : a study in theory-mediated measurement /
George E. Smith and Raghav Seth.
Description: New York, NY : Oxford University Press, 2020. |
Series: Oxford studies in philosophy of science |
Includes bibliographical references and index.
Identifiers: LCCN 2020008760 (print) | LCCN 2020008761 (ebook) |
ISBN 9780190098025 (hardback) | ISBN 9780190098049 (epub)
Subjects: LCSH: Brownian movements. | Atomic theory.
Classification: LCC QC184 .S65 2020 (print) | LCC QC184 (ebook) | DDC 530.4/75—dc23
LC record available at https://lccn.loc.gov/2020008760
LC ebook record available at https://lccn.loc.gov/2020008761

Contents

Preface

This study began as an undergraduate honors essay by Seth in 2011, the topic his choice out of a handful I had suggested. He had previously read Perrin's Nobel Prize lecture as a response to Duhem's objections to molecular science and, in reviewing my suggestions, found van Fraassen's "Perils of Perrin, in the Hands of Philosophers" raising issues that he had not considered earlier. I agreed to the topic under the stipulation that he not focus on either the philosophic literature that van Fraassen was challenging or on Perrin's *Atoms* on which that literature has relied so heavily, but instead go back to the original papers appearing in scientific journals by Perrin and those by Einstein, Rutherford, Planck, and Millikan, in particular, cited in his book. I also suggested that to put the issue of molecular reality as of 1905 into historical context, he turn to the original editions of Wilhelm Ostwald's *Outline of General Chemistry* (1889–1890) and the 1899 edition of O. E. Meyer's *The Kinetic Theory of Gases*, the book everyone else was citing back then as *the* comprehensive review of the experimental evidence on the topic.

Seth's essay, *Measurement and the Discreteness of Matter: Evidence for Discreteness and the Extension of Molecular Motion in the Brownian Movement*, was awarded highest honors as he graduated in 2012. It and the research that went into it yielded a number of discoveries that my familiarity with the philosophic literature had given me no way of anticipating: the extraordinary quality of Perrin's theory-mediated measurements of Brownian motion itself, and with them how much he was able to establish about this phenomenon entirely independently of molecular theory; his tendency not to present the experimental results of others with the same exactitude as he presented his own; the no less extraordinary theory-mediated measurements of molecular magnitudes achieved by Rutherford and his collaborators during the same 1908–1913 period, including the first well-founded values for the fundamental charge and the most direct measurement of Avogadro's number through collecting a sufficient count of α-particles to form a gas; the systematic gap between Perrin's values for molecular magnitudes derived from Brownian motion, on the one hand, and, on the other, the values published between 1908 and 1913 for the same magnitudes on the basis of α-particle emissions (by Rutherford and others), blackbody radiation (earlier by Planck, and then by others), and of course ionization (by Millikan); and, finally, the fact that Ostwald's well-known refusal to grant the reality of molecules before 1908 had nothing to do with "metaphysical" scruples, but only with the failure before then to achieve well-founded values for molecular magnitudes, as amply displayed in Meyer's book.

I had been preoccupied with evidence coming out of theory-mediated measurement for some 30 years at the time I posed the topic as an option for Seth. I had initially focused on J. J. Thomson's agreeing measurements of the mass-to-charge ratio of cathode rays and of photoelectric and thermionic emissions and how they had revolutionized our conception of electricity, but had then come to focus on such measurements in Newton's *Principia* and the subsequent history of celestial mechanics. From 1996 until his terminal illness in 2013, I was fortunate to have many hours of discussion with the physicist Kenneth Wilson on the history of the recommended values for the fundamental constants, starting with Raymond Birge's landmark paper of 1929 and continuing through the publication of CODATA 2006. Our focus in these discussions was with the nature of the evidence coming out of increasingly precise determinations, independently of one another, of values for individual constants that, according to theory, are tightly interlinked with one another. Yet I had never found a way to organize what I had learned over these years about theory-mediated measurement and the evidence coming out of it to present to an audience wider than a classroom.

That is, I had never found a way before Seth's essay. By virtue of the approach he had taken, going back to the original papers and then assessing each independent measurement of a molecular magnitude on its own terms, he had co-incidentally revealed how the approach to evidence from measuring interlinked constants that Wilson and I had been discussing first emerged. Equally, provoked by van Fraassen's denial that Perrin had established the existence of molecules, Seth's analyses of what Perrin had in fact measured in Brownian motion independently of molecular theory revealed how those measurements provided an ideal exemplar for explicating theory-mediated measurement and evidence deriving from it. Knowing that Seth was going on to medical school and a career as a physician, I asked if I could join him as a coauthor in transforming his essay into a paper, with increased emphasis on these two themes.

Taking Seth's essay to be the initial version of this book, I drafted the second while a visiting professor in the Philosophy Department of Stanford University during the first seven months of 2013. There I benefited from discussions on the topic of Brownian motion and van Fraassen with Michael Friedman, and once-removed with Bill Demopolous (with whom Michael had already been engaged in continuing discussion on this topic). The new version that emerged, though too long to be a publishable paper, included a reasonably close draft of what have now become Chapters 4 through 6, the core of the book.

On my return to Boston, Seth and I had to decide what to do with this draft, cut it into one or two publishable papers or expand it into a book. Our decision to do the latter was based on two considerations: the draft of the core had the fault of cutting too many corners in an effort to limit its length; and that core would gain enormously from a more expansive discussion of the nineteenth-century developments forming the historical context of the efforts to measure molecular magnitudes between 1905 and 1913. We had both come to realize while I was at Stanford that the many editions between 1893 and 1921 of Walther Nernst's textbook in physical chemistry, *Theoretical Chemistry from the Standpoint of Avogadro's Rule and Thermodynamics*, were at least as important as Ostwald's book in defining the historical context of those efforts between 1905 and 1913; and that book impressed upon us the importance to our topic of the discoveries in the late 1880s which solidified the emerging discipline of physical chemistry, especially the Nobel Prize–winning discoveries on solutions by J. H. van't Hoff and Svante Arrhenius.

So, we decided in 2013 that we would turn the draft into a book and, in order to do so, we would have to conduct a substantial amount of further historical research. Seth took responsibility for the further research on the history of Brownian motion in the nineteenth century, and I took primary responsibility for the history of developments pertaining to molecular reality in the second half of that century. We made a draft of the seven present chapters part of my seminar on "Scientific Realism" in the fall semester of 2017 (with Seth in attendance, in spite of the demands of his final year in medical school). The discussions in that seminar, especially points made by Professors Jody Azzouni and Teru Miyake, led to a number of refinements to the book in the following months, including its "Postscript on the Realism-Instrumentalism Debate." We completed those revisions during the first six months of 2018.

The discoveries that Seth made in 2011–2012, listed in the second paragraph of this Preface, form the backbone of the book. By all rights he should be listed as the first author; it is only at his insistence that he is not. We owe our gratitude to several others beyond those already acknowledged, most notably Amelia Perkins for preparing the Bibliography and putting the notes in their present form, Jody Azzouni for convincing us to submit the manuscript to Oxford University Press, Peter Ohlin of OUP for his encouragement, and three anonymous reviewers for the respect they showed for such an unusual manuscript.

Speaking personally, my contributions to this book would never have been possible without Ken Wilson and Sam Schweber, both of whom passed away while it was underway. I have acknowledged Ken's influence in the preceding. Back in 1996, Sam took it upon himself to make sure that Ken and I got to know one another, thereby initiating our 16 years of one-on-one discussions on an array of topics. After Ken died, Sam continued to be a constant sounding board for me on all matters pertaining to modern physics until he too passed away in the summer of 2017. It is difficult to explain to others how much these two enriched my life. In the deepest gratitude, I dedicate my efforts on this book to the two of them.

—GES, August 2019

Phil 116 on the Philosophy of Science is a course offered at Tufts University that draws an eclectic mix of not just philosophy students, but students from the physics, biology, and the chemical and mechanical engineering departments. When I took this course with Professor Smith in the spring of 2011, the first assignment was to reconstruct the logical design of J. J. Thomson's 1897 measurement of the mass-to-charge ratio of the constituents of cathode rays. The prompt provided the framework for this reconstruction in the form of a conditional followed by a bi-conditional—$p \rightarrow (q \leftrightarrow r)$—to identify background assumptions mediating the law-like relation between the deviation of cathode rays in a magnetic field and the mass-to-charge ratio of the constituents of these emissions. This same logical format is the cornerstone here of Chapter 4.

Working on this book has provided me an extended opportunity to continue learning from Professor Smith. The lengths to which he has gone in identifying lacunae in the evidential reasoning presented in this book at times made me wonder whether he had set out to disprove that Perrin's results on Brownian motion established the particulate nature of matter. Our answer, it turns out, is that Perrin's and others' convergent microphysical measurements provided grounds to take the particulate conception of matter to be true, subject to further testing and revision, as it constitutively entered ongoing research. The book has matured far beyond the content of my senior thesis, and it has been an exciting exploration as much as it has been an education. The lasting insight for me working on this book has been Professor Smith's example of what it takes to undertake a proper investigation.

There are certain insights by Professor Smith that I would be remiss not to mention here. My favorite ones include the principal conclusions reached about Brownian motion independent of molecular-kinetic theory toward the end of Chapter 4, the logical design of Perrin's Brownian motion experiments to get there, and the identification of convergence across phenomena as the only evidence for weak-link premises connecting primary measured quantities—the mean kinetic energy per granule per degree Kelvin in Brownian motion, the fundamental charge in gas ions, the number of α-particles collected in radioactive decay, and the mean kinetic energy per resonator in blackbody radiation—to Avogadro's number for each case in Chapter 6. I am still struck in particular whenever I go through the analysis on continuity with molecules in Chapter 5 that necessitates sustained, rapidly changing, highly localized random fluctuations in the liquid surrounding the granules based on features of Brownian motion established by Perrin's experiments. I hope readers will enjoy these and other insights in the pages ahead.

The opportunity to work on this project materialized for me after I transferred to Tufts as a second-year undergraduate. I was prompted by a growing desire to learn more about philosophy, and, on a personal note, my father Sudhir's constant encouragement was pivotal for me in maintaining my focus to refine my goals and interests. His example of what hard work consists of continues to inspire me daily.

—RS, September 2019

Bibliographical Notice

Even though the period covered by our study ends with the summer of 1913, we have included several quotes from and other citations of Jean Perrin's *Les atomes*. Doing so presented a problem. The book was published in 1913, but by 1914 it was already in its fifth edition, and new editions continued at least through the eleventh in 1921. Two English translations by D. L. Hammick were published with the title *Atoms*, the first of the "fourth revised edition" in 1916 and the second of the "eleventh edition" in 1923. The Ox Bow Press English reprint in 1990 is of the 1923 translation, but nothing in it indicates this, and its frontispiece displays the cover of the original 1913 edition, as if it were a translation of it; worse, some passages in it are not to be found in the French editions dating from 1913 and 1914, potentially creating confusion among those relying on it alone. Our study has relied on the original and the fifth French editions and the two English translations; all our citations are to them. A further complication arose during the course of our efforts when CNRS (Centre national de la recherche scientifique, founded by Perrin) issued a centennial version of the 1913 edition, but with entirely different pagination. Insofar as this version is the most readily available in French now, whenever citing the original 1913 edition we have listed page numbers in it as well. Similarly, because the Ox Bow Press reprint is the most readily available English edition, we have generally listed page numbers in it, but to protect the reader from mistaken impressions, we have added corresponding page numbers in the 1916 edition and, when appropriate, to the 1913 or 1914 French editions too.

Our study has focused more on Perrin's publications before *Les atomes* than on it, but the availability of the Ox Bow Press English reprint and the extent to which the philosophic literature has relied on it has led us to cite it most often in spite of how far removed in time it was from the original. Including parallel citations to the prior editions whenever they include corresponding passages hopefully will safeguard against drawing mistaken conclusions about whether they do or do not date from 1913.

A related problem arose with the publications during the 1900–1908 period of the other key figure on Brownian motion, Albert Einstein. Everything he published pertinent to the topic appeared in those years. Most of it has been available in a single volume in English in a 1956 reprint by Dover of the 1926 translation of the 1922 German volume edited and annotated by R. Fürth. Meanwhile, the original papers in German have appeared in the second volume of *The Collected Papers of Albert Einstein*, with new English translations in an accompanying volume. We relied primarily on the Dover edition while doing our research, and hence our citations are largely of it. We have tried to compensate for this in citations and the bibliography by listing each of his papers pertaining to Brownian motion in its original appearance, adding then its pagination in the German and English *Collected Papers* and the Dover edition.

Three books that we have frequently cited were important to our study precisely because of changes in them over the course of their multiple editions: Wilhelm Ostwald's *Grundriss der allgemeinen Chemie*—in English translations, *Outlines of General Chemistry*; Walther Nernst's *Theoretische Chemie vom Standpunkte der Avogadroschen Regel und Thermodynamik*—in English translations, *Theoretical Chemistry from the Standpoint of Avogadro's Rule and Thermodynamics*; and James Jeans's *The Dynamical Theory of Gases*. Here too our choices on which editions we cite require some comment.

Ostwald's *Grundriss* went through five editions: 1889, 1890, 1899, 1909, and 1917, with English translations of the first (1890, James Walker), second (1895, James Walker), and fourth (1912, W. W. Taylor) of these. We have consulted all of the German editions, though the last only cursorily, but have customarily cited only the 1895 and 1912 translations because of their availability in English. No substantive changes in the text, nor even in the pagination, occurred from the 1889 to the 1890 edition, so that nothing is lost in our relying on the 1895 translation in the case of these two. In our citations of the 1912 translation we have added corresponding page numbers in the 1909 German original whenever they have any potential bearing on the point we are making.

The first edition of Nernst's *Theoretische Chemie* appeared in 1893 and the fifteenth in 1926, with six of them translated into English. We have consulted the 1893, 1898, 1900, 1909, and 1921 editions in German, and English translations of the 1893 edition (1895, Charles Skeele Palmer), the 1903 (fourth) edition (1904, Robert A. Lehfeldt), the 1909 (sixth) edition (1911, H. T. Tizard), and the 1921 ("eighth–tenth") edition (1923, L. W. Codd). Our citations are mostly from the 1895, 1911, and 1923 English editions, occasionally listing as well the corresponding pages in German. Nernst licensed his translator to add a few paragraphs on Perrin's results on Brownian motion that had not been available at the time the 1909 edition went to press; we have expressly marked these, in both our text and citations. Because of its frequent editions and the central position Nernst himself maintained not only in the field of physical chemistry, but after 1900 in physics as well, the changes in his book from one edition to the next provide a chronicle of the changing standing of molecular theory over the three decades they subtend.

Sir James Jeans's *Dynamical Theory* appeared in four editions, in 1904, 1916, 1921, and 1925—the last reissued by Dover in 1954. All of these concern the theory of gases, in most places under the assumption of spherical molecules. Our citations are entirely from the first three editions, always serving to highlight changes in the text reflecting experimental advances, especially those made in the 1905–1913 period.

One further work that we have relied on heavily requires comment, the second (1899) edition of O. E. Meyer's *Die kinetische Theorie der Gase*. Insofar as its first (1877) edition dates from a significantly earlier stage in the development of evidence for kinetic theory than the second, all our citations are to the second edition. An English translation of it was published in 1899 too, within months of the German original. Our citations are mostly to it, but because the translator (Robert E. Baynes) in some places modified equations and converted values into different units, we have in some places cited pages in the original German as well.

—GES, RS

Brownian Motion and Molecular Reality

1
Introduction

1.1. Brownian Motion and Molecular Reality: The "Lore"

Jean Perrin received the Nobel Prize in Physics in 1926 "for his work on the discontinuous structure of matter, and especially for his discovery of sedimentation equilibrium." In his presentation speech on the occasion of Perrin's receiving the prize from the King of Sweden, Carl Wilhelm Oseen said that the "object of the researches" that gained Perrin the prize "was to put a definite end to the long struggle regarding the real existence of molecules;"[1] and Perrin chose as the title for his Nobel Lecture the next day "Discontinuous Structure of Matter,"[2] emphasizing the contribution his research on Brownian motion had made toward that conclusion, though covering other contributions as well. Subsequent "lore" (to use Bas van Fraassen's word for it) has tended to credit Perrin's Brownian motion research as having made *the* decisive step in ending "the long struggle regarding the real existence of molecules." Looking at the situation from the perspective of 1926, however, with so much diverse evidence having accumulated over the preceding 30 years, the Nobel committee chose to be more guarded.

Perrin's Nobel in physics was one of three that the committee awarded in that year to research pertaining to colloid chemistry and Brownian motion. Richard Adolf Zsigmondy was awarded the delayed prize in chemistry for 1925 for "his demonstration of the heterogeneous nature of colloid solutions and for the methods he used, which have since become fundamental in modern colloid chemistry."[3] Those methods included the ultramicroscope which he had developed in 1902 in conjunction with Henry Siedentopf of Carl Zeiss AG—that is, the innovation in microscopy that opened the way to the advances in research on Brownian motion made by Perrin and others in the next few years. And Theodor Svedberg was awarded the 1926 prize in chemistry for his research, contemporaneous with Perrin's, on Brownian motion and his more recent development of the ultracentrifuge, the topic he chose for his Nobel Lecture. In his presentation speech for Svedberg, H. G. Söderbaum cites both him and Perrin for having experimentally confirmed the Einstein-Smoluchowski theory of Brownian motion, and then remarks,

> Should it now be true that the movement of particles suspended in a liquid, which we can observe with the aid of our extremely highly magnifying

Brownian Motion and Molecular Reality. George E. Smith and Raghav Seth, Oxford University Press (2020). © Oxford University Press. DOI: 10.1093/oso/9780190098025.001.0001.

> instruments, can be explained *only* as a result of the movement of molecules beyond the limits of human vision, then this provides visual evidence for the real existence of molecules and consequently for that of atoms, evidence which is all the more remarkable as not so long ago an influential school of scientists declared these particles of matter to be unreal fictions representing an obsolete viewpoint of science.[4] [italics added]

The "influential school of scientists" in question, as Söderbaum went on to make clear, was the "energists," for whom a chief spokesman was Wilhelm Ostwald.

The "lore" about Brownian motion and the reality of molecules does indeed have its historical roots in the reactions of certain major figures in science to Perrin's (and Svedberg's) announced results on Brownian motion at the time of their publication. And, indeed, the most notable among these reactions was Ostwald's complete reversal on the status of the "molecular hypothesis" in the fourth (1909) edition of his textbook, *Grundriss der Physikalische Chemie*:

> I am now convinced that we have recently become possessed of experimental evidence of the discrete or grained nature of matter, which the atomic hypothesis sought in vain for hundreds and thousands of years. The isolation and counting of gas ions, on the one hand, which have crowned with success the long and brilliant researches of J. J. Thomson, and, on the other, the agreement of the Brownian movements with the requirements of the kinetic hypothesis, established by many investigators and most conclusively by J. Perrin, justify the most cautious scientist in now speaking of the experimental proof of the atomic nature of matter. The atomic hypothesis is thus raised to the position of a scientifically well-founded theory, and can claim its place in a text-book intended as an introduction to the present state of our knowledge of General Chemistry.[5]

The earlier editions of the book had employed the "atomic conception" as a "convenient mode of representation" of a wide range of physical and chemical experimental results while underscoring places where it seemed not to work. Elsewhere, as we shall see, Ostwald had been as outspoken as Söderbaum indicates in questioning whether the atomic conception could ever amount to anything more than a convenient mode of "as-if" representation.

Ostwald's about-face on atomism at the time Perrin's Brownian motion results were first surfacing has led to his being the scientist most often cited in claiming that those results were decisive. He was not, however, the only major figure in the world of chemistry to conclude that atomism was coming to be established at the time Perrin was publishing his results. In the sixth (1909) edition of his *Theoretische Chemie: von Standpunkte der Avogadroschen Regel und der*

Thermodynamik, Walther Nernst, who was all along more favorably disposed to the molecular hypothesis than his former mentor Ostwald, lists a series of values of Avogadro's number N_0 obtained via theory-mediated inferences from different phenomena: 100×10^{22} from different approaches within kinetic theory; 83×10^{22} from cloud droplets forming around ions; 62×10^{22} from alpha-particle radiation; and 62×10^{22} from blackbody radiation.[6] On the basis of these numbers he concludes, "In any case we may regard the molecular dimensions as established with a remarkable degree of certainty, so that the atomistic conception begins to lose its hypothetical character."[7]

Nernst also introduced for the first time a short section on Brownian motion in his sixth edition, but with no mention of Perrin's research. That section instead cites Svedberg's (1906) and Seddig's (1908) tests of Einstein's theory of it to conclude,

> It can therefore be taken as proved that the Brownian movement is caused directly by the heat motion of the molecules. From this point of view, we must assume that a particle suspended in a liquid receives impulses on all sides from the rapidly moving molecules of the liquid.[8]

The section ends with a remark to the same effect as the one quoted at the end of our preceding paragraph, though here giving a reason different from converging values of Avogadro's number:

> In the face of these *ocular* confirmations of the kinetic theory of the molecular world, we may well acknowledge that the theory begins to lose its hypothetical character.[9]

The English translator of Nernst's 1909 edition did not alter the section on Brownian motion, but with the author's permission did add a paragraph on Perrin's determination of Avogadro's number just before the list of values given in the preceding and added that value, 71.0×10^{22}, to the list, while leaving the final sentence in the chapter on molecular sizes—". . . begins to lose its hypothetical character"—intact.[10]

In later editions Nernst still did not mention Perrin in his section on Brownian motion, though he changed the phrasing of the final sentence to read ". . . we may well acknowledge that the theory has lost its hypothetical character."[11] The chapter on molecular sizes, however, was thoroughly recast, with a two-page subsection on the values of N_0 obtained from different aspects of Brownian motion by both Perrin (from 1909 and 1910) and Svedberg (from 1911). While noting that Svedberg's values for N_0 using the same approaches as Perrin were more than 10 percent lower, Nernst concluded,

> The relatively good agreement of the values of N_0 obtained by these various methods from experiments with suspensions is an important fresh proof of the correctness of the theory of thermal motion; the accuracy of the results obtained by Perrin and Svedberg in the work mentioned above is very remarkable, since it would appear very difficult to obtain precise measurements of N_0 by such methods.[12]

Nernst's overall conclusion in the chapter is that "the problem of measuring accurately the size of the individual atoms has been solved; *a fortiori*, no doubt can remain as to their actual existence."[13] His preferred value for N_0, 6.064×10^{23}, however, is taken from a Millikan article of 1917, not from the Svedberg and Perrin results for Brownian motion, for which he cites a mean value of 6.4×10^{23}, "with a probable error of a few percent."[14]

Ostwald, in his mid-fifties when he did his about-face on atomism, and his former assistant Nernst, 10 years younger, had for some time been *the* authoritative voices in physical chemistry.[15] That the mathematician and physicist Henri Poincaré was no less authoritative at the time is why he was asked in 1912 to close a series of lectures organized by the French Society of Physics. In it, he too proclaimed, "The atom of the chemist is now a reality," following a review of what he took to be the key evidence:

> The kinetic theory of gases has acquired, so to speak, unexpected props. New arrivals have modeled themselves upon it exactly; these are on one side the theory of solutions and on the other the electronic theory of metals. The molecules of the solute, as well as the free electrons to which metals owe their electrical conductibility, behave like gas molecules in the enclosed spaces in which they are contained. The parallelism is perfect and can be pursued even to numerical coincidences. In that way, what was doubtful becomes probable; each of these three theories, if it were isolated, would seem to be merely an ingenious hypothesis, for which it would be possible to substitute other explanations nearly as plausible. But, as in each of the three cases a different explanation would be necessary, the coincidences which have been observed could no longer be attributed to chance, which is inadmissible, whereas the three kinetic theories make these coincidences necessary. Besides, the theory of solutions leads us very naturally to that of the Brownian movement, in which it is impossible to consider the thermal disturbance as a mere figment of the imagination, since it can be seen directly under the microscope.
>
> The brilliant determinations of the number of atoms computed by Mr. Perrin have completed the triumph of atomism. What makes it all the more convincing are the multiple correspondences between results obtained by entirely different processes. Not too long ago, we would have considered ourselves

> fortunate if the numbers thus derived had contained the same number of digits. We would not even have required that the first significant figure be the same; this first figure is now determined; and what is remarkable is that the most diverse properties of the atom have been considered.[16]

Poincaré then points out that this closing of one question has led to a host of new questions insofar as "each new discovery in physics reveals a new complexity of the atom."[17] We will return to this point later.

From our vantage point now, after a century of accumulating evidence on the details of molecules and atoms, these pronouncements seem to reflect the "good sense" of the scientific community at the time: *exercise caution toward theories that explain phenomena by positing hypothetical entities, but only until truly decisive evidence emerges warranting their acceptance.* The intervening century of evidence may be distorting our judgment, however. Consider the other major hypothetical entity posited during the nineteenth century, the ether—that is, the medium in which the transverse waves constituting light propagate. Mary Somerville summarizes the situation in the 1830s as the last opposition to the transverse wave theory of light was dying out:

> Now it is contrary to all our ideas of matter to suppose that two particles of it should annihilate one another under any circumstances whatever; while on the contrary, two opposing motions may, and it is impossible not to be struck with the perfect similarity between the interferences of small undulations of air or of water and the preceding phenomena [of optics]. The analogy is indeed so perfect, that philosophers of the highest authority concur in the supposition that the celestial regions are filled with an extremely rare, imponderable, and highly elastic medium or ether, whose particles are capable of receiving the vibrations communicated to them by self-luminous bodies, and of transmitting them to the optic nerves, so as to produce the sensation of light. The acceleration in the mean motion of Encke's comet, as well as of the comet discovered by M. Biela, renders the existence of such a medium almost certain.[18]

Nearly four decades later, Maxwell was no less adamant about the reality of the ether. After reviewing the unsuccessful attempts to empirically establish anything specific concerning the ether, he nevertheless concluded:

> Whatever difficulties we may have in forming a consistent idea of the constitution of the aether, there can be no doubt that the interplanetary and interstellar spaces are not empty, but are occupied by a material substance or body, which is the largest, and probably most uniform body of which we have any knowledge.[19]

At the very least we might ask how the analogy between Brownian motion and molecular motion comprised more definitive evidence in 1910, or so, than the analogy between the many aspects of wave motion in the media of air and water, most notably the cyclic exchange between potential and kinetic energy, and corresponding aspects of wave motion of light in the postulated medium labeled the ether during the nineteenth century.

Some have challenged whether it did comprise more definitive evidence. Consider, for example, the following assessment of the evidence by Stephen Brush, the foremost historian of kinetic theory:

> For the next few years, he [Perrin] seems to have devoted much of his time to popularizing the significance of his work on Brownian motion, in particular the idea that atoms have now been proved to exist. In this he was surprisingly successful. In fact, the willingness of scientists to believe in the "reality" of atoms after 1908, in contrast to previous insistence on their "hypothetical" character, is quite amazing.
>
> The evidence provided by the Brownian-movement experiments of Perrin and others seems rather flimsy, compared to what was already available from other sources. The fact that one could determine Avogadro's number and the charge on the electron by one more method seems hardly sufficient to justify such profound metaphysical conclusions. Several independent methods of determining these parameters had been known since 1870 or before, to say nothing of the many successes of kinetic theory in predicting the properties of gases.[20]

Writing at almost the same time as Brush, Eyvind Wichmann, in the Introduction to his textbook *Quantum Physics*, does not mention Brownian motion at all in dismissing the resistance to accepting the reality of atoms and molecules that it overcame:

> the reader is urged to ponder the remarkable understanding of natural phenomena which was achieved during the nineteenth century on the hypothesis that matter is made of atoms. On this assumption we can understand the basic fact of chemistry, namely that a given chemical compound always consists of certain basic chemical elements in fixed, definite proportions, characteristic of the compound. . . . As further evidence in favor of the atomic hypothesis we point to the success of the *kinetic theory of gases*, developed in particular by J. C. Maxwell and L. Boltzmann. This theory could explain many properties of gases on the hypothesis that a gas in a container is a swarm of molecules moving randomly inside the container, incessantly colliding with each other and with the

> walls of the container. The kinetic theory could furthermore be used to estimate Avogadro's number, $N_0 = 6.02 \times 10^{23}$. . . .
>
> In view of such evidence for the existence of atoms it is hard to understand a certain school of thought, which persisted until the turn of the century, and which rejected the atomic hypothesis on the grounds there was no *direct* (!) evidence that matter is made of atoms.[21]

In other words, Brush questions what contribution Brownian motion truly made at the time, and Wichmann questions whether there was then any need whatever for Brownian motion to contribute something beyond what was already available.

One question prompted by these two retrospective views is: What did Ostwald, Nernst, and Poincaré think was still needed at the beginning of the twentieth century in order for the molecular hypothesis, in Ostwald's words, to be "raised to the position of a scientifically well-founded theory," or, in Nernst's, to "lose its hypothetical character"? In other words, what shortcoming did they see in the evidence at the time when Perrin and Einstein began their efforts on Brownian motion? A way to get at this question is to ask just what Ostwald, Nernst, and Poincaré and their like took the new standing of the molecular hypothesis to amount to when it lost its hypothetical character and became a scientifically well-founded theory. The answer to this question is less obvious than one might think. Granted, authorities were telling the scientific community that the molecular hypothesis should now be taken to be true, and hence criticizing it and proposing alternatives to it were no longer in order. Surely, however, it retained a provisional status at least insofar as it remained subject to reconsideration in the light of new evidence in the future. But then what difference did its new standing make to the science itself?

(Lest readers be misled by the quotations from Wichmann and, perhaps less so, Brush, we should pause here to emphasize that molecular-kinetic theory, for all its explanatory accomplishments, still had no solution for the problem Maxwell had singled out from the outset in 1860: the inability to reconcile the experimental values for the specific heats of gases with a fundamental principle of the theory, the equipartition of energy across all degrees of freedom of molecules. We will present this problem properly in Chapter 2. Suffice it to say here that by 1901 it had become notorious, if only because of Lord Rayleigh's "The Law of Partition of Kinetic Energy" of 1900 and Lord Kelvin's response in "Nineteenth Century Clouds over the Dynamical Theory of Heat and Light" of 1901.[22] Wichmann notwithstanding, the intractability of this problem alone gave those at the time who continued to view molecules as merely hypothetical solid empirical grounds for doing so, quite independently of any positivistic philosophical scruples.)

1.2. Two Issues

The earlier editions of Ostwald's *Grundriss der Physikalische Chemie* and Nernst's *Theoretische Chemie* had both throughout presented chemistry and pneumatic physics from the point of view of the atomic-molecular hypothesis. Ostwald, in more than one place, had included (in a contrasting font) cautionary notes like the following:

> Such a broad and far-reaching agreement of the empirically determined abstract laws with the atomic hypothesis formed to explain the cause of these laws justifies us in expecting still further concordances between this hypothesis and experience. In fact, all our chemical experience harmonises with the atomic theory and finds in it an easy mode of expression, so that in what follows we shall always employ it. But once for all be it understood that the atomic hypothesis is only a mode of picturing to ourselves what we know of the behavior of substances. What the "real" nature of matter is, is to us a matter of complete ignorance, as it is of complete indifference.[23]

The 1909 edition drops all admonitions of this sort and adds material on such newly developing topics as Brownian motion, electrons, and radioactivity. To a remarkable extent, however, the material on all other topics, including the kinetic theory itself, changes little from the earlier edition. So, it isn't as if chemistry or pneumatic physics underwent some sort of transformation with the new standing of molecular theory.

If we are to understand what contribution Brownian motion made, we had best be clear about what it was that it made a contribution to. This is one of the two principal issues of this monograph: ***What did the new standing of atomic-molecular theory at the end of the first decade of the twentieth century amount to***? How did putting "a definite end to the long struggle regarding the real existence of molecules" affect science?

It is scarcely enough to say that Ostwald and others who had agreed with him had finally come to *believe* in the reality of atoms and molecules—in words uncomfortably reminiscent of religious conversion. Lots of chemists and physicists had for some time apparently believed in the reality of atoms and molecules, at least insofar as they presupposed them in their research. So, what difference did still others coming to believe as they had make to the science? Indeed, what difference do the beliefs of individual scientists generally make to their science? Many scientists had believed in the reality of the ether during the second half of the nineteenth century. Maxwell's two-volume *Treatise on Electricity and Magnetism* had throughout presented the phenomena of electricity and magnetism from the point of view of the ether hypothesis. Nevertheless, the denial of the reality of the ether three or so decades after his *Treatise* was published made virtually no difference to his dynamical theory of the electromagnetic field. Was more at stake in coming to believe in the reality of atoms and molecules?

One way to get at what the new standing of the molecular hypothesis amounted to for Ostwald, Nernst, and Poincaré is to consider the differences they saw the Brownian motion results making. They single out three aspects of the results. First, Brownian motion itself—that is, the motion of discrete granular particles in liquids—is a visible fact under microscopes. Second, contrary to a widespread view before 1900, the observed motions of the granules are in fact entirely consistent with kinetic theory. Third, the values for Avogadro's number that Perrin inferred from different sorts of measurements he made of Brownian motion showed notably less variance than the values that had been estimated from kinetic theory itself, and they agreed well with values inferred from phenomena independent of kinetic theory during the first dozen years of the twentieth century.

This last point suggests that Brownian motion was providing experimental access to the atomic-molecular realm insofar as having a reliable value for Avogadro's number meant that more reliable values for several other parameters of atomic-molecular theory could then be obtained. That suggestion is wrong, however. What Perrin determined through his theory-mediated measurements was in every case a mean kinetic energy—translational or rotational—of his granules. He could then calculate how many granules having such a mean kinetic energy are needed to match the total energy, *RT*, in a mole of gas at the corresponding temperature. *This number, however, is not in and of itself Avogadro's number. It can be taken as a measure of Avogadro's number only under the further assumption that molecules exist and have the same mean translational kinetic energy at the temperature in question as the Brownian motion granules have.*

In other words, a gap remained between the measurements made for Brownian motion and the value then claimed for the molecular parameter. A similar gap remained between the observable motions of the granules versus unobservable molecular motion, as well as between the consistency of the observed motions with kinetic theory versus the consistency of kinetic theory with what actually occurs microphysically. What was it about the specifics of the measured results for Brownian motion that licensed inferences bridging these gaps?

Various commentators on "scientific realism" among philosophers of science have weighed in on this question over the last few decades: Clark Glymour,[24] Nancy Cartwright,[25] Wesley Salmon,[26] Deborah Mayo,[27] Peter Achinstein,[28] Penelope Maddy,[29] and more recently, Stathis Psillos[30] and Alan Chalmers.[31] The issue addressed by all of them, in contrast to the issue we posed earlier, concerns how the results from Brownian motion gave scientists warrant to believe in the reality of atoms and molecules. With the notable exceptions of Mayo and, following her, Chalmers, all of them focused primarily on the values Perrin derived for Avogadro's number and the general agreement between those values and values that were being derived from phenomena other than Brownian motion. The philosophical problem, as they see it, is to identify the principle of scientific reasoning that allows the results on the motions of visible particles to confirm the existence of hypothetical particles. That is, they take for granted that the historical inference from the announced findings concerning Brownian motion to the reality of molecules was valid. Their task is to enunciate the principle licensing this inference.

Interestingly, the eight of them have not reached accord on what that principle is. This may reflect nothing more than the fact that they are doing philosophy, that is, trying to formulate with care a principle that not only applies instructively to the case of Perrin, but that holds generally of evidence in science, especially evidence bearing on the existence of theoretically posited entities. Scientists have the advantage when drawing inferences from evidence of not having to enunciate the precise general principles licensing the inferences in a philosophically unobjectionable form.

Bas van Fraassen has offered a very different view of Perrin's achievement in his "The Perils of Perrin, in the Hands of Philosophers"[32]—a view that is remarkably close, at least at first glance, to claims that Ostwald himself expressed in print on the question of atomism before and even after his "conversion." Van Fraassen argues that philosophers have misrepresented the guarded attitude of scientists toward atomism before Perrin. Their worry was not over the reality of hypothesized molecules, but over the absence of adequate *empirical grounding* of kinetic theory—that is, the absence of "theoretical connections" between the significant parameters of kinetic theory and parameters whose values could be measured in experiments to allow adequately precise values of the former to be empirically determined. What Perrin did was to employ theoretical relationships connecting two key parameters of kinetic theory, *Avogadro's number* and *mean kinetic energy* of molecules, to Brownian motion parameters that could be measured in his extraordinary experiments. The theoretical relationships in question were obtained by extending (van Fraassen's word is "enriching") kinetic theory with further assumptions to the effect that "visible particles comprising a dilute emulsion will behave like molecules in a gas" in certain respects.[33] As Perrin emphasized, the values of those two key parameters then allowed values of other parameters of kinetic theory, as well as various molecular magnitudes, to be determined.[34]

Ostwald said that the atomic hypothesis had been "raised to the position of a scientifically well-founded theory." Quite rightly, van Fraassen insists that a theory is not scientifically well founded in the eyes of scientists unless it is empirically well grounded. This was the transformation, he contends, that we should see Perrin's results for Brownian motion as having effected at the time.

Van Fraassen emphasizes more than once that Perrin's measurements were "relative to kinetic theory." In particular, "Perrin begins his theoretical work in a context where the postulate of atomic structure is taken for granted."[35] In other words, the indirect, theory-mediated measurements through which Perrin empirically grounded kinetic theory presupposed kinetic theory and hence could not have provided evidence for the reality of molecules independently of the molecular hypothesis. That nevertheless, van Fraassen says, does not lessen the evidential significance of Perrin's success in finding ways for Brownian motion to ground kinetic theory.

Van Fraassen gives three requirements for a theory to be grounded, two of them taken from Herman Weyl, and the third adapted from Clark Glymour.[36] First, conditions must be available in which definite values of its theoretically significant parameters can be determined by means of measurement; usually this involves indirect measures that presuppose "theoretically posited connections" between the parameters and other parameters amenable to more direct measurement. Perrin obtained values for Avogadro's number based on four different such connections involving three different Brownian motion phenomena; in each case his results in fact had more claim to having determined a definite value for it than any of the prior estimates obtained on the basis of kinetic theory. Second, measured values for any parameter obtained through different theoretically posited connections must agree with one another, at least to within the limits of experimental precision. Perrin's three different phenomena yielded values for Avogadro's number in impressively good agreement with one another, especially when compared with the lack of agreement among the values obtained from alternative approaches based on kinetic theory prior to him. Third, there must be an alternative possible outcome for the same measurements to have refuted the theory on the basis of the same theoretically posited connections.[37] As we shall see in the following, all four of Perrin's approaches might have yielded results thoroughly in conflict with kinetic theory. These three requirements, taken together, make grounding anything but trifling. In meeting them, Perrin's results really did put kinetic theory on a different standing, even though the values he obtained for Avogadro's number may have only been "relative to" that theory.

The present study of Brownian motion began in response to van Fraassen. One reason for focusing on him and saying little about the rest of the philosophical literature is that, instead of worrying about what scientists believed about molecules, he has offered an answer to the issue we posed earlier: *What did the new standing of atomic-molecular theory at the end of the first decade of the twentieth century amount to*? His answer, moreover—that this theory finally became empirically well grounded—is at the very least a *prima facie* plausible minimal answer. As such, it puts the burden of argument on anyone who contends that there is more to the answer than he is giving, in particular more pertaining to the reality of molecules. We claim to have met that burden, but by taking an approach different from any in the existing literature.

Save for the more recent articles by Psillos and Chalmers, all the "realist" literature we cited preceded van Fraassen's account of grounding and hence had no reason to try to meet this burden of argument. Still, our most important reason for focusing on van Fraassen is that his answer centers on theory-mediated measurement and the conditions under which it can produce strong evidence. This is what our analysis of Perrin's Brownian motion measurements will be preoccupied with in the following. Our view is that the burden of argument van Fraassen has imposed can be met only through a careful assessment of how different elements of theory entered into those measurements.

Given what we said earlier about a gap between the measurements Perrin made for Brownian motion and the values he then claimed for Avogadro's number, we are not simply going to reject van Fraassen's claim that those values were "relative to" kinetic theory. But exactly where and how kinetic theory entered into Perrin's measurements is not so straightforward. Perrin employed four theoretical relationships that licensed inferences from values of parameters that he could measure comparatively directly in his experiments to values for the mean kinetic energy of the granules in Brownian motion and then, with a further step, to values of Avogadro's number. One of these relationships he seems to have derived before he became aware of Einstein's efforts on Brownian motion, and the other three came from Einstein. Einstein specifically derived his relationships from molecular-kinetic theory. Perrin is less clear whether he is deriving his relationship from kinetic theory or just appealing to kinetic theory in motivating them. Either way, van Fraassen can make a *prima facie* case that kinetic theory was being extended to cover Brownian motion. Whether the measurements mediated by those relationships would turn out to be well-behaved might, from this point of view, then be claimed to provide a hypothetico-deductive test of kinetic theory, perhaps even a test stringent enough to have strongly supported the reality of molecules.

In fact, however, the theoretical relationships involving the mean kinetic energy of the granules can *all* be derived entirely independently of *molecular*-kinetic theory and hence independently of any question of the reality of molecules. The measurements were still mediated by those relationships, but they were theory-mediated measurements of aspects of Brownian motion taken unto itself, measurements that yielded results that at the time had to be (and in fact were) regarded as extraordinary in their own right. Viewing Perrin's efforts this way still leaves us with the question of how the gap between the visible granules and invisible molecules is to be bridged, but it clarifies just what that gap amounts to, and it opens up possibilities for bridging it beyond those considered by van Fraassen.

This brings us to the second principal issue of this monograph: ***What did the research by Perrin and others establish about Brownian motion* independently *of the molecular hypothesis and kinetic theory?*** The answer to this, we claim, will put us in a better position to answer our other principal issue, and with it the question of whether the contribution made by the research to the new standing of atomic-molecular theory involved anything more than that theory's becoming, in van Fraassen's sense, "empirically grounded."

We said earlier that our study of Brownian motion began in response to van Fraassen. We did not foresee that careful analysis of Perrin's efforts would add an entirely new dimension to it—for his determinations of the mean translational kinetic energy of granules in Brownian motion turned out to be an exceptionally instructive exemplar of theory-mediated measurement. He, like Einstein, realized from the outset that the motions of the granules seen through a microscope are not at all the actual motions, and consequently any determinations of their velocities and hence their mean translational kinetic energies would have to come from such measurement. He introduced one, and Einstein two other relationships linking their mean translational kinetic energy to quantities that could be determined from observation, namely the vertical concentration gradient of the granules, their mean square displacements over intervals of time, and their rate of diffusion in solutions.[38] Perrin then pursued complementary experimental results to cross-check the assumptions underlying these three relationships, thereby freeing them from depending on molecular theory. As a consequence he was able not only to determine values for the mean translational kinetic energies of the granules independently of molecular theory, but also to establish experimentally several other conclusions about Brownian motion, including that these energies vary with temperature, but not with such other factors as granule size and mass or the liquid substrate in which they move.

Our analysis of Perrin's efforts in Chapter 4 will accordingly at the same time be an analysis of the requirements theory-mediated measurement needs to meet in order to yield results that have a full standing on their own. It will also make clear how Brownian motion in itself did not provide experimental support for the further presupposition Perrin adopted in order to obtain his values for Avogadro's number from his kinetic energies, and hence did not provide these values with such a standing. Evidence supporting his further presupposition had to come from complementary theory-mediated determinations of Avogadro's number that did not depend on it. We shall examine the complementary determinations to which Perrin appealed for this purpose in Chapter 6, in the process extending our analysis of theory-mediated measurement beyond that in Chapter 4. This, we hope, explains our subtitle, for our monograph is as much a study of theory-mediated measurement during the period in question as it is of Brownian motion in relation to molecular reality.

1.3. Challenges to Our Second Issue

Asking what the research by Perrin and others established about Brownian motion *independently* of the molecular hypothesis and kinetic theory is likely to seem so wrong-headed to many of our readers for them to see little reason to continue. Wasn't the whole point of the research on Brownian motion in the first decade of the twentieth century to *test* kinetic theory? Consider Einstein, in particular. In his initial paper on the subject in 1905 he derives a prediction from the "molecular-kinetic theory of heat" that "bodies of microscopically-visible size suspended in a liquid will perform movements of such magnitude that they can easily be observed in a microscope." After acknowledging that he does not have enough information to say whether "the movements to be discussed here are identical with the so-called 'Brownian molecular motion,'" he states the point of his derivation:

> If the movement discussed here can actually be observed (together with the laws relating it that one would expect to find), then classical thermodynamics can no longer be looked upon as applicable with precision to bodies even of dimensions distinguishable in a microscope: an exact determination of actual atomic dimensions is then possible. On the other hand, had the prediction of this movement proved to be incorrect, a weighty argument would be provided against the molecular-kinetic theory of heat.[39]

So, shouldn't the principal issue be, *how stringent of a test did the confirmation of Einstein's predictions (and various related results) provide of the molecular-kinetic theory of heat?* And this can scarcely be answered by focusing on what aspects of Brownian motion were established independently of the molecular hypothesis and kinetic theory.

We could, of course, postpone this and the other challenges we consider in this section until the end of our monograph. Our main reason for not doing so is that Alan Chalmers and Stathis Psillos have published responses to van Fraassen's article in which they clearly do not consider our approach of focusing first on Perrin's results for Brownian motion independently of molecular-kinetic theory. Both of them insist that van Fraassen's mistake is not to regard those results as first and foremost tests of that theory. Moreover, in arguing this, both suggest a couple of other respects in which Perrin's results can be recognized for what they are only by taking into consideration from the outset their relationship to kinetic theory. Our point here is that this relationship is not nearly so straightforward as Chalmers and Psillos take it to be, and hence the issue of just what Perrin's results showed about Brownian motion needs to be addressed first if that relationship is to be understood.

Turning first to Chalmers, while he insists that Perrin's results should be viewed as a test of molecular-kinetic theory, he is also careful to point out that any such test can only be of some "partition" of the theory and not the whole theory. He then argues that what Perrin's efforts did were to test "the central and most basic assumptions of the kinetic theory," namely the random character of molecular motion and the equipartition of energy among the degrees of freedom of the molecules.[40] Chalmers grants that Perrin's measurements of the mean kinetic energy of the granules in Brownian motion at most required only the assumption that these two features hold for the granules; and in fact the measurements did not require these two to be assumed at all insofar as Perrin provided clear evidence independent of molecular-kinetic theory that both in fact hold for the granules. So, as Chalmers concedes in his book, "the experimental results against which Perrin tested the predictions of the kinetic theory were established independently of the theory itself."[41] We fully agree that Perrin's results showed that the Brownian motion of granules exhibits the two central features attributed to molecules by nineteenth-century kinetic theory. But precisely because those results in no way presupposed that theory, we do not see how it is possible to assess the *stringency* with which they tested its "central and most basic assumptions." This explains why we think the issue of the stringency of any such test should not be a principal issue, but instead should be postponed until at least the second of our principal issues, if not the first as well, has been fully addressed.

In fairness to Chalmers, he recognizes our concern and consequently states with care the sense in which Perrin's results comprised a test: "The kinetic theory predicts a range of phenomena associated with Brownian motion borne out by experiment. To this extent, we have a powerful argument from coincidence in favour of the theory,"[42] that is, "an argument to the effect that it would be a coincidence were the range of evidence to be the case and the theory false nonetheless."[43] Such an appeal to coincidence sounds to us as a variant of inference to the best explanation, a topic to which we shall turn in a couple of paragraphs. What we do not see is how it helps toward assessing the stringency with which Perrin's results comprised a test.[44] The conclusions we shall ultimately put forward in this monograph are in a great many respects close to those Chalmers reaches in both his reply to van Fraassen and his book. They will not, however, be based on any claim to the effect that Perrin's results comprised an especially stringent test of claims derived from molecular-kinetic theory, or even from the two central assumptions of that theory.

A different way in which Perrin's results have been construed as a decisive test of molecular-kinetic theory, and an argument from coincidence as well, invokes the agreement between his four determinations of Avogadro's number and the various determinations of it independently of Brownian motion. For example, Stathis Psillos, in his reply to van Fraassen, claims that "the role of Perrin's work was to show that a certain way to calculate *N* (based on a certain theoretical prediction of it) could provide a *decisive* test in favour of the atomic conception"[45]—a test that he subsequently says consisted of a comparison of the prediction "with already known calculations of *N* based on kinetic theory."[46] In much the same vein, Chalmers, in response to van Fraassen's claim about Perrin's results having grounded kinetic theory, says, "I suggest it is more appropriate to see Perrin's determinations of *N* as a means to an end, namely testing the claims of kinetic theory. It was to this end that Perrin dramatized the significance of the agreement of his first measurement of *N*, via the density distribution, with previous estimates of that number."[47] The "previous estimate" to which Chalmers is referring here is, however, to a value that Perrin himself had calculated for argon from van der Waals formula and then published in his monograph of 1909.[48] This was scarcely an established value from the prior literature against which Perrin could have been testing his determinations.

As we shall subsequently see, between 1908 and 1912 roughly 10 determinations of Avogadro's number were published that had no tie to Brownian motion and were close enough to Perrin's four determinations to constitute, at least *prima facie*, the notable agreement which so many philosophers like Psillos and Chalmers have stressed. Our Chapter 6 will be devoted to an analysis of just what this agreement was evidence for. None of these 10, however, was based on kinetic theory, and none of them predates Perrin's first published determination of 1908. Table 1.1 presents a listing of prominent values for both Loschmidt's (the number of molecules in a cubic centimeter of gas at standard conditions) and Avogadro's number in the literature prior to 1905, some of them based on kinetic theory, and others not. These numbers differ far too much from one another to support any suggestion that a comparison of Perrin's determinations of *N* with them constituted a test of molecular-kinetic theory. In fact—and this is important historically—Perrin's results have claim both to being seen as and to being the first experimentally well-founded values for *N*. We will return to this point at length in Chapter 6.

We shall explain in Chapter 2 the bases for the different values of Avogadro's number in Table 1.1. Planck's 1900 value, however, merits an aside here because it underscores the want of a reliable value for Avogadro's number before 1908. Planck's value is the only one of those listed in Table 1.1 that lies within 20 percent of the values Perrin published in 1908 and 1909. (Unlike any of Perrin's values, it happens also to have turned out to be the first experimentally derived value within 3 percent of our current value, $60.2214076 \times 10^{22}$.) Planck obtained

Table 1.1 Typical Published Values of Loschmidt's and Avogadro's Numbers before 1905

Loschmidt's number	Avogadro's number	Author	Year	Source
[2.0×10^{18}]	[4.5×10^{22}]	Loschmidt	1865	From viscosity, mean free path: size
$\leq 6 \times 10^{21}$	[$\leq 1.3 \times 10^{26}$]	W. Thomson	1870	From four estimates of size
19×10^{18}	[4.3×10^{23}]	Maxwell	1873	Following Loschmidt
5×10^{19}	[11×10^{23}]	van der Walls	1873	From mean free path and his coef. *B*
5.4×10^{19}	[12×10^{23}]	Nernst	1893	Following Maxwell
10^{20}	[22×10^{23}]	Richarz	1894	From kinetic theory, as per Drude 1900
20×10^{18}	[4.5×10^{23}]	J. J. Thomson	1898	From ion $e = 6.5 \times 10^{-10}$ esu (and Faraday's law)
21×10^{18}	[4.7×10^{23}]	J. J. Thomson	1898	Quoted w/o citation, from "viscosity"
61×10^{18}	[13.7×10^{23}]	O. E. Meyer	1899	From mean free path, following van der Waals
2.1×10^{19}	[4.7×10^{23}]	Drude	1900	From conduction in solids
2.76×10^{19}	61.75×10^{22}	Planck	1900	From black-body radiation *k*
3.9×10^{19}	[87×10^{22}]	Wilson, J. J. Thomson	1903	From $e = 3.1 \times 10^{-10}$ esu (water drop)

As of 2018, Loschmidt's number = $2.686780111 \times 10^{19}$, Avogadro's = $60.2214076 \times 10^{22}$, and $e = 4.80320471 \times 10^{-10}$ esu. Numbers enclosed in brackets were not stated in the publications cited, but were instead calculated here from values stated in them.

his value from his equation for the energy distribution in blackbody radiation as a function of frequency and temperature:

$$u = \frac{8\pi h \upsilon^3}{c^3} \frac{1}{e^{h\upsilon/kT} - 1}$$

where his values of the two constants, *h* and *k*, were determined from experimental observations of the variation of *u* with *υ* at different temperatures. In the second of his papers on the topic in 1900, he offered a theoretical derivation of this equation based on Boltzmann's theory of entropy, a derivation in which the constant *k* was interpreted as the energy per molecule per degree of temperature—that is, what we now call Boltzmann's constant $k = R/N$, the molar gas constant divided by Avogadro's number. Then, as a "further test of the reliability" of his derivation, he calculated from his value for k the values of Loschmidt's and Avogadro's numbers listed in Table 1.1, as well as a value for the "quantum of electricity e" (4.69×10^{-10} esu), comparing each to a value from the literature (e.g., Thomson's 1898 value, 6.5×10^{-10} esu).[49] The paper then concludes,

> If the theory is at all correct, all these relations should be not approximately, but absolutely, valid. The accuracy of the calculated number [for e] is thus essentially the same as that of the relatively worst known, the radiation constant k, and is thus much better than all determinations up to now. To test it by more direct methods should be both an important and a necessary task for further research.[50]

In other words, Planck was calling for more reliable independent values for N and e than were available as of 1900 to confirm his *derivation* of his blackbody law, specifically for its claim that the coefficient of T in it is in fact Boltzmann's constant in statistical thermodynamics.

This is ironic in more respects than one. A dozen years later Planck cited the good agreement between the value for k measured from blackbody radiation and values for N and e obtained between 1908 and 1912 as compelling evidence for a key claim in his 1900 derivation, though not for the full derivation itself:

> Probably the most direct support for the fundamental idea of the hypothesis of quanta is supplied by the values of the elementary quanta of matter and electricity derived from it [that is, my radiation formula]. When twelve years ago, I made my first calculation of the value of the elementary electric charge and found it to be 4.69×10^{-10} electrostatic units, the value of this quantity deduced by *J. J. Thomson* from his ingenious experiments on the condensation of water vapor on gas ions, namely 6.5×10^{-10} was quite generally regarded as the most reliable value. This value exceeds the one given by me by 38 percent. Meanwhile the experimental methods, improved in an admirable way by the labors of *E. Rutherford, E. Regener, J. Perrin, R. A. Millikan, The Svedberg* and others, have without exception decided in favor of the value deduced from the theory of radiation which lies between the values of *Perrin* and *Millikan*.[51]

During those same 12 years, Planck's original 1900 derivation had been outspokenly challenged by Einstein, who had offered his own derivations of Plank's formula in 1906, 1907, and 1911; and he proposed still other derivations of it in the next dozen years, most notably in 1916–1917, before finally lending his support to Bose's 1923 derivation.[52] The central issue surrounding Planck's blackbody law for the first quarter of the twentieth century was thus not the equation itself, nor the occurrence of the Boltzmann constant in it, but what constituted a proper theoretical derivation of it.

A still further irony is that even as late as 1921, Perrin himself expressed misgivings about Planck's value, assigning to it a probable error of "more or less 5 percent" and remarking,

> In spite of the importance of these results, we cannot conceal the fact that Planck's theory presents great difficulties. It will be (and already has been) profoundly modified, and we shall certainly see Planck's postulates replaced by others more comprehensible and in more accurate agreement with experiment.[53]

Perrin's misgivings notwithstanding, Planck's call in 1900 for independently determined values of molecular constants as a source of evidence in support of his derivation calls attention to a further topic of our study. Theory-mediated measurement had been a part of physics for a long time, but before 1900 not so centered on a web of constants that are theoretically interlinked to one another. As the second of our quotes from Planck indicates, during the period from 1905 to 1913 complementary theory-mediated measurements of interlinked molecular constants became a primary source of evidence for microphysical theory. It has remained so ever since. Perrin was no less a champion of this new dimension of theory-mediated measurement than Planck was. Chapter 6 will examine the different aspects of evidence arising from such measurements, and the final section of Chapter 7 will consider the period from 1905 to 1913 within the history of theory-mediated measurement.

Returning to Table 1.1, we take it to dispense with any suggestion that a comparison of Perrin's values for Avogadro's number in 1908 and 1909 with values already in the literature comprised some sort of test of kinetic theory. A related, yet different reason for why Perrin's results cannot be properly considered in isolation from kinetic theory is that the whole point of research on Brownian motion at the time was to show that molecular-kinetic theory is the only plausible candidate for explaining it. Psillos stresses this in his reply to van Fraassen more than Chalmers does in his, but one need only turn to Perrin himself, or for that matter Einstein, to see that the research was simply a means to this end. We do not dispute this. That Brownian motion persists even years on end was a mystery calling for explanation. Moreover, Perrin's results eliminated the two considerations that had been standing in the way of molecular motion providing the explanation, namely (1) the apparent granule velocities, as observed in the microscope, being far too low to be consistent with the mean molecular velocities required by kinetic theory, and (2) questions about whether the granule motion can truly be random stemming from the controversy over whether the second law of thermodynamics is merely statistical. As Chalmers notes more than once, however, the Brownian motion results did nothing to dispel the long-standing principal problem with kinetic theory, the inability to reconcile the specific heats of gases with the equipartition of energy.

We also do not dispute the historical importance of kinetic theory being shown to provide the only plausible explanation of Brownian motion. More so than chemists, physicists at the time put a good deal of stock in the extent to which an overarching theory could explain known phenomena. This likely explains why they had not been all that concerned by Ostwald's refusal to grant the existence of molecules and atoms, for his "energetics" alternative was offering little in the way of the sort of explanation they wanted, especially when compared with molecular-kinetic theory. Indeed, the capacity of this theory to provide such explanations is why Ostwald nevertheless employed it throughout the earlier editions of his textbook as offering helpful pictures. The importance physicists attached to such explanations also undoubtedly explains our earlier quotes from Brush and Wichmann questioning why the results on Brownian motion following Einstein's 1905 paper made much difference, for Brownian motion was not something physicists had shown much interest in before that as something further that molecular-kinetic theory could explain. The theory was already explaining so much.

Although it is speculation on our part, a case can be made—not just in response to Brush and Wichmann—for a quite specific reason why establishing that molecular-kinetic theory can explain Brownian motion was historically so important at the time. In 1907 Einstein made a radical proposal concerning the specific heats of solids: energy is quantized, and degrees of freedom have energy thresholds that have to be crossed before the equipartition principle extends to them.[54] This led Nernst to begin exploring the specific heats of gases at extreme low temperatures, and subsequently to initiate and then organize what became the first Solvay Conference, in 1911. The topic of the Conference was how to incorporate quanta within physics, with Nernst choosing Einstein's "On the Present State of the Problem of Specific Heats" to conclude it. In the middle of the Conference, Perrin had presented his "The Proofs of Molecular Reality." Unlike almost all the other papers at the Conference, it contained nothing about quanta, but everything Perrin had claim to having established about Brownian motion.

To pursue a solution to the problem of specific heats in a way inconsistent with classical physics was, on its face, an extraordinary way of saving the molecular-kinetic theory of heat unless, of course, the reality of molecules and most of the core of the theory was already beyond dispute. The combination of Ostwald's capitulation and that theory's offering the only viable explanation of all aspects of Brownian motion gave clear reason for taking this much to no longer being under legitimate dispute. Nernst seems to have seen the situation in this way, and judging from the response of others at the Conference to Perrin's paper, so did they. Our suggestion, then, is that it was in this specific respect that Perrin could legitimately be said even at the time, and not just in 1926, to have "put a definite end to the long struggle regarding the real existence of molecules."

If we are thus willing to grant that those at the time saw Perrin's results as having shown that the molecular-kinetic theory of heat offered the only plausible explanation of Brownian motion, then why are we nonetheless insisting on limiting our attention initially to those results in and of themselves, independently of that theory? We in a way gave our answer on the first page of this chapter when we italicized the word "only" in the quotation from Söderbaum's Nobel Presentation Speech for Svedberg. That the molecular-kinetic theory of heat offered the only plausible explanation of Brownian motion after Perrin's results does not entail that it is the only explanation. That the existence of an etherial medium offered the only plausible explanation for Fresnel's and Young's transverse light waves in the 1840s and Maxwell's electromagnetic waves in the 1870s did not entail that it was the only explanation. Decades after Maxwell, his electromagnetic field offered an alternative medium that no one had foreseen for explaining transverse light waves. Inferences to the only plausible explanation have the same limitation as inferences to the best explanation generally, namely the failure to exclude alternatives not yet conceived of. It was because of this shortcoming that the phrase "inference to the best explanation" first appeared in the philosophical literature in a book entitled *Fallacies*.[55]

One way to lessen worries about an inference to an only plausible explanation is to marshal evidence that the phenomena in question, in and of themselves, determine certain specific features of the proposed explanation that are then so strongly supported to amount to a constraint on any possible alternative explanations. In the case of Newtonian gravity, for example, the dominant motions of the planets determined specific values of the centripetal tendencies toward them and the Sun, values that were subsequently so strongly supported by further nuances they cause in those motions for Einstein to insist on preserving them within his alternative to Newton's theory.[56]

More pertinent to the concerns at hand, this is precisely what did not happen in the case of the ether: the phenomena providing the evidence for the transverse wave character of light and for electromagnetic waves did not begin to determine any specific features of the proposed luminiferous ether.[57] One might propose Perrin's determinations of Avogadro's number as having done it in the case of the molecular-kinetic explanation of Brownian motion, but this number seems not to have entered into the explanation itself in any way. Or, at any rate, so van Fraassen claims, and hence it would amount to begging the question to suppose otherwise from the outset, instead of waiting until Perrin's results for Brownian motion have been examined in and of themselves.

Instead of Avogadro's number, a seemingly more promising specific feature of the molecular-kinetic explanation dictated by the phenomena of Brownian motion is the very character of the motion of the granules itself, extremely high frequency changes in speed and direction, displaying a Gaussian distribution, that persist without any apparent limit of time. This suggestion is a nuanced variant of the point emphasized by so many in one way or another, namely that the motion displayed by the granules is simply molecular motion writ large enough to be visible. Psillos offers a version of such a proposal in his reply to van Fraassen, starting with a quote from Perrin's *Atoms*:

> The objective reality of the molecules therefore becomes hard to deny. At the same time, molecular movement has not been made visible. The Brownian motion is a faithful reflection of it, or better, it is a molecular movement in itself, in the same sense that infra-red light is still light.[58]

Psillos then goes on to argue,

> Perrin's point here is precisely that size does not matter, but causal role does! Like Pasteur before him, Perrin did place the molecules firmly in the laboratory, grounding their causal role and offering experimental means for their detection and the specification of their properties—even though, the molecules did not become visible. This is of great significance because it becomes clear that Perrin's argument should be compelling for anyone who does not take it that strict naked-eye observability is a necessary condition for accepting the reality of an entity. It should also be compelling for anyone who thinks that continuity of causal role is a criterion for accepting the reality of an entity—irrespective of whether some instances of this entity are observable, while others are not. Recall Perrin [*sic*] claim that the movement of the Brownian particles was a "faithful reflection" of the molecular movement, since the Brownian particles *were* large molecules.[59]

Though not always so rhetorical, appeals of this sort to Brownian motion being a visible "reflection" or causal extension of molecular motion can be found throughout the literature, both then and now.

Nevertheless, the idea that Brownian motion is molecular motion writ large enough to be visible is, on two different counts, rhetorical excess. To begin with, Brownian motion occurs only in emulsions sufficiently dilute for there to be, in contrast to molecules, negligible interaction among the granules themselves. Beyond this, consider the number of molecules within one granule-radius of the surface of each granule—that is, the number of molecules in a spherical shell surrounding the granule of twice the size of the granule. Save for some special studies (including of granule rotation) in which he used granules of a 10- to 15-micron diameter, Perrin's quoted results were for granules ranging from 0.212

microns in radius to 0.52. In the case of a 0.212 radius granule, the number of molecules in such a shell surrounding it would, by Perrin's own values for Avogadro's number, exceed 8 million even were it a gas at standard conditions instead of a liquid, while in the case of a 0.52 radius granule the number would exceed 120 million. The picture of granules interacting with individual molecules is thus a fiction. Whatever the action of the millions of molecules in the immediate vicinity of a granule may be when they abruptly change its motion, it is not remotely akin to the action of molecules on one another of the molecular-kinetic theory of heat.

To conclude this defense of our approach, none of what we have said has been meant to argue that Perrin's results for Brownian motion did not produce decisive evidence for the reality of molecules or for its explanation by the molecular-kinetic theory of heat. Rather, our point has been that the only proper way to assess the evidence is first to consider Perrin's results in and of themselves, independently of molecular theory. Even more so, it is the only proper approach to assessing whether the evidence achieved is anything more than van Fraassen has claimed for it, namely a "grounding" of molecular-kinetic theory. Any other approach to responding to van Fraassen, so far as we can see, is sure to beg the question.

1.4. Why Focus So Narrowly on Perrin?

As we noted at the outset, Perrin received the Nobel Prize in Physics in 1926 for his work on Brownian motion, but then so too, at least in part, did The Svedberg in the same year, and surely not just because he happened to have been a member of the Royal Swedish Academy of Sciences that awards the Nobel Prizes in the sciences. Svedberg's initial publications on Brownian motion preceded any of Perrin's. Moreover, not only did his *Die Existenz der Moleküle* appear a year before Perrin's *Les atomes*, but in it, according to Mary Jo Nye, "Svedberg claimed priority for himself as the first individual to present quantitative confirmation of the molecular-kinetic theory of Brownian movement."[60] We have already seen some evidence supporting this claim, namely in Nernst's citing Svedberg's (1906) and Seddig's (1908) tests of Einstein's theory in the sixth edition of his book, with no mention of Perrin; and Nye reports that Ostwald "first wavered in his unmitigated hostility to the atomic-molecular hypothesis in a 1908 review of" Svedberg's thesis on colloidal solutions.[61] Shouldn't we therefore be focusing not only on Perrin's research, but at the very least on Svedberg's as well, if not also that of others like Seddig and Henri who conducted experimental research on Brownian motion during the first decade of the twentieth century?

Our answer to why we are going to focus on Perrin has three parts, one pertaining to Svedberg's research on Brownian motion when compared with Perrin's, the second centered on Perrin's standing in the community of physicists at the forefront of research at the end of the first decade of the twentieth century, and the third reflecting what it is that we are, and are not, trying to do in this monograph.

Two considerations were responsible for the surge in experimental research on Brownian motion in the middle of the first decade of the new century: the announcement of the ultramicroscope in 1903 and the publication of new theories, principally by Einstein in 1905 and 1906 and by Smoluchowski in 1906. Apparently unaware of these theories, Perrin began his experimental investigations of Brownian motion in 1905, but did not publish any results until four papers in 1908. Svedberg, who turned 21 only in 1905, published his first prominent paper on the subject, on tests of Einstein's theory, in 1906, followed by the publication of his Inaugural Dissertation in Sweden in 1907. So, his priority claim quoted earlier was legitimate. As we shall see, however, Einstein published a somewhat outspoken criticism of this 1906 paper a year later, and, once Perrin became familiar with them, he expressed definite reservations in 1909 about Svedberg's results. Moreover, the results reported in Perrin's 1908 papers not only covered the same ground that Svedberg had, but added an important entirely new result, singled out in his Nobel Prize award of 1926, on the vertical density variation of Brownian motion granules. By the end of 1908, then, Perrin had at least begun to establish his laboratory as the leading center for experimental research on the topic.[62]

In 1909 Perrin published a 114-page monograph in *Annales de Chimie et de Physique* entitled "Mouvement brownien et réalité moléculaire" that presented not only the evidence for molecular reality coming out of the research on Brownian motion, but also the growing body of determinations of Avogadro's number arising from other phenomena. The monograph was published as a book that same year in France, and in English and German translations in 1910. Its conclusion opens with the statement, "I think I have given in this Memoir the present state of our knowledge of the Brownian movement and of molecular magnitudes." It provided Perrin with his own priority claim, namely the first to present in systematic, comprehensive form the evidence derived from research on Brownian motion in support of molecular reality, in the process laying out how it related to evidence complementing it from other sources. Svedberg's book to this end was published three years later.

Perrin's monograph was almost certainly responsible for the singular standing he had acquired by 1911. We need a short detour, however, to clarify his claim to such standing. In 1910 Nernst managed to persuade the industrialist Ernest Solvay to sponsor a conference in Belgium of leading physicists, by invitation only, on the topic of "the theory of radiation and the quanta." Nernst's draft of the invitation, to go out in Solvay's name, states the purpose of the conference, "to elucidate certain current questions of the kinetic theory" and the concerns prompting it:

> It appears that we find ourselves at present in the midst of an all-encompassing re-formulation of the principles on which the erstwhile kinetic theory of matter has been based.
>
> On the one hand, this theory leads to a logical formulation—which nobody contests—of a radiation formula whose validity is contradicted by experiments; on the other, there follow from the same theory certain results on the specific heat (constancy of the specific heat of a gas with the variation of temperature, the validity of Dulong and Petit's law up to the lowest temperature), which are also completely refuted by many measurements.
>
> As Planck and Einstein in particular have shown, these contradictions disappear if one places certain limits (doctrine of energy quanta) on the motions of electrons and atoms in the case of their oscillations around a position of rest. But this interpretation, in turn, is so-far removed from the equations of motion of material points employed until now, that its acceptance would incontestably lead to a far-reaching reformation of our erstwhile fundamental notions.[63]

The Conference, held at the end of October 1911, was attended by 22 physicists (not including Svedberg) at the cutting edge of research. Twelve of them, by invitation, prepared papers that were circulated beforehand, with the proceedings, published both in French and in German, containing these papers plus the discussions following their presentations at the conference. All but one of these 12 pertain to the concerns Nernst expressed. The exception was Jean Perrin's "Rapport sur les preuves de la réalité moléculaire," which, even including its exceptionally short following discussion, occupied 99 pages in the published proceedings, by far the longest of any of them.[64]

That Perrin was chosen to summarize the present state of the evidence for molecules shows his singular standing at the time. His paper at the conference is in its own right of singular importance for the purposes of this monograph. Two years later he turned it into his book for a general audience, *Les atomes*. Unlike the book, however, the paper itself was addressed to as discriminating an audience as he could have imagined at the time. It rightly does have claim to being the most authoritative statement of the evidence for the reality of molecules coming out of both Brownian motion and other research as of 1911. It will therefore be at the center of our evaluations of this evidence in Chapters 4 through 6. This has the further virtue of freeing us from having to address the rhetorical excesses of *Les atomes*.

Separating rhetoric from substance is not our only reason for relying less on his *Les atomes* than on his Solvay Conference paper. Already by 1914, *Les atomes* was in its fifth revised edition in French, with chapters on quantum theoretical consequences for the rotation of molecules and conjectures on the structure of atoms, though with no mention of Bohr. The now most widely read English translation, specifically of the 1921 eleventh edition, includes a section on "Bohr's laws" and a 1921 Appendix that presents the Bohr model in some detail. To rely on *Les atomes* accordingly is to rely on a moving target.

Largely because of Bohr's three-part paper in 1913 and contemporaneous with it a series of papers by the Braggs announcing their initial results on the structure of crystals, that year marks a revolution in the history of research in microphysics. Before then, the detailed structure of atoms and molecules had remained a matter of conjecture. Bohr initiated not only theory-mediated measurements of details of atoms, but ones employing quantum principles departing radically from the principles of classical physics that Perrin and others had employed in the theory-mediated measurements involving Brownian motion. On our view, the question of the "reality of molecules" took on a totally different guise once measured values were being put forward for aspects of the structure of atoms, molecules, and crystals. A still further virtue of our giving Perrin's Solvay Conference paper priority over his subsequent book is that its approach to the question of the reality of molecules not only falls strictly within the limits of classical physics, but also had to take even the question of the shape of molecules as totally open.

This brings us finally to the third part of our answer to why we are focusing on Perrin, namely what we are, and are not, trying to do in this monograph. We are not trying to give a history of research on Brownian motion during the first decade or so of the twentieth century. Those wanting such a history can find a far more thorough presentation of it in Mary Jo Nye's book than they will find here. Our goal is to conduct a careful, critical assessment of the evidence coming out of this research, both in its own right and in its relation to other evidence pertaining to the reality of molecules before the advent of the Bohr model. Readers can rightly demand that our presentation and assessment of this evidence be historically accurate, and that we take into consideration missteps and misleading results within this research that Perrin either did or should have himself taken into consideration. However, going into all the missteps and misleading results—as any thorough history should—would, we think, do more to hamper than to foster our goal.

1.5. Organization of the Monograph

The remainder of the monograph consists of six chapters. The issue of what Perrin's research established about Brownian motion independently of the molecular hypothesis is the subject of Chapter 4. Chapters 5 and 6 then consider, *as much as we can from the point of view of someone writing a review article at the time*, how the results for Brownian motion might have been seen as bearing on molecular theory. Specifically, Chapter 5 will consider this for the results on Brownian motion in and of themselves; and Chapter 6 will consider it in the context of other contemporaneous determinations of Avogadro's number that were independent of those from Brownian motion in liquid solutions. Chapter 7 will then return to the issues of what the new standing of atomic-molecular theory amounted to and what Perrin, and others at the time publishing their values for Avogadro's number, contributed to it, including in both cases whether it exceeded what van Fraassen has claimed; as noted earlier, this chapter will end with an analysis of how the advances in the theory-mediated measurement of molecular magnitudes in the period from 1905 to 1913 fit into the history of such measurement.

Needless to say, we cannot explain either what Perrin's research contributed about Brownian motion or how the standing of atomic-molecular theory changed as a consequence of it and other research during the decade without first spelling out what was known about Brownian motion at the time Perrin began working on it, and what the standing of atomic-molecular theory was at that time. Chapter 2 lays out what had and had not been established in atomic-molecular and molecular-kinetic theory at the end of the nineteenth century, as seen by such figures as Ostwald, Nernst, and O. E. Meyer. Chapter 3 then does the same for Brownian motion, which was almost entirely ignored by the principal figures working on molecular-kinetic theory before 1900.

Even though most of the interest in Perrin's research on Brownian motion over the last half century has been in the "instrumentalist-realist" debate literature, Chapters 4 through 6, the core of our monograph, will largely ignore this literature. As our subtitle says, this is a study in theory-mediated measurement, more precisely in a particular episode of such measurement between 1905 and 1913. Our central concern in these chapters will be with what evidence can—and cannot—legitimately be extracted from the measurements in question. The principal overlap with the "instrumentalist-realist" literature is in Chapter 6 insofar as it too considers the evidence arising from Perrin's determinations of Avogadro's number, both in their own right and in combination with the several other theory-mediated determinations of it in the same period entirely independently of Perrin's. But even in that chapter we will be taking a somewhat different approach, focusing on the theoretical presuppositions that had to be made to bridge the gap from macroscopic measurements to values for the number of molecules in a mole.

The emphasis in the instrumentalist-realist debate on these determinations of Avogadro's number, moreover, has unfortunately obscured the importance of other theory-mediated measurements derived from Brownian motion. The invention of the ultramicroscope in 1902 was only part of the reason for the breakthrough in research on Brownian motion that occurred at the time. No less important was the realization by Einstein and Perrin that the direct observation of the velocities of the granules was not reliable, and hence that some indirect—that is, theory-mediated—means was needed for determining them and the corresponding kinetic energies. Our question about the evidence that can—and cannot—legitimately be extracted in the case of these measurements, we contend, is no less important than the much discussed question of the evidence from the complementary measurements of Avogadro's number.

Save then for an assessment in Chapters 5 and 6 of van Fraassen's claims about the contribution Perrin's results made, the three core chapters of our monograph will largely ignore the back-and-forth that has been occurring over the last half century within the "instrumentalist-realist" dispute. Our goal is not to take or to defend some stance on this dispute, but rather to provide a thorough analysis of the evidence in the Perrin episode that we hope will make at least that part of the dispute more tractable. The "Postscript on the Realism-Instrumentalism Debate" at the end of the monograph is intended to serve this aim by suggesting some lessons pertinent to the dispute that might be drawn from our study. As such, it, unlike the rest of the monograph, is addressed to a narrow audience. Its last section, however, returns once again to the topic of theory-mediated measurement, examining why various results from such measurement have historically managed to persist through theory-change and what implications this has for a question of wider interest, namely which scientific results have the strongest claim to being permanent.

Notes

1. Oseen (1926), n.p.
2. Perrin (1926).
3. The Nobel Prize in Chemistry 1925. NobelPrize.org. https://www.nobelprize.org/prizes/chemistry/1925/summary/.
4. Söderbaum (1926), n.p.
5. W. Ostwald (1912), p. vi; the Preface is dated November 1908.
6. Nernst (1911), p. 437 [(1909), p. 438]. We have converted all the values listed to molecules per mole from the molecules per cubic millimeter at standard conditions listed by Nernst, in the process rounding the values to two significant figures. His kinetic theory value is a mean of different values using van der Waals's approach to determining the fraction of the volume occupied by molecules from deviations

from Boyle's law and the Clausius-Mossotti approach to determining this fraction from the dielectric constants of gases, each combined with Maxwell's approach to determining mean free paths from the gas viscosity. Nernst's droplet formation value is the mean of values reported by J. J. Thomson in 1898, Harold Wilson working under Thomson in 1903, and R. A. Millikan and M. Begeman in 1908. His alpha-particle radiation value came from Ernst Rutherford and Hans Geiger in 1908, though adding that Regener had obtained a somewhat greater value; and his blackbody radiation value came from the first edition of Max Planck's *The Theory of Heat Radiation* (1906).

7. Ibid., p. 437.
8. Ibid., p. 206f [(1909), p. 211].
9. Ibid., p. 207 [(1909), p. 212].
10. Ibid., p. 437.
11. Nernst (1923), p. 224.
12. Ibid., p. 510. Nernst speaks of "fresh proof" because of the conclusion he drew in his discussion of Brownian motion earlier in the book from Svedberg's confirmation of Einstein's theory.
13. Ibid., p. 505.
14. Ibid., p. 514. The mean of the three values Nernst lists for Perrin is 6.87×10^{23}. Nernst cites Millikan (1917) in giving his preferred value, but in fact the value Millikan gives there (p. 16) is $6.062 \times 10^{23} \pm 0.006 \times 10^{23}$, not the value given by Nernst.
15. Two long-time colleagues of Ostwald's, Jacobus van't Hoff and Svante Arrhenius, had won the first and third Nobel Prizes in Chemistry ever awarded, in 1901 and 1903, respectively, and so they had claim to being authoritative too. They had not, however, produced textbooks in physical chemistry used throughout the world, as Ostwald and Nernst had.
16. Poincaré (1963), p. 90; for an explanation of what Poincaré was referring to in the first paragraph of this quotation, see sections 2.6 and 2.7 of Chapter 2. In another lecture from the same year (1912), Poincaré stresses the agreement of Perrin's value for Avogadro's number with other totally independently obtained values for it, especially from radioactive decay, as the grounds for why "at present we are going towards atomism," and away from Duhem's goal of a "thermodynamics free of hypotheses and exclusively based on experience"; see Bub, Demopolous, and Frappier (2012).
17. Ibid.
18. Somerville (1840), p. 194.
19. Maxwell (1878), p. 775.
20. Brush (1976), p. 697f.
21. Wichmann (1967), p. 6.
22. Strutt (1900) and W. Thomson (1901).
23. W. Ostwald (1895), p. 7.
24. Glymour (1975).
25. Cartwright (1983), pp. 82–85.
26. Salmon (1984), pp. 213–227.

27. Mayo (1986), (1988), and (1996), pp. 214–250.
28. Achinstein (2001), pp. 243–265.
29. Maddy (2001), (2007).
30. Psillos (2011a), (2011b), (2014).
31. Chalmers (2009), (2011).
32. van Fraassen (2009).
33. van Fraassen (2009), p. 20, where van Fraassen is quoting Achinstein (2001), p. 246.
34. Perrin (1908d), (1910), pp. 9–15.
35. van Fraassen (2009), p. 23.
36. See, as well, van Fraassen (2012).
37. This requirement is quoted from van Fraassen (2012), p. 783, save for the substitution of the word "theory" where van Fraassen employed "hypothesis," referring to a hypothesis added to "enrich" a theory.
38. The other relationship that Einstein introduced and Perrin used concerned the mean *rotational* kinetic energy of the granules; the relationship in this case is between the observable mean square rotational displacements and the mean rotational kinetic energy, where again the rotational motion is too complex for the rotational velocities to be determined through observation. In contrast to the three cases noted in the text, any relationship between the mean rotational energies of the granules and the mean rotational energies of the molecules of the substrate seems to lack any causal basis at all.
39. Einstein (1956), p. 1f. Einstein's reference to "classical thermodynamics" is to contrast it with Maxwell-Boltzmann statistical thermodynamics.
40. Chalmers (2011), p. 723; in a footnote earlier in the article (p. 715) he credits Mayo with convincing him "of the importance of the notion of partitioning."
41. Chalmers (2009), p. 243. He concedes the point in his response to van Fraassen as well: "None of these conclusions required appeal to the kinetic theory," (2011), p. 723.
42. Chalmers (2011), p. 724.
43. Ibid., p. 714.
44. Psillos (2014), in his reply to van Fraassen, offers a guarded Bayesian approach to assessing the stringency of such coincidence. His own caution in putting it forward gives us reason not to consider it here.
45. Psillos (2014), p. 159.
46. Ibid., p. 154.
47. Chalmers (2011), p. 726.
48. Perrin (1910, p. 17f), where he first cites a value for N of 45×10^{22} from van der Waals formula for oxygen and nitrogen, then questions the reliability of this value on the grounds that molecules of these elements are "assuredly not spherical," and then gives his calculated value of 62×10^{22} for argon, which he takes to be a better candidate for being spherical. As Chalmers says, Perrin subsequently (p. 46) says that the values for N he obtained from the density distribution of various kinds of granules "agree with that which we have foreseen for the molecular energy," noting that "the mean departure does not exceed 15 percent, and the number given by the equation of Van der Waals does not allow of this degree of accuracy." In a long footnote in his (1908d),

Perrin lists a value for *N* of 60 × 10^{22} that he attributes to the theory of van der Waals among a wide range of other values in the literature at the time, but he does not give a source, nor does he claim any notable agreement between his values and it.

49. Planck (1967), esp. p. 89. Planck, by the way, misstates the value for Avogadro's number in Meyer (1899b). Meyer, in fact, derived a value of 640 trillion molecules in a milligram of hydrogen (1899a, p. 337; 1899b, p. 335) from the value of Loschmidt's number he had obtained for air (p. 333), listed in our Table 1.1.
50. Planck (1967), p. 90.
51. Planck (1991), p. vii.
52. See Kuhn (1978) for an account of the first dozen years of the controversy, and the key papers by Einstein, in English translation: (1989b), pp. 214–224; (1993), pp. 426–437; (1997). For a discussion of Bose's paper and Einstein's involvement in it, see Pais (1982), pp. 423–439.
53. Perrin (1990), p. 156. This remark was not in the first, fourth, or fifth editions of *Les atomes*.
54. Einstein (1989b), pp. 214–224. For an extended discussion, see Rogers (2005).
55. The notion of inference to the best explanation appears to have first been put forward in Sidgwick (1884), pp. 250 and 271ff, where it was presented as a fallacy for the reason we give. It was subsequently picked up by John Henry Wigmore, with attribution to Sidgwick, in the first volume of his (1913); there, however, it was treated not as a fallacy, but as the cornerstone of Wigmore's analysis of evidence in a courtroom, especially circumstantial evidence. Several years ago Smith confirmed in conversation that Gil Harman was unaware of this background when he advanced the notion in his (1965).
56. See Smith (2014).
57. This claim requires a minor qualification, for the speed of light, as a transverse wave, should be sufficient to determine a characteristic quantity of any isotropic elastic luminiferous ether, namely the square root of its shear modulus divided by its density. The problem was that, according to the theory of wave transmission in an elastic medium, there must be a longitudinal pressure wave, as well as a transverse shear wave, and its velocity must be greater than that of the transverse wave. The absence of any such faster longitudinal wave made the inference from the velocity of light to this putative characteristic quantity of the ether open to question. See, for example, Stokes (1880), esp. pp. 124ff. (Stokes's paper originally appeared in 1845.)
58. Perrin (1990) and (1916), p. 105; (1913), p. 150f; (2014), p. 201 f. The remainder of Perrin's paragraph is worth quoting: "From the point of view of agitation, there is no distinction between nitrogen molecules and the visible molecules realised in the grains of an emulsion, which have a gramme-molecule of the order of 100,000 tones." In a footnote, Perrin added, "Of course, such grains are not *chemical* molecules."
59. Psillos (2014), p. 158f.
60. Nye (1972), p. 146.
61. Ibid., p. 151.

62. The points made in this paragraph will be revisited in more detail in subsequent chapters, where individual references will be given. They are covered with proper historical care in Chapter 3 of Nye (1972).
63. Quoted from Mehra (1975), p. 6.
64. de Broglie and Langevin (1912), pp. 153–251. The next longest, Sommerfeld's, runs 60 pages followed by 20 pages of discussion.

2
The Historical Background
Molecular Theory as of 1900

The quotations from Ostwald and Nernst in the preceding chapter have stressed that a variety of research, other than that on Brownian motion, was producing revolutionary experimental results affecting the molecular hypothesis during the decade and a half between J. J. Thomson's "discovery" of the electron in 1897 and Perrin's *Les atomes* of 1913. Ostwald singled out cloud chamber results on the charge of ions, in combination with results on Brownian motion, in announcing his conversion in the preface to the 1909 edition of his *Grundriss der allgemeinen Chemie*. Beyond this, he added a full chapter on "conduction in gases and radioactivity," whereas neither topic received any mention in the preceding 1889, 1890, and 1899 editions. Nernst listed results on radioactive decay and black-body radiation, along with the ones from cloud chambers, in the 1909 edition of his *Theoretische Chemie vom Standpunkte der Avogadroschen Regel und der Thermodynamik*; again, none of these topics received mention in his 1893 or 1898 editions. Each of these new research areas was yielding values for Avogadro's number well within 20 percent of Perrin's, and hence his results for Brownian motion cannot be assessed in isolation from these other breakthroughs.

While all of this research had an impact on the molecular hypothesis, most of it, in contrast to the research on Brownian motion, had nothing as such to do with kinetic theory. Thomson had identified the discrete carrier of negative electric charge from its signature mass-to-charge ratio—first in cathode rays in 1897, and then in photoelectric and thermionic emissions in 1899. His evidence that the charge itself is discrete and universal came from measurements of the charge of ions under the assumption of one ion per droplet formed in rapid expansion in cloud chambers. In his initial experiments of 1898 he sought a cross-check for this assumption through the best estimates then available for Avogadro's number.[1] Faraday's universal value for the total electricity needed to release a given quantity of a monovalent element in electrolysis had been established half a century earlier; dividing this number by the value of Avogadro's number gives the charge per molecule released. Thomson's student John Townsend shortly afterward managed to confirm the identity between the charge per ion and the charge per molecule of hydrogen in electrolysis, independently of the value of Avogadro's number.[2] After that, dividing Faraday's constant by the charge per ion

Brownian Motion and Molecular Reality. George E. Smith and Raghav Seth, Oxford University Press (2020). © Oxford University Press. DOI: 10.1093/oso/9780190098025.001.0001.

in cloud chambers—and subsequently by the charge of the electron—became a way of obtaining values for Avogadro's number cited by Ostwald and Nernst. Values thus obtained were independent of kinetic theory. The only role kinetic theory had played in reaching this method was in Townsend's confirming experiment, which presupposed Maxwell's theory of diffusion.

Values for Avogadro's number obtained from radioactive emission of α-particles were also obtained from dividing Faraday's constant by the fundamental unit of charge. Rutherford's efforts on α-particles culminated in a pair of papers he coauthored with Geiger that appeared in 1908.[3] The discreteness of α-particle radiation had been established by the scintillations the particles produce. The first of these papers established the rate of α-particle production (3.4×10^{10}) per second from 1 gram of radium, in the process confirming the count obtained from scintillations; the second used this value to obtain a value for the charge per α-particle (9.3×10^{-10} esu), with strong evidence that the particles amount to helium ions of positive charge twice that of the fundamental unit of charge. The value of Avogadro's number thus obtained (62×10^{22}) was therefore again totally independent of kinetic theory, and at most only loosely tied to molecules.

Two further revolutionary proposals were put forward by Einstein while Perrin was in the process of developing his experiments, the quantization of light in his 1905 paper on the photoelectric effect and the quantization of energy in his 1907 paper on the specific heats of solids.[4] Although neither of these papers gave results for Avogadro's number, they were as much at the forefront as the new research giving such results at the first Solvay Conference of 1911. Both of those Einstein papers open with a discussion of Planck's blackbody radiation law, which (as we noted in Chapter 1) not only was at the forefront of the Solvay Conference, but also was offering its own value for Avogadro's number. In particular, Einstein's discussion of Planck's law in the 1907 paper concludes that "molecular-kinetic theory of heat must be modified in order to be brought into agreement with the distribution of blackbody radiation."[5] So, at least in Einstein's eyes, if not to the same extent in Planck's, the blackbody radiation value for Avogadro's number was not independent of kinetic theory to the same extent as the values obtained from ionization and radioactivity.

The extent to which kinetic theory entered into the value for Avogadro's number inferred by Planck from his blackbody radiation formula is nevertheless not straightforward. As noted in Chapter 1, his formula has two constants whose values can be obtained from radiation data: one that became known as Planck's constant h, and the other, Boltzmann's constant k from statistical thermodynamics—within kinetic theory, the mean kinetic energy per molecule per degree Kelvin. Planck's value for Avogadro's number (again, 62×10^{22}) then came from dividing the gas constant R, the energy in a mole of an ideal gas per degree Kelvin, by k. The tie between Planck's formula and kinetic theory at the time was less than transparent because of the way *k* was introduced in the derivation of the formula, namely not through any direct appeal to molecular activity, but through Planck's novel construal of Boltzmann's probabilistic analysis of entropy. Indeed, in the paper in which he announced the formula, Planck appealed to the general agreement between the value he had obtained for Avogadro's number and the values from kinetic theory listed by O. E. Meyer as evidence supporting the physical interpretation of the formula set forth in his derivation of it.[6] That derivation, however, was controversial.[7] Einstein's 1907 paper was one of several that offered alternative derivations over the next decade, and it was not the last of his alternative derivations. Perrin himself continued to express discomfort with Planck's derivations of his formula as late as 1921.[8]

So, while other approaches were yielding values of Avogadro's number during the first decade of the twentieth century, none of them had so clear and direct a tie to kinetic theory as those based on Brownian motion. Each of the others was providing evidence for the discontinuous nature of something—electrical charge, radioactive emission, and energy (or something related to it) in radiation—but not as such for the discontinuous nature of matter itself, much less the motions of molecules. Einstein's 1905 paper on the photoelectric effect had proposed the discontinuous nature of light, and his 1907 paper, the discontinuous nature of energy driving the vibration of atoms in crystals; but, in contrast to the intent of his 1905 paper on Brownian motion, neither of these as such provided evidence for the discontinuous nature of matter. The efforts on Brownian motion during the decade thus had a more direct bearing on the issue of the discontinuous nature of matter than any of these others. We are therefore going to examine them in the following independently of the other efforts yielding values of Avogadro's number that were going on at the time, returning only in Chapter 6 to the relation between the contribution Brownian motion made to the change in the standing of the molecular hypothesis and the contribution made by these others.

We had best pause to clarify how we will be using such terms as "molecular hypothesis" and "kinetic theory." The term "molecule" we will use to refer to a chemical molecule, that is, a body or particle with its own dynamical integrity compounded out of chemical atoms. By "kinetic theory" we mean broadly that heat is a form of motion within matter, with the question left open of what precisely is in motion. In his 1875 lecture before the Chemical Society, "On the Dynamical Evidence for the Molecular Constitution of Bodies," for example, Maxwell adopted the term "particle" for these bodies in motion interacting with one another, thus leaving open to empirical investigation their relation to molecules.[9] By "molecular-kinetic theory," a phrasing we have taken from Einstein's 1905 paper on Brownian motion,[10] we will generally mean kinetic theory in which the particles in motion are indeed the chemists' molecules. When discussing Maxwell-Boltzmann kinetic theory, however, "molecular-kinetic theory" will not always include the chemical component, leading us in those contexts to a distinction between it and "chemical-molecular theory." By "the molecular hypothesis" we refer to the generic claim that molecules formed of atoms exist, and hence with it that atoms too exist. We will sometimes use "atomic-molecular hypothesis" to include the further claim that atoms retain their individual integrity when they are constituents of a molecule. All of these distinctions were commonplace in the late nineteenth-century literature.

Our reason for these choices is that one could maintain the molecular hypothesis in the nineteenth century without insisting on the kinetic theory of heat, and one could maintain the latter, and its reformulation as statistical mechanics, without insisting that the particles in motion are chemical molecules, or even that such molecules exist. As evidence that we are not being anachronistic, we can appeal to many sources from the period, some of which will appear in passing in what follows. To give one example, the physicist A. W. Rücker, in his Presidential Address to the British Association for the Advancement of Science in 1901, referred to "the three chief conceptions which for many years had dominated physical as distinct from biological science," namely "the theories of the existence of atoms, of the mechanical nature of heat, and of the existence of the ether"; a page later, he cites the doubts that had arisen about all three, noting that "the mechanical theory of heat is closely bound up" with atomic theory.[11]

Focusing in subsequent chapters exclusively on Brownian motion first, our assessment of Perrin's results from the first decade of the twentieth century will put us in a position to answer how Brownian motion contributed to the new standing of molecular-kinetic theory. We cannot even begin to answer this question, however, without understanding the status of molecular-kinetic theory before Perrin. At the very least, we must ask whether something was truly missing in this theory, or were criticisms of it merely a matter of polemics? Both Steven Brush and John Nyhof maintain that philosophical prejudices against admitting unobservables were the real reasons behind resistance to the kinetic theory.[12] On top of this, recall the quotation from Wichman in the previous chapter, that the kinetic theory was too abundantly successful for Brownian motion to make any difference whatsoever to the standing of molecular-kinetic theory. If we are to claim otherwise, we need to answer the following: *What was still missing in molecular-kinetic theory in the first decade of the twentieth century*? To answer this question, in this chapter we track the development of kinetic theory in the later decades of the nineteenth century. Parallel to this development was the emergence of the new hybrid discipline of physical chemistry and its efforts to measure chemical magnitudes through physical phenomena—a topic of sections 2.6 and 2.7.

Equally, of course, we cannot begin to understand what Brownian motion contributed to the new standing of molecular-kinetic theory at the time without understanding why it had not done so earlier. Several philosophers and historians have taken "background information" on Brownian motion to be vital to an account of the impact it ultimately had on the standing of molecular-kinetic theory.[13] Brownian motion, which had been observed as early as 1765,[14] had a rich and extensive history prior to Einstein and Perrin. While this early observation was of little consequence, Robert Brown's discovery of the thoroughgoing prevalence of Brownian motion in 1827 propelled research into the nature and cause of the movement. In particular, research in the latter half of the nineteenth century was by and large devoted to Brownian motion in relation to kinetic theory. This was exactly the motivation behind Perrin's experiments in the first decade of the twentieth century; yet, despite their having the same intended purpose, none of the earlier efforts made any real difference to the status of molecular-kinetic theory. Why was that? Said another way, *What was missing in the earlier research on Brownian motion that led to its making little to no impact on molecular-kinetic theory*? Since the answer to this question turns out to have so little relevance to the question of this chapter, we are relegating to Chapter 3 the historical background to the efforts on Brownian motion covered in subsequent chapters.

2.1. On the "Hypothetical" Status of the Molecular Hypothesis

Brownian motion research cannot be understood independently of the prevailing attitude toward the molecular hypothesis at the time. Judging from the studies cited by van Fraassen,[15] one might question whether there even was a *prevailing* attitude, for not only did different people have different attitudes, but the reasons they offered involved differing considerations. Rather than review these here, we shall rely on the editions of two physical chemistry texts that were the most authoritative during the 1890s, Ostwald's *Grundriss* (1890) and Nernst's *Theoretische Chemie* (1893), both of which had appeared in English in 1895. They both present atomic-molecular theory as a *hypothesis that was yet to be established*. The fact that Ostwald was a chemist, and Nernst, though trained as a physicist, was writing as a physical chemist, gives valid reason to question whether their assessments of the molecular hypothesis were less favorable than those of most physicists. The second edition of Meyer's *Kinetische Theorie der Gase* (1899), a standard text in both German and English on the topic at the time among physicists, was more favorably disposed toward the molecular hypothesis than were either of the texts in physical chemistry. Nevertheless, Meyer did not insist otherwise than that the hypothesis had yet to be established. Insofar as the efforts of both Perrin and Einstein on Brownian motion had the express purpose of finally establishing the reality of molecules, they too must have thought it not yet established. What we need to make clear is why it could still be regarded as only a hypothesis in 1905. Taking the specific reasons for this given by Ostwald and Nernst, with Meyer's text providing a balance against bias in theirs, will enable us to do this.

(Rather than explain here why these three books are uniquely appropriate for the purposes of the chapter, we have added an appendix at the end of our monograph giving background on the three and our grounds for taking them to be authoritative. We encourage readers not familiar with them to read this appendix before continuing.)

The answer to the question of why the hypothesis was not yet established depends on which part of it one considers. The atomic hypothesis was introduced early in the nineteenth century as a way of explaining such fundamental phenomenological findings of post-Lavoisier chemistry as the laws of fixed and multiple proportions. Avogadro's hypothesis added a way of explaining the further law of combining volumes discovered by Gay-Lussac. Ostwald's *Grundriss*, which teaches chemistry by recounting its nineteenth-century history, opens with this core of the atomic-molecular hypothesis and the related determination of relative atomic and molecular weights—which Ostwald emphasizes in no way presuppose the reality of atoms.

Those laws and weights provided the basis for a huge amount of progress in chemistry in the form of definitive chemical formulas for a large range of compounds over the course of the nineteenth century. All of those results, as Ostwald concedes, remained entirely consistent with this core of the hypothesis. It continued to offer a way of explaining—or at least conceptualizing—all of them. At the same time, however, none of that progress required atoms and molecules to exist. Stated more precisely, *the atomic-molecular hypothesis did not enter constitutively into any of those results.*[16] So, nothing in the research was giving chemists decisive reasons to view it as anything more than a convenient, yet dispensable heuristic.

Worse, the hypothesis was failing to help where chemists most needed it, namely on what happens when a compound—that is, according to the hypothesis, a molecule—forms, and the associated question of why certain compounds form and others do not. The large array of formulas for compounds led to valence theory and the periodic table and hence a step toward answering such questions, especially so for compounds of carbon. As Ostwald immediately points out, however, valence theory introduced new questions as well as answers: valence increases from 1 to 4 in the first four groups of Mendeleev's table, but "from this point on the elements have mostly more than one valency, one towards chlorine, oxygen, etc. as we proceed, and one towards hydrogen, which decreases at the same rate."[17] The periodic table itself was not purely explanatory, for blanks in it led to the "discovery" of several new elements. The questions it raised about multiple valences, however, called attention to how little information atomic theory was offering about the atoms themselves or how they manage to be selective in the compounds in which they enter. The problem was not the invisibility of atoms and molecules; it was the dearth of chemical phenomena that gave experimental access to features of atoms and molecules beyond relative weights, numbers of atoms of different kinds in molecules, and valences.

The one area of chemistry that did appear to be offering some experimental access to the "constitution of molecules" was organic chemistry, that is, compounds of carbon. The phenomenon of isomerism—Berzelius's term—had initiated the need to include some sort of structural aspect in chemical formulas; and various organic reaction phenomena had then supported the introduction of "structural formulas" that went well beyond the need to account for isomerism. The example shown in Figure 2.1 for acetic acid is taken from Ostwald's book, as is the following quotation.

> Such a formula is capable of rendering evident a great many different relations. It shows that one of the hydrogen atoms behaves differently from the others, because it is united with oxygen to form hydroxyl, while the others are all united to carbon. Further, the two oxygen atoms behave differently, the hydroxyl oxygen being more easily attacked and removed than the other. Lastly, the two carbon atoms have different functions: the one, united to two oxygen atoms, easily passes into carbonic acid, the other splits off as methyl, CH_3. All these relations just derived from the formula exist in fact; the structural formulae thus fulfill in a high degree the claim of being both functional and constitutional formulae.[18]

Nernst, more favorably disposed toward the molecular hypothesis than Ostwald, emphasizes the promise displayed by this functional aspect of structural formulas in a footnote:

> In fact, the study of organic chemistry is not only a study of the compounds of carbon, but also fully as much a special treatise on the *pure science of reaction*

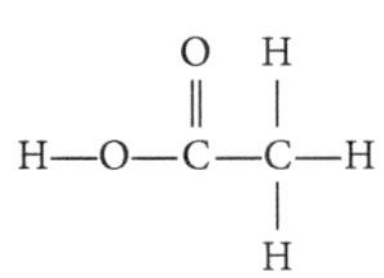

Figure 2.1. Ostwald's example of "Structural Formula" specifically for acetylene.
Source: W. Ostwald (1895), p. 198

> *chemistry*, which is perhaps the chief basis of the so-called "graphic formulae." Certainly, when we shall have had the kinetic and dynamic side of organic chemistry as well developed as is the static side at present, we shall have even less occasion to apologize for the speculative nature of the atomic theory.[19]

In the paragraph immediately following the preceding quotation, in small print, Ostwald characteristically insists that structural formulas are to be regarded as merely formally representing established facts concerning chemical valency, "serving to aid the mind in comprehending and retaining the relations experimentally found."[20] He says nothing more there about his reasons for not attaching more significance to structural formulas. Nernst, by contrast, wants to attach added significance to structural formulas even while acknowledging that the molecular hypothesis is just a hypothesis:

> Therefore it must be assumed that certain forces are exerted between the atoms in the molecule, which determine the relative positions of the atoms; also that the relative positions of the atoms can vary with the mode of union of the atoms with each other.
>
> *The differences, shown in the physical and chemical behavior of metameric compounds, must thus be ascribed to differences in the constitution of the molecule.*
>
> It cannot be denied that in attempting to develop the atomistic science from this standpoint, and to frame more definite conceptions regarding the arrangement of the atoms in the molecule, we go into a region of a purely hypothetical nature—a region which can only be reached by a leap of a bold phantasy. The usual requirement customary in the careful study of nature, of abstaining from such an attempt, does not appear to be justified on these grounds; for such a requirement, on the one hand, would amount to a refusal to obtain many obvious conceptions regarding many important phenomena, which neither the experimental nor the theoretical student can explain in any other way; and, on the other, it would not harmonise with the fundamental principle of the method of natural science which commands us to follow out, to the ultimate, such a practical and fruitful hypothesis as the atomic hypothesis is well known to be.[21]

Why then does it still have only the marginal status of a hypothesis? Because experiment has yet to provide a basis for "framing more definite conceptions of the arrangement of atoms in the molecule." Nernst's very next sentence states what is most missing: "*At present scarcely anything definite is known regarding either the nature of the forces which bind the atoms together in the molecule and which hinder them from flying apart in consequence of the heat of motion, or regarding their laws of action.*"[22] We have italicized this sentence because we shall be returning to it repeatedly; it recurs in all subsequent editions of Nernst's book, right through 1921.

In the portion of the preceding quote that he italicized, Nernst refers to the "physical behavior" of compounds "that must be ascribed to differences in the arrangements of the atoms in the molecule." What he is referring to, above all else, is "optical isomerism"—that is, what we now call the phenomenon of chirality discovered by Pasteur in 1848 and much explored within the domain of organic chemistry thereafter. Nernst cites the phenomenon as pointing to the need to add a spatial element to structural formulas, citing van't Hoff's "creation of stereochemistry." Among the examples of the latter he considers is "the question as to the directions in which the four valences radiate from the carbon atom," concluding, with a limited qualification, that "*we are forced to regard the tetrahedron arrangement of the valences in space as correct.*"[23]

Ostwald similarly invokes experiments on the rotation of the plane of polarization as establishing aspects of the spatial arrangement of atoms within molecules:

> Now Le Bel and van't Hoff showed almost simultaneously (1874) that all optically active substances that could rotate the plane of polarisation in the non-crystalline state contained an "asymmetric" carbon atom, *i.e.* one whose valences were satisfied by four atoms or radicals of different kinds. . . . Whatever we may think of the assumption of the tetrahedral arrangement of the valencies—an assumption which has of late proved very useful in another field—the fact is at least very remarkable, that hitherto no optically active substance has been discovered which does not, in the above sense, possess an asymmetric carbon atom.[24]

Ostwald's general practice in the book was to add a cautionary note, usually in fine print, about not taking talk of atoms and molecules too literally when the text itself seemed on its face to be doing so. His discussion of optical isomerism does not include any such cautionary note. The nearest to a qualifier he adds is to note, in a couple of places, the ongoing state of this research.

The qualifier Ostwald includes in the preceding quotation—"whatever we may think of the assumption of the tetrahedral arrangement of the valencies"—points to the sense in which the atomic-molecular hypothesis nevertheless continued to have the status of merely a hypothesis as of the 1890s, not just within chemistry, but within physical chemistry as well. If (1) the existence of atoms is granted (whatever they may be in themselves), as well as (2) their having valences (whatever these may be), and (3) those valences have a spatial distribution with respect to the atom, then, (4) insofar as the carbon atom has four valences, if one assumes (consistent with the chemical evidence) (5) "that the four valences of carbon are like each other in every particular,"[25] then there are only two possible spatial arrangements of these valences, in a plane at 90 degree angles with respect to one another or in a tetrahedral arrangement; chemical phenomena can then select between these two options.[26] This is one of several examples that Ostwald and Nernst respectively cite in which the empirical world is answering a question under the assumption that atoms exist and combine with one another to form molecules. Notice, however, that in answering this question the empirical world is saying nothing about whether atoms exist, what they are like, what valences are and how they have spatial orientations, or about the forces that bind atoms together into molecules.

Putting to one side for the moment the technical sense he assigns to it, van Fraassen's word "grounding" seems on the mark for describing what is missing here. During the second half of the nineteenth century the atomic-molecular hypothesis was enriched in ways that enabled it to explain—indeed, at the time to be the only proposal for explaining—an increasingly wide range of chemical and physico-chemical phenomena. In this respect it was becoming an increasingly promising and fruitful hypothesis. But the foundational aspects of the hypothesis—what atoms are, how they selectively combine with other atoms to form molecules, and what holds these molecules together over a wide range of circumstances—remained purely a matter of speculation, and hence not yet "grounded" in any chemical or physico-chemical phenomena that would allow the speculations to be replaced with empirically grounded substance.

Both Ostwald and Nernst have chapters devoted to the "mutual relations of atoms within molecules"[27]—Nernst's entitled "The Constitution of Molecules" and Ostwald's, "Theory of Chemical Composition"—focused on the enrichment of molecular theory during the second half of the nineteenth century. The quotations from Nernst's chapter given in the preceding display his view of these developments. Two quotations from Ostwald's chapter, one from its beginning and the other from its end, give his view. First, the second paragraph of the chapter, in fine print:

> This problem [of the mutual relations of the atoms within the molecule] is of course purely hypothetical, for the existence of the atoms themselves is only a hypothesis. But based on this hypothesis a theory of great fertility has been developed, and—apart from metaphysical scruples about the atomic theory, which for the most part rest on misapprehension—there lies in the results already obtained a guarantee that further advancement of the theory will lead to results equally serviceable. This advance may not be made by assuming any arbitrary forces or properties, but must come directly from the fundamental

> assumption that the atoms are finite, and, for the same element, similar bodies. But as every finite quantity of matter occupies a position in space which is definable with regard to other material particles, the question as to the relative position (or motion) of the atoms in the molecule is scientifically justified, and must be put sooner or later by the atomic theory.[28]

The chapter proceeds to present the enrichments to molecular theory that we have summarized. Its final paragraph, ending Part I of the book, is not in fine print:

> In these geometrical considerations we see a necessary and therefore well-justified stage of the development of the theory of chemical structure on the basis of the doctrine of valency. The ideas, it is true, are of a purely hypothetical character; in this, however, they do not differ from the generally accepted atomic and molecular theories themselves. Every hypothesis must be specialised as far as the facts require. As soon as the formulae written without respect to the geometrical relations became incapable of representing all the observed phenomena, an extension of their signification was rendered necessary, and this was made by taking the geometrical relations into account. The chief object of such hypothetical generalisations—that of simplifying the representation and of inciting fresh research—has in the case of the geometrical formulae already been fully attained. Not only have old facts been made clear through it for the first time, but a considerable number of new substances have been searched for at its instance—searched for and in most cases found.[29]

As we noted in Chapter 1, in the subsequent (1909) editions of the two books we have been considering, both Ostwald and Nernst announce that the atomic-molecular hypothesis no longer has the status of being just a hypothesis. Our goal in the preceding discussion has been to clarify what they meant by insisting it had only the status of a hypothesis, at least within chemistry and physical chemistry, in their preceding editions. Their reasons had nothing as such to do with atoms and molecules not being visible, or any other such "metaphysical scruples"—to use Ostwald's phrasing. Neither, however, makes clear in the books we have been considering what further advances might end the hypothetical status of atomic-molecular theory within their discipline. So far as we have found, Nernst never states what specifically is needed for any hypothetical "explanatory" theory to cease having this status. Ostwald, however, did, though in the abstract and hence not in a fashion suitable for a textbook in physical chemistry. We shall consider what he said in 1907 in this regard in section 2.4, after we examine the status of the atomic-molecular hypothesis within kinetic theory.

2.2. Kinetic Theory as a Means of Gaining Access to Molecules

A promise of gaining experimental access to the molecular realm was just what Maxwell singled out in defending the kinetic theory and its molecular conception of matter in a lecture before the Chemical Society in 1875, entitled "On the Dynamical Evidence of the Molecular Constitution of Bodies."[30] Maxwell defines "molecule" in this lecture as "that small portion of the substance which moves as a lump during the motion of agitation," adding, "this is a purely dynamical definition, independent of any experiments on combination."[31] So, the evidence referred to in his title does not have any bearing on what atoms and valences are, nor on the forces holding atoms together. He nevertheless talks at one point about molecules as compounds of atoms, for one of the topics he raises (with which we will deal in section 2.3) is the motion of atoms within molecules. The evidence he presents to the chemists, however, pertains to matter being composed of molecules in the minimal sense stated in his definition. The term "minimal" here is appropriate, given the spirit in which he presents it to an audience that he seems to have assumed was sympathetic with skepticism toward molecular hypotheses:

> Of all the hypotheses as to the constitution of bodies, that is surely the most warrantable which assumes no more than that they are material systems, and proposes to *deduce from the observed phenomena* just as much information about the conditions and connections of the material system as these phenomena legitimately furnish.
>
> When examples of this method of physical speculation have been properly set forth and explained, we shall hear fewer complaints of the looseness of the reasoning of men of science, and the method of inductive philosophy will no longer be derided as mere guess-work.[32] [italics added]

The kinetic theory we learn in school explains the gas laws as arising from the mean translational kinetic energy of molecules, in the process explaining too what is actually measured by the devices we use to determine values of pressure and temperature. But it does more than that. It identifies conditions under which the "ideal" gas laws would hold exactly, opening the way to using the measured deviations of real gases from the ideal—deviations that Regnault had shown differ markedly from one gas to another—to reach conclusions about some aspects of the molecules themselves. In particular, Maxwell cites Clausius's virial theorem:

$$pV = \frac{2}{3}T - \frac{2}{3}\Sigma\Sigma\left(\frac{1}{2}Rr\right)$$

where T is the kinetic energy of the system, r is the distance between a pair of molecules, R is the force between them, and the double sum indicates summation over every pair of molecules in the system.[33] Measured deviations from the ideal gas law thus have the promise of giving information about the forces between molecules.

Notice that kinetic theory would be entering constitutively into any inferences about forces thus drawn from the measured deviations. Demanding that the inferred values for the forces be well-behaved would consequently amount to a test of kinetic theory. We shall be spelling out in some detail what "well-behaved" amounts to in Chapters 4 and 6, and then again in the last section of the Postscript on the Realism-Instrumentalism Debate. Suffice it here to point out that relationships between accessible and less accessible parameters of the sort Maxwell invoked allow measurements of the former to answer questions about the values of the latter. Any such answer, however, is required to be reasonably determinate and unequivocal to be an answer at all. When this requirement is not met, the relationship authorizing the theory-mediated measurement becomes suspect.

Unfortunately, the promise in the case of the preceding relationship proved much slower to be realized than Maxwell anticipated. The reason is transparent from the equation: not the summation over all pairs of molecules, which can be handled statistically, but the unknown distances r between the molecules. Efforts instead focused on obtaining a "real" gas law directly from the data on deviations from Boyle's law at different temperatures. By 1900 more than 30 distinct real gas laws had been proposed.[34] The most promising of these at the time—and still the best known—was put forward by van der Waals in his doctoral thesis of 1873; while still a curve-fit, it had the virtue of having a specific physical interpretation:[35]

$$\left(p + \frac{a}{V^2}\right)(V - b) = RT$$

where the constant b represents the portion of the volume taken up by the molecules and a/V^2, the effects of the forces among the molecules. This formula too had the promise of yielding information about the molecules themselves, and it was widely appealed to in discussions of the transition from gas to liquid; yet its "constants" for each individual gas turned out not to be constant, but to vary with temperature, and hence inferences drawn from deviations from Boyle's law using it remained less than conclusive.[36]

Those deviations, as represented by van der Waals's formula, provided one of three kinds of phenomena that Meyer considers in his 1899 review of the efforts to determine kinetic-theory-mediated values for the sizes of molecules and, with them, Loschmidt's number, that is, the number of molecules in a cubic centimeter of gas at standard conditions. All three approaches relied on a theoretical relationship, derived by Clausius and refined by Maxwell, for the mean free path of *spherical* molecules at standard conditions:

$$L = \frac{1}{\pi\sqrt{2}}\frac{V}{ns^2}$$

where s is the diameter of the molecule or, following van der Waals, of its "sphere of action," and n/V is the number of molecules per unit volume, which by Avogadro's law is the same for all gases at the same conditions.[37] Values for L were derived from the viscosity and diffusivity of gases. Values for s were obtained from (1) the "condensation coefficient," that is, the ratio of the volume taken up by the gas molecules when condensed into a liquid to their volume in the gas phase; (2) the indices of refraction of gases and hence their dielectric capacities; and (3) the values of b in the van der Waals formula. Table 2.1 lists values for both Loschmidt's and Avogadro's numbers obtained via the mean-free-path

Table 2.1 Kinetic-Theory-Based Values of Loschmidt's and Avogadro's Numbers from 1865 to 1900

Loschmidt's number	Avogadro's number	Author	Year	Source
[1.83×10^{18}]	[4.1×10^{22}]	Loschmidt	1865	From mean free path and condensation coef.
$\leq 6 \times 10^{21}$	[$\leq 1.3 \times 10^{26}$]	W. Thomson	1870	From four estimates of size
19×10^{18}	[4.3×10^{23}]	Maxwell	1873	Following Loschmidt
5×10^{19}	[11×10^{23}]	van der Waals	1873	From mean free path and his coef. *b*
5.4×10^{19}	[12×10^{23}]	Nernst	1893	Following van der Waals
61×10^{18}	[13.7×10^{23}]	O. E. Meyer	1899	From mean free path and dielectric capacity
[$<2.4 \times 10^{18}$]	[$<5.9 \times 10^{22}$]	O. E. Meyer	1899	Following Loschmidt
[6.8×10^{18}]	[1.5×10^{23}]	O. E. Meyer	1899	Following van der Walls

As of 2018, Loschmidt's number = 2.686780111 × 1019 and Avogadro's = 60.2214076 × 1022. Numbers enclosed in brackets were not stated in the publications cited, but were instead calculated from values stated in them.

formula by various authors from 1865 to 1900, some of them published and others entailed by the published values for *s*.[38]

More important than the extent to which the values in the table differ from one another is the absence of any indication of a tendency to converge toward a preferred range of values from the time Loschmidt performed the first such determination of *s* to the second (1899) edition of Meyer's text. This absence of a tendency to converge stands in contrast to values for Avogadro's number yielded by Brownian motion, the charge of the electron, and blackbody radiation between 1908 and 1913. This absence of a tendency to converge before 1905 is what Poincaré was referring to in the remark at the end of the passage we quoted in Chapter 1: "Not too long ago, we would have considered ourselves fortunate if the numbers thus derived had contained the same number of digits. We would not even have required that the first significant figure be the same."[39]

The source of the large discrepancies among the values listed in Table 2.1 stems not so much from differing values for the mean free path *L*, but from the lack of any consistency at all among the values obtained for *s*. We have added the last two rows of values, attributed to Meyer though not stated by him, to drive this point home. As we noted earlier, Meyer's second edition was the most authoritative text on the experimental basis of kinetic theory in both German and English at the end of the nineteenth century. It is divided into three parts, the first on molecular motion, the second on mean free paths, and the third, consisting of a single chapter, "On the Direct Properties of Molecules." Most of this last chapter is devoted to reviewing the three different approaches to determining *s* listed earlier, with repeated cautions that *s* is being taken to be a diameter of a sphere or "sphere of influence," in spite of numerous reasons not to think that molecules are generally spherical.[40]

Meyer's values for *s* were invariably larger than 1×10^{-7} cm in the approach based on the condensation coefficient, in the range of 0.8×10^{-7} to 0.4×10^{-7} cm in the approach based on van der Waals equation of state, and generally less than 0.2×10^{-7} cm in the approach based on dielectric capacity. Meyer's published value for Loschmidt's number used a value of 0.2×10^{-7} cm, notably less than the values for *s* from the other two approaches, resulting to more than an order of magnitude difference in Loschmidt's number. Even though all the values were of the general order of 10^{-8} cm, the claim that these were three different approaches to determining values for the same quantity *s* was open to obvious doubt. Doubting this, moreover, was tantamount to doubting whether any phenomena were enabling kinetic theory to gain theory-mediated access to either the sizes or shapes of molecules.

The primary goal of not only Meyer's efforts, but the efforts of the others cited in Table 2.1, was to obtain at least approximate values for the sizes of molecules, which of course were not thought to be the same for all gases. The further step of deriving values for Loschmidt's and Avogadro's numbers, though of some importance in their own right, was more like an afterthought. In his groundbreaking paper of 1865, Loschmidt did not bother to calculate values for the constant that came to be named after him from his values for the sizes, but others following him did, usually by selecting some representative values for L and s. Although it generally went unmentioned, values obtained for Loschmidt's number n for different combinations of values for L and s—that is, for different gases—served to provide a cross-check on the reliability especially of the values for s insofar as the values derived for n, by virtue of Avogadro's law, should be the same for all gases. Indeed, one way to assess the reliability of any of the approaches to determining s and L was to check whether the different combinations of values obtained for different gases were nevertheless yielding more or less the same value for n.

Nernst was the one individual among those cited in Table 2.1 who presented such a cross-check, giving a table of values for Loschmidt's number for five different gases based on the van der Waals approach to s, the approach he ended up adopting for s in his 1893 text. He did not bother to include a corresponding cross-check for the other approach he considered, based on dielectric capacity, but he did provide a table comparing the values for s obtained by the two approaches for different gases, putting us in a position to carry out the same cross-check for this other approach. Table 2.2 displays the values for Loschmidt's number for his five gases based on these two different approaches to determining s.[41] One thing obvious from the table is why Nernst chose not to rely on dielectric capacity for determining s. Another conclusion to draw from it, though a little less obvious, is that the values for Loschmidt's number that were being derived from the mean-free-path formula depended not only on what approach was being taken to s, but also on what gases were being taken into consideration—something that according to kinetic theory should make no difference. Consider how much more compelling the determinations of s for different compounds would have been had all the values for Loschmidt's number, even just with the van der Waals approach, fallen within a range of ±10 percent rather than the ±27 percent in Table 2.2. At the very least, the variation raises obvious questions about the across-the-board assumption of sphericity.

The idea behind kinetic theory being a means for gaining experimental access to the molecular realm arose from its potential to supply more relations like *temperature varies as the mean kinetic energy of molecules* that link molecular to macroscopic parameters in a way that allows values of the latter to determine values for the former. Such theory-mediated measurement is what we take Maxwell to have intended when he proposed, in the spirit of Newton, "to deduce from the observed phenomena just as much information about the conditions and connections of the material system as these phenomena legitimately furnish."[42] In Part II of his book Meyer adduces evidence that the relation between viscosity

Table 2.2 "Measured" Values of Loschmidt's Number for Select Gases, as per Nernst (1893)

Gas	Mean Free Path $L \times 10^6$ cm	$n \times 10^{19}$ via a van der Waals Approach for *s*	$n \times 10^{19}$ via Dielectric Capacity Approach for *s*
Nitrous oxide	6.8	5.8	9.3
Carbon dioxide	6.8	6.9	15.8
Ethylene	5.8	5.8	6.2
Ethyl chloride	3.7	4.5	0.16
Sulphur dioxide	4.8	3.9	0.15
	Mean:	5.4	6.3

As of 2018, the CODATA recommended value for Loschmidt's number was $2.686780111 \times 10^{19}$.

and mean free path derived from kinetic theory provides such a measure, namely through the cross-check of using values for the mean free path to calculate diffusion coefficients in good agreement with measured values.[43] If the three different approaches to determining values of *s* had been in remotely as good agreement with one another, there would have been evidence for kinetic theory having supplied means for measuring a parameter giving sizes of molecules. If, for that matter, even one of the three approaches had met the cross-check of yielding values for Loschmidt's number across a range of gases in such good agreement with one another, there would have been evidence that its relation derived from kinetic theory was doing so. As matters stood as of 1900, however, kinetic theory was yielding at best only a rough order of magnitude for *s*, in the range of 10^{-7} to 10^{-8} cm, raising questions about just how physically meaningful *s* even was.

As we have already indicated on occasion, questions were being raised at the time about the physical meaning of *s* independently of these considerations. It is supposed to be representing the diameter of a sphere in the formula given earlier for the mean free path, but few at the time seemed to have taken seriously the idea that molecules in general are spherical in shape. What the proponents of kinetic theory were looking for were relationships between macroscopic phenomena—deviations from Boyle's law, viscosity, diffusion, liquefaction, etc.—and motions of molecules interacting among one another. Kinetic theory offered the possibility of providing such relationships independently of the internal structure of molecules. It could not do so, however, without making some further assumption concerning shape, for the motions of molecules must depend on their interaction with one another, and any such interaction has to depend on some aspect of shape. Not merely for reasons of mathematical tractability, the one salient option was a sphere.[44]

(Although it dates from after 1905, we should note in passing that in his 1909 memoir Perrin says that he has obtained a value of 62×10^{22} for Avogadro's number (that is, 2.8×10^{19} for Loschmidt's) by applying the van der Waals approach to argon—this, after announcing a value of 45×10^{22} via the same approach for oxygen and nitrogen and remarking that "this choice of substances is not a very suitable one, since it necessitates the consideration of *molecular diameter* for molecules assuredly not spherical."[45] He does not give the values he used for L, b, or s in his calculation for argon in 1909. (He does give a value for s of 2.85×10^{-8} cm for argon in his 1911 Solvay Conference paper, which together with Meyer's 9.9×10^{-6} value for the mean free path of argon[46] does indeed yield his 62×10^{22}.)[47] We are uncertain how much significance should be attached to the fact that his value of 62×10^{22} from argon is so close to the values he was about to give from Brownian motion, not to mention how close it was to our current value. Perrin himself expressed caution in 1909:

> I should add that it is not easy to estimate the error possibly affecting this number, because of the lack of rigour of the equation of Clausius-Maxwell and of that of Van der Waals. Unquestionably an uncertainty of 30 per cent. will not be a matter for surprise.[48]

Based on what we have said earlier, such caution was fully appropriate. Nevertheless, near the end of his 1909 memoir Perrin lists a value for Avogadro's number of 60×10^{22} obtained via "the viscosity of gases" together with "the exact law of compressibility,"[49] with no qualifier attached. Perhaps he was letting its agreement with his own values persuade him of the trustworthiness of this value. One would be more impressed with it had he carried out a parallel calculation for helium and obtained a value close to the one he obtained for argon.)

We concluded in the preceding section that, while the molecular hypothesis was becoming increasingly promising and fruitful within chemistry and physical chemistry, foundational aspects of it—what atoms are, how they selectively combine with other atoms to form molecules, and what holds these molecules together over a wide range of circumstances—remained matters of speculation. The corresponding foundational questions for the hypothesis that gases consist of molecules in motion concern the number, masses, sizes, and velocities of molecules and the mechanics of their interaction with one another. These questions did not remain so much matters of speculation in the 1890s as the preceding were, but only because the assumption that molecules are either spheres or have definite spheres of interaction allowed pertinent numerical values to be derived from macroscopic phenomena. These numerical values, however, were not remotely well-behaved enough to justify this assumption, and in its absence the foundational questions associated with the molecular-kinetic hypothesis remained open. Put in other terms, van Fraassen was fully justified in claiming that experimentally derived values for molecular magnitudes derived from an assumption of sphericity had not grounded molecular-kinetic theory.[50]

2.3. The Problem Posed by the Specific Heats of Gases

In fact, the situation with kinetic theory in 1900 was much worse than this. So far, we have been considering only the initial form of kinetic theory in which the molecular parameters are mean quantities—mean translational kinetic energy, mean free path, etc. As Maxwell remarked in the lecture we cited, when only mean quantities are involved, all the molecules might just as well have the same velocity. Kinetic theory expanded into statistical mechanics when he and Boltzmann proposed the distribution named after them of velocities in a gas. The statistical or probabilistic analysis of entropy, which we alluded to earlier in conjunction with Planck, initiated a vigorous controversy concerning whether the second law of thermodynamics, rather than being deterministic, allows at least the remote possibility of decreases in entropy. This controversy was still at the forefront in 1900, and as such was one of the reasons some individuals were opposing molecular theory. We shall postpone this worry until section 2.5, for a fundamental postulate underlying the Maxwell-Boltzmann distribution was posing a more important *empirical* problem for statistical mechanics, namely the principle of equipartition of energy.

A phenomenon discovered in the early years of the century that we have not yet noted is the distinction between the isothermal expansion of a gas, conforming to Boyle's pV = constant, and its adiabatic expansion in which pV^{γ} = constant, where γ is the ratio of the specific heat of a gas at constant pressure to its specific heat at constant volume. On Clausius's formulation of kinetic theory, γ is related to the fraction of energy added to a gas that increases the translational kinetic energy of molecules:

$$(\gamma-1)=\frac{2}{3}\times\frac{\text{translational kinetic energy}}{\text{intramolecular energy + translational kinetic energy}}$$

Under the Maxwell-Boltzmann equipartition of energy—that is, the equipartition of energy among all the degrees of freedom of a molecule—Clausius's relation transforms into

$$\gamma = \frac{\textit{number of degrees of freedom} + 2}{\textit{number of degrees of freedom}}$$

Now, regardless of its internal structure, a molecule has three translational degrees of freedom in space. So, measured values of γ, which came to be readily obtained from the speed of sound in gases, had the promise of yielding information about the further degrees of freedom that molecules have beyond those three.

From the time Maxwell noted this formula, in 1860, it yielded ill-behaved theory-mediated values for the number of degrees of freedom. Those values should be integers, yet for many gases they clearly were not. The distinctive emission and absorption spectra display a multitude of frequencies; assuming these correspond to vibrations within molecules, they imply a huge number of degrees of freedom. And the number of degrees of freedom should be a configurational feature of molecules that is independent of temperature, yet for some gases γ, and hence by implication the number of degrees of freedom, was found to vary with temperature. In his 1875 lecture Maxwell called anomalies with γ "the greatest difficulty which the kinetic theory has yet encountered,"[51] adding quite presciently:

> On the physical side, however, it [the Maxwell-Boltzmann distribution] leads to consequences, some of which, being manifestly true, seem to indicate that the hypotheses are well chosen, while others seem to be so irreconcilable with known experimental results, that *we are compelled to admit that something essential to the complete statement of the physical theory of molecular encounters must have hitherto escaped us.*[52] [emphasis added]

A quarter century after Maxwell said this, the situation, if anything, had become worse. The research in the intervening years did yield some encouraging results: values for γ at least close to 5/3 for the noble gases (which had been discovered in the 1890s) and mercury, and room temperature values for several presumably diatomic molecules close to 7/5. But no integral number for the degrees of freedom matched, for example, the 1.359 for chlorine and the 1.265 for carbon dioxide at room temperature, and the variation of γ with temperature had been confirmed for a range of substances.[53] In January 1900 Lord Rayleigh, who had long been a defender of the Maxwell-Boltzmann equipartition principle, published a review article in which he offered a derivation of equipartition from seemingly elementary principles of mechanics. His penultimate paragraph opens with the conclusion that this derivation drove him into: "We are here brought face to face with a fundamental difficulty, relating not to the theory of gases merely, but rather to general dynamics."[54] The final paragraph of the article merits quoting in full:

> What would appear to be wanted is some escape from the destructive simplicity of the general conclusion relating to the partition of kinetic energy, whereby the energy of motions involving larger amounts of potential energy should be allowed to diminish in consequence. If the argument, as above set forth after Maxwell, be valid, such escape must involve a repudiation of Maxwell's fundamental postulate as practically applicable to systems with immense numbers of degrees of freedom.[55]

Within a matter of weeks after Rayleigh's article appeared, Lord Kelvin responded to Rayleigh in a lecture published in an expanded version in 1901, "Nineteenth Century Clouds over the Dynamical Theory of Heat and Light."[56] One "cloud" was to reconcile the aberration of light and the results from the Michelson-Morley experiment with ether theory and the motions of bodies through the ether; the other, to which the majority of the published version was devoted, was to reconcile measured values of specific heats with kinetic theory. A review of the data leads Kelvin to conclude that "there is, in fact, no possibility of reconciling the Boltzmann-Maxwell doctrine [of equipartition] with the truth regarding the specific heats of gases."[57] Kelvin ends by quoting the two paragraphs of Rayleigh's article preceding the one we just quoted and the opening of that paragraph, "What would appear to be wanted is some escape from the destructive simplicity of the general conclusion." To this Kelvin responds,

> The simplest way of arriving at this desired result is to deny the conclusion [of equipartition]; and so, in the beginning of the twentieth century, to lose sight of a cloud which has obscured the brilliance of the molecular theory of heat and light during the last quarter of the nineteenth century.[58]

To deny equipartition amounted to denying Maxwell-Boltzmann statistical mechanics. That still left the core of kinetic theory, to which Kelvin remained committed, but with no way of drawing inferences about the configuration of molecules from the specific heats.

Maxwell, Rayleigh, and Kelvin were all outspoken defenders of the molecular hypothesis. So too was the young James Jeans, who in 1901 published a paper, "The Distribution of Molecular Energy," exploring the possibility that the difficulties with the specific heats were stemming from dissipation of molecular energy into the ether.[59] The conjectural character of the paper is made clear from the outset:

> As there is not sufficient known about the constitution of a molecule to enable it to be completely specified as a dynamical system, the paper is limited to the consideration of two imaginary types of molecules.[60]

It should already be apparent that this paper made no historically notable contribution. Its value for us lies in the aspects of his imaginary molecules that Jeans had to stipulate, for they bring out especially clearly just what the ignorance of "configurational" specifics at the time amounted to. Jeans needed the shape of his molecules in order to state equations for what happens when they collide with one another. His real problem, however, lay in how molecules can contain energy beyond that of their translational motion, that is, the energy Clausius had labeled "intra-molecular." Jeans took that to involve distinct modes of vibration within a molecule, vibration that somehow interacts with the ether to produce the lines observed in the spectrum. Much of his paper concerned this interaction.

Our point is that the lack of specificity within the molecular hypothesis involved more than the size and shape of molecules. No less so than in its application to chemistry, its application to gases called for details about the internal structure of molecules. Tying these two applications together were the spectra exhibited by different substances. More than a half century of data had accumulated on these spectra, and their "fingerprint" character had made them indispensable to chemistry. Without question, they were indicating something central to the microphysics of matter, but no one could figure out what that was. While the molecular hypothesis explained an impressive range of chemical and pneumatic phenomena, as it stood it was offering no explanation of spectral phenomena.[61]

Ostwald and Nernst, as the preceding section showed, had several reasons for regarding the status of the molecular hypothesis still to be that of a hypothesis in the 1890s. In the Preface to his seminal abstract formulation of statistical mechanics in 1901, J. Willard Gibbs cited a single reason alone as sufficient:

> Moreover, we avoid the gravest difficulties when, giving up the attempt to frame hypotheses concerning the constitution of material bodies, we pursue statistical mechanics as a branch of rational mechanics. In the present state of science, it seems hardly possible to frame a dynamic theory of molecular action which shall embrace the phenomena of thermodynamics, radiation, and of the electrical manifestations which accompany the union of atoms. Yet any theory is obviously inadequate which does not take account of all these phenomena. Even if we confine our attention to the phenomena distinctively thermodynamic, we do not escape difficulties in as simple a matter as the number of degrees of freedom of a diatomic molecule. It is well known that while theory would assign to the gas six degrees of freedom per molecule, in our experiments

> on specific heat we cannot count more than five. Certainly, one is building on an insecure foundation, who rests his work on hypotheses concerning the constitution of matter.[62]

2.4. Ostwald on How a Hypothesis Can Become Something More

In the 1895 English translation of Ostwald's *Grundriss,* he ends his account of the kinetic theory of gases by briefly noting the difficulties it was having with the specific heats, but with no hint at all that those difficulties weighed heavily against the theory:

> Theoretical investigations have been undertaken to determine the energy necessary for the intramolecular work in compound molecules, but they have not yet met with much success. Not only the number but the nature of the atoms has a great influence on the conditions, as may be seen from the fact that the molecular heats of gases containing the same number of atoms in the molecule have been found different. A mathematical expression for this influence has not yet, however, been given.[63]

Having told the reader at the outset that the molecular hypothesis should be taken as nothing more than a convenient way of visualizing abstract information, and not as something that can be true or false, the book never presents arguments for rejecting it outright.

To find his reasons why he still regarded it as only a hypothesis before the Brownian motion results, we need to turn to a 1907 article defending his "energetics" alternative to it:

> When expressions or notations of magnitudes which can not be observed and measured and for which we can substitute *no definite and empirically determinable value*, occur in a formula by means of which some physical relations are to be represented, we have to deal with an hypothesis. For the task of the exact sciences is to establish the reciprocal relation of measurable and demonstrable quantities, or in other words to find the mathematical forms or functions by which these quantities are interrelated, so that one of them can be calculated when the others are given. In order to establish experimentally such a functional relation it is therefore necessary to measure singly all variable or constant magnitudes which appear in such an equation. No other means is known by which to establish whether the protothetically assumed functional exists or not. As long as a single magnitude appears which is not susceptible to measurement,

> we can not consider the assumed condition as proven. Then, too, such an equation is useless, for since it expresses the condition of a magnitude which is not susceptible to measurement, it makes a statement about a thing which has no influence on, nor significance for, science or life. *For not being susceptible to measurement is only another expression for the fact that nothing at all depends upon this thing. If anything did depend upon it this dependence would be one way to experience something about the thing, and it would be measurable.*[64] [italics added]

Ostwald is not rejecting theory-mediated measurement here, for his *energy* can be measured only through its relation to observable quantities. What he is requiring is that each quantity entering into a physical relation among quantities be measurable independently of that relation.

We will return to this quotation in Chapter 5 when we consider what it was about the Brownian motion results that led Ostwald to change his view of the molecular hypothesis. Notice, however, as we remarked in Chapter 1, how close his requirements are to van Fraassen's requirement of grounding, which itself supplements requirements put forward two decades after Ostwald's article by Weyl.[65] Ostwald's requirements are also at least in the same spirit as the point we made earlier about molecular theory serving only to "explain" phenomena without entering *constitutively* into ongoing research and hence being dispensable without affecting the results of the research. All of these requirements are pointing to a distinction scientists draw in practice between a claim that remains a hypothesis no matter how many phenomena it "explains" and, to use Ostwald's words from the quotation with which we began this monograph, "a scientifically well-founded theory."

Still, given the extent to which Ostwald relies on atoms and molecules as a way of conceptualizing principles of chemistry, one might well ask what a systematic presentation of chemical knowledge without reference to them might look like at the time. Ostwald himself has given us the answer with a book published in 1907 entitled *Prinzipien der Chemie: Eine Einleitung in alle chemischen Lehrbücher*, or *The Fundamental Principles of Chemistry: An Introduction to All Text-books of Chemistry* in its English translation of 1909. Ostwald states the goal of the book in its Preface:

> The extraordinary development of the experimental side of the science of Chemistry has in some measure thrust into the background the work which has been on the methodical side. This work has not been wholly lost, but an examination of the various theories which have been advanced in the past few decades, especially in organic chemistry, indicates that most of them were born of the necessity of a day and that they were ephemeral in their influence.

A desire for generalization is an important and justified one, and the only reason for the unsatisfactory outcome of all these theories is to be sought, in my opinion, in a fundamental error. Hypothetical assumptions were used in their development. Hypotheses were set up in each case with special reference to the phenomena to be explained, and I believe that the right way has been obscured in many instances.

Indication of the right way is given by the experience of other sciences which are older and simpler than Chemistry and which have therefore already attained the necessary ripeness. Mathematics, Geometry, and Mechanics began an examination of their fundamental principles years ago, and a firm foundation has now been set up for each of these sciences. The present time is decidedly philosophical in its trend, and in the past few years this task has been taken up with renewed vigour. The results of this labour form a valuable and fruitful portion of modern scientific knowledge. The end sought is the discovery of final truths and the relations between them, and these, when found, give safe foundations for further investigation. This does not mean setting up analogies and hypotheses, but the careful analysis of concepts and indication of the general facts of experience from which they are derived. . . .

The present book has for its object the presentation of the actual fundamental principles of the science of Chemistry, their meaning and connection, as free as possible from irrelevant additions.[66]

As promised, the book eschews claims about individual elements and compounds, focusing instead on generalizations within chemistry, like the law of combining weights and the law of gas volumes, stated respectively in terms of *combining weights* and *molar weights*. These generalizations extend to topics that would seem especially to lend themselves to statements in terms of atoms and molecules, such as isomerism, valences, and ions. With isomerism, Ostwald introduces a distinction between composition and constitution, where the latter concerns whether factors experimentally forming a compound consist of elements or radicals. This in turn enables him to characterize first valences, and with these, principles of electrolysis. The sole mention of atoms or molecules in the book is in a pair of footnotes, the first giving a historical account of the relation between his "combining weights" and Daltonian "atomic weights,"[67] and the second quoted here:

It is necessary to differentiate carefully between conclusions like these and hypothetical investigations of so-called molecular size among gaseous, liquid, and solid substances. By this is meant the size of hypothetical molecules, which are defined as the smallest amounts of substances which can exist independently. Various considerations, all of which contain a larger or smaller number

> of arbitrary assumptions, or other uncertainties, have been applied with the hope of reaching conclusions about molecular sizes. Such processes lead to the assumption that the weights of molecules must be proportional to the molar weights, as calculated from the laws for gases and solutions.[68]

What is remarkable about the book is the range of generalizations within chemistry that it can capture without any reliance on the concepts of atoms and molecules. Ostwald's changing his position with regard to the latter had nothing to do with their explanatory power.

In saying that scientists in general draw a distinction between hypotheses and "scientifically well-founded" theoretical claims, we are not saying that they all draw it in quite the way Ostwald did. At the heart of this distinction is a distinction, perhaps of degree rather than kind, between having and not yet having been able to establish *definite, empirically determinable values* (to use Ostwald's phrasing) of the key unobservable magnitudes of a theory. Meyer's 1899 text consists in its entirety of a review of the extent to which this had by then been achieved. Part 2 of the book presents values of *mean free path* and *collision frequency*, generally to three significant figures, for an impressive array of gases derived from viscosity and then gives evidence from diffusion for the validity of these values. Part 3 of the book, consisting of a single chapter of the same name as the Part, "On the Direct Properties of Molecules," culminates with numerical estimates of values for molecular *sectional area, volume, number, spacing*, and *absolute* and *specific weight*. He opens this portion of the chapter with the following paragraph:

> Although we may not pretend to see an exact evaluation of the size of the molecular diameter in the value 0.2 [$\times 10^{-7}$ cm], which we have assumed as a mean, yet it seems justifiable to suppose that this number may serve as an approximately correct estimate. It is therefore not lost trouble, and it is more than a play with figures, if we calculated the sectional area and the volume of a molecule from this estimate of its diameter. We shall be conscious that the calculation can give us only approximate values, since we must once more introduce the assumption of a spherical figure, which is not strictly correct.[69]

A page later, he prefaces his value of 61×10^{18} for Loschmidt's number with the remark, "I shall content myself again with an approximate estimate."

So, even though Meyer was not about to say that kinetic theory was not at the time "a scientifically well-founded theory," he was clearly honoring a distinction between having and not yet having been able to establish *definite, empirically determinable values* of the key unobservable magnitudes of kinetic theory. Ostwald, needless to say, was thoroughly familiar with Meyer's book at the time he wrote his 1907 article. We shall return to the distinction we are here calling attention to in the last three chapters of this monograph, using our account of the Brownian motion and attendant results in an effort to clarify it further. For, rhetoric aside, this distinction is at the heart of the change of status of molecular theory that occurred between 1905 and 1913.

2.5. A Further Dimension of the Dispute over Molecular-Kinetic Theory

None of the three textbooks on which we have been focusing in this chapter mentions the controversy over whether the second law of thermodynamics is deterministic or, as Boltzmann had argued, probabilistic, admitting of *possible*, though extremely low probability, exceptions. Ostwald's book devoted an entire chapter to the second law, emphasizing the difference between the total energy of a system Q and the free energy Q'—in his terminology, the amounts of heat entering and leaving a working system; yet he never introduces Clausius's proposed new parameter, entropy:[70]

$$S = \frac{Q - Q'}{T}$$

Meyer, while devoting both an entire chapter and an Appendix to the "applicability of the calculus of probability to kinetic theory" and hence to Maxwell's distribution, never mentions even so much as the distinction between total and free energy, much less the second law of thermodynamics. Nernst is the sole one of the three to single out *entropy*, though only briefly in his initial chapter, remarking that it is "a function [that is, the preceding equation] which is of especial advantage in considering adiabatic processes."[71] (His "third law" of thermodynamics, which turned S from a relative into an absolute magnitude, dates from the period at the center of our monograph, 1905–1913.)

That entropy is so little mentioned in the three books from the 1890s on which we have been focusing is less surprising than those accustomed to the way we now view the second law might think. This law, as Clausius's originally stated it, asserted simply that heat does not flow from cold to hot. Even after Clausius subsequently introduced the term *entropy* and stressed the distinction between reversible and irreversible processes in the statement of the law, few followed him, especially so among chemists and those in the vanguard of the emerging discipline of physical chemistry.[72] Part of the reason for this was "confusion"—Helge Kragh and Steve Weininger's choice of words—about what entropy amounted to physically. In particular, how to measure it at all was less than clear, in part because Clausius's definition of the entropy difference between two states A and

B involved an integral along a path that "corresponds to a reversible transformation from A to B."[73]

Confusion over the concept of entropy and the importance of irreversible processes was at the heart of an extended correspondence in the early 1890s between Max Planck, a champion of entropy and irreversibility, and Ostwald, who denied the centrality of both in thermodynamics.[74] At the time, Planck had been drawn into the research within physical chemistry sparked by van't Hoff's theory of osmotic pressure and Arrhenius's theory of electrolytic dissociation—topics covered in the next two sections of this chapter. Prior to this, he had been focusing on the mechanical foundations of thermodynamics, especially in regard to the second law, following his 1879 dissertation on the two laws. One example of the problem he saw in reconciling the reversibility implied by the first law and the irreversibility implied by the second is his conclusion at the end of an article in 1882:

> In conclusion I should like to call explicit attention to a previously known fact. Consistently developed, the second law of the mechanical theory of heat is incompatible with the assumption of finite atoms. It can therefore be foreseen that the further development of the theory will lead to a battle between these two hypotheses in which one of them will perish. An attempt to predict the conflict's outcome with precision at this time would be premature. Nevertheless, a variety of present signs seems to me to indicate that atomic theory, despite its great success, will ultimately have to be abandoned in favor of the assumption of continuous matter.[75]

Little wonder that Boltzmann's analysis of entropy ultimately had such an influence on Planck.

In sum, Boltzmann was responding to a real need, yet not at the time to a widespread demand, in his analysis of entropy, and hence the controversy over his proposal was more confined at the time than it would have been had entropy assumed the central place in thermodynamics that it subsequently came to have. (This will help to explain both Perrin's discomfort with Planck's derivation of his blackbody law and some of the reactions to Perrin's Brownian motion findings by such defenders of Boltzmann as Einstein and Ehrenfest when we come to Chapters 5 and 6.)

Given the extraordinarily low probability of Boltzmann's suggested exceptions to the second law, one might well conclude that the dispute he precipitated at the time over whether such exceptions are even in principle possible was purely philosophical. However much some at the time thought it was, it was not. In particular, it was not about whether nature is at root somehow stochastic, for no one was denying that the laws of mechanics are entirely deterministic, including at the molecular level. Rather, the issue was about what the quantity S amounted to at the molecular level, and hence was not entirely independent of the issue of whether there really is a molecular level at all. Put even more accurately, the issue was about what the measurement of the quantity S amounts to at the molecular level, in parallel to the claim, central to kinetic theory from its outset, that what the measurement of temperature amounts to is a physical averaging over time and space of the kinetic energy of the molecules.

A further role of theories of underlying processes and mechanisms in physics, besides explaining phenomena, is to determine the physics involved in measuring quantities. Because measurements in physics are themselves physical processes, physics cannot help but include its own theory of measurement. The concern here extends beyond questions about the physics of the complicated instruments used in measurement, their calibration, and their potential sources of error to questions about what the magnitudes assigned to any quantity represent physically. Molecular-kinetic theory from its inception had supplied answers to such questions in the case of temperature and pressure, both of course in terms of the mean kinetic energy of molecules. Boltzmann had supplied his answer in the case of entropy.[76] The history and details of the development of his answer by means of his H-theorem need not concern us here. Its key feature is that entropy of a gas varies as the logarithm of the probability of the state of its molecules. What the dispute over the second law of thermodynamics was about at its core was whether his answer to what entropy amounts to molecularly is correct.

Boltzmann's answer—indeed, his proof of his H-theorem—was inextricably tied to Maxwell's distribution of the velocities in a gas. It was through this link that probabilities entered, namely by virtue of its giving the probability for any molecule having a velocity within a range $v + \Delta v$, with the probability corresponding to the proportion of the molecules having a velocity within this range. Indeed, one consequence Boltzmann immediately draws from his H-theorem is that any other distribution of velocities within a gas "must approach that of Maxwell."[77] The justification for talk of probabilities lies in part in the assumption of equiprobability of certain outcomes when molecules collide with one another in Maxwell's derivation of the distribution from principles of mechanics, but no less so in the fact that the distribution in question is the same as the one Gauss had put forward for chance errors. Meyer expresses the latter point in italics:

> Maxwell's distribution, the theoretical foundation of which rests on the calculus of probabilities, agrees exactly in form with another law which is also founded on this calculus. *The possible values which the components of the molecular velocities can assume are distributed among the molecules in question according*

> *to the same law as the possible errors of observation are by the method of least squares distributed among the observations.*[78]

Accordingly, the question of whether Boltzmann's analysis of what entropy amounts to at the molecular level is also inextricably tied to the question of whether Maxwell's distribution is correct.

While Meyer says nothing in his book about Boltzmann's analysis of entropy, his book does develop a 50-page chapter on "Maxwell's law," supplemented by a nearly 50-page mathematical appendix. His chapters and mathematical appendices on "Molecular Free Paths" and "Viscosity of Gases" note the contrast between Clausius's equation for the mean free path, "obtained on the assumption of equal speeds among the molecules" and Maxwell's, obtained on the assumption of his distribution: where Maxwell has a value of $1/\sqrt{2}$ in his equation for the mean free path, given earlier in section 2.2, Clausius had a value of 3/4.[79] Insofar as this equation forms the basis for the analysis of viscosity within kinetic theory, and (as we noted earlier) Meyer confirms that theory by means of a cross-check provided by diffusion coefficients, one might think this difference could have provided evidence one way or another bearing on Maxwell's distribution. It did not, for the simple reason that the evidence in support of the accounts of viscosity and diffusion was not sufficiently precise to discriminate between the two values, 0.707 and 0.75.

Ostwald too noted the difference between Clausius's assuming equal speeds among molecules and Maxwell's distribution of speeds in his chapter "The Kinetic Theory of Gases," remarking (in fine print), "Instead of the expression 'kinetic energy' we should, strictly speaking, substitute 'average kinetic energy' in the above deductions. This of course does not in any way alter the general results arrived at." Returning then to the main text, he adds, "This result, that the different molecules of a gas must have different velocities, *abstract as it appears on account of our inability to measure these velocities,* is nevertheless of great value for the right comprehension of many chemical processes"[80] [italics added]. His concession of a distribution of velocities should not be taken, however, as an endorsement of Maxwell's specific distribution. In fact, the Maxwell distribution is nowhere else in the book ever mentioned. Ostwald seems to have regarded it as a still further hypothesis attached onto the molecular hypothesis, with no means at all for empirically assessing whether it is correct.

In sum, the dispute over the second law of thermodynamics is best thought of as a dispute over Boltzmann's analysis of what *entropy* amounts to at the molecular level. And, unlike what kinetic theory claims about the sort of statistically aggregated quantities *temperature* and *pressure* amount to at the molecular level, under Boltzmann's analysis *entropy* is not a simply aggregated quantity insofar as it depends on the specific character of the distribution of the velocities among the molecules, and not just its mean. As Ostwald notes, under the molecular-kinetic theory of heat, the velocities of the molecules surely are non-uniform. Nothing empirical, however, was available at the time to bring evidence to bear on the question of whether this non-uniformity matches Maxwell's distribution. For that matter, nothing empirical gave clear evidence of any motions in nature exhibiting the chance-like behavior entailed by the mathematical form of the Maxwell distribution.

Consequently, regardless of the degree to which the molecular hypothesis and the molecular-kinetic theory of heat still had the status of hypotheses in the 1890s, the status of the Maxwell distribution, and with it Boltzmann's analysis of entropy, had this status even more so. It goes without saying that we have stressed this point here precisely because Brownian motion in the hands of Perrin had such a crucial bearing on it.

2.6. A Major Development in Support of the Molecular Hypothesis

Before we turn to Brownian motion, we need to note one further factor bearing on kinetic theory in the historical background to Perrin's efforts—in some ways the most important factor insofar as it was a development intermediate between Brownian motion in liquids and the kinetic theory of gases. Some stunning results emerged during the 1880s on non-electrolytic dilute solutions, yielding what is now known as the van't Hoff equation for osmotic pressure.[81] Sugar dissolved in water, in contrast to salt, is a typical non-electrolytic dilute solution. Osmotic pressure is the pressure exerted on a semi-permeable membrane, that is, a membrane through which the solvent, but not the dissolved substance, flows freely. The advance opening the way to the stunning results consisted of ways of measuring osmotic pressure, both "direct" ways and, to greater precision, "indirect" ones.[82]

The results in question unfolded in a four-step sequence. The first step was the discovery that the relation between osmotic pressure and the volume of the dissolved substance at a fixed temperature has the same form as Boyle's pV = a constant. The second step was the discovery that the constant in this relationship between pressure and concentration varies with temperature in the same way as Gay-Lussac's pV/T = a constant. The third was the discovery that, once the concentration is expressed as moles (that is, gram molecules) of the dissolved substance per unit volume, that constant is independent of the solvent and the dissolved substance—that is, it seemed to be a universal relationship, much as the ideal gas law was thought to be before Regnault exposed the deviations from it for real gases. The fourth was the realization that the value of this constant for non-electrolytic solvents is, to at least two significant figures, the same as *R*, the universal gas constant. In other words, the (ideal) gas law holds for osmotic pressure in dilute liquid solutions, or as Nernst puts it (in italics): "*the osmotic pressure is exactly the same as the gas pressure which would be observed if the solvent were removed, and the dissolved substance were left filling the same space in the gaseous state at the same temperature.*"[83]

Nernst presents these results expressly in terms of molecules—for example, "in both cases the amount of pressure is determined by the number of molecules, dissolved or gaseous, contained in unit volume"—and then turns to four related results, all stated in italics: (1) "*the relative lowering of the vapour tension experienced by a solvent on dissolving a foreign substance, is equal to the quotient obtained by dividing the number of dissolved molecules* n, *by the number of molecules* N, *of the solvent*"; (2) "*the relative lowering of solubility is equal to the number of molecules of the dissolved substance, divided by the number of molecules of the solvent*"; (3) "*if one dissolves in any selected solvent, equi-molecular quantities of any selected substances, the freezing-point is lowered the same degree in all cases*"; and (4) "*the lowering of the freezing-point is proportional to the molecular proportions of the dissolved substance.*"[84] All these results, of course, could in principle have been stated in terms of fractions of moles instead of number of molecules. Nernst's choice of phrasing underscores the extent to which the results on solutions naturally lent themselves to molecular-kinetic interpretation.

In the two chapters following his chapter "Osmotic Pressure"—"Vapour Pressure of Solutions" and "The Freezing Point of Solutions"—Ostwald too in a couple of cases states the further results for solutions in terms of molecules—e.g., "*The relative lowering of vapour pressure of any solution is equal to the ratio of the number of molecules of the dissolved substance to the number of molecules in the solution.*"[85] But he does not do so for the results on osmotic pressure. The nearest he comes is when he presents the fourth step in the sequence, asking "if Avogadro's law holds for solutions, i.e. if all other substances besides sugar exhibit the same value for *R* when molecular quantities are considered." His answer, in the affirmative, is a qualified yes, adding, "the solutions of the most diverse substances have the same influence, if these substances be contained in the solutions in the ratio of their molecular weight."[86] Nowhere that we have found in his book does he acknowledge that the results for solutions provide any important evidence for molecular-kinetic theory. As late as 1907 he remarked,

> What conclusions had been drawn from the mechanistic hypothesis, for instance, that heat consisted of a motion of atoms? Nothing definite at all; for the kinetic hypothesis which Bernoulli evolved on the condition of gases rests upon a large number of other hypotheses, as is at once apparent from the fact that it has no application to liquid and solid states. With regard to the particular

> character of the supposed motions, the mechanistic hypothesis provided no direct information, and since these hypothetical motions had to have some magnitude and direction, many questions arose which had no empirical meaning, i.e., pseudo-problems (*Scheinprobleme*), to use a happy term of Mach's,—problems of such remarkable character that even if their solution were made possible by some supernatural powers we would be unable to utilize it since it has no reference to observable magnitudes.[87]

Why Ostwald did not see the results on solutions as providing some sort of application of kinetic theory to the liquid state is not clear to us.

During his discussion of osmotic pressure, Ostwald does develop one point (all but the first sentence in fine print) that, though an aside here, will be important later, when we consider Perrin's results for the diffusion of granules in a liquid:

> It has been shown by many other measurements, both direct and indirect, that the law of osmotic pressure holds for all dissolved substances no matter what their nature.
>
> From this fact, that the particles of a substance in solution exercise a pressure on the neighboring pure solvent, the above-mentioned phenomenon of diffusion becomes comprehensible. The particles are driven by this pressure into the solvent, and as the pressure is proportional to the concentration, the action will only cease when the same concentration has everywhere been reached.
>
> The extreme slowness with which the diffusion takes place, notwithstanding the magnitude of the pressures, is remarkable. The cause of this is to be found in the friction which the particles experience in moving through the liquid, and in the very small size of the particles themselves. If we reduce a stone, which would fall several hundred feet in a few seconds, to a very fine powder, this will remain under similar circumstances suspended in the air for hours, although gravitation acts upon it with exactly the same force as before. When, therefore, we consider the extraordinary minuteness of the molecules we begin to understand why such very slow progress is made in spite of the great pressures at work.[88]

What is most notable here is not his mention of molecules, nor even his remark about the vertical suspension of a fine powder versus what Perrin will be saying, but Ostwald's complete lack of mention of Maxwell's probabilistic account of diffusion.

We have said enough about Ostwald's view of the results on solutions; let us return to the results themselves. The product RT gives a characteristic energy for a given amount of gas at a given temperature. With the temperature in absolute units and the amount of gas in appropriate chemical units[89]—usually moles, a term Ostwald introduced—R amounts to a fundamental constant, notwithstanding the small deviations of real gases from the ideal gas law.[90] That a mole of every gas has the same characteristic quantity of energy RT, up to a high level of approximation, *regardless of the chemical composition of the gas*, amounted to a cornerstone discovery in the history of physics and chemistry. Its explanation, according to kinetic theory, lies in the fact that the mean translational kinetic energy of the molecules of all gases would be the same at any given temperature, regardless of their size, shape, and mass, were it not for small forces between the molecules that vary from one kind of molecule to another. That is what made the mean translational kinetic energy of molecules the fundamental parameter of the kinetic theory of gases.

But now consider how stunning to some it must have been in the 1880s to have discovered that every mole of any non-electrolytic dissolved substance in a dilute liquid solution has this same characteristic quantity of energy RT up to a high level of precision, regardless of the nature of the substance or solvent. We say "stunning," but of course it had to have been highly satisfying as well to the supporters of molecular-kinetic theory, for they then had evidence that in liquids too the microphysical quantity governing temperature is the mean translational kinetic energy of molecules.

Van't Hoff's equation was what Poincaré was referring to in his remark in the quotation we gave in Chapter 1:

> The kinetic theory of gases has acquired, so to speak, unexpected props. New arrivals have modeled themselves upon it exactly; these are on one side the theory of solutions and on the other the electronic theory of metals.[91]

Such "unexpected props" that kinetic theory acquired—that is, the phenomena it explained beyond those for which it was constructed to explain—many at the time, and now too, would have cited as more than adequate grounds for accepting the theory and thereby granting the reality of its posits. So, what was missing? The answer should be obvious by now: *no way of determining definite values for its central quantity*, the mean translational energy per individual posited molecule, *and hence no independent way of confirming that, at any given temperature, it is the same regardless of the substance.*

Van't Hoff's equation inductively generalized results from experiments involving dissolved substances like sugar of comparatively low molecular weight. A natural question was whether it extends to solutions of substances of indefinitely large molecular weight, even to colloidal solutions and emulsions. Nernst's *Theoretische Chemie* has a chapter on colloids giving experimental results before 1893.[92] As expected from the known very slow diffusion rates of colloids, their osmotic pressures were quite low, raising worries about how accurately they were being measured. Lacking chemical formulas, the "molecular" weights (in excess of 1,000) had to be inferred from the measured osmotic pressures. The experimental uncertainties were thus too great to reach conclusions, one way or the other, about the applicability of van't Hoff's equation to emulsions. But then, as we shall see in Chapter 4, that is where Brownian motion and Jean Perrin enter the story.

2.7. Ions in Solutions: van't Hoff, Arrhenius, Ostwald, and Nernst

The preceding account of van't Hoff's discovery is misleading in one respect. It is stated from the perspective of the mid- to late 1890s, not from that of the mid- to late 1880s when it unfolded. In particular, in the preceding section we have repeatedly restricted his discovery to *non-electrolytic* dilute solutions, a restriction not made by van't Hoff until after he had learned of advances that the comparatively unknown Svante Arrhenius had made in Sweden. Although the distinction between electrolytic and non-electrolytic solutions did not have much bearing on the Brownian motion research, it was indirectly important to some of the determinations of Avogadro's number that Perrin came to cite as convergent with his determinations from Brownian motion. It also had a major bearing on the influence J. J. Thomson's research had on Ostwald's finally abandoning his insistence that molecules are merely hypothetical. So, even though its importance to our monograph is somewhat indirect, we are going to briefly summarize the extraordinary research that brought the four figures listed in the title of this section together in the years between 1887 and 1890, and with it secured the full establishment of the new discipline of physical chemistry.[93]

In his initial papers on osmotic pressure, published in 1886, van't Hoff had included an additional purely empirical coefficient, i, in his key equation, $pV = iRT$, instead of the $pV = RT$ form discussed in the preceding section. This was to accommodate the exceptions revealed in his experiments to the established principle, stated in Ostwald's words, "if one takes chemically comparable or 'molecular' quantities of gases the constant R of the gas equation is independent of the nature of the substance."[94] Continuing with Ostwald's phrasing, "This proved right in a great many cases but not in other, very important cases. The exceptions were mostly known substances like the salts and acids in inorganic chemistry."[95] The extent to which the value of i deviated from unity in the case of these exceptions depended on the dissolved compound.

Unbeknownst to van't Hoff, Arrenhius had written a dissertation in Sweden on electrical conductivity in solutions and electrolysis, including Faraday's law of electrolysis, with the results published in that country in 1884. Ostwald responded with great interest to this work, largely because it revealed a "*correlation between the electrical conductance and the capacity for chemical action.*"[96] Again in his words, "For a number of years W. Ostwald had been pursuing the question how the so-called magnitude of affinity or chemical affinity of different substances is to be measured."[97] One of Arrhenius's conclusions was that "coefficients of affinity are proportional to the electrical conductances of the solutions concerned."[98] The prospect of this relationship providing him a basis for an indirect measure of affinities had led Ostwald to conduct a series of experiments testing its validity. "The result was a brilliant confirmation of Arrhenius's theory."[99] Among other things, this had established an affiliation between the two of them by 1887. It was early in that year that Arrhenius encountered van't Hoff's paper.

Initially in a letter, then subsequently in articles, Arrhenius appealed to his theory of electrical conductances in solutions to claim that the values of *i* greater than unity result from dissociation of electrolytic solutes into separate, equal, but oppositely charged ions. That some, but not all, solutes naturally dissociate into such separate, charged ions, with the fraction of the total solute thus dissociating depending on the level of dilution, was central to his account of electrical conductance. In his letter and subsequent articles, Arrhenius showed that values for van't Hoff's *i* could be derived from a simple formula involving this fraction and the amount of solute. That is, Arrhenius provided evidence ranging over a number of solutes that, in the context of his theory of dissociation, van't Hoff's *i* was not a mere empirical coefficient, but was instead derivable from measured values of two parameters governing electrical conductances in solutions.[100] This formed the basis for the distinction between $pV = RT$ in the case of non-electrolytic solutions and $pV = iRT$ in the case of electrolytic ones. Moreover, in tying van't Hoff's discoveries about osmotic pressure in solutions to Arrhenius's discoveries about electrical activity in them, the latter's proposal was pointing to a unified, comprehensive "theory of solutions."

In order to explain how this proposal played out at the time, we need to give some details of the context in which it was put forward. That year, 1887, was also the year in which the 34-year-old Ostwald completed his massive two-volume *Lehrbuch der allgemeinen Chemie* and, in September, assumed the position of professor of Physical Chemistry and director of the Physical-Chemistry Laboratory of Leipzig University. In this capacity he hired 23-year-old Walther Nernst as an assistant to oversee the one of three laboratories under him devoted to physical chemistry. Though Nernst's training was in physics, not chemistry, the work he had done in electromagnetic experimental physics made him especially suitable for the position in Ostwald's laboratory, given the burgeoning developments in ionic dissociation theory from Arrhenius's response to van't Hoff earlier in the year.[101] The year 1887 was, moreover, the year Ostwald and van't Hoff initiated the first journal devoted exclusively to "physical chemistry," *Zeitschrift für physikalische Chemie*. Arrhenius's proposal in response to van't Hoff's theory first appeared in print in the first volume of the journal later that year, and several further developments pertaining to the new theory of solutions appeared in its subsequent volumes over the next few years. Erwin Hiebert was not exaggerating when he claimed, "The *annus mirabilis* of physical chemistry was 1887."[102]

Arrhenius's dissociation theory met with substantial resistance when it appeared in print.[103] Of course, as we have emphasized earlier through quoting Nernst, virtually nothing was known about what holds a chemical compound together. Nevertheless, the claim that a varying fraction of some, but not all, chemical compounds dissociate—spontaneously, as it were—into oppositely charged ions of their constituent elements seemed incompatible with the concept of securely bound compounds. Here again, Ostwald, ever first and foremost an experimentalist, took it upon himself and his laboratory to develop conclusive evidence in support of the new theory of solutions formed by the combination of Arrhenius's and van't Hoff's theories. The results of the ensuing experiments, published in a series of articles in his journal, responded point by point to the opposing arguments.[104] To give one example, theory required C in the following formula to be a constant in the case of any binary electrolyte across all dilutions:

$$\frac{m^2}{(1-m)v} = C$$

where m is the ratio of the conductance to the maximum conductance and v is the dilution. To quote Ostwald, "the quantity C proves to be constant over such a wide range that the dissociation formula was never tested nor could be tested over such a wide range for gaseous compounds."[105]

Much of the resistance to the new theory, as Ostwald saw it, centered on the difficulty of "accepting free ions, which one has to conceive as nothing but a property of the element concerned."[106] A paper in 1889, coauthored by Ostwald and Nernst, "Über freie Ionen," presented a range of experimental results that did much to end this resistance. Nernst continued to do research supporting the dissociation theory and its implications after he left Leipzig for Göttingen in 1890. Nernst's efforts on the new theory of solutions can be seen in the notably greater extent to which the first (1893) edition of his *Theoretische Chemie* treats the topic than even the second (1890) edition of Ostwald's *Grundriss* does. Arrhenius singled out Ostwald and Nernst, as well as van't Hoff, in his 1903 Nobel Lecture for the experimental support they had provided for his theory.[107]

Arrhenius received the third Nobel Prize in Chemistry in 1903 "in recognition of the extraordinary services he has rendered to the advancement of chemistry by his electrolytic theory of dissociation." The penultimate paragraphs of the Presentation Speech on the occasion are worth quoting in full:

> In the time of Berzelius this notion [*viz.* electrical charges contained in the constituents of reacting substances] rested on a qualitative basis only, whereas Arrhenius's theory determined it quantitatively, thus allowing it to be treated mathematically. In his doctor's thesis, twenty years ago, Arrhenius had deduced from this principle all known laws governing chemical changes, but despite this the new theory was very little understood. It so conflicted with current ideas as to disprove them. According to this theory, for instance, common salt, sodium chloride, when dissolved in water splits up to a varying extent, in other words it is *dissociated* into its constituent parts which are diametrically opposed but charged with electricity, i.e. into *ions of chlorine and sodium*, the only chemically effective substances in a solution of common salt. The theory also claimed that when an acid and a base react upon one another, water is the primary product and salt the secondary, and not reversely, as was then generally believed. Ideas so contrary to those at that time could not be accepted immediately. A struggle lasting more than ten years and an enormous number of new experiments were required before the new theory was accepted by everyone. During this long battle over Arrhenius's theory of dissociation tremendous advances were made in chemistry and ever closer links were established between chemistry and physics—to the benefit of both sciences.
>
> One of the most important consequences of Arrhenius's theory was the completion of the great generalizations for which the first Nobel Prize for Chemistry was awarded to **Van't Hoff**. Without the support of Arrhenius's theory that of van't Hoff would never have gained general recognition. The names of Arrhenius and Van't Hoff will go down in history of chemistry as marking the modern period of this science and it is for this reason that the Academy, despite the fact that the experimental basis of the theory of dissociation belongs to physics, did not hesitate to award the Nobel Prize for Chemistry to Arrhenius.[108]

With this we end our case that the development of the new theory of solutions combining van't Hoff's theory of osmotic pressure and Arrhenius's theory of electrolytic dissociation, coupled with the experimental efforts, with Ostwald and Nernst at the forefront, confirming it, secured the full establishment of the new discipline of physical chemistry in 1887 and the years immediately following.

We ended the preceding section with the remark, "this is where Brownian motion and Jean Perrin enter the story." In a sense he was already part of the story before he began his research on Brownian motion.[109] In the year van't Hoff received the inaugural Nobel Prize in Chemistry, Perrin was initiating research in his newly created physical-chemistry laboratory at the Sorbonne, focusing on osmosis, ion transfer, and electrical charge at surfaces. He published his paper "Les hypotheses moléculaires" in that year, expressing support for the hypothesis as a physical chemist. In the year Arrhenius received his Nobel Prize, Perrin published his text, *Traité de Chemie Physique. Les Principes*.[110] By 1905 he was focusing on electricity in colloid solutions, a natural entrée into his research on Brownian motion.[111] Surely, therefore, he must have seen this latter research as a continuation of the efforts culminating in the theory of solutions for which those two had received the Prize.

What we have not done so far is to make clear what the name "physical chemistry" designated beyond research on the edge between the two disciplines. The remark in the preceding quotation about how Arrhenius had gone beyond Berzelius—"quantitatively, thus allowing it to be treated mathematically"—underscores what seems to us the key feature. After Lavoisier introduced the "balance sheet," measurement of weights had been central to chemistry, but other measurements, like freezing points, had not found connections to chemical parameters. The work we have been describing in this and the preceding section of this chapter, by contrast, centered on indirect—that is, theory-mediated—measurements licensed by relationships tying parameters of prime interest to chemists to parameters drawn from physics, like electrical conductance. This was Ostwald's vision all along, finding parameters from physics that would enable measurement of such quantities central to chemistry as affinities and "reaction velocities." In conjunction with the standing that physical chemistry had gained from the success of the new theory of solutions, Ostwald produced not only his 1896 thousand-page history of electrochemistry heralding the contribution such measurements had made ever since Volta introduced his cell in 1800, but even before this, his 1893 *Manual of Physico-Chemical Measurements* instructing a wide audience of chemists on the basics of carrying out such measurements.[112]

Because theory-mediated measurement is a central topic of the present monograph, a single example of how matters had changed following 1887 will help prepare the way for later chapters. As we presented it earlier, Arrhenius's contribution to van't Hoff's theory was to show that the values of a purely empirical parameter, i, which bordered on being a fudge-factor, could be derived from measured values of two parameters governing electrical conductivity in solutions. This is not, however, the way in which Nernst presented the result in the chapter on electrolytic dissociation in the 1893 edition of his *Theoretische Chemie*. Rather, he turned the result on its head, so to speak: "The degree of the dissociation, or the value of the coefficient of dissociation, by which is meant the ratio of the dissociated molecules to the whole number of molecules, can be directly obtained in two independent ways; viz. from the osmotic pressure and from the conductivity."[113] This does more than just allow the quantity of primary interest for chemistry to be measured from two distinct quantities of physics; it announces a pair of *convergent* theory-mediated measures for this quantity, in the process pointing to the evidential implications for the theory of free ions as well as the relationships licensing these measures. The evidential implications of convergent theory-mediated measures will come up repeatedly in Chapters 4 through 6.

Nernst's reference to the dissociated and whole numbers of molecules in the preceding quotation underscores how natural it was at the time to speak of dissociation as molecules splitting into charged ions. Three pages earlier, he had presented the equation, $HCl = H^+ + Cl^-$, to draw two conclusions: "the ions of hydrochloric acid are thus monatomic" and (in italics) "*an atom of hydrogen and an atom of chlorine are electrically equivalent*."[114] So, just as van't Hoff's theory had provided support for molecular-kinetic theory (especially after Boltzmann had derived the law of osmotic pressure from the kinetic theory of gases[115]), Arrhenius's theory provided correlative support for the atomic-molecular theory of chemistry. An obvious question, therefore, is why the empirical evidence for Arrhenius's theory did not end the hypothetical standing of that theory. The answer is also obvious. Insofar as what atoms amount to and how they combine to form enduring molecules was still almost entirely a matter of conjecture, to an even greater extent was how an atom could be or become electrically charged. Consider, for example, the sort of proposals that were made in response to J. J. Thomson's 1897 conclusion that cathode rays consist of negatively charged particles of an extraordinarily small mass-to-charge ratio—twists of ether becoming attached to molecules.[116] Ostwald's phrasing in his explanation of the reluctance of others to accept free ions that we quoted earlier says a good deal: "which one has to conceive as nothing but a property of the element concerned."

This point can be made in another way that will better prepare the reader for Chapter 6. Immediately after Nernst draws the two conclusions, the second in italics, quoted in the preceding paragraph, he draws a third, again in italics: "*chemically equivalent quantities of the most various ions are set free from the most various solutions, by the same quantity of electric current*," adding that this is "simply the fundamental law of electrolysis, advanced by Faraday and demonstrated most exactly by experiment."[117] The value of the constant that became known as Faraday's was firmly established: 96,540 coulombs are required to yield a mole of any monovalent substance.[118] Nernst's statement suggests that this number is simply the product of the number of molecules in a mole—that is, Avogadro's number—and the characteristic charge of a free ion. The dissociation theory had not provided any way of determining the latter number, nor any way of counting the number of free ions. Had a reliable value for Avogadro's number been established, a value for the charge could be inferred by taking Faraday's constant to be this product. Equally, a determination of the charge could yield a value for Avogadro's number that could be compared with other values. Best of all, determinations of each of these values independently both of each other and of Faraday's constant could eliminate any need for conjecture about whether indeed Faraday's constant is the product of the two.

As we shall see in Chapter 6, this is where the research on the electron initiated in 1897 in J. J. Thomson's laboratory in Cambridge, and a decade later extended in Rutherford's laboratory in Manchester and Millikan's in Chicago, enters the story. The important point for now is that Arrhenius's research had linked electrolytic dissociation to Faraday's constant in much the way, though not to the same extent, that van't Hoff's had linked osmotic pressure to the gas constant. These two macrophysical constants will come to anchor, so to speak, the efforts to gain experimental access to the molecular realm through theory-mediated measurements in the 1905 to 1913 period on which we will be focusing in Chapters 4 through 6.

2.8. By What Authority Still Only a Hypothesis?

Although we have not surveyed the literature to verify it, we are confident that a large majority of the research community of physicists had ceased questioning the atomic and molecular composition of matter, if not before, then surely after the unqualified experimental demonstration of van't Hoff's claim that the gas law extends to liquid solutions. Judging from the Presentation Speech on the occasion of the inaugural Nobel Prize in Chemistry being awarded to van't Hoff in 1901—"As a result of his investigations in the fields of atomic and molecular theory, van't Hoff has made the most important discoveries in theoretical chemistry since Dalton's time"[119]—the same seems by then to have become true as well of the research community of physical chemists and, though perhaps not yet to the same extent, even that of experimental chemists. We feel confident in saying this because scientists then, as much as now, placed great weight on the sort of considerations expressed in our quote from Wichmann in Chapter 1, that is, on "the remarkable understanding of natural phenomena which was achieved during the nineteenth century on the hypothesis that matter is made of atoms."[120] Indeed, that atoms, molecules, and molecular-kinetic theory were providing such *understanding* is attested to by Ostwald's reliance on them in his explications of phenomena within physical chemistry throughout the pre-1900 editions of his *Grundriss*. Granting that such a large fraction of the relevant research communities had ceased questioning the reality of molecules by 1900, however, obligates us to end the chapter with a response to the question, *By what authority did the reality of atoms and molecules and their relation to heat still have only the status of conjectured hypotheses in 1900?*

The answer is *not* the personal authority of Ostwald and Nernst by virtue of their standing in the field. If that were the answer, the last part of the Wichmann quote in Chapter 1 would be more legitimate than it is—"In view of such evidence for the existence of atoms it is hard to understand a certain school of thought, which persisted until the turn of the century, and which rejected the atomic hypothesis on the grounds there was no *direct* (!) evidence that matter is made of atoms."[121] Rather, the authority consisted of the three book-length review articles we have relied on so heavily in this chapter, by Ostwald on the overall state of the field of physical chemistry, by Nernst on the state of theoretical chemistry, and by Meyer on the state of the kinetic theory of gases.[122]

The raison d'être of any review article is to identify what questions within the topic the available evidence has settled, what questions it has yet to settle, and what factors are standing in the way of making progress on the latter. The whole point of any review article is to provide the research community—those working within the topic area, those whose work is on its periphery, and even those at some remove from it—a balanced, thorough, objective assessment of the state of the evidence at the time. Failure to do so is a disservice rather than a service to the field. We are not denying that prominent figures have sometimes failed to meet these standards. We know of no one, however, complaining that the three books in question misrepresented the state of the evidence; and judging from the international standing of these books for over a decade, and from our own comparison of the assessments of the evidence in the three, Ostwald, Nernst, and Meyer met those standards.

As we have seen during the course of this chapter, the limitations of the evidence at the time consisted of having not yet managed to experimentally establish specific features of atoms, how they combine to form molecules, specific features of those molecules, and how they interact with one another. This, however, is fully consistent with a large fraction, even a substantial majority, of those engaged in research in physics and chemistry at the time having ceased questioning the atomic and molecular composition of matter. Their research in no way depended on any specifics about atoms or molecules, nor on any details about the motions of the latter! The value of atomic theory and the molecular-kinetic theory of heat for them was to provide a unified way of conceptualizing—van Fraassen's word may be even better, *picturing*—a wide range of phenomena without having to concern themselves with any specifics or details. The latter, and any difficulties in experimentally establishing them, they could leave to those at the cutting edge of microphysics to work out. Precisely because nothing in their research depended on specifics of atoms and molecules, atoms and molecules could remain amorphous for the purposes to which they were putting them. But that is tantamount to saying that the results they were obtaining in their research in no way constitutively presupposed that atoms and molecules even exist.

Contrast this with the situation of those on the cutting edge of microphysical research. The aim of that research was to establish specifics of the microphysical realm and their links to macrophysical phenomena, that is, links determining the differences specific microphysical details make to specific features of various macrophysical phenomena. Consequently they could not help but be confronting all the time limitations of the evidence pertaining to atoms, molecules, and their interactions of the sort we have noted in this chapter. This, moreover, would have made them appreciate the extent to which any microphysical details that were being taken to have been established—for example, the mean molecular velocity at any given temperature—were actually attained from their having been postulated as part of chemical atomic theory or molecular-kinetic theory in the first place—for example, that 2/3 of the mean molecular kinetic energy of a gas is equal to the product *RT*. The difficulties they were having in experimentally establishing, in contrast to postulating or hypothesizing, specifics in the microphysical realm had to have constantly been calling their attention to the fact that they had not yet gained the sort of experimental access to this realm needed to do so. But that in turn would have constantly reminded them that, so far as their research was concerned, the atomic and molecular composition of matter still had the status of a hypothesis.

The contrasts we are drawing in the preceding paragraphs showed up, somewhat strikingly, at the 1901 Nobel Award ceremony for van't Hoff. In his presentation speech, C. T. Odhner singled out the results summarized in the preceding section of this chapter, saying that van't Hoff had "proved that gas pressure and osmotic pressure are identical, and *thereby* that the molecules themselves in the gaseous phase and in solutions are identical"[123] [emphasis added]. In discussing this result in his following Nobel Lecture, van't Hoff's reasoning was more careful:

> It was found namely that with sufficiently dilute solutions the osmotic pressure was the same as the gas pressure—i.e. the pressure which the dissolved substance would exert as a gas. To some extent this is obvious: Just as one imagines the gas pressure *P* to arise as a result of the movement of molecules and of their collisions with walls (Fig. 1), so can one imagine the osmotic pressure *p* to arise as a result of the collisions of the dissolved molecules with the semi-permeable membrane (Fig. 2) surrounded by the solvent (denoted by shading).
>
> However, *independently of an anyway hypothetical cause of this pressure*, it was found that under the same circumstances, i.e. with the same number of

> molecules in the same volume and at the same temperature, the pressures also were the same.[124] [italics added]

It goes without saying that, even had a reliable value for Avogadro's number been available at the time, the only means they had of verifying that the number of molecules in a given volume was the same was in terms of the fractions (by weight) of a mole of the relevant substances. In other words, van't Hoff's extraordinary finding in no way depended on the existence of molecules, however ideal they were for understanding, and picturing, his result. Odhner would have reflected the state of the field better had he instead said, ". . . and thereby that the molecules themselves in the gaseous phase and in solutions are identical, taking for granted that molecules exist and effect pressure on walls and membranes by collisions with them."

(Another way to draw the contrast between the large fractions of the research communities that had ceased questioning the atomic and molecular composition of matter and those still regarding it a hypothesis is that the former were at some remove from the details of the evidence at the time, while the latter were quite close to them. More often in the history of science this difference works in the opposite way: those close to the evidence see it as having established some exceptional new advance in theory, while those more distant from it are unwilling to accept the result because of difficulties in reconciling it with the way they conceptualize phenomena.[125])

One of the two main issues of this monograph is, *What did the new standing of atomic-molecular theory at the end of the first decade of the twentieth century amount to*? To do justice to this issue will require a clear understanding of what the prior standing amounted to. Saying that the theory still had the standing of a hypothesis before then is too vague to be instructive. So too are the statements by Ostwald and Nernst in their 1909 updates of their book-length review articles, that is, in the new editions of their books: "The atomic hypothesis is thus raised to the position of a scientifically well-founded theory" and ". . . the theory begins to lose its hypothetical character." The fact that the theory did *not* still have the standing of a hypothesis in so many eyes at the end of the nineteenth century all the more demands that we spell out what its standing as a hypothesis in the eyes of those closest to the details of the evidence amounted to. Fortunately, the contrasts drawn in the immediately preceding paragraphs put us in a position to do this.

Consider first what the hypothetical standing of the molecular-kinetic theory amounted to. The theory arose as an explanation of what was then taken to be *the* gas law, and following Clausius it had the promise of explaining such phenomena as the systematic deviations from that law discovered by Regnault.

Absent, however, determinations of the sizes and shapes of molecules and whatever forces there are between them, those promises took the form of purely qualitative, hypothetical explanations. Contrast now the situation in the second half of the nineteenth century with a fulfillment of that promise in the form of experimental determinations of their sizes and shapes that might then have allowed the forces between them to be derived from specific patterns in the deviations from the gas law for each kind of gas (as Maxwell had hoped) and, with those forces, then derivations of the individual systematic deviations over the full range of temperatures and densities. Notice how the latter development would support counterfactuals specifying how the deviations from the ideal gas law would change if the size of the molecules in the gas were 10 percent greater, or how the mole density would change in specific circumstances as a consequence of molecular dissociation or from the spontaneous combining of, for example, pairs of NO_2 molecules to form nitrogen peroxide (N_2O_4) molecules as the temperature is lowered.[126]

Similarly, the statistical mechanical formulation of the molecular-kinetic theory gave generic qualitative explanations of the phenomena of viscosity and its independence of density in gases, yet absent any determination of the sizes and shapes of molecules of different kinds, it offered only a hand-waving explanation of the specific differences in viscosity from one gas to another or the specific systematic variation of viscosity with temperature in each gas. The inability to determine the sizes and shapes of molecules of different kinds resulted in the theory of viscosity and diffusion being a theory about hypothetical spherical molecules, and even then, efforts to determine their sizes had failed to yield determinate values. In short, the molecular-kinetic theory of heat, even with everything that could be said in its favor, remained a theory about hypothetical discrete entities.[127]

The contrast we have drawn here, while emphasizing a different aspect of the situation, is not unrelated to the critique of molecular theory Ostwald lodged in his 1907 article that appeared in translation in the *Monist*: "When expressions or notations of magnitudes which can not be observed and measured and for which we can substitute no definite and empirically determinable value, occur in a formula by means of which some physical relations are to be represented, we have to deal with an hypothesis."[128] Nor is it unrelated to van Fraassen's contention that the shortcoming in molecular theory before Perrin was its lack of what he has labeled "grounding." Our point is that, absent "definite and empirically determinable" values of the magnitudes that distinguish one kind of molecule from another, *the theory has no way of relating specific macroscopic contrasts to specific microphysical details that govern those contrasts*. Pursuing such relations among specific macroscopic variability and specific microphysical detail amounts to a way of testing the generic qualitative explanations that give reason to take a theory seriously in the first place. Often it is the only real way to test the theory. Thus our point is tantamount to saying that the molecular-kinetic theory of heat had yet to be truly tested.

The same can be said of the atomic-molecular theory of chemical combination. Dalton had put it forward, in response to experimental results on multiple oxides of nitrogen, to explain the "laws" of fixed and multiple proportions. Eighty years later, however, no explanation was yet forthcoming for what specifically it is about nitrogen and oxygen atoms that explains why two of their compounds, NO and $NO_{2,}$ and only these, are stable.[129] Especially important to the history of the gas law was the phenomenon of NO_2 gas transforming as the temperature is lowered into a mixture of it and nitrogen peroxide, N_2O_4. (When this occurs, the molar density of the gas changes, yet its density does not, so that the preferred statement of the gas law has to be in terms of the former.[130]) The obvious molecular explanation of this is that the number of molecules in the gas decreases when pairs of NO_2 molecules combine to form an N_2O_4 molecule, and this decrease then alters the relationship between temperature and pressure in the gas. This, however, does not begin to explain what it is about NO_2 molecules that leads to their forming N_2O_4 molecules when, for example, pairs of H_2O molecules do not join to form stable H_4O_2 molecules, nor why the phenomenon occurs in the conditions it does. Absent the relevant specifics of NO_2 molecules, explaining the formation of nitrogen peroxide in terms of pairs of these molecules joining to form a larger molecule provides nothing more than a relatively crude, qualitative way of picturing what is happening.

Needless to say, we could go on indefinitely with examples like these of contrasting chemical phenomena for which atomic-molecular theory at the time could offer no specific explanation. The reason why such examples abound was given by Nernst in the sentence we quoted earlier that immediately followed his reasons for pursuing atomic theory notwithstanding its hypothetical standing: "At present scarcely anything definite is known regarding either the nature of the forces which bind the atoms together in the molecule and which hinder them from flying apart in consequence of the heat of motion, or regarding their laws of action."[131]

The distinction we have been invoking between generic qualitative explanations of phenomena, on the one hand, and specific quantitative explanations of detailed features of them, on the other, is less widely acknowledged than it ought to be. The success of the atomic-molecular and molecular-kinetic hypotheses in explaining a wide range of phenomena generically and qualitatively had provided a unified way of conceptualizing and hence, in a sense, understanding them. This, we agree, gave reason for *believing* in atoms, molecules, and their relation to heat. But it did not give reason for concluding that experimental efforts had succeeded in gaining access to atoms and molecules. Gaining experimental access to atoms, molecules, and their motions, moreover, was a prerequisite for establishing specific features of them that could then lead to success in quantitatively explaining distinct aspects of those phenomena. And, one would naturally conclude, evidence that experimental access had been gained to atoms, molecules, and their motions would have amounted to difficult-to-dispute evidence that they exist.

The requirement that we are saying had not yet been met as of 1900 should not be confused with Wichmann's claim that what was being demanded was "*direct* (!) evidence that matter is made of atoms." Wichmann's italics and exclamation point were presumably intended to underscore the absurdity of demanding direct observational evidence of unobservables. We have throughout the preceding paragraphs used the phrase "experimental access" expressly to contrast with visual access, aided or unaided. We have no visual access to electric current, witness whereunto physicists had the flow of electric charge in current going in the wrong direction, from positive to negative, until a paper by J. J. Thomson in 1900 called everyone's attention to the asymmetry of charge.[132] Experimental access had nonetheless been gained to electrical current decades earlier, following Ampère's discovery of the law named after him and his proposing the "galvanometer" (his word) incorporating this law into an instrument to provide a theory-mediated means for measuring its magnitude with precision. Achieving experimental access to unobservables requires combinations of experiment and theory, especially theory that opens the way to theory-mediated measures of quantities. The three book-length review articles we have been citing make clear that, despite all the nineteenth-century efforts, the requisite combinations had yet to be found in order to secure experimental access to atoms and molecules.

Our answer earlier to the question posed in the heading of this section—*by what authority were the atomic-molecular and molecular-kinetic hypotheses still only hypotheses?*—was these three book-length review articles. While not incorrect, that answer is misleading in one respect. Review articles cannot help but reflect to some extent or other the personal viewpoints of their authors. In the present instance, Nernst's book was free of the warnings in small print in Ostwald's not to take atoms and molecules as anything but a helpful means for conceptualizing matters, and Meyer's book was free of the reminders in Nernst's that atoms and molecules remain hypothetical. Nevertheless, once adjustments are made for their different dates of publication, the three books do not disagree with one another on any point that we have examined concerning the results of experiments and measurements. Perhaps, therefore, a more precise answer to the question of this section is, *by the authority of the state of the evidence at the time.*[133]

Notes

1. J. J. Thomson (1898). Thomson, by the way, used a value for the number of molecules in a cubic centimeter at standard conditions, not a value of Avogadro's number; the value he used, 60×10^{18} (equivalent to an Avogadro number of 134×10^{22}), came from Meyer (1899b). For a fuller account of this history, see Smith (2001b).
2. Townsend (1900).
3. Rutherford and Geiger (1908a), (1908b).
4. In his (1978), Thomas Kuhn argued, we think compellingly, that Planck had not proposed quantization of energy itself in his derivations of his blackbody radiation formula in either 1900 or 1905, and hence Einstein's 1907 paper has claim to being the inception of modern quantum theory. Regardless, the 1907 paper, which included an alternative derivation of Planck's formula, and Debye's subsequent refinement of Einstein's approach to specific heats in 1910 seem to have weighed most in leading Nernst to initiate the first Solvay conference of 1911. See Barkan (1999).
5. Einstein (1989b), p. 218; his 1905 paper cited in the text can be found in the same volume (1989), pp. 86–103.
6. Planck (1967), especially p. 89. As we mentioned in a note in Chapter 1, Planck misstates the value for Avogadro's number in Meyer's *The Kinetic Theory of Gases*. Meyer, in fact, derived a value of 640 trillion molecules in a milligram of hydrogen (1899b, p. 335) from the value of Loschmidt's number he had obtained for air (1899b, p. 333). We shall review Meyer's values later in this chapter.
7. See, again, Kuhn (1978) or, for a briefer account, Hund (1974), pp. 22–40. An important source of the controversy was the way Planck had counted alternative distributions of quanta of energy, namely treating these quanta as indistinguishable from one another in the manner that a quarter century later became known as Bose-Einstein statistics. Planck adopted such counting not on any theoretical basis, but simply to be able to derive the known form of the blackbody formula. In a footnote, in response to criticism of the derivation by Wien, Planck remarked—ironically in the context of our monograph—"From the same point of view one should also declare the kinetic theory of gases to be unsatisfactory since nobody has yet proved that the atomistic hypothesis is the only one which explains irreversability" (Planck, 1967, p. 88).
8. Perrin (1990), p. 156.
9. Maxwell (1875), p. 430.
10. Einstein (1956), p. 1). The phrase in the German title is "*molekularkinetischen Theorie der Wärme*."
11. Rücker (1901), p. 427f.
12. Nyhof makes this thesis clear in the first two lines of his paper abstract: "Towards the end of the 19th century there were those who wished to see the kinetic theory abandoned. This paper attempts to show that this reaction was *primarily due to philosophical objections* rather than the result of scientific difficulties encountered by the kinetic theory" [emphasis added] (Nyhof, 1988, p. 81). Brush similarly maintains: "In retrospect it seems clear that the criticisms of kinetic theory in this period were

motivated not primarily by technical problems, such as specific heats of polyatomic molecules but, rather by a general philosophical reaction against mechanistic or 'materialistic' science and a preference for empirical or phenomenological theories as opposed to atomic models" (Brush, 1976a, p. 245).

13. The "background information," which draws on Brownian motion research from the latter half of the nineteenth century, essentially amounts to the rejection of external causes for an explanation of the movement. Achinstein (2001) appeals to the background information in assigning a prior probability greater than ½ for the claim that "chemical substances are composed of molecules." Psillos, in all three of (2011a), (2011b), and (2014), makes the same appeal to suppose the prior probability of the atomic hypothesis as "not very low." In formalizing their respective arguments using Bayes's Theorem, they both assume an ad hoc prior probability, which ultimately depends on the "background information" on Brownian motion. Mary Jo Nye also credits background research on Brownian motion—in particular, the efforts of Léon Gouy—for definitely establishing and demonstrating the main characteristics of Brown's phenomenon. See Nye (1972), p. 28.
14. Riedi (2013).
15. See, in particular, Clark (1976); Gardner (1979); Nyhof (1988); and de Regt (1996).
16. Ostwald makes this clear as he recounts the results in question, never, however, denying the utility of the hypothesis as a way of picturing chemical substances. More recently, Alan Rocke has made the same point by showing how all the talk of atomic and molecular weights amounted to nothing more than a system of "equivalent" weights that could do all the work they did. See his (1984), and also Ostwald (1907–1909) for direct confirmation.
17. W. Ostwald (1895), p. 37. Ostwald, by the way, notes that other versions of the periodic table had been put forward at roughly the same time differing from Mendeleev's.
18. W. Ostwald (1895), p. 198. For an extended discussion of the advent of structural formulas and their historical significance, see Klein (2003); also, Chapter 9 of Chalmers (2009) discusses these formulas in the context of nineteenth-century chemical formulas in general and their relevance to the molecular hypothesis.
19. Nernst (1895), p. 237, n. 1.
20. W. Ostwald (1895), p. 198. The whole of this paragraph in fine print reads as follows:

> Now, as to the meaning of such formulae, there are two things which must be carefully kept distinct—on the one hand, the chemical valency, on the other hand its representation by the so-called structural formulae. The first is of actual material significance, it is founded on the observation that the atoms may have their capacity of combination satisfied by a certain number of equivalents of other atoms or radicals, no matter what the nature of these atoms or radicals may be. The representation of this fact by structural formulae, in the first instance at least, is purely formal; it only serves to aid the mind in comprehending and retaining the relations experimentally found, and shows whether the postulates of the theory of valency are fulfilled.

Without mentioning Ostwald, Nernst (1895), p. 251, acknowledges this sort of view of structural formulas:

> By means of the methods described, it is possible to ascertain with great certainty the constitution of a legion of carbon compounds the number of which increases every day. These constitutional formulae so obtained, have much significance in the eyes of those who would regard them as established, *entirely independently of all molecular and theoretical speculations*; for these formulae represent to a large degree, and in a very condensed form, their "facies" in many empirical relations, and he who is skilled in reading this symbolism, learns very much regarding the nature of the compound, from what is expressed through the formula. [emphasis added]

This is the opening paragraph in a section entitled "The Limit of the Validity of Structural Formulae," which goes on to introduce the need for van't Hoff's new stereochemistry to supplement structural formulas.

21. Nernst (1895), p. 237. Nernst's footnote quoted in the text and cited in note 19 is attached to the end of this quote.
22. Ibid.
23. Ibid., p. 253f.
24. W. Ostwald (1895), p. 103.
25. Nernst (1895), p. 253.
26. This reasoning is a summary of what Nernst says; ibid.
27. W. Ostwald (1895), p. 192.
28. Ibid.
29. Ibid., p. 204.
30. Maxwell (1875).
31. Maxwell (1875), p. 430.
32. Ibid., p. 420. The relation of this passage to Newton's *Principia* should not go unnoticed.
33. Ibid., p. 421.
34. See Partington and Shilling (1924), p. 29ff, for a listing of more than 50 alternative gas laws that had been proposed between Regnault's work in the 1840s and 1924.
35. See Kipnis, Yavelov, and Rowlinson (1996) for a history of van der Waals's "law" and its relation to some other real gas laws.
36. For example, see W. Ostwald (1895), pp. 84–87, where formulas expressing values for the critical volume, critical pressure, and critical temperature are given in terms of van der Waals's two constants, allowing the formula to be tested by obtaining the values for the "constants" from gases and then comparing inferred with measured values of the critical parameters.
37. The mean-free-path equation used by Loschmidt (1865) was in fact $L=\frac{3}{4\pi}\frac{V}{ns^2}$, which assumes uniform velocity across molecules. In contrast, Maxwell's correction for his velocity distribution changes the factor $\frac{3}{4}$ to $\frac{1}{\sqrt{2}}$, yielding the corrected $L=\frac{1}{\sqrt{2}\,\pi}\frac{V}{ns^2}$, as cited here. This factor difference gives different equations for the molecular diameter—in the case of Loschmidt, $s = 8\,\epsilon\,L$ (where ϵ is the condensation coefficient of a gas) versus Maxwell's equation $s = 6\sqrt{2}\epsilon\,L$. Accordingly, Loschmidt's

value for the molecular diameter s of 9.69×10^{-8} cm underestimates the molecular radius since the corrected diameter under Maxwell's velocity distribution, using Loschmidt's own data, comes out to 10.3×10^{-8} cm. This disparity in the value of s then gives, again using Loschmidt's own data, n as 1.82×10^{18} (under the assumption of uniform velocity) versus n as 1.52×10^{18} (under the assumption of Maxwell's velocity distribution). Given the lack of well-behaved results across the board for molecular parameters based on kinetic theory in the latter half of the nineteenth century, this small numerical difference is historically trivial. That said, we include the uniform-velocity-based value 1.82×10^{18} for n in our Table 2.1 to maintain consistency with the derivation from Loschmidt's original paper of 1865.

38. Table 2.1 is based on results from the following publications: (1) Loschmidt (1865); (2) W. Thomson (1870), p. 553; (3) Maxwell (1873), p. 300; (4) van der Waals (1873), p. 105; (5) Nernst (1895), p. 352; and (6) Meyer (1899b), pp. 319–333. What came to be known as Loschmidt's number in the wake of his 1865 paper is a measure of the number of molecules per unit volume, while what came to be known as Avogadro's number, apparently from Perrin so calling it, is a measure of the number per unit molecular mass. Loschmidt's number multiplied by the volume occupied by a mole of gas at standard conditions—22,400 cm^3 to three significant figures—equals Avogadro's number.
39. Poincaré (1963), p. 90.
40. See Meyer (1899b,) pp. 299–352, which covers not only the determinations of the sizes of molecules and the corresponding values for Loschmidt's number, but also arguments that molecules more likely have "a shape that is flat, and not spread out on all sides into space . . . at least for the gaseous state" (p. 309).
41. Nernst (1895), p. 352, actually tabulated values for a related quantity, L^3x^2, where x "represents the volume of the molecules," taken to be spherical; this quantity too should be invariant across different gases insofar as the formula Nernst then gives for Loschmidt's number has it equal to $1/(319\, L^3x^2)$.
42. Maxwell (1965), p. 420.
43. See Meyer (1899b), p. 275. His assessment of the comparison of calculated and observed values merits quoting in full:

    > The agreement of the theory with experiment is of course not entirely perfect, but is sufficiently good for us to see in it a further important argument for the correctness of the hypotheses of kinetic theory. For it remains, indeed, a striking and remarkable circumstance that the coefficient of diffusion may be calculated from that of viscosity with so near an approximation.

    The values Meyer cited for mean free paths from viscosity in Chapter 7, by the way, implied as well values for molecular collision frequencies for different gases, of the order of 10^9 per second; see pp. 191–200.
44. The seeming indispensability of assuming a spherical shape is underscored by both Enskog and Chapman adopting it in their respective efforts in 1916–1917 on non-uniform gases, as can be seen in Chapman and Cowling (1939).
45. Perrin (1910), p. 17f.
46. Meyer (1899b), pp. 193–197.

47. Perrin (1912a), p. 161.
48. Perrin (1910), p. 18. He added a caution in 1911 as well (1912a, p. 162), though somewhat more weakly stated.
49. Perrin (1910), p. 90.
50. van Fraassen (2009), p. 19 and elsewhere.
51. Maxwell (1875), p. 430.
52. Ibid., p. 428. The remark was prescient because what was escaping everyone turned out to be quantum theory. The appreciation Richard Feynman expresses for the situation physicists found themselves in with the problem of specific heats from 1860 until the emergence of quantum theory is striking; see Feynman, Leighton, and Sands (1964), pp. 40–8ff.
53. See Meyer (1899b), pp. 115–145, for a full review of the state of specific heat theory as of 1899, including extensive lists of measured values; and Partington and Shilling (1924) for a compendious review 25 years later of measured values of γ and their variation with temperature.
54. Strutt (1900), p. 117.
55. Ibid., p. 118.
56. W. Thomson (1901).
57. Ibid., p. 18.
58. Ibid., p. 40.
59. Jeans (1901).
60. Ibid., p. 397.
61. Strikingly, Ostwald does not discuss spectra in any of the editions of his *Grundriss* we have examined. Nernst, however, does mention spectra in the first edition of his *Theoretische Chemie*, remarking (1895, p. 169), "As far as we can judge at present, of all the physical properties, the consideration of the *spectra of the elements* offers altogether the best way to attack this problem" of the "nature and intimate behavior of atoms."
62. Gibbs (1948), p. ixf.
63. W. Ostwald (1895), p. 76.
64. W. Ostwald (1907), p. 500f. The English translation is of an article that had been published in German earlier that year in the opening number of the journal *Rivista di Scienza*. The quoted passage culminates in Ostwald's proposal to distinguish between "hypotheses" and what he calls "prototheses":

    > A protothesis, therefore, is [an assumption] set up at the beginning of an investigation and disappears at its close if the work is successful; while a hypothesis is established when the work can be carried no farther. For this reason it is customary that in the production of scientific work the different prototheses which the investigator has employed are for the most part not mentioned at all, for it has been usual to communicate only those assumptions which the investigation has finally proved to be correct or at least approximate. Silence is kept with regard to unsuccessful prototheses just as the scaffolding is removed after the building is completed. Only very rarely, as for instance in Kepler's reports of his astronomical researches, do we learn a little about the unsuccessful prototheses. On the other hand the hypotheses in the narrower sense occupy a

large place in the literature. Because inasmuch as they have reference to scientifically inaccessible things they can be neither proved nor refuted, an endless pro and con is the usual result (p. 499f).

Ostwald's "protothesis" is thus just another term for what we here, and Smith elsewhere, call "working hypotheses" that enter constitutively into empirical research.

65. Weyl (1949), pp. 121–122; the original German edition was published in 1927.
66. Ostwald (1907–1909), p. vf. The Preface is dated September 1907, a mere 14 months before the Preface to the new edition of his *Grundriss* in which he announces, "The atomic hypothesis is thus raised to the position of a scientifically well-founded theory, and can claim its place in a text-book intended as an introduction to the present state of our knowledge of General Chemistry."
67. Ibid., p. 272.
68. Ibid., p. 289.
69. Meyer (1899b), p. 332. Calling 0.2×10^{-7} cm a "mean" is misleading insofar as in no sense whatever does it correspond to an average of a collection of values.
70. This is Nernst's formulation of the definition of entropy, not Clausius's; see Nernst (1895), p. 18.
71. Ibid., 18.
72. Kragh and Weininger (1996).
73. Ibid., p. 93.
74. See Deltete (2010), (2012).
75. Planck (1882), p. 474; quoted in translation from Kuhn (1978), p. 23. The first three chapters of Kuhn's book lay out the long struggle Planck had with the second law and entropy, as does Planck's (1949).
76. A quote from Boltzmann (1995), p. 73, shows we are not misrepresenting his intent: "Since by now we have learned the physical meaning of all the other quantities, we shall deal with the physical meaning of the quantity *H*."
77. Boltzmann (1995), p. 55.
78. Meyer (1899b), p. 44.
79. Ibid., p. 162.
80. W. Ostwald (1895), p. 65.
81. See Atkins and de Paula (2010), p. 173. J. H. van't Hoff (1852–1911) received in 1901 the first Nobel Prize ever awarded "in recognition of the extraordinary services he has rendered by the discovery of the laws of chemical dynamics and osmotic pressure in solutions." Both Ostwald and Nernst devote a chapter to the results in their texts that we have been citing; W. Ostwald (1895), pp. 127–130; and Nernst (1895), pp. 116–150. We have relied primarily on Nernst's account, first because it is more extensive, covering details of the measurement of osmotic pressure that Ostwald treats in a single sentence, and second because Nernst goes on to consider as well colloidal solutions, that is, solutions in which Brownian motion occurs.
82. See Nernst (1895), pp. 119–134, and (1904), pp. 132–146.
83. Nernst (1895), p. 137, and (1904), p. 149.
84. Nernst (1895), p. 138f.
85. W. Ostwald (1895), p. 132.

86. Ibid., p. 129.
87. W. Ostwald (1907), p. 497.
88. W. Ostwald (1895), p. 128.
89. Ostwald explains why the unit for quantity must be *mole*, or some comparable *chemical* unit, and hence for density must be number of moles per unit volume in his (1895), pp. 59–60: in some cases the chemical composition, and hence the number of moles, of a given mass of gas changes as the pressure or temperature changes—itself a fundamental discovery.
90. Ostwald remarks, the gas laws "must be looked upon as ideal limiting cases, to which the behavior of existing gases approximates more or less closely, without ever actually attaining them" (1895, p. 41).
91. Poincaré (1963), p. 90. The reference to the electronic theory of metals is to the Lorentz-Drude "sea of electrons" theory of the conduction of electricity in gases, proposed in 1900 and immediately amended that year by J. J. Thomson to exclude the positively charged electrons they had included. That theory had a complex subsequent history, at the center of Braun et al. (1992).
92. Nernst (1895), pp. 340–346 and (1904), pp. 408–415.
93. We are relying on two accounts of this research, one by W. Ostwald himself in the final chapter of his 1896 two-volume history of electrochemistry, republished in (1980); the other the opening chapter of Servos (1990), pp. 3–45. We have turned as well to Hiebert (1982) for how physical chemistry became established in 1887; and to Barkan (1999), for information relating to Nernst's perspective of the developments during the years in question. We have also consulted Laidler's (1985), though its focus on the role of research on reaction rates in the origins of physical chemistry makes it less germane to our concerns here.
94. W. Ostwald (1980), p. 1084.
95. Ibid.
96. Ibid., p. 1071, italics Ostwald's.
97. Ibid., p. 1075.
98. Ibid.
99. Ibid., p. 1076.
100. Ibid., pp. 1086ff, for a comparison of the values for *i*, obtained in the two different ways for roughly 100 different substances.
101. See Barkan (1999), p. 32ff.
102. Hiebert (1982), p. 102.
103. Ibid., p. 104.
104. See W. Ostwald (1980), pp. 1093–1106.
105. Ibid., p. 1100.
106. Ibid., p. 1101.
107. Arrhenius (1903), pp. 45–58, especially p. 57f.
108. Törnebladh (1903).
109. See Nye (1972), Ch. 2.
110. Perrin (1903).
111. Perrin (1905).

112. See W. Ostwald (1894).
113. Nernst (1895), p. 312.
114. Ibid., p. 310.
115. See Hiebert (1982), p. 107.
116. Responses to Thomson of the sort to which we are alluding can be found in Fitzgerald (1987) and Sutherland (1899).
117. Nernst (1895), p. 311.
118. See W. Ostwald (1890b), p. 272.
119. Odhner (1901).
120. Wichmann (1967), p. 6.
121. Ibid.
122. A recent example of a particularly impressive review article of book length is Weinberg's (2008).
123. Odhner (1901). His very next sentence was, "As a result of this the concept of the molecule in chemistry was found to be definite and universally valid to a degree hitherto undreamed of."
124. van't Hoff (1901).
125. For an example of how complicated the relationships can be between the views held about a hypothesis by those close to the details of the evidence and the views held by those more remote from the details, see Oreskes (1999) and the companion book of reminiscence, Oreskes (2001).
126. "Nitrogen peroxide" is the designation Ostwald employed for what is now known as dinitrogen tetroxide. According to him, it was among the first cases of an apparent change in the gas constant *R* with temperature, specifically the formation of N_2O_4 as the temperature drops below roughly 21°C. See W. Ostwald (1912), pp. 306–309.
127. We should note again not only that van Fraassen stresses the reliance on hypothetical spherical molecules in his (2009), but also that the much-heralded independent efforts by Chapman and Enskog on the mathematical theory of non-uniform gases in 1916–1917, even after molecules themselves had ceased having the standing of a hypothesis, presupposed spherical molecules.
128. W. Ostwald (1907), p. 500.
129. This has turned out to be a more difficult challenge than one might have thought. As late as the 1970 edition of his *General Chemistry*, Linus Pauling's answer to why NO and NO_2 are the only stable oxides of nitrogen was, "*probably* . . . the resonance of the odd electron between the two or three atoms of the molecule stabilizes [them]" [Pauling (1970), p. 286, italics added].
130. See W. Ostwald (1912), pp. 306–309.
131. Nernst (1895), p. 237. This sentence, by the way, remained word-for-word the same in the 1909 edition of his book. We will return to this point in Chapter 7.
132. J. J. Thomson (1900), p. 138ff.
133. We owe a debt of gratitude to Jody Azzouni, along with Teru Miyake, for pressing us to add what became this section of the chapter, and to the other members of Smith's 2017 seminar on "scientific realism" for helping to see what its content needed to be.

3
The Historical Background
Brownian Motion as of 1905

Jean Perrin was thoroughly immersed in the study of colloids at the time of the revival of interest in colloid chemistry following the innovation of the ultramicroscope by Zsigmondy and Siedentopf in 1903.[1,2] Perrin had first turned to colloids at the end of the nineteenth century to study their gross electrical properties, a professional trajectory well documented by Mary Jo Nye.[3] The more than hundredfold increase in resolution offered by the ultramicroscope over the light microscope offered greater access in particular to the motion of the constituent granules in colloids. Zsigmondy was the first to employ the greater magnification of the ultramicroscope to study granular motion, in his solo 1905 publication *Zur Erkenntnis der Kolloide*, describing with wonder the motion of small gold particles previously beyond resolution:

> The small gold particles no longer float, they move—and that with astonishing rapidity. . . . They hop, dance, jump, dash together, and fly away from each other, so that it is difficult in the whirl to get one's bearings. . . .
>
> This motion gives an indication of the continuous mixing up of the fluid, and it lasts hours, weeks, months, and, if the fluid is stable, even years.
>
> Sluggish and slow in comparison is the analogous Brownian movement of the larger gold particles in the fluid, which are the transition forms to ordinary gold that settles.[4]

Perrin was similarly thrilled as he likened the view of granules through the ultramicroscope to directly perceiving a swarm of bright stars on a dark background.[5] Writing in 1905, he took this as evidence for the discontinuous structure of the liquid substrate of colloids, leading to his more ambitious project of harnessing Brownian motion to provide evidence for the discontinuous structure of matter.

The year 1905 also marked a notable leap in the theoretical development of Brownian motion, with Einstein publishing his first paper on the topic in the *Annalen der Physik*.[6] Einstein presented a theoretical treatment of the movement of small particles in a stationary liquid, venturing that it might be identical with the so-called Brownian molecular motion. By 1906 Siedentopf had confirmed as much to Einstein via personal correspondence. The technical means of studying

Brownian Motion and Molecular Reality. George E. Smith and Raghav Seth, Oxford University Press (2020). © Oxford University Press. DOI: 10.1093/oso/9780190098025.001.0001.

Brownian motion with greater resolution using the ultramicroscope thus serendipitously aligned with the theoretical framework laid down beginning in 1905. Seen in the context of this historical snapshot, Perrin came on the scene at just the right time for an experimental investigation of Brownian motion aimed at establishing the discontinuity of matter.

While this historical snapshot is an adequate starting point to understand Perrin's efforts on Brownian motion, the history of Brownian motion belongs in the nineteenth century. Much of its history was characterized by a disconnect between experimental work and theoretical considerations that led to erroneous characterizations of granular movement. These, at best, called for dramatic ad hoc additions or modifications to the molecular-kinetic theory that was emerging during the century, simply to explain the essentials of the movement; and, at worst, they were at such odds with a kinetic theory of heat that both Maxwell and Boltzmann—the two giants of nineteenth-century statistical mechanics—never even proposed a thermal molecular origin to Brownian motion, much less undertook a proper theoretical investigation of the phenomenon. This chapter tracks the experimental history of Brownian motion in an effort to identify the mischaracterizations that arose when experimental work and its interpretation were divorced from fundamental theoretical considerations.

The first half of the chapter details how the characterization of Brownian motion and investigation into its cause became a collective, even if disjointed, scientific enterprise. Section 3.1 outlines the promise surrounding granular motion as a means of accessing features of the motion of atoms and molecules—an expectation that formed around Brownian motion soon after Robert Brown published the first thoroughgoing description of it as a bona fide phenomenon in 1828. Sections 3.2 and 3.3 identify how this project ran into trouble in the nineteenth century due to disjointed measurements of Brownian motion across experimentalists lacking a theoretical foundation. Disparate measurements focusing on different features of granular movement were all taken under the umbrella term of *granular velocity*, which only exacerbated the problem that the concept of granular velocity in the nineteenth century had nothing to do with the true motion of granules. The assumption that perceived granular motion is the same as, or even close to, true granular motion remained unchallenged in the nineteenth century, leading to the so-called granular velocities being taken at face value. This not only led to a false disparity between the features of granular movement and expectations from the kinetic theory of heat; it misdirected those thinking about a kinetic theory of Brownian motion in the nineteenth century toward ad hoc additions and modifications to this theory.

The second half of the chapter details the genesis of the approach to indirectly measure granular movement—a reconsideration that allowed the natural world to deliver on the promise of Brownian motion. Section 3.4 tracks the approach of Perrin and Einstein in particular with their key insight to establish theoretically how true granular motion leads to perceived granular motion and relations among granules (and not the other way around). Section 3.5 summarizes how Brownian motion progressed from the nineteenth century as a bona fide phenomenon in itself to the first decade of the twentieth century, when it was finally poised to deliver on the promise of establishing the rudiments of the kinetic theory of heat.

3.1. The Promise of Brownian Motion

Robert Brown was the first to announce the thoroughgoing prevalence of the indefinitely continuing motion of particles, or *granules*, that comprise a dilute emulsion or colloidal solution. In the summer of 1827, he used a simple microscope (fitted with a double convex lens 1/32 of an inch in focal length) to observe grains from the pollen of *Clarkia pulchella* immersed in water.[7] Brown discovered that these grains were "evidently" in persistent complex motion and then identified the same behavior elsewhere, proceeding systematically from organic to inorganic material. Some of the materials in which he identified the movement included pollen from other living and dried plants, living and dead animal tissue, and inorganic material including window glass, various minerals, and even a fragment from the Sphinx. Brown thus not only evidenced the thoroughgoing persistence of what came to be known as Brownian motion; he gave reason to seek a general explanation for the phenomenon.

Given the universality of Brownian motion irrespective of its liquid substrate, a molecules-in-motion or kinetic explanation for it was a leading contender from the very beginning, with the expectation that the atomic-molecular hypothesis and development of a kinetic theory of heat in particular would make sense of all its details. A rigorous kinetic theory of heat came on the scene three decades after Brown's observations, and, as we noted in Chapter 2, it was developed into a molecular-kinetic theory over the course of the latter half of the nineteenth century. These efforts, however, did not connect molecular motion to granular motion; in fact, so far as we have found, none were devoted to even providing a framework within which to place and quantify granular motion. Absent any guidance from theory, and absent any formal link from molecules to granules, why then did some form of kinetic theory remain a leading contender to explain Brownian motion throughout the nineteenth century? Part of the reason for this was the indefinite persistence of Brownian motion even after controlling for other potential causes of granular agitation; and the other part, we submit, was "causal-eliminative"—that is, the lack of an adequate explanation found through other physicochemical phenomena.

A word is needed here about just what the challenge was in pursuing a cause of Brownian motion. The term "Brownian motion" is now synonymous with motion that is exceptionally irregular, with individual granules observed to be constantly changing direction. But this was so even before the advent of the ultra-microscope and, as Zsigmondy's testament quoted earlier makes clear, far more so afterward. So, any cause had to produce such irregularity in the motions of the individual granules. No less challenging, however, was the fact that Brownian motion was observed to persist indefinitely, for months and even years. So, any cause had to be no less persistent, yet also consistent with the principles of conservation of momentum and energy, the latter of which finally emerged during the 1840s. The idea that gases, and liquids too, consist of discontinuous elements in constant motion—an idea that went back to Daniel Bernoulli in the prior century—was thus an obvious candidate to which to turn.

Enthusiasm for a kinetic explanation thus surrounded Brownian motion almost immediately after Brown turned it into a bona fide phenomenon persisting universally and indefinitely in granules below a certain size. This enthusiasm is well exhibited in the following remarks by the physicist David Brewster,[8] writing just months after Robert Brown's 1828 publication; while well aware of the lack of any inroads into a kinetic explanation for Brown's microscopic observations, he was not short on optimism that Brownian motion would provide access to, or at the very least be tied to, the inner workings of atoms and molecules:

> If these motions resisted every method of explanation, it is the last supposition in philosophy that they are owing to animal life; and in future times, when the science of molecular organization shall be further advanced, it will be viewed in the same light as the opinion of Kepler, that the planets themselves were living animals, swimming in the ethereal ocean of the heavens. . . . Why should not the molecules of the hardest solids have their orbits, their centres of attraction, and the same varied movements which are observed in planetary and nebulous matter? The existence of such movements has already been recognized in mineral and other bodies. . . . A mineral body will, in the course of time, part with some of its ingredients, or take in others, till it has become a new mineral, and has entirely lost its personal identity. . . . In these changes the molecules must have turned round their axes, and taken up new positions within the solid. . . . Before another century passes, the laws of such movements will probably be determined; and when the molecular world shall thus have surrendered her strongholds, we may look for a new extension of the power of man over the products of inorganic nature.[9]

Indeed, Brewster's expectation gets fulfilled with Maxwell's, Boltzmann's, and Gibbs's development of statistical mechanics late in the nineteenth century. Until

Einstein's reformulation of Boltzmann's approach at the beginning of the twentieth century,[10] however, Brownian motion did not come close to receiving an adequate kinetic description, and the putative import of Brownian motion for molecular theory remained open to objections.

We have already noted that, despite the failure of kinetic theory to provide an explanation for Brownian motion throughout the nineteenth century, it remained an enduring contender to explain it. While part of the reason for this was the indefinite persistence of granular motion that naturally lent itself to an explanation in terms of heat motion of atoms and molecules, no less important was the failure to find any sustainable explanation elsewhere. Other physical and chemical explanations for Brownian motion were wide-ranging—including electrical effects, surface tension, capillary action, osmosis, microscopic liquid currents, and unequal heating of the liquid through radiation[11]—but none of these could quantify granular motion through an underlying cause any more than the kinetic hypothesis could. There was essentially thus only causal-eliminative[12] support for the kinetic picture as the cause of Brownian motion in the nineteenth century, and the idea of accessing the molecular realm through granular motion was accordingly as much just one possibility out of many for some as it was a promise sure to be delivered for others.

With a description of Brownian motion thus up for grabs, criticism of the possibility of a kinetic explanation was just as warranted as Brewster's optimism. Such a sentiment is found in the following note by Reverend John Henry Conybeare, professor at Oxford, to the Reverend Dr. Buckland of Christ Church Oxford:

> Touching Brown's theory that all matter consists of live mites, I don't believe a word on't. . . . If you suspend particles of matter in a fluid for microscopical observation, a thousand circumstances may generate motion, and to this I attribute his facts. . . . Biot, if I remember, in the optics of his Nat. Phil., had some curious speculations on the subject. He states it to be possible that solids may be composed of systems of moving molecules, representing in small what the planetary systems do in large. I would only add one supposition more; that these molecules are inhabited, and have philosophers among their populations who, having observed the motions of some half-dozen molecules in their neighbourhood and ascertained their laws, believe they have developed the system of the universe.[13]

Beyond the notion that Brownian motion was not due to animal life, the Reverend's scathing critique of the kinetic explanation had little in common with Brewster's optimism. Brown's 1827 research had necessitated a physical or chemical picture of Brownian motion, but the real work was still to be done. The contrasting opinions of his contemporaries were pre-figurations of the push and pull between rival explanations in years to come.

We discussed in Chapter 2 the failure of the molecular-kinetic theory in the nineteenth century to deliver on the promise of accessing the molecular realm. Recall in particular the example of Jeans's 1901 paper on "The Distribution of Molecular Energy" that began by outright supposing two entirely imaginary types of molecules in attributing the specific heats problem to dissipation of molecular energy into the ether.[14] The situation was not dissimilar for Brownian motion, where similarly ad hoc suppositions about atoms and molecules were being employed. For instance, the physicist Christian Wiener (1863)—the first to report on the effect of granular size on the magnitude of granular movement ("*Grösse der Bewegung*")—assumed vibrating corporeal and ether atoms as the cause of the heat motion of bodies within which irregular impacts led to Brownian motion. Several twentieth-century authors writing about the episode have retrospectively credited Wiener as the first to conceive the cause of Brownian motion in the kinetic theory of heat.[15] His vibrating corporeal and ether atoms picture, however, was a far cry from the kinetic theory picture of equiprobable translational motion of uncoordinated material atoms and molecules; and, as was the case with Jeans in 1901, Wiener's 1863 speculation made no historically salient contribution.

Another notable kinetic proposal came from Joseph Delsaulx (1877), a Belgian Jesuit, who reflected that the apparently homogenous pressure on a vessel wall requires that the dimensions of the vessel be incomparably greater than the mean free path of a molecule between two consecutive collisions; this condition, he notes, does not hold for particles and bubbles in Brownian motion:

> We are therefore justified in attributing the oscillatory motion of little gas-bubbles imprisoned in a liquid, to the velocity of translation of the molecules, varying in intensity and direction from one to another, and giving rise at the same instant, by default of the total communication of pressure, to unequal propulsions upon the different points of the envelope.[16]

In retrospect, we can confirm that Delsaulx had the right idea; historically, however, his proposal amounted to little more than conjecture for he was unable to formalize it into a theoretical framework that could be tested; the closest he came to doing so was speculating on the parameters that might influence Brownian motion:

> The tension of the gas, the mass of the molecules, and the coefficient of resistance of the surrounding liquid, will determine, in each particular case, the conditions of volume for which the Brownian motion ceases or is produced.[17]

As we shall see in section 3.3, Marion Smoluchowski's 1906 inquiry into Brownian motion[18] begins with the same insight as Delsaulx's 1877 proposal that pressure variations do not cancel out at the scale of granules in Brownian motion. He dug deeper, however, and combined probability theory with Stokes's law to provide an experimentally verifiable framework for Brownian motion. Smoluchowski's successful inquiry prompted Arnold Sommerfeld to remark in 1917 that "his name will, forever, be associated with the first flowering of atomic theory"[19] whereas Delsaulx's informal proposal, albeit correct, made no historically salient contribution.

The kinetic proposal remained a contender for explaining Brownian motion in the nineteenth century in large part because the persistence of the motion remained otherwise unexplained. Granular agitation seemed to persist despite the removal of many outside influences. Such support for a kinetic theory of Brownian motion owed much in particular to the work of Léon Gouy, who published three notable papers on Brownian motion in 1888, 1889, and 1895. In fact, in Perrin's first 1908 publication on Brownian motion, he gave Gouy foremost credit for establishing that the eternal agitation was an essential property of the surrounding fluid medium.[20] In his first 1888 "Note sur le mouvement brownien," Gouy already suggested that Brownian motion is due to molecular heat-motion in liquids after identifying that the irregular granular motion could not be from currents, vibration, or temperature differences. He acknowledged, however, that the proposal of a thermal molecular origin of Brownian motion was still far from being an explanation of Brownian motion rooted in the kinetic theory of heat:

> The Brownian movement shows us, certainly not the movements of molecules, but something very near it, and gives us direct and visible evidence of the exactness of the current hypothesis on the nature of heat. If one adopts these views, then the phenomenon, the study of which is far from being terminated, takes on assuredly an importance for molecular physics of the first order.[21]

Gouy understood that experimental and theoretical work on Brownian motion was far from complete, yet at the same time saw its singular importance in offering an empirical inroad into molecular physics.

Gouy's next two papers added to the causal-eliminative support for a kinetic theory of Brownian motion. The 1889 paper in particular was more systematic in approach, with additional results that granular movement decreases in intensity (*intensité*) with increase in granular size and viscosity of the surrounding medium.[22] Gouy reached the limit of his inquiry in his final 1895 paper, entitled "Le mouvement brownien et les mouvements moléculaires," which begins by reflecting on how, as he put it, certain scientific issues that touch on the limits of two sciences do not always receive the attention they merit. After reiterating the need for further work on the essentials of granular movement that he had evidenced as indefinite, irregular, and internal to the surrounding medium, Gouy concluded this final paper with his own marked expectancy for a kinetic theory of Brownian motion in the future:

> Experimental and theoretical concepts are still lacking. . . . We can at least conclude that Brownian motion gives us what is missing in the kinetic theory of matter: direct experimental evidence. Without doubt, we do not see and will never see the movement of molecules; but we see at least something that results directly and comes necessarily from the internal agitation of bodies. It is therefore very desirable that this phenomenon, long neglected as an unimportant accident, becomes the object of attention of physicists and remains within the sphere of their studies; I firmly trust that, through their efforts, we will enter more and more deeply into the intimate knowledge of the properties of matter, already so rich, and full of promise for the scientific and industrial development of humanity.[23]

Brewster's naïve optimism had been replaced by Gouy's informed optimism; the promise of Brownian motion was fully alive among some at the turn of the twentieth century.

So much then for how a thermal molecular-based explanation for Brownian motion endured through the nineteenth into the twentieth century, underscored by Gouy's conviction in the concluding remarks to his investigations that it was poised to deliver *une preuve expérimentale directe* for the kinetic theory. Accompanying Gouy's optimism, however, was also his dismay that Brownian motion had been so long neglected by physicists. A question this section raises, accordingly, is why no leading nineteenth-century physicist pursued a kinetic theory of Brownian motion or tried to find a place for it within the molecular-kinetic theory of heat that was at center stage in physics during the second half of the nineteenth century.

Gouy suspected that the origin of Brownian motion within the purview of biology was responsible for the physicists' disinterest in the phenomenon. While that may have been part of the reason, the major reason seems to have been the mischaracterization of granular movement in the nineteenth century that misdirected the larger scientific community away from a thermal molecular origin of Brownian motion. In particular, measurements of granular velocities were reported at many orders of magnitude less than what was called for on the basis of the kinetic theory of heat. The next section tracks how these so-called granular velocity measurements were characterized by nineteenth-century experimentalists; this will then put us in a position to better identify further shortcomings of announced granular velocities—namely, inconsistent measurements across experimentalists—that likely did more harm than good in bringing Brownian motion into the purview of physics.

3.2. Measuring Granular Velocities: A Failure in Experimentation

So-called granular velocity measurements were the major and, as far as we have determined, the one clear quantitative discrepancy between Brownian motion and kinetic theory in the nineteenth century. In his 1907 dissertation on the theory of colloidal solutions, Svedberg begins the section on the determination of granular motion entitled "Bestimmungen der Bewegungskonstanten" by presenting a table of theretofore known velocity measurements, starting with J. Regnauld in 1857 to Zsigmondy in 1905.[24] While Svedberg cautions that these previous determinations were imprecise, he presents them all under the title of *Geschwindigkeit*, corresponding to the speed or velocity of the granules.

In the same year that Svedberg published his dissertation, Einstein published the following admonition concerning granular velocity measurements:

> Since an observer operating with definite means of observation in a definite manner can never perceive the actual path traversed in an arbitrary small time, a certain mean velocity will always appear to him as an instantaneous velocity. But it is clear that *the velocity ascertained thus corresponds to no objective property of the motion under investigation*—at least, if the theory corresponds to facts.[25] [emphasis added]

Einstein began his note by specifically citing Svedberg's first publication on Brownian motion (1906) that prompted him to stress, once and for all, that the

Beobachter und Zeitpunkt der Beobachtung	Durchmesser der Teilchen in μ	Geschwindigkeit der Teilchen in μ sek	Temperatur in C-Graden	Natur der Teilchen

Figure 3.1. Headings from Svedberg's table that compiled half a century of disjointed research into the characterization of granular movement under the umbrella-term of "*Geschwindigkeiten*."

Table 3.1 later in this section distinguishes how the various descriptions were actually characterized by the authors historically. Source: Svedberg (1907), p. 134

apparent velocities in Brownian motion in no way correspond to the path actually traversed by granules.[26]

In contrast to Einstein's cautionary remarks, Svedberg's table of *Geschwindigkeiten* has been irrelevant to any account of the import of Brownian motion in establishing the kinetic theory or the molecular hypothesis. This is not surprising, however, for why would the results of so-called velocity measurements be important when they are, in effect, optical illusions! This is obvious to us now in retrospect, but without the benefit of hindsight and Einstein's now well-known 1907 caution, the situation was entirely different at the end of the nineteenth century. The pursuit of apparent granular velocities had been an enterprise going back to the mid-nineteenth century, and as late as 1907 Svedberg believed that he had devised a method to better isolate the apparent *Geschwindigkeiten*. The current section will track the history of these apparent granular velocity measurements. This will set us up for section 3.3 on the consequent garden path of various modifications that were being suggested to the basic tenets of the kinetic theory in an effort to explain away the discrepancy with the announced granular velocities.

Taking the most rudimentary picture of kinetic theory first worked out by Waterston in an unpublished 1845 draft and in print by Krönig in 1856,[27] the formula for molecular velocity jumps out from one of the most basic tenets of kinetic theory: that pressure is the transfer of molecular momentum. The formula was firmly established by Clausius in 1857 and remained unchanged through Meyer's 1899 *Kinetische Theorie der Gase*. Perrin's *Les atomes* of 1913 introduced kinetic theory using the same formula, and this formula for molecular velocity is the quantitative introduction to kinetic theory even in contemporary physics textbooks.[28] Using only classical mechanics, therefore, we get:

$$pV = \frac{1}{3}NmG^2$$

In unit volume, i.e., $V = 1$, this simplifies to:

$$p = \frac{1}{3}NmG^2$$

where p is the pressure, N is the number of molecules in unit volume, m is the molecular mass, and G^2 is mean squared molecular velocity. Since $Nm = \rho$, where ρ is the density, the substitution of ρ reduces the formula to three variables:

$$p = \frac{1}{3}\rho G^2$$

By measuring p and ρ, one can now estimate G^2 for various gases, which is exactly how molecular velocities became the first absolute values calculated for molecular magnitudes, with root mean square velocities for oxygen, nitrogen, and hydrogen first appearing in Clausius's 1857 paper as 461, 492, and 1,844 meters per second, respectively.[29] These values for all three gases remained the same to two significant figures in Meyer's 1899 text, where root mean square values were tabulated for 31 different gases.[30]

The arithmetic mean value Ω of molecular velocities is actually less than the value G of root mean square velocities. (In any distribution, the mean of two different numbers a and b is always less than the root of the mean square of these numbers a^2 and b^2.) This distinction gets resolved by a further transformation of the preceding formula based on Maxwell's law of velocity distributions, where:

$$\Omega = G\sqrt{\frac{8}{3\pi}}$$

The relationship now becomes:

$$p = \frac{\pi}{8}\rho\Omega^2$$

and mean molecular velocities can be calculated, again using just pressure and density measurements. A notable feature of this derivation was that it did not require any assumptions about molecular shapes or intermolecular forces; as Perrin remarked in *Les atomes*, "this is a mechanical problem, into which no physical difficulties enter."[31]

To be sure, the basic tenets of kinetic theory—that pressure is the transfer of molecular momentum and that temperature is proportional to mean molecular translational kinetic energy—were still assumptions in the nineteenth century, but the formula for molecular velocity was a byproduct of the basic framework that emerged from these tenets without further assumptions. The possibility of a confirmation of molecular velocities by alternative means was thus paramount to firmly establishing the basic tenets of the kinetic theory of heat. If the indefinite incessant Brownian motion of granules was indeed from the heat motion of atoms and molecules in thermal equilibrium in a colloid, the granules themselves would maintain a mean kinetic energy. Herein ostensibly was an appeal of trying to measure granular velocities since directly ascertaining and quantifying the movement of visible granules—and confirming it against indirectly measured results for invisible atoms and molecules—would not just be an alternative measurement of molecular velocities, but importantly a determination independent of the assumptions of kinetic theory. The measurement of granular velocities and from there the velocities of atoms and molecules in the surrounding medium thus promised a direct verification that was different from other outstanding microphysical measurements at the time.

Confirmation of implied molecular velocities by comparing them to the visible motion of granules hinged a priori on the seemingly straightforward task of tracking a granule's movement over a given time. None of the experimentalists in the nineteenth century even came close to appreciating, however, the degree to which this task was practically impossible. The predominant reason for this oversight seems to have been that none of them actually engaged with the emerging kinetic theory of heat to identify how features of molecular motion would influence and constrain perceived granular motion. It would be unfair to suggest that those nineteenth-century experimentalists should have known what Einstein announced in 1907, that a granule changes speed and direction over 10 million times per second. Nonetheless, there was certainly a disconnect between experimental work and theoretical considerations when the question of whether granular velocities are even ascertainable in principle was never seriously raised in the nineteenth century, even after apparent granular velocities turned out to be at odds with those from the kinetic theory by many orders of magnitude.

The problem of measuring granular velocities was essentially the problem of ascertaining true granular motion, i.e., the path actually traversed by a granule. The suggestion that this might be an impossible challenge to meet was part of the physics literature for most of the time that experimentalists continued pursuing and publishing granular velocity results based on perceived motions. The key insight that atoms and molecules in a gas are undergoing an extraordinary number of collisions and consequent changes in speed and direction per second was already established by Clausius and Maxwell by 1860, with Maxwell's published estimate that year that molecules of air at 60°C undergo on average over 8 billion collisions per second.[32] Meyer opened Part II on "The Molecular Free Paths and the Phenomena Conditioned by Them" of his 1899 textbook by noting the same idea of many millions of collisions for molecules per second, with molecular motion "proceeding tumultuously hither and thither in straight zigzags" and the molecules thus executing "such a motion that the best representation of it is that of grains of corn shaken in a closed box."[33] These contentions, however, never became part of the discussion surrounding Brownian motion in the nineteenth century; apparent granular velocities continued to be taken at face value. Ironically, this ended up posing challenges to the basic framework of the kinetic theory of heat that should have informed the characterization of granular movement in the first place.

Without kinetic theory mediating the interpretation of apparent granular velocities, then, measurements based on perceived granular motion ended up imposing false constraints on any kinetic theory of Brownian motion, a topic we discuss in further detail in section 3.3. The other fundamental problem with the measurement of granular velocities in the nineteenth century was inconsistent experimental descriptions; for our purposes, the issue was inconsistency in what the term *granular velocity* meant to different experimentalists. The remainder of this section tracks these various characterizations to better identify how Svedberg's 1907 table, taken under the umbrella term *Geschwindigkeit der Teilchen*, was actually a collection of different quantities idiosyncratic to each experimentalist's different take on what he took perceived granular motion to be.

Regnauld's 1857 measurements, appearing under "verbal-reports" of an 1858 meeting of the Society of Pharmacy of Paris, were the earliest quantitative attack on Brownian motion.[34] At 500 magnification, Regnauld observed that spherules 1/4,000 mm in radius were displaced five times their diameter, or 1/400 mm, in *oscillation* (Fr., "*oscillation*"). This notion of oscillatory motion portrayed granules as having ordered movements with characteristic amplitudes, which was a far cry from the stochastic uncoordinated movement that we now understand Brownian motion to be. This first mischaracterization of Brownian motion as oscillatory is one that kept resurfacing in subsequent literature, but Regnauld was not necessarily responsible for this trend.

Even within the trend of mischaracterizing Brownian motion as oscillatory, different experimentalists' reports on what the essentials of the movement consisted of were idiosyncratic to each of them. The real issue was that these choices were fundamentally arbitrary, far removed as they were from theoretical consideration. Publishing his results at the other end of the era of measuring granular velocities, Zsigmondy in 1905 reported that Brownian motion consisted of a *combined motion of translation and oscillation*. We began the current chapter by quoting Zsigmondy's wonder upon viewing the Brownian motion of small gold particles through the ultramicroscope; he continued by describing the motion of the particles:

> The smallest particles which can be seen in the hydrosols of gold, show a combined motion, consisting of a *motion of translation* by which the particle moves from 100 to 1000 times its own diameter in one sixth to one eighth of a second, and a *motion of oscillation* of a considerably shorter period. . . . I must expressly call attention to the fact that this motion of the little gold particles just described, differs in many respects from the typical Brownian movement.[35] [emphasis added]

Zsigmondy goes on to reference the observations of the German physicist Otto Lehman on the fat globules in milk as the typical Brownian motion, "vibrating unsteadily about a mean position seldom reached"; he then contrasts this with and again reiterates the distinct translational motion that he has observed using the ultramicroscope:

> In contradistinction to this typical Brownian movement about a mean position, the motion of the little gold particles is continuous, a quick moving gold particle for instance, after a series of speedily executed zig-zag moves, rushing across the illuminated field of view, almost as if it were a living thing, and vanishing; whereas larger particles may for the most part be conveniently viewed for a long time before the slow current of the fluid removes them from the field of view.[36]

Here, then, is another interpretation of granular movement: Zsigmondy's *translational motion for ultramicroscopic particles*, distinct from the typical Brownian motion about a mean position.

(The current section preoccupies itself with the specific methodological issue in the measurement of granular velocities of lack of a consistent characterization of granular movement across experimentalists; however, each study discussed in this section had further shortcomings when it came to controlling for other variables that could influence granular movement. This was especially important for establishing the thermal molecular origin of Brownian motion since it involved continued granular agitation even after the removal of gross currents, gradients, or energy exchanges with influences outside the closed colloidal system. Zsigmondy's mention of the "slow current of the fluid," that is, the medium of his gold suspensions, underscores this problem of gross variables that had to be controlled for in any proper experimental verification of a kinetic theory of Brownian motion.)

Further into his 1905 text, Zsigmondy again elaborates, in a paragraph entitled "Turbid Solution of Gamboge," that bright gamboge particles exhibit the slow, trembling motion characteristic of Brownian motion versus gold particles that have motion of a quite different nature.[37] For Zsigmondy, particles below a certain size, undetected before the advent of the ultramicroscope, graduated to an entirely new level of Brownian motion. Between the *typical oscillatory motion* of large granules and the *translational motion* of smaller granules, Zsigmondy chose the latter motion of translation as representative of Brownian motion for gold particles 0.035 to 0.006 μ in diameter. In his tables, he called this motion the *free path* of particles. Despite all this (somewhat arbitrary) nuance, Zsigmondy's results were given the same description as Regnauld's of *Geschwindigkeiten* in Svedberg's 1907 table on granular velocity measurements.

Possibly influenced by Zsigmondy, Svedberg himself endorsed a twofold description of granular movement in his first paper on Brownian motion.[38] For Svedberg, however, the motion of oscillation—a ceaseless vibration around a central position—was the true ("*eigentlichen*") Brownian motion, and he saw this characteristic motion ("*Eigenbewegung*") of particles as a characteristic property ("*Eigenschaft*") of colloidal solutions. It was entirely appropriate, then, that the electric arc lamp used by Svedberg to illuminate particles did not reach the lower limit of visibility (6 μ); according to him, "a study of the Zsigmondy translational or progressive movement was thereby excluded."[39] Thinking that he had thus isolated oscillatory Brownian motion from the circumstantial progressive motion, Svedberg attempted measurement of the essential movement.[40]

Taking Brownian motion to be oscillatory, Svedberg superimposed granular movement on a linear motion to obtain a sinusoidal curve. He studied the movements of platinum organosols suspended in a fluid with known linear velocity H, essentially treating Brownian motion as simple harmonic motion in translation. Looking at this movement through an ocular scale, he then estimated the amplitude A and the wavelength λ of the curve. Although the wavelength was merely contingent on the linear fluid velocity, it allowed Svedberg to determine the time period t of one oscillation cycle:[41]

$$t = \frac{\lambda}{H}$$

Next came the most dramatic transformation, where Svedberg calculated the "absolute velocity" h ("*absolute Geschwindigkeit*") of the particles:

$$h = \frac{4A}{t}$$

Though Svedberg did not include his own absolute velocity results in his 1907 table of velocity measurements, his approach further shows what an oversimplification it was to gather the various characterizations of granular movement under the umbrella-term of *Geschwindigkeiten*.

(Despite what appear to be very obvious shortcomings in Svedberg's characterization of granular movement, his name at the time was often repeated next to Perrin's for confirming the Einstein-Smoluchowski theory of Brownian motion. Wilhelm Ostwald credited both Perrin and Svedberg in securing results for Brownian motion satisfactorily in agreement with the kinetic theory of heat and, in fact, he only referenced Svedberg for results on granular movement that agreed with theory close enough to be taken as "fairly satisfactory proof of the kinetic nature of heat";[42] H. G. Söderbaum, in his 1926 presentation speech for Svedberg's Nobel Prize, similarly cited both him and Perrin for having experimentally confirmed the theory of Brownian motion.[43] Why did Svedberg receive such credit when his characterization of granular movement as oscillatory was nowhere near the mean displacement of stochastic granular motion described by Einstein? Although we're not going to take this question head on, it is a fascinating artifact in the history of Brownian motion. It has been considered by other authors writing on the episode.[44] Kerker writes that Svedberg's work continued to persist in the colloid chemical literature partly because Svedberg himself continued insisting, years after his initial publications on Brownian motion, that his 1906 results were the first quantitative confirmation of the kinetic theory of Brownian motion. Furthermore, it seems that, other than Einstein and Perrin, those at the time crediting Svedberg did not actually take a careful look at the suppositions that Svedberg imposed on granular motion in his papers.

Before returning to the primary goal of this section of listing the various characterizations of granular motion, it would be remiss for us not to mention the suppositions Svedberg employed to link his measurements to Einstein's theory. In his second 1906 paper on Brownian motion, he does what Maiocchi calls a double empirical translation—all without discussion!—of translating the term ξ, which in Einstein's theory represents a *mean* of linear displacements, to his amplitude A of a *non-random* sinusoidal oscillation, and translating τ, which in the theory is *any* arbitrary time period of observation of sufficient length, to his t, the *non-arbitrary* period of oscillation.[45])

At the end of his second 1906 paper on Brownian motion, Svedberg plays off his results for granular velocities against Ramsay's 1892 results; however, the basis for the comparison was only superficial—that is to say, it was a comparison of granular velocities only in name—because Ramsay believed that granular movement was in no sense regular, quite unlike an ordered oscillatory movement.[46] Ramsay's characterization of granular movement as a "skipping motion" was particularly vague, though the other predominantly translational-based characterizations of Brownian motion were not much more precise; recall from section 3.1 that Christian Wiener referred to his measurements as the size of the movement ("*Grösse der Bewegung*"), and Felix Exner in 1900 took himself to be measuring the *Geschwindigkeit,* but without accounting for the smaller trembling movements of granules.[47] In fact, no two experimentalists had the same description for what they took granular velocity to be, and this inconsistency kept nineteenth-century measurements of granular movement from ever taking off as a unified enterprise—certainly not one where granular velocity results by different experimentalists could be played off against one another to identify what features of the measurement process might make a difference to the results for apparent granular velocities.

In the six studies mentioned in this section on granular velocity measurements, we come across six different ways of ascertaining the true granular path (Table 3.1). Given this inconsistency across experimentalists in characterizing granular movement, nineteenth-century efforts to quantify it started off on the wrong foot, never to be steered right into a unified project, and ultimately abandoned as a collective mistake. Einstein's 1907 admonition of measuring so-called granular velocities was in response to Svedberg, but it applied equally to all these other efforts as well. Without the foundation of a kinetic theory of Brownian motion to

Table 3.1 Alternative Characterizations of the Granular Motions

Author and Date	Description of the So-called Granular Velocity ("*Geschwindigkeit*")
J. Regnauld 1857	"*Oscillations des corps*"
Chr. Wiener 1863	"*Grösse der Bewegung*"
W. Ramsay 1892	"*Skipping motion*"
F. Exner 1900	"*Geschwindigkeit*" without accounting for "*zitternden Bewegung* [trembling movements]"
R. Zsigmondy 1905	"*Zurückgelegter Weg* [distance traveled]" as well as "*Brownschen Bewegung um eine Mittellage*"
T. Svedberg 1906	"*Eigentlichen* [true] *Brownschen Bewegung*" of "*vibration um eine Mittellage*"

guide the experimentalists' efforts, their choice on what counted as true granular motion was ultimately arbitrary.

This was part of the bigger problem of the experimentalists having no framework within which to identify what features of granules and the surrounding medium to account for, what to control for, and what features of the system to systematically vary in order to establish law-like relations. Given that no two experimentalists shared the same description for granular velocity, it would have been difficult even to judge the quality of the experimental work. This may have been part of the reason for the absence of theoretical work by the greater scientific community on the thermal molecular origin of Brownian motion. Granular velocity measurements thus fittingly epitomized Marian Smoluchowski's criticism on the history of Brownian motion: uncertainty due to inaccurate experimental data, ultimately due to faulty and mathematically imprecise theories.[48]

While the inconsistent characterization of granular velocities was precluding meaningful comparison of granular movement across different studies (or the possibility of testing and trying to replicate studies given their imprecise characterizations), the complete disconnect of the interpretation of granular velocities from the kinetic theory of heat was turning out to be a more serious problem for the thermal molecular origin of Brownian motion. Granular kinetic energies based on velocity results were many orders of magnitude smaller than molecular kinetic energies derived from the kinetic theory of heat from 1857 onward. The question of whether the true granular path is even ascertainable in principle was never raised by the nineteenth-century experimentalists to address this disparity, and apparent granular velocities continued to be taken as true granular velocities in the interim. The situation was primed to lead down a garden path[49] where the concept of simple thermal equilibrium between molecules and granules—or something close to it—was just an untenable foundation for a kinetic theory of Brownian motion, given the results for granular velocities. One way out of this was to abandon the thermal molecular origin of Brownian motion; the other was to modify the fundamental framework of the kinetic theory of heat to incorporate the erroneous granular velocity results. The latter was the approach taken by most of the experimentalists, imposing false constraints on the kinetic theory of heat. It is the subject of the next section.

3.3. Explaining Granular Velocities: A Failure in Theory-Mediation

In his second paper on Brownian motion in 1889, Gouy stated that granular speeds were about a hundred millionth the speed of molecules.[50] How much smaller exactly did this make the kinetic energy of granules than that of molecules? Answering this question requires an estimate for molecular kinetic energies, and on the kinetic theory of heat, molecular kinetic energy per temperature increment is equal to the quotient R/N_0 where R is the molar gas constant in units of energy per temperature increment per mole and N_0 is Avogadro's number, i.e., the number of atoms or molecules per mole. While the gas constant R was well-established based on agreeing stable measurements from different gases, kinetic-theory mediated estimates for N_0 in the nineteenth century were not consistent with one another; nor, as we discussed in section 2.2 of Chapter 2, did the discrepant results indicate a tendency to converge toward a preferred range of values. Even though the value of N_0 was unclear, there was nevertheless an order of magnitude estimate of what derived molecular kinetic energies at given temperatures should be. In his first Brownian motion paper in 1908, Perrin took 7.0×10^{23} for N_0 to obtain an estimate of molecular kinetic energy of 3.43×10^{-14} ergs at room temperature. Although we postpone a detailed discussion of Perrin's comparison until Chapters 4 and 5, we will use his molecular kinetic energy value here to illustrate the disparity between molecular and granular kinetic energies based on velocity measurements.

The disparity between molecular and granular kinetic energies misdirected a kinetic theory of Brownian motion in two distinct ways. One of these was simply the rejection of a kinetic theory of Brownian motion; in fact, the disparity posed by granular velocity measurements may well have contributed to the greater scientific community's inattention toward Brownian motion throughout the nineteenth century. The other form of misdirection was insisting on a kinetic theory of Brownian motion despite the disparity at hand, which necessitated ad hoc additions and modifications to kinetic theory. This latter point can be phrased in terms of our earlier description of granular velocity measurements as a garden path in Brownian motion research. It is the main topic of this section.

Fortunately, the garden path paved by granular velocities did not take hold in the greater scientific community, but hindsight might be obscuring our view of the fact that perceived granular velocities were representative of the state of Brownian motion research at the turn of the century. Around the same time that Lord Kelvin published his "Nineteenth Century Clouds over the Dynamical Theory of Heat and Light" in 1901 and questioned the theorem of equipartition, everyone engaged in Brownian motion research was pursuing one or another form of a kinetic theory of Brownian motion that questioned other basic tenets of the kinetic theory of heat. This garden path development was stopped in its track by the end of the first decade of the twentieth century by the work of Einstein and Perrin, but in the meantime granular velocity results were demanding the kinetic theory to mold itself in order to explain a gross misrepresentation of the true molecular impact on granules.

Before diving into the proposed modifications to the kinetic theory of heat prompted by the kinetic energy disparity of granular velocities, we had best recap what the basic tenets of kinetic theory were in the first place. We began section 3.2 by noting how molecular velocities came out of the identification of pressure as the transfer of molecular momentum. The other central tenet of kinetic theory was the proportionality between absolute temperature and the mean translational kinetic energy of molecules, which, along with pressure as the transfer of molecular momentum, connected the Boyle-Mariotte and Gay-Lussac laws of bulk phenomena to the underlying translational motion of molecules. The assumption common to both these tenets was *independent molecular motion* in straight lines until redirected by striking against other molecules or against an impermeable surface. Clausius started with this very assumption in his mathematical investigations of 1857 to establish that the absolute temperature is proportional to the *vis viva* of the translational motion of molecules.[51] Contrast this with the proposal of *coordinated molecular motions* at the microscopic level of granules that was endorsed by most of those working on Brownian motion at the turn of the century.

Maxwell began his 1860 "Illustrations of the Dynamical Theory of Gases" by emphasizing the fecundity of the assumption of rapid molecular motion in straight lines that he goes on to describe as equiprobable in any direction and thus independent of the motion of other molecules:

> So many properties of matter, especially when in the gaseous form, can be deduced from the hypothesis that *their minute parts are in rapid motion*, the velocity increasing with the temperature, that the precise nature of this motion becomes a subject of rational curiosity. . . . M. Clausius has determined the mean length of path in terms of the average distance of the particles, and the distance between the centres of two particles when collision takes place. We have at present no means of ascertaining either of these distances; but certain phenomena, such as the internal friction of gases, the conduction of heat, through a gas, and the diffusion of one gas through another, seem to indicate the possibility of determining accurately the mean length of path which a particle describes between two successive collisions.[52] [emphasis added]

This paper made groundbreaking contributions on the velocity distribution of molecules and on the mean-free-path theory of transport coefficients. This was just the beginning of the refinement of kinetic theory, but our point here is not to catalog those successes.[53] Our mention again of the beginnings of kinetic theory is rather to emphasize that the assumption of independent molecular motions was entrenched in the development of the kinetic theory of heat from the very beginning. It was the basis for the two pillars of the kinetic theory of heat—of pressure as the transfer of molecular momentum and the proportionality between absolute temperature and the mean translational kinetic energy of molecules. The major proposal of coordinated molecular motions to explain away the disparity between molecular and granular kinetic energies was directly undermining this assumption and, therefore, the foundation of the kinetic theory.

On the kinetic theory of heat, the key link between granules and the molecules comprising the surrounding liquid is that they are in thermal equilibrium. The expectation was that their kinetic energies would be in a 1:1 ratio. As it turned out, Perrin's finally establishing the thermal molecular origin of Brownian motion also established that this ratio was at least close to unity. There was, however, no a priori guarantee that this identity would actually hold from molecular kinetic energy to granular kinetic energy. For a link between granules and molecules, the kinetic theory of heat minimally required only a proportionality between the two. This less demanding standard still included that the kinetic energy of granules should vary proportionally with the temperature. This was tested by Felix Exner (1900), who plotted squared velocities—representing the *vis viva* ("*lebendigen Kräfte*") of the granules—as ordinates at nine different temperatures as abscissa for granules 0.7 μm in diameter (Figure 3.2).

If granules participate in the thermal motion of molecules in the liquid, the curve in Figure 3.2 would correspond to a straight line that goes through absolute zero. Although a linear regression curve fits Exner's data with a coefficient

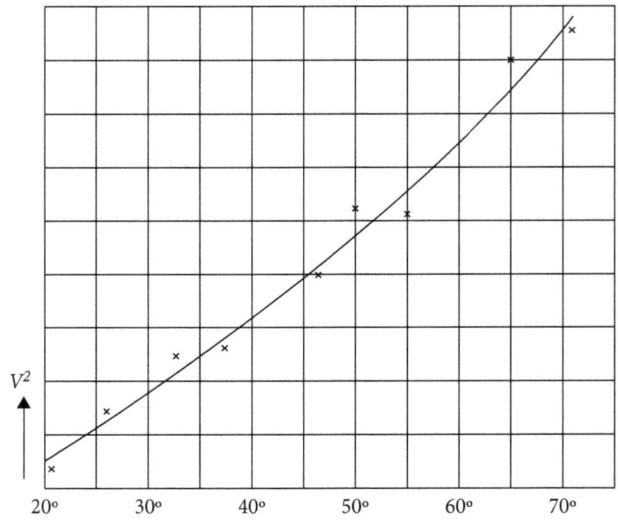

Figure 3.2. Felix Exner's plot for squared granular velocities at different temperatures.
Source: Exner (1900), p. 856

of determination of 0.98, the line extended crossed the abscissa at about −20°C instead of at absolute zero.

Exner's results thus in no way corroborated the fundamental features of the kinetic theory of heat, and he responded to this incongruity by proposing a slow ascent ("*langsam aufsteigend verläuft*") of *vis viva* from absolute zero. He concluded that the simple assumption, that the particles are analogous to the molecules of the liquid, is inadequate:

> This model does not give data consistent with molecular motion (and only considers material points while neglecting molecular forces and friction), but it may still be conceivable that, if one starts from the hypothesis of a relation between the movement of molecules in the liquid and that of suspended particles, the visible movements and their corresponding metrics will be of real use for the clarification of the internal movements of a liquid.[54]

A remarkable departure from the central tenet of kinetic theory on the proportionality between temperature and kinetic energies notwithstanding, Exner still seems to maintain that a more intricate model of molecular motion would explain Brownian motion.

While Exner did not discuss coordinated molecular motions, his concept of a slow ascent of *vis viva* from absolute zero until about 20°C and a linear relationship between temperature and granular kinetic energy from there onward was an entirely ad hoc requirement that he imposed on a kinetic theory of Brownian motion. Exner thus did not see his experiments as a refutation of a kinetic theory of Brownian motion, even though they imposed arbitrary requirements not independently supported by the kinetic theory of heat. Commenting on the episode, Maiocchi has maintained that Exner's experiment did not seem crucial to anybody;[55] this seems to have been true for the greater scientific community. While it was encouraging that not many followed Exner's proposal on a garden path of modifications to the kinetic theory to incorporate a slow ascent of granular kinetic energies until about 20°C, this also underscores the problem that Brownian motion had to a great degree been turned into a marginal phenomenon in physics at the turn of the twentieth century. Little wonder, then, that Meyer's *The Kinetic Theory of Gases* (1899) makes no mention of Brownian motion in all of its 455 pages and that neither Maxwell nor Boltzmann undertook a proper theoretical investigation into the thermal molecular origin of Brownian motion. Compare this to the attention given to the ratio of specific heats anomaly, with Maxwell calling it the greatest difficulty encountered by the kinetic theory and Boltzmann's ad hoc postulation of five degrees of freedom instead of six for diatomic molecules to explain experimental data.

A problematic assumption in some accounts of how molecular motion leads to a weakened granular velocity was that a large number of molecular collisions would cancel each other out, for the most part, to produce a diminished impact of molecular motion on granular motion. This was endorsed by Gouy in his second paper on Brownian motion (1889) where, after estimating that granular speeds were about a hundred millionth the speed of molecules, he cited the law of large numbers as explanation for how Brownian motion was a *weak* and *distant* result of the heat motion of molecules.[56] He expressed the same viewpoint in his final Brownian motion paper of 1895, namely that the effect of a large number of molecules, not generally in the same direction, would counteract and neutralize in part.[57]

Ironically, misapplication of the law of large numbers was also the cornerstone of one of the more effective criticisms of a kinetic theory of Brownian motion in the nineteenth century by the German botanist Karl Von Nägeli. In 1879 he calculated that the speed that would be imparted to a typical dust particle by the elastic collision of a single gas molecule is 0.002 mm/sec.[58] On the kinetic theory, molecular impacts in opposite directions are equally likely, and part of Nägeli's argument was his misunderstanding that a large number of molecular impacts would more or less have a net neutralizing effect. He further elaborated that the imparted velocity would be even less in liquids due to higher viscous drag and greater intermolecular cohesive forces among the liquids' molecules.

This naïve view that the large number of simultaneous collisions on a granule would effectively cancel each other out was not resolved until Smoluchowski (1906) used simple combinatorics to show that, given equally likely impacts in the positive and negative directions, the velocity imparted (v_i) by N simultaneous impacts would typically be:

$$v_t \approx \sqrt{N} \times v_{mol}$$

where v_{mol} is the velocity imparted by an individual molecule.[59] In the interim, however, the law of large numbers was being used by the proponents of a kinetic theory of Brownian motion as an excuse for the several orders of magnitude difference between molecular and granular kinetic energies, just as much as it was being used by Nägeli to reject a thermal molecular origin of Brownian motion.

Proponents of a kinetic theory of Brownian motion in the nineteenth century seemed to have accepted Nägeli's misapplication of the law of large numbers wholesale insofar as none of them questioned that granular motion was a weakened result of molecular motion. Given that they accepted the gap between molecular and granular kinetic energies as a genuine disparity, misapplication of the law of large numbers was perhaps a welcome explanation for why granular velocities were so much smaller compared to expectations based on the kinetic theory. This balance between the two problems, however, did not make the situation any better, for it forced experimentalists into supposing intricate models of coordinated molecular motions as the cause of Brownian motion. We quoted Gouy's first publication from 1888 toward the end of section 3.1 with his mention of how Brownian motion shows us not the movement of molecules, but something very close to it. In a footnote to this remark, Gouy speculated that molecular movements might be coordinated at magnitudes around one micron to produce Brownian motion.[60] He did not elaborate further on this claim, which became a forgettable footnote to his otherwise influential and lasting work on Brownian motion.

Like Gouy, Ramsay also endorsed the coordinated motion of molecules that "must exist in complex groups of considerable mass, and of some stability" for them to lead to granular movement.[61] Such coordinated movements, however, were in opposition to one of the most fundamental assumptions of kinetic theory of the independence of translational motion across molecules that had not only led to Maxwell's velocity distribution, as discussed at the beginning of this section, but was later generalized through Boltzmann's H-theorem into the Maxwell-Boltzmann distribution as the only equilibrium state of a system. Rejecting the independence of molecular motions was thus paramount to rejecting much of the progress achieved independently in the kinetic theory of heat, while rejecting coordinated molecular movements at the microphysical level left no good alternatives for a kinetic theory of Brownian motion. The details of granular kinetic energies implied by velocity measurements in the nineteenth century were ultimately either a rejection of or a rejection by the classical kinetic theory of heat, and the greater impact of this push and pull seems to have been the relegation of Brownian motion to a marginal phenomenon in physics until after the turn of the century.

Although none of the garden paths threatened by granular velocity measurements was actually pursued across the greater scientific community, the problems they posed for the kinetic theory of heat surely discouraged many in that community from pursuing a kinetic theory of Brownian motion in the nineteenth century. We have so far discussed two distinct garden paths to explain away the erroneous results: (1) that granular kinetic energies were several orders of magnitude smaller than molecular kinetic energies at the same temperature, and (2) that granular kinetic energies do not vary linearly with absolute temperature. A third result that threatened a separate garden path was that granular kinetic energies were not the same for different sizes and types of granules at the same temperature and thus not universal in the way that molecular kinetic energies are. This is underscored by the ratio between granular and molecular kinetic energies, with ratios for Svedberg, Ramsay, and Zsigmondy, respectively, as 1:30,000, 1:1,000,000, and 1:500,000,000. These granular kinetic energy ratios varied by over four orders of magnitude among themselves, in contrast to the universality of mean translational kinetic energy of molecules.

Briefly, the universality of mean translational kinetic energy across molecules also came out of Clausius's 1857 paper, with him combining Avogadro's hypothesis with the *vis viva* of gases at the same temperature and pressure.[62] Maxwell, in 1860, then deduced the equality of mean translational kinetic energy across molecules by proving that for two systems of particles moving in the same vessel, the mean *vis viva* of each particle will become the same in the two systems after many impacts.[63] The universality of mean kinetic energy was thus no less entrenched in the kinetic theory than the proportionality between absolute temperature and the mean translational kinetic energy of molecules. It was an equally significant link between the macrocosm and the microcosm, for it linked the universality of the characteristic energy RT of molar amounts of gases to the universality of mean translational kinetic energy of their constituent atoms and molecules. Add to this the discovery that molar amounts of nonelectrolytic substances in dilute solutions have the same characteristic energy RT as gases[64] and the expectation of accessing the microcosm through colloidal substances naturally followed. Nineteenth-century results on kinetic energies across granules, however, were nowhere close to matching the expectation of universality of kinetic energy across substrates, and this further reinforced the seeming gap between molecular motion and granular motion.

Our goal in this section has been to show the misdirection that can ensue when experimental work proceeds uninformed by theoretical consideration. The ensuing confusions can range from false constraints on a theory imposed by inaccurate experimental data and equally a false distance between theory and experimental results when the latter are not modeled in the right way. The kinetic theory of Brownian motion in the nineteenth century was subject to both these problems, neither of which was resolved until granular velocity measurements were identified as a collective mistake. So long as the granular velocity measurements were seriously entertained, it was impossible to establish the true nature of the thermal equilibrium between molecules and granules. Initially even Perrin, in the same paper where he credited the ultramicroscope for proving the discreteness of colloidal substrates, accepted that the smallest measured granules have kinetic energies about 100,000 times smaller than the kinetic energy of the molecules of a liquid or gas at the same temperature.[65] Exner had expressed no less hesitation in 1900 that his granular velocity measurements of roughly 3 micrometers per second be taken as a genuine result while at the same time acknowledging estimates for molecular velocities based on the kinetic theory at several hundreds of meters per second.

One response to these problems was to reject or at the very least withhold judgment on the thermal molecular origin of Brownian motion. This seems to have been the path taken by the greater scientific community, including the two giants of kinetic theory in the nineteenth century, Maxwell and Boltzmann. The other response, embraced by all the nineteenth-century experimentalists working on Brownian motion, was to search for the theoretical relationship between molecular motion and granules that imparts to the latter a kinetic energy many orders of magnitude smaller than the former. This was the kind of relationship that Exner hoped would clarify the internal movements of a liquid, and just like the various descriptions of granular motion presented under the umbrella term of "*Geschwindigkeiten*" (see Table 3.1), the experimentalists offered various ad hoc proposals for the cause of Brownian motion that would still somehow be based on the heat-motion of the surrounding liquid (see Table 3.2).

Fortunately, the additions and modifications to the molecular-kinetic theory that were being implied by erroneous granular velocity measurements never progressed beyond the rudimentary proposals of Table 3.2.[66] Any ideas of further developing these proposals were mooted by Einstein and Smoluchowski's theoretical work in the first decade of the twentieth century (not to mention Einstein explicitly pointing out the impossibility of ascertaining granular velocities by direct observation). That said, the proximity was close enough for Haas-Lorentz (1913) to adopt the mistaken retrospective view that Felix Exner's call

Table 3.2 Alternative Approaches to Explaining Away *prima facie* Discrepancies between Brownian motions and the Motions Called for by the Molecular-Kinetic Theory of Heat

Author and Date	Rudiments of Various Heat-Motion Based Explanations of Brownian Motion
J. Regnauld 1857	"*Radiation solaire*"
Chr. Wiener 1863	Vibration of attracting corporeal atoms and repulsive ether atoms ("*Körperatome* und *Aetheratome*")
J. Delsaulx 1873	Pressure fluctuations due to unequal molecular impact on the surface of granules since dimensions of granules not considerably greater than estimates for the mean free path of molecules
L. Gouy 1888	Molecular movements in liquids coordinated in part ("*partie coordonnés*") at magnitudes around 1 micron
W. Ramsay 1892	Coordinated molecular motion in "complex groups of considerable mass, and of some stability"
F. Exner 1900	Slow ascent ("*langsam aufsteigend verläuft*") of granular kinetic energy from absolute zero and linear relationship from roughly 20°C onward
R. Zsigmondy 1905	(1) Manifold causes with kinetic theory of fluids of prime importance, and (2) for the smallest particles, electrical attraction to ions of the liquid ("*. . . elektrisch geladenen Goldteilchen mit den ionen der flüssigkeit*")

for a renewed hypothesis was a "carefully marked expectancy" of Einstein and Smoluchowski's theories.[67] In point of fact, while Einstein and Smoluchowski's theories erased the earlier efforts from the collective memory, there is a lesson to be learned from the unfounded efforts. *Brownian motion did not find a place within kinetic theory in the nineteenth century because observational concepts having to do with granular motion were divorced from theory in the first place.* Even the most rudimentary observational concept only gains meaning through a theoretical construct.[68]

As we noted earlier, the requisite theoretical construct for Brownian motion was first provided by Einstein in 1905 in his paper on the movement of small particles suspended in a stationary liquid, as demanded by the kinetic theory of heat. Perrin came on the scene at just the right time to match Einstein's theoretical ingenuity with his experimental finesse, exhibiting the perfect union of the theoretical with the experimental in his verification of sedimentation equilibria, the discovery credited to him in the second half of his Nobel Prize citation.

3.4. The Problem Reconsidered; The Promise Delivered

Einstein began his scientific career taking for granted the molecular constitution of matter. His first two published papers dealt with intermolecular forces governed by a universal function of distance between molecules.[69] Although he later characterized the two papers as "worthless," they were the beginning of his efforts at building atomism as the conceptual foundation of physics. One part of this was to extend the framework of kinetic theory beyond gases to electrons and heat radiation. To this end he pursued a dynamics basis for the thermodynamic laws in three papers on the foundations of statistical mechanics from 1902 to 1904. The last paper in particular dealt with energy fluctuations in a physical system at equilibrium. Einstein believed that, while such fluctuations were not large enough to be observed in a classical ideal gas, black-body radiation is one physical system that experience suggests would exhibit observable energy fluctuations.[70] Seeking such comparisons with theory was especially important to him given his criticism of Boltzmann's *Gastheorie* that it placed too little value on comparison with reality.[71] Here lay the other part of Einstein's approach to atomism and his major aim "to find facts which would guarantee as much as possible the existence of atoms of definite finite size."[72]

The quest for facts concerning the microcosm led to Einstein's development of a new theoretical method for determining Avogadro's number and molecular radii in his PhD thesis.[73] Although the thesis appeared in print after Einstein's first paper on Brownian motion, Abraham Pais gives a convincing argument that Einstein's results on Brownian motion were in part byproducts of his thesis work.[74] Einstein marked April 30, 1905, as the completion date of his thesis, and his first paper on Brownian motion was received by the *Annalen der Physik* 11 days later, on May 11, 1905. The thesis made no reference to Brownian motion, but it is where Einstein worked out the fluctuation-dissipation theorem relating diffusion to viscosity for the first time. The derivation hinged on relating the osmotic pressure of solute particles—based on van't Hoff's theory of osmotic pressure, assuming discrete solute particles in a liquid composed of discrete molecules—to the liquid's viscous force on the particles—based on the hydrodynamic theory of Stokes assuming a continuous fluid medium. This daring union of microphysical and macrophysical laws saw the first explanation of a mesoscopic system where the novel phenomenon of pressure fluctuations, absent in bulk solids, made an appearance.

Einstein, however, was not the first to announce this relationship; William Sutherland communicated it over a year before Einstein to the tenth Congress of the Australasian Association for the Advancement of Science held from January 6 to January 13, 1904, in Dunedin, New Zealand. This account was published in the Congress proceedings in early 1905,[75] followed by the publication in *Philosophical Magazine* of an expanded account by Sutherland, dated March 1905.[76] His extended equation contains a term for slip at the solute-solution interface such that:

$$D = \frac{RT}{N}\frac{1}{6\pi a\zeta}\frac{1+3\zeta/\beta a}{1+3\zeta/\beta a}$$

where D is the diffusion coefficient, R is the ideal gas constant, T is the temperature, N is the number of molecules in a gram-molecule (that is, a mole), a is the radius of the presumed spherical solute particle, ζ is the viscosity of the liquid, and the new term β is the coefficient of sliding friction to account for slip between the diffusing molecule and the solution. If $\beta = \infty$, i.e., if there is no slip at the boundary, we recover the usual form of the equation originally derived by Sutherland and later by Einstein:

$$D = \frac{RT}{N}\frac{1}{6\pi a\zeta}$$

Einstein, like Sutherland in his presentation at the Congress meeting, assumed no slip to derive this equation outright and not as a special case of his extended equation. Both ways, we agree with Pais that the fluctuation-dissipation theorem relating the diffusion coefficient to viscosity should properly be called the Sutherland-Einstein relation.[77]

With this relationship at their disposal, Sutherland and Einstein did very different things—the former sought out particular results of large molecular masses and the latter sought out universal results, including the measurement of N. In fact, after deriving the relationship, Sutherland made no reference to N again. He noted that at the same temperature, aD would be a constant and consequently $B^{1/3}D$ should also be a constant where B is the volume of the molecules in a gram-molecule of solute. From available data on the diffusion of gases and nonelectrolytes in water, Sutherland inductively generalized to the equation:

$$10^6 B^{1/3} D = b + \frac{k}{B^{2/3}}$$

where b and k are constants for a given solvent at a given temperature determined as 21 and 220, respectively, for water at 16°C. When $B^{1/3}$ is large, $10^6 B^{1/3} D = 21$ is approximately correct. Sutherland used this relationship to determine B for egg albumin as 27,000 from its diffusion coefficient and from there, the molecular mass of egg albumin as 32,814 from its minimum molecular formula and the limiting volumes of its constituent gram-atoms.[78]

In contrast, Einstein took data on the molecular weight and viscosity of a sugar solution to calculate the hydrodynamically effective radius of sugar molecules and the universal *N*. In his words, ". . . *N* must show itself to be independent of the nature of the solvent, of the solute and of the temperature, if our theory is to correspond with the facts."[79] The theory consisted of a second equation for the increase in viscosity of a solution due to the presence of very small rigid spheres:

$$\frac{k^*}{k} = 1 + \varphi$$

where k^* is viscosity of the solution, k is the viscosity of the solvent, and φ is the total volume of the spheres suspended in unit volume. Substituting for φ:

$$\frac{k^*}{k} = 1 + \frac{N\rho}{m}\frac{4\pi}{3}a^3$$

where ρ is the mass of the dissolved solute per unit volume and m is its molecular weight. With two equations and two unknowns, Einstein obtained 2.1×10^{23} for N and 6.2×10^{-8} cm in his thesis for the hydrodynamically effective radius of sugar molecules. Using better data, Einstein gave 4.15×10^{23} for N in a supplement to his thesis dated January 1906 that appeared in the *Annalen der Physik*, and he revisited his thesis one last time in 1911 to make a correction to his change in the viscosity equation such that:

$$\frac{k^*}{k} = 1 + 2.5\varphi$$

Using the same data as before, N now came out as 6.56×10^{23}. Einstein was thus interested not only in calculating N, but also in finding progressively more precise values for it, as he did with each correction to his account of the transport and intrinsic viscosity of invisible solute particles. This effort was tied to his quest to establish the existence of atoms of definite finite size through the application of the transport relation for invisible solute particles to visible granules in Brownian motion.

Not knowing much about Brownian motion upon the completion of his thesis, Einstein sent a paper to the *Annalen der Physik* on the movement of microscopically visible particles suspended in a stationary liquid with an introductory remark that, "It is possible that the movements to be discussed here are identical with the so-called 'Brownian molecular motion.'"[80] He again derived the fluctuation-dissipation theorem, though with two seemingly trivial changes. First, in relating dynamic equilibrium between the internal force acting on the particles and their osmotic pressure, he construed the pressure gradient as $K\nu$ where ν is the number of suspended particles per unit volume. In his thesis, he equated the pressure gradient to $K(\rho/m)N$ where ρ is the mass of solute per unit volume and m is its molecular weight. Second, he determined the diffusion current as the product of the diffusion coefficient and a particle number density gradient, whereas he used a particle mass density gradient in the thesis. In the end, these differences wash out to give the same equation relating diffusion to viscosity, but there is clearly an emphasis in Einstein's first paper on Brownian motion on discrete particles in dynamical motion instead of on mass properties like molecular weights.

The truly novel part of Einstein's paper, however, was deriving the diffusion equation on the assumption of independence between particle motion and relating it to the mean square displacement per unit time of particles:

$$\xi^2 = 2D\tau$$

where ξ^2 is the mean of squared displacements, D is the diffusion coefficient from before, and τ is the time period of observation. This relationship, along with the Sutherland-Einstein relation, in principle allowed a determination of N using just a microscope fitted with an ocular scale and a stopwatch; we return to the significance of calculating N from this method in Chapter 4, but in the meantime note the importance Einstein attached to such a determination:

> It is to be hoped that some enquirer may succeed shortly in solving the problem suggested here, which is so important in connection with the theory of Heat.[81]

Jean Perrin was the inquirer ready to take on this challenge, and in the same year as Einstein's publication, of which he was initially unaware, he was searching for ways to access Brownian motion and measure the properties of suspended granules.

Perrin's intrigue with granules, the discrete constituents of colloidal solutions, turned into a project for experimental study soon after the invention of the ultramicroscope. The ultramicroscope zoomed further into turbid media to reveal the discontinuous structure of colloids, Perrin describing the discrete granules as swarms of bright stars on a dark background. He discussed these findings on the discontinuous structure of colloids in the second part of his paper on contact electrification and the electrical properties of colloidal solutions. The first part appeared in volume 2 (1904) of *Journal de Chimie Physique*, where Perrin discussed the action of H^+ and OH^- ions as the fundamental agents of electrification. The second part appeared in volume 3 (1905) of the same journal, and although Perrin focused there on the rules of contact electrification—in relation to the electrical potential, size, and coagulation of granules—he also proposed several ways to tie granular activity to molecular motion. He coined the term "granules" for the discrete constituents of colloids and laid out his expectation in them of a direct verification of molecular-kinetic theories. The verification, as Perrin put it, was to establish that granules are stirred and suspended by a mechanism similar to that which stirs and suspends a molecule of sugar in a sugar solution.[82]

Perrin right away searched in Brownian motion for continuity and parallels with the molecular realm, yet he also entertained the erroneous granular velocity measurements in his first publication on this relation in 1905. In describing the motion of granules 4 to 5 microns in diameter, he remained noncommittal between using the terms *displacement* and *amplitude* for the magnitude of the movement. Moreover, it appears that Perrin did not yet appreciate the impossibility of measuring granular velocities, still believing that the kinetic energy of a granule was much smaller ("*très inférieure*") than the kinetic energy of an isolated molecule.[83] This misapprehension had vanished by early 1908 as he jotted his first ideas for producing uniform emulsions of same-sized granules by fractional centrifugation in his lab notebook. His first entry reads, "Langevin-Einstein hypothesis: each granule is assimilable to molecule (same),"[84] for which he was now seeking experimental verification. Perrin knew he could not return to velocity measurements for this verification, however, and neither did he need to, as the idea of indirectly measuring granular movement through the vertical distribution of granules was already in 1905 evolving in Perrin's mind.

During his first speculations on granular movement, Perrin already reasoned that granular diffusion, measurable or not, is theoretically certain as a consequence of their irregular movement in Brownian motion. His conceptual understanding was not far off from Einstein's or Smoluchowski's in that, even though any granule remains for some time in the vicinity of the same point, the perfectly irregular movements do not cancel out and hence one must simply wait for the displacement of the granule to become large.[85] This process of a true diffusion ("*diffusion veritable*") must surely play a role in maintaining a vertical distribution of granules, and Perrin's musings on this merit quotation, as they were the sparks that set him on the path of eventually providing a full explanation of the vertical distribution:

> On reflecting on this diffusion, one sees above all that it cannot have the effect of rigorously equalizing the distribution of granules. It is only in the same horizontal plane that this distribution becomes and remains uniform. But the concentration of granules must increase as this plane is taken to a lower level (at least if the granules are denser than the liquid in which they are immersed).
>
> *Thus, diffusion has the effect of establishing, then of maintaining invariable, a certain mean distribution of granules, such that an element of equal volume contains all the more (granules) as it is at a greater depth in the liquid.*

> This explains why, for example, a colloidal solution of copper ferrocyanide, left alone for a long time, is always darker in the lower region, *and by no means due to precipitation.*[86]

Perrin continued that, if the intensity of gravity were, say, a thousand times greater, the inequality in vertical distribution would be even more pronounced. This is why centrifuging a solution of copper ferrocyanide for several hours almost completely discolors the upper layers. He concluded by noting that Gouy had also shown by thermodynamic reasoning that a solution of copper sulfate should be more concentrated at the bottom than at the top; and, that in centrifuging several salt solutions, Lobry de Bruyn had produced and measured large concentration differences between the upper and lower layers of a previously homogenous solution.[87]

One sees in Perrin's comments his appreciation for gravity as the other factor affecting concentration differences across horizontal layers of a colloidal solution or a true solution. Later, he would dig deeper into the osmotic pressure of granules as a mechanism in granular diffusion and construe the equilibrium distribution of granules across vertical layers as a balance between this pressure and the effective gravitational force. By the time of Perrin's 1908 publication of "L'agitation moleculaire et le mouvement brownien," he brought these two forces together into a well-defined theoretical construct to go along with well-defined experimental features. In doing so, Perrin succeeded where other approaches had been unfounded or imprecise. He systematically explained the properties of granules in Brownian motion and, in the process, got well-behaved results that offered links to the microstructure of matter.

These links, of course, did not amount to literally seeing atoms and molecules—a feat arguably accomplished by images of atoms in the 1950s obtained using the field ion microscope. Yet, evidence for the discreteness of matter did not require such a narrow and contingent form of verification for, as Pais has pointed out, particles smaller than atoms were "seen" much earlier; electrons, for instance, were perceived as tracks in a cloud chamber.[88] In relation to Brownian motion, he says:

> Today, we say that Brown saw the action of the water molecules pushing against the suspended objects. But what we see in Brownian motion is as dependent on theoretical analysis as is the statement that a certain cloud chamber track can be identified as an electron.[89]

Fast-forward a hundred years and the situation is not very different for the discovery of the Higgs Boson from the detection of its decay signature. Evidence for the discrete structure of matter similarly appealed to the signature of atoms and molecules detected in the movement of granules.

In short, the characterization of Brownian motion itself is of paramount importance in saying anything about its cause. Furthermore, when Perrin came to concentrate on the nature of Brownian motion, there was a great deal of confusion not just about the cause of the motion, but about the motion itself. Einstein seems to have been the first to provide a basis—in his case, theoretical—for ending this confusion. Perrin seems to have been the first to do so experimentally, characterizing Brownian motion for the first time in the way we have now come to understand it. We focus on this characterization in the next chapter. This brings us to the second principal preoccupation of this book: *What did the research by Perrin and others establish about Brownian motion* independently *of the molecular hypothesis and kinetic theory?* Answering this question will then put us in a position to say something about implications of Brownian motion for the molecular-kinetic theory in Chapters 5 and 6.

3.5. The History in Retrospect

As we remarked in passing earlier, the second edition of Meyer's *The Kinetic Theory of Gases* of 1899 (the year he died) contained no mention at all of Brownian motion. More to the point, insofar as it was about gases, neither did either of the other two "book-length review articles" that we relied on in Chapter 2, the 1889–1890 editions of Ostwald's *Grundriss* and the 1893 edition of Nernst's *Theoretische Chemie*. The latter did contain a chapter on colloidal chemistry, contrasting it with the chemistry of "crystal" solutions, with the focus on such matters as osmotic pressures and changes in the freezing points, but with no mention of the motion of the colloidal constituents. Ostwald, by contrast, elected not to include anything but a passing reference to colloids not only in the cited edition, but also in the 1899 edition of his *Grundriss*. This was not because he was unaware of the current research on colloids, for the fourth book of the 1890 second edition of his *Lehrbuch der Allgemeinen Chemie* contains an extended discussion of the contrast between colloid and other solutions, which Nernst's discussion mirrors; but here too there is no mention at all of Brownian motion.[90]

This absence of attention to Brownian motion in major books pertaining to molecular chemistry and kinetics, moreover, was not confined to ones written in German; the first 1904 edition of Jeans's *The Dynamical Theory of Gases* also contains no mention of it, though it does briefly consider liquid as well as gaseous solutions in a chapter on "Aggregation and Dissociation."[91] By contrast, the very first page of the second, 1916, edition of Jeans's book—and in the two subsequent editions as well—features the following paragraph, to which we shall return later:

> The second hypothesis, the identification of heat with molecular motion, is that with which the Kinetic Theory of Matter is especially concerned. This hypothesis was long regarded as pure conjecture, incapable of direct proof, and probable just in proportion to the number of phenomena which could be explained by its help. In recent years, however, the study of Brownian movements has provided brilliant visual demonstration of the truth of this conjecture, and the actual heat-motion of molecules—or at least of particles which play a role exactly similar to that of molecules—may now be seen by anyone who can use a microscope.[92]

Of course, Jeans's pronouncement, for all its flourish, was simply echoing the emphasis put on Brownian motion in the 1909 editions of Nernst's *Theoretische Chemie* and Ostwald's *Grundriss*, which included a new section on "Microchemistry" featuring a chapter on colloids. The once recondite topic of Brownian motion had come to occupy center stage.

Quite a change, accordingly, took place in a very short period of time in the attitude toward Brownian motion of those producing authoritative books on molecular chemistry and kinetics. One question this raises is why it took place in such a short time. No less does it raise a question about why the change was so extreme, from no mention at all in presenting the kinetic theory of heat to its becoming heralded as the most compelling evidence for it. Regardless of how one views Jeans's assessment of the hypothetical status of molecules before Perrin's and Svedberg's results on Brownian motion, he was surely correct that all the evidence for molecules in motion before these results was indirect and hence in large part a matter of surmise. Insofar as no one had pretended otherwise, it does seem odd that none of the principal advocates for the molecular-kinetic theory of heat before Einstein—to add still one more example, Boltzmann in his *Lectures on Gas Theory* from the late 1890s—seems to have given any thought to the possibility of Brownian motion offering a source of strong evidence in support of the theory. Moreover, independently of this theory, one would think that the persistence of Brownian motion would have drawn the attention of the champions of standard thermodynamics like Ostwald, for as he remarks in his 1909 edition, "These movements are in apparent contradiction to the Second Law."[93] Viewed in retrospect, therefore, the history of research on Brownian motion raises questions about why the shift was so extreme at least as much as questions about why it was so abrupt.[94]

Both the timing of the shift and much of its abruptness are to be explained by the innovation of the ultramicroscope. We opened this chapter with the quote from Zsigmondy announcing how much more remarkable the motions of the gold particles that had become visible through the ultramicroscope were than the motions of the larger gold particles that he had previously taken to exemplify Brownian motion. The full title of the 1905 book where this statement appeared is *Zur Erkenntnis der Kolloide: Über irreversible Hydrosole und Ultramikroskopie*. An indication of how much attention it attracted was its appearance in English as soon as 1909. Zsigmondy was himself a colloid chemist and became involved in the ultramicroscope project in the first place in order to gain visible access to colloid particles that were not resolvable under an ordinary microscope, yet were clearly making a difference.[95] The innovation immediately turned colloid chemistry into a prominent research area, and independently of this the device itself drew a great deal of attention, if not excitement. Indeed, the quote with which we opened the chapter and the paragraphs surrounding it appear to have been historically significant, for Svedberg's research on Brownian motion, readers will recall, followed Zsigmondy in employing gold particles.

The increased access to the particles engaged in Brownian motion nevertheless did not immediately resolve the question of whether this motion should be construed as oscillatory or translational. Svedberg had construed the motion as oscillatory in his paper announcing confirmation of Einstein's formula, prompting the latter's 1907 paper intended "to make it somewhat easier for the physicists who study this problem to interpret their observational data and to compare them with theory."[96] The very question of whether the motion is oscillatory or progressive, of course, disappears once one takes it to be truly random, for the seeming prevalence of oscillation is simply an (at-times) circumstantial consequence of any random-walk process. That this question was arising at all in nineteenth-century research on Brownian motion might have been seen as evidence not only that the motion itself is Gaussian, but that for this very reason it was a potential source of evidence in support of the Maxwell-Boltzmann account of molecular motion. That it was not seen in this way by those engaged in research on Brownian motion is symptomatic of how difficult it was at the time to conceptualize random-walk processes. We shall see this difficulty again at the end of the next chapter with C. W. Wolfgang Ostwald's seeming inability to appreciate the by-then confirmed Gaussian distribution of the displacements of granules in Brownian motion and its implications for diffusion. One cannot help but wonder how much more the attention of someone like Boltzmann would have been drawn to Brownian motion if someone doing research on it had proposed that it exemplifies the sort of random motion called for by him and Maxwell.

Not being able to appreciate what came to be known as a random-walk process was surely not the only reason no one had proposed this. Even after the advent of the ultramicroscope, those doing research on Brownian motion were relying on what they were seeing as the basis for the conclusions they were putting forth. That Brownian motion is Gaussian is not something that one can simply see, even under an ultramicroscope, much less the microscopes that were being used before it. Moreover, as we have emphasized throughout the chapter, this reliance on vision had led to values for the speeds of granules that could not be reconciled with those called for by the molecular-kinetic theory of heat.

Giving up on the idea that vision is our most reliable source of information is not so easy. Perhaps no one at the time was even aware that what Huygens had done when he resorted to pendulum-based theory-mediated measurements of the strength of surface gravity a couple of centuries earlier was to recognize the limits of direct vision. The necessity of doing so in the case of Brownian motion was the primary message Einstein was trying to convey in the paper prompted by Svedberg's paper; the second of the three points he develops in it ends with a single-sentence paragraph: "It is therefore impossible—at least for ultramicroscopic particles—to determine $\sqrt{v^2}$ by observation."[97]

Readers might naturally assume that the subtitle to our monograph—"A Study in Theory-Mediated Measurement"—refers to the role such measurement played in advancing research into the microphysical realm. It of course does so; but it no less refers to the indispensable role of theory-mediated measurement in the extraordinary advances Perrin made in research on Brownian motion in and of itself—that is, independently of anything it has to do with microphysics—over the course of the years between 1905 and 1913. Spelling out how it did this is a central goal of the next chapter.

Notes

1. Siedentopf and Zsigmondy (1903).
2. For an excellent historical account of the development toward the ultramicroscope, see Cahan (1996). Cahan notes the novelty of Zsigmondy's conception of the ultramicroscope in utilizing the Faraday-Tyndall diffraction cone, yet stresses that this innovation was facilitated within the scientific, technical, industrial, and economic context at the Zeiss complex in Jena. In particular, the Zeiss microscope used by Zsigmondy and Siedentopf was an advanced but nonetheless traditional instrument that had already been developed and refined by Abbe over the latter half of the nineteenth century.
3. Nye (1972), pp. 85–91.
4. Zsigmondy (1909), pp. 134–135.
5. Perrin (1905), p. 55.
6. Einstein (1905). The English translation of this essay is entitled "On the Movement of Small Particles Suspended in a Stationary Liquid Demanded by the Molecular-Kinetic Theory of Gases." This, along with Einstein's other essays on Brownian motion, are collectively presented in English translation in Einstein (1956). The page

numbers we cite in this chapter will be to this book. Both originals and translations of all the papers in this book, plus Einstein's dissertation, are available in Einstein (1989a) and, in English, Einstein (1989b).

7. Brown (1828).
8. Brewster was co-editor of the *Edinburgh New Philosophical Journal* until 1826, at which point he parted ways with Robert Jameson. Jameson took sole editorship of the journal and Brewster started a rival journal edited by himself, called the *Edinburgh Journal of Science*. Brown's microscopic observations from the summer of 1827 were published in the third quarterly publication of the 1828 *Edinburgh New Philosophical Journal*. Brewster's remarks on Brown's observations soon followed in a note dated December 18, 1828, in the *Edinburgh Journal of Science*. Brewster (b. 1781–d. 1868) was a prominent figure in the debate over wave versus particle theories of light, relentlessly defending the latter; he also produced the first comprehensive biography of Isaac Newton.
9. Brewster (1828), pp. 219–220.
10. Renn (1997). Renn argues that Einstein's reinterpretation was motivated by his greater preoccupation with specific problems in physics, like radiation and the electron theory of metals, and establishing atomism as a general foundation in physics. The key difference is the equipartition of energy for arbitrary material bodies in thermal equilibrium, which played an only secondary role in Boltzmann's work, but became the centerpiece of Einstein's 1905 theoretical treatment, in his words, of the so-called Brownian molecular motion.
11. For a more detailed historical account of these efforts, see Nye (1972), pp. 9–13, 21–29. Also Brush (1968), pp. 7–10.
12. This background information of precluding external causes for an explanation of Brownian motion is the starting point of several accounts of the Perrin episode, viz. Achinstein (2001) and Psillos (2011a, 2011b, 2014). Both philosophers use this background information to assign a prior probability to the discreteness of matter. In particular, Achinstein assumes a prior probability greater than ½ for the claim that chemical substances are composed of molecules, and Psillos assumes that the prior probability of the atomic hypothesis "is not very low." Although the two then utilize Bayes's Theorem in different ways, they must both assume the ad hoc prior probability to avoid the base-rate fallacy in their respective arguments.
13. This quote is from the *Journal of the Royal Microscopical Society*, vol. 10, 1890, p. 120. The society purchased a number of the *Philosophical Magazine and Annals of Philosophy* for 1828, wherein Brown's microscopical observations were first published (pp. 161–173). At the beginning of Brown's article, they found an MS letter addressed to Reverend Dr. Buckland and signed J.H.C., understood to be Reverend John Henry Conybeare from a relative of Dr. Buckland.
14. Jeans (1901).
15. See Kerker (1974), p. 766; also see Freundlich (1926), p. 341.
16. Delsaulx (1877), p. 4.
17. Ibid.
18. Smoluchowski (1906), p. 577.

19. A. Sommerfeld (1917), p. 533. English translation taken from Piasecki (2007), p. 1624.
20. Perrin (1908a), p. 968.
21. Gouy (1888), p. 563.
22. Gouy (1889).
23. Gouy (1895), p. 7.
24. Svedberg (1907), pp 134–135.
25. Albert Einstein (1956), p. 67, in translation of Einstein (1907).
26. Ibid., p. 63.
27. Daub (1971).
28. See the chapter on kinetic theory, for instance, in the textbook used for introductory physics at Tufts University: Giancoli (2008), pp. 477–478.
29. Clausius (2003), p. 131.
30. Meyer (1899b), pp. 57–58.
31. Perrin (1990), p. 56.
32. Maxwell (1860), p. 32.
33. Meyer (1899b), p. 151.
34. Regnauld (1858), p. 141.
35. Zsigmondy (1909), pp. 134–135.
36. Ibid., pp. 137–138.
37. Ibid., p. 194.
38. Svedberg (1906).
39. Ibid., p. 853.
40. See Kerker (1976) for how Svedberg developed his theoretical characterization of Brownian motion. Kerker's paper is essential for a reading of Svedberg's work on Brownian motion, starting with the first paper of 1906 into Svedberg's 1907 dissertation, especially for the clarity with which he identifies the various assumptions in Svedberg's construction—namely, construing granular motion as oscillatory and then studying it as a simple harmonic motion—and how tenuous the link was from Svedberg's simple harmonic granular motion to Einstein's root mean square displacement of granules.
41. We've used t to represent the time period of oscillation instead of Svedberg's notation τ in his original paper to distinguish it from τ in Einstein's displacement formula that is the mainstay of section 4.4.
42. Ostwald (1912), p. 585.
43. Söderbaum (1926).
44. See, in particular, Kerker (1976), pp. 212–216, and Maiocchi (1990), pp. 269–274.
45. Maiocchi (1990), p. 271.
46. Ramsay (1882), p. 301.
47. Exner (1900), p. 845.
48. Smoluchowski (1906b), p. 577.
49. Smith in print—(2002) and (2010)—has referred to a garden path as pursuing an extended body of research that ultimately has to be thrown out because it was predicated on a fundamentally mistaken working hypothesis; or, put another way, it is

the risk of making successive approximations to a theory to deal with second-order phenomena that might turn out to be artifacts stemming from errors in the original theory. This is particularly relevant in the case of inductively generalizing a law-like relationship beyond the scope of its application; for a discussion in the context of Newton's law of gravity and his law of universal gravitation, see Smith (2002), pp. 50–55. We use the concept of a garden path here to refer to the theoretically misinformed measurement of apparent granular velocities that imposed false constraints on the kinetic theory of heat that should have informed the measurement of granular velocities in the first place.

50. Gouy (1889), p. 105.
51. Clausius (2003), p. 115. This is the same paper in which Clausius relates the ratio of specific heats to the fraction of the total energy of a gas that increases the translational kinetic energy of molecules, pp. 131–134.
52. Maxwell (2003), p. 149.
53. For an account of the major players in the development of kinetic theory, see Brush (1976a). In Table 1.1-1, pp. 6–7, Brush provides a list of 39 important papers, along with their subject matter, that catalogs the progressive development of a general kinetic theory and the various attempts of relating it to experimental phenomena.
54. Exner (1900), p. 847.
55. Maiocchi (1990), p. 262.
56. Gouy (1889), p. 105.
57. Gouy (1895), p. 7.
58. Nägeli (1879); see calculation on p. 395.
59. Smoluchowski (1906b), pp. 583–584. While Smoluchowski used this calculation to show Nägeli's error, he noted that it did not represent the correct result because each collision would not change the granular velocity by the same amount as the formula supposes. His actual analysis of granular movement assumed a granular velocity based on the equipartition theorem and considered the change in direction due to random molecular impacts.
60. Gouy (1888), p. 563.
61. Ramsay (1892), p. 90.
62. Clausius (2003), p. 123.
63. Maxwell (2003), pp. 156–157.
64. See the concluding pages of section 2.5 in Chapter 2.
65. Perrin (1905), pp. 58–59.
66. In contrast to Table 3.1, we have not mentioned Svedberg in this table of various proposals for the heat-motion based explanation of Brownian motion because he presented his results within the framework of Einstein's derivation of the formula for mean-squared-displacements. It is another matter that Svedberg misrepresented Einstein's mean-squared-displacement by replacing it with his amplitude and then further imposed his own notion of "absolute velocity" ("*absolute Geschwindigkeit*") to obtain granular velocities for comparison with past values when his notion of granular velocity had no place within Einstein's derivation and interpretation of his formula. Svedberg nevertheless considered that what he was taking in his results to be

rough agreement with Einstein's formula was evidence for granular movement as a manifestation of the general molecular movement of matter. Svedberg thus did not have a novel heat-motion-based proposal to explain Brownian motion in the way of Wiener's "corporeal and aether atoms" or Ramsay's "coordinated molecular motions," and hence does not belong in the table.

67. G. L. de Haas-Lorentz (1913), p. 12.
68. See Herman Weyl's chapter on "Methodology," especially the sections on "Measurement" and "Formation of Concepts" in Weyl (1949), pp. 121–122 and 139–151; the original German edition was published in 1927.
69. In both papers, Einstein proposed a universal molecular energy function in analogy with the inverse-square force of gravity. The first paper on the thermodynamics of liquid surfaces dealt with liquid molecules, and the second paper on the thermodynamics of electrolysis dealt with electrolytic solutions. See (1) Einstein (1901) and (2) Einstein (1902). Abraham Pais's (1982) outstanding biography of Einstein has been essential in informing us of Einstein's scientific endeavors and interests leading up to his *annus mirabilis*. In particular, Pais informs that Einstein deemed his first two papers as "worthless" in a letter to J. Stark dated December 7, 1907 (p. 57), and that Einstein's fluctuation-dissipation relationship between diffusion and viscosity had been independently discovered by William Sutherland in Australia at around the same time as Einstein (p. 92).
70. Einstein (1904), p. 361.
71. Einstein, letter to Mileva Marić, April 30, 1901, in Einstein (1987), Doc. 102.
72. Einstein (1979), p. 44; translation, p. 45.
73. Einstein (1956), pp. 36–62; originally in *Annalen der Physik*, vol. 19, 1906, pp. 289–306, with corrections in *Annalen der Physik*, vol. 34, 1911, pp. 591–592.
74. Pais (1982), pp. 88–92.
75. Sutherland (1904).
76. Sutherland (1905).
77. Pais (1982), p. 92.
78. Sutherland (1905), p. 784f.
79. Einstein (1956), p. 60.
80. Einstein (1956), p. 1.
81. Ibid., p. 18.
82. Perrin (1905), p. 58.
83. Ibid.
84. This is taken from Bigg (2011), p. 162, where she cites Perrin's lab notebook: Jean Perrin, "*tube de crookes*," laboratory notebook, pp. 1–3. Jean Perrin file, Archives of the Académie des Sciences, Paris.
85. Perrin (1905), p. 60.
86. Ibid., pp. 60–61.
87. Ibid., p. 61.
88. Pais (1982), p. 86.
89. Ibid.
90. Ostwald (1891), p. 151.

91. Jeans (1904).
92. Jeans (1916), p. 1.
93. Ostwald (1912), p. 484.
94. We should note that we are not the first to raise such questions. Stephen Brush opens "A History of Random Processes" with the same points we have made (1968, p. 1):

> It is surprising that Brownian movement played almost no role in physics until 1905, and was generally ignored even by the physicists who developed the kinetic theory of gases, though it is now frequently remarked that Brownian movement is the best illustration of the existence of random molecular motions. Moreover, the existence of Brownian movement itself has been interpreted to imply that the Second Law of Thermodynamics does not have absolute validity on the microscopic level; yet this argument was almost completely overlooked at the time when the validity and consequences of the Second Law were being hotly disputed in the 1890s.

95. Cahan (1996).
96. Einstein (1956), pp. 63–67; (1987a), p. 399f; and for the English (1987b), pp. 229–231.
97. Einstein (1956), p. 66; (1987b), p. 231.

4
Perrin's Brownian Motion Experiments

Brownian motion refers to the indefinitely continuing complex motion of particles, or *granules*, in a dilute emulsion or colloidal solution. Perrin's experiments were designed to yield theory-mediated measurements of aspects of that motion from features of it that could be observed in a sufficiently high-powered microscope. By the *logical design* of these experiments, we mean the assumptions made and the step-by-step derivation from them of the theoretical relationship mediating between the indirectly measured quantities and the features observed. As we indicated in Chapter 1, the logical design of Perrin's experiments can be construed in two very different ways. On the one hand, the assumptions in question can expressly be about molecules, and the mediating relationships amount to an extension of molecular-kinetic theory to Brownian motion. On the other hand, the assumptions can be limited strictly to claims about the granules themselves, adopted as provisional *scaffolding*, to use Ostwald's term,[1] to enable conclusions to be drawn *about Brownian motion* itself.

Insofar as the conclusions that were ultimately drawn from Perrin's experiments were about molecules, there cannot be anything out-and-out wrong with construing their logical design as an extension of kinetic theory. Our alternative construal nonetheless has two notable advantages for purposes of assessing just what Perrin's evidence showed. First, his experiments yielded conclusions about Brownian motion itself that were beyond dispute, at least far more so than the conclusions about molecules. In other words, *the strongest evidence coming out of the experiments concerned Brownian motion itself, separately from any claims at all about molecules*. Second, construing the logical design in this narrower way can bring out precisely what gap remained between the conclusions pertaining to the visible realm of the granules and the further conclusions about the non-visible realm of molecules. In this chapter we focus on the first of these two, postponing the second until Chapters 5 and 6.

As we indicated in Chapter 1 when explaining our subtitle, we shall be taking the occasion of presenting Perrin's efforts on Brownian motion to bring out some logical intricacies of theory-mediated measurement. In particular, we shall be showing how his "well-behaved" results provide evidence in and of themselves for the validity of his measurements and hence for the theoretical presuppositions on which they are indispensably predicated. Again, as noted there, Perrin's efforts on Brownian motion are especially instructive in this

Brownian Motion and Molecular Reality. George E. Smith and Raghav Seth, Oxford University Press (2020). © Oxford University Press. DOI: 10.1093/oso/9780190098025.001.0001.

regard. His experimental approach (and independently of him, Einstein's proposed approach as well) recognized from the outset that, even though Brownian motion can be watched through a microscope, the quantity of primary interest, the mean square velocities of the particles, cannot be determined visually. The only way to get at this quantity was through indirect, theory-mediated measurement in which a visually accessible quantity serves as a proxy for the quantity of interest. We shall examine the four theory-mediated measurements he carried out each in turn and then collectively, stressing how his results in and of themselves supported his claim to have measured mean kinetic energies in Brownian motion. Section 4.9 will then consider his efforts within the broader history of the practice of theory-mediated measurement, in the process stating in a general form what it is for results from such measurements to be "well-behaved" and how their being so in itself amounts to evidence for their validity.

4.1. Some Preliminaries

Perrin's theory-mediated values were derived from three distinct Brownian motion phenomena: (1) the vertical gradient in the concentration of granules as a consequence of gravity, referred to as Perrin's "discovery of sedimentation equilibrium" in his Nobel Prize citation; (2) the lateral or translational displacement of granules versus time and, related to this yet yielding a separate determination, the diffusion of granules, which was the primary phenomenon for which Einstein's theory of Brownian motion (and Smoluchowski's) gave predictions; and (3) the angular or rotational motion of granules, a further phenomenon for which Einstein's theory gave predictions.

In every case the principal values announced by Perrin were for Avogadro's number—that is, the number of *molecules* per mole, and hence a molecular parameter that involves more than just the motion of granules. In every case, however, Perrin obtains values for Avogadro's number by dividing (algebraically, not numerically) the characteristic energy per mole *RT* of gases and dilute solutions by 2/3 of the mean kinetic energy of the granules.[2] *His principal measured values for Brownian motion itself were therefore for mean kinetic energies of granules.* The quotient of *RT* and these values can be interpreted in two ways: as the number of granules needed for their total (translational) kinetic energy to match 3/2 of the *RT* values for a mole of gas or dissolved substance; or as the number of molecules in a mole, *assuming* that their mean translational kinetic energy at any given temperature is the same as that of the granules in Brownian motion.

That said, it will often be easier in the rest of this chapter to work with Perrin's claimed values for Avogadro's number rather than the corresponding values for the mean kinetic energy of granules. This is because Avogadro's number is (presumably) a constant, allowing different values for it to be compared immediately with one another, in contrast to the mean kinetic energies of granules, which vary with the temperature of the emulsion. So, whenever we cite Perrin's values for Avogadro's number in the rest of this section, all we mean is the quotient of RT and 2/3 of a mean kinetic energy of granules. In other words, *the values we cite for Avogadro's number in this chapter should always be taken as mere surrogates for values of the mean kinetic energies of the granules.* To help remind readers of this, we shall be distinguishing between our symbol for Avogadro's number, N_0—the number of molecules in a mole of gas at standard conditions—and the symbol N, for the number of granules needed for their mean translational energy to sum to $3RT/2$—that is, for what Perrin called "Avogadro's number" in his statement of the results of his Brownian motion experiments.

Spelling out their logical design is just the initial step in any proper logical analysis of experiments. A second step is to assess the relationship between the data generated in them, on the one hand, especially the scatter, and on the other, the summary statements announcing the results.[3] Perrin's data in any one experiment generally had a good deal of scatter, which he then distilled into a single value. Insofar as Brownian motion is stochastic, such scatter was to be expected, and his announcing results in terms of mean values was fully appropriate. The question can still be raised, however, whether the specific results he announced were properly warranted by his data—or, in other words, whether his "reduction" of the data was proper. We are largely going to ignore that question here, most of the time taking his announced results at face value without commenting on the data behind them. We are likewise largely going to ignore a third step of any proper logical analysis of his experiments, instead simply granting him that the range of variations of the manipulable parameters covered by his experiments was sufficient to warrant his inductively generalized statements of his results.

One reason for focusing almost exclusively on their logical design in our review of Perrin's experiments is that others at the time appear not to have quibbled all that much with his statements of their results and his inductive generalizations of them.[4] The main reason, however, is that the two philosophic issues we singled out in Chapter 1 concern what his experiments *ideally* could have shown, and these issues turn on their logical design, not the specifics of his execution of them.

Even though we will be doing only part of the task, the point of view we are going to try to maintain in the next three subsections is that of someone writing a review article on Brownian motion just before or soon after the first Solvay Conference late in 1911—in particular, before the publication of the Bohr model in 1913. Before we shift to that point of view, however, we should make one remark about Perrin's results that did not become unmistakably clear until after 1913: his announced values for Avogadro's number were consistently

10 percent and more too high. In other words, the values for the mean kinetic energies of Brownian motion granules were 10 percent and more lower than the subsequent firmly established values for the mean translational energies of molecules at the same temperatures. The source of this discrepancy, which extends across all of the phenomena out of which he obtained his values, is not at all obvious.[5] It might even be from macroscopic features of granules that molecules lack. Other than in Chapter 6, we are going to ignore this *retrospectively* glaring discrepancy; it is of negligible significance so far as Brownian motion itself is concerned. But that does not mean that it is philosophically uninteresting or unimportant.

4.2. Perrin's Vertical-Gradient Experiments

That the atmosphere becomes more rarified at higher altitudes had been recognized since Pascal, if not before. Perrin attributes the calculation of the variation of barometric pressure with altitude to Laplace, but in fact it goes back to Halley in 1686:[6] the pressure (and hence by Boyle's law the density) incrementally decreases in a geometric progression in relation to an arithmetic progression of increments of height. In other words, ignoring any temperature variation, the rule is:

$$\ln\left(\frac{p_0}{p}\right) = \ln\left(\frac{\rho_0}{\rho}\right) \propto h$$

where the subscript designates a reference value, say at the Earth's surface. Because the granules comprise only a constituent of an emulsion, however, the pertinent atmospheric phenomenon is the one singled out by Maxwell in the lecture of 1875 to which we have been referring:

> It follows from this, that if several gases in the same vessel are subject to an external force like that of gravity, the distribution of each gas is the same as if no other gas were present. This result agrees with the law assumed by Dalton, according to which the atmosphere may be regarded as consisting of two independent atmospheres, one of oxygen and the other of nitrogen; the density of the oxygen diminishing faster than that of the nitrogen, as we ascend.[7]

The corresponding rule for the variation of the partial pressures of the distinct constituents (assuming they act as ideal gases), and hence their relative concentrations, is:

$$\ln\left(\frac{p_0}{p}\right) = \ln\left(\frac{n_0}{n}\right) = \frac{Mgh}{RT}$$

where n is volumetric concentration and M is the molecular weight of the constituent.[8]

Maxwell's point in singling out this phenomenon, by the way, was that within kinetic theory it is a consequence of, and hence explained by, the claim that the molecules of the different constituents of a gaseous mixture at uniform temperature all have the same mean translational kinetic energy. As we are about to see, the profound result of Perrin's vertical gradient experiments was an analogous claim for the granules of different kinds in dilute emulsions. The difference is that the claim was not merely theoretical in Perrin's case. He turned Maxwell's reasoning upside down, obtaining it inductively from the phenomenon through theory-mediated measurements of the mean translational kinetic energies of a wide range of granules.

Two considerations lie behind the theoretical relationship enabling Perrin to measure these kinetic energies. First, for the granules to remain in equilibrium vertically, the fraction of their total weight not supported by buoyancy in any thin vertical slice of the liquid must be supported by a pressure difference across that slice:

$$p - p' = \frac{n}{V} m_{gran} g_{eff} (\delta h)$$

where n/V is the volumetric concentration of the granules in the slice, m_{gran} is the mass of the individual granules, and g_{eff} is the effective acceleration of gravity, that is, the acceleration of gravity reduced by the ratio of the fluid density to the density of the granules, or $g(1 - \rho_{fluid}/\rho_{gran})$. Second, the impact of the granules on any solid surface produces a pressure on the surface; in particular, their impact on a semi-permeable membrane through which the fluid flows freely produces a pressure at least akin to an osmotic pressure:

$$p_{gos} V = \frac{2}{3} n W$$

where W is the mean translational kinetic energy of the granules, and the permanence of the Brownian motion justifies taking their impact on any solid surface to be perfectly elastic. Combining these two equations to eliminate (n/V) then gives:

$$p - p' = \frac{3}{2} \frac{p_{gos} m_{gran} g_{eff}}{W} (\delta h)$$

Now, the difference in pressure across the thin slice has to support both the weight of the fluid and the weight of the granules, so that $(p - p')$ in this equation represents only a fraction of that total pressure difference. *If* the pressure associated with the motion of the granules truly is an osmotic pressure, and hence a partial pressure in the fluid, then $(p - p')$ must be the same as δp_{gos}. Making this substitution and rearranging gives:

$$\frac{(\delta p_{gos})}{p_{gos}} = \frac{3}{2}\frac{m_{gram} g_{eff}}{W}(\delta h)$$

Notice that the assumption we have made, that a difference in the pressure associated with the motion of the granules is the difference in pressure supporting their buoyant weight, does not assume anything about the molecular realm. When Perrin makes the step we have made in this paragraph, he invokes molecular-kinetic theory;[9] we have not.

Any difference in the osmotic pressure associated with the granules of a dilute emulsion even over a thick vertical slice is far too small to measure reliably—at least at the time, if not still.[10] So, a further step is needed before W is the only quantity in the relationship that cannot be measured directly. Assuming that W is uniform across the slice, any vertical difference in the pressure associated with the motion of the granules must stem from a difference in their volumetric concentration, n/V. In other words,

$$\frac{(\delta n)}{n} = \frac{3}{2}\frac{m_{gran} g_{eff}}{W}(\delta h)$$

where n represents the number of granules (per unit area) in a thin slice and δn, the difference in that number in the adjacent thin slice. With the slice taken thin enough, this amounts to a differential equation. If W is constant throughout the emulsion and hence independent of h, the solution of this equation is,

$$\ln\left(\frac{n_0}{n}\right) = \frac{3}{2}\frac{m_{gran} g_{eff}}{W} h$$

So, in principle, if n can be measured at (any) two heights in the emulsion, an indirect measurement of W becomes possible. Furthermore, the assumption that W is constant throughout the emulsion can be cross-checked by verifying that the vertical variation of n really does form a geometric progression over an arithmetic progression of heights, a check Perrin did indeed consistently carry out.[11]

A minor practical problem remains: the measurement of the mass of the individual (presumed identical) granules. This Perrin solved by measuring their density *en masse* and taking them to be spherical. Replacing m_{gran} then yields the equation governing Perrin's measurements of the mean kinetic energy of the granules:

$$\ln\left(\frac{n_0}{n}\right) = \frac{2\pi a^3(\rho_{gran} - \rho_{fluid})g}{W}h$$

where a is the radius of the granules and g is now just the acceleration of gravity. As we indicated, Perrin himself does not quote values for W, but instead for the number of granules N required for 2/3 of their total translational kinetic energy to match the characteristic energy RT of a mole of gas at the temperature T. The governing equation for this surrogate for W is,

$$\ln\left(\frac{n_0}{n}\right) = N\frac{4\pi a^3(\rho_{gran} - \rho_{fluid})g}{3RT}h$$

Put this way, whether the values obtained for N represent the value of Avogadro's number N_0—that is, the number of molecules in a mole—is an open question.

Perrin's Nobel Prize citation for "the discovery of sedimentation equilibrium" was not for this last formula expressing the equilibrium distribution. Unbeknownst to Perrin, Einstein too had derived this formula, indeed with the same goal in mind.[12] The citation was for the pains and ingenuity required of him in developing experimental techniques that enabled the "observable" quantities in the formula to be determined and, needless to say, the well-behaved values for N and hence W that he then obtained. The pains and ingenuity included getting emulsions with uniform, spherical granules (uniformity through fractional centrifuging); counting through microscopes the number of granules, which are in motion, at horizontal levels a couple of microns thick, separated from one another by only a matter of a few microns ("some thousands of readings are required if some degree of accuracy is aimed at"[13]); and determining the radii of granules no bigger than a fraction of a micron (which especially called for precision insofar as the term in the formula is their radius cubed). Perrin used three methods for determining the radii; the cross-check these three provided for one another yielded an "extrapolation from the law of Stokes,"[14] which then served as the basis of the method he primarily relied on. The pride Perrin displayed in summarizing all these efforts two decades later in his Nobel Prize lecture was entirely appropriate.[15]

Our concern here, however, is not with the painstaking details of the experiments he and his cohorts carried out, but with the results they obtained for N, and hence W, and what those results showed about Brownian motion. Perrin reported six series of measurements in his 1909 monograph, covering a range of granule types and sizes. For the initial series, when he was still developing the requisite experimental techniques, he concluded only that the value for N lies between 50×10^{22} and 80×10^{22} (for granules with $a = 0.14\mu$). For the remaining five, his announced values (for granules with $a = 0.30\mu$, 0.29μ, 0.45μ, 0.52μ, and 0.212μ, respectively) were 75, 65, 72, 70, and 70.5, all times 10^{22}. He regarded the last as most accurate:

> The mean radius of the granules of the emulsion employed was found equal to 0.212μ, by counting 11,000 granules of a titriated emulsion, and to 0.213μ by application of the law of Stokes. The difference of density between the material of the granules and the intergranular water was 0.2067 at 20°, the temperature to which the measurements refer. 13,000 granules were counted at different heights (direct observation through a needle-hole), and it was verified that the distribution was quite exponential, each elevation of 30μ lowering the concentration to about half of its value.[16]

Ignoring any differences in temperature across the five series, Perrin's values for N (the mean of which happens to match his 70.5 preferred value) correspond to measured values for W that vary from their mean by less than 8 percent.

Subsequent experiments conducted by Perrin's colleague, M. Bruhat, showed that the effect of temperature on the rate of rarefaction of the granules conforms with Gay-Lussac's law,[17] implying that the measured values for W vary linearly with absolute temperature, at least within experimental accuracy. Incorporating such a linear dependence on temperature into the equation for inferring values for W gives,

$$\ln\left(\frac{n_0}{n}\right) = \frac{2\pi a^3(\rho_{gran} - \rho_{fluid})g}{W_T T}h$$

where W_T is the mean translational kinetic energy of the granules per degree of absolute temperature, or in other words, $W = W_T \times T$. Put this way, to the extent that Perrin's measured values for N showed that it is constant across all dilute emulsions, they also showed that W_T is constant across all dilute emulsions. That is, *at any given temperature, the mean translational kinetic energy of granules is the same in all dilute emulsions*—at least within experimental accuracy over the ranges of the manipulable variables covered in the experiments.

This was the primary conclusion about Brownian motion itself that Perrin's vertical-gradient experiments revealed. Of course, it, together with the value equaling 3/2 of that of the Boltzmann constant $k = R/N_0$, was what Perrin had hoped for. Notice, however, that nothing in the design of the experiments, as we have construed them, assured or even predicted either of these. The experiments were designed to answer a question about the values of the mean translational kinetic energy of the granules and how it varies from one emulsion to another. The answer the experiments gave might very well have turned out differently.

4.3. Perrin's Vertical-Gradient Results Re-Examined

In Chapter 2 we concluded that molecular theory, while offering explanations for a wide range of phenomena, was generally not entering constitutively into the research on those and related phenomena; and when it did, as in the attempts to infer molecular dimensions from viscosity, diffusion, and the deviations from Boyle's law, the results were not "well-behaved." By contrast, the equations for N and W_T, and the assumptions required to derive them, which are not molecular, entered constitutively into Perrin's measurements, and those measurements yielded reasonably well-behaved results. *Precisely because those theoretical elements entered constitutively into his measurements, the demand that the results be well-behaved comprised a test of them.* We have already given one example of this with the assumption that W is constant over the height of the emulsion in any one experiment made in integrating the differential equation for a single slice: if W fails to remain constant over the height, the variation of n is not going to be logarithmic. The constancy of W over what turned out to be very small heights, however, was not a particularly problematic assumption, and it was ultimately susceptible to cross-check by the constancy of the measured values of W_T over the full array of experiments.

But now consider the more problematic assumption that the fraction of the vertical pressure increase in the fluid caused by the weight of the granules, δp_g, is identical with δp_{gos}, that is, with a vertical difference in the pressure exerted on the wall of the container from impact of the granules. Perrin's justification for it, through an appeal to van't Hoff's law, appears to presuppose that the osmotic pressure in dilute solutions itself arises from the impact of discrete particles, specifically the molecules of the dissolved substance, on semi-permeable membranes.[18] To presuppose that is to presuppose a kinetic theory claim of just the sort we are here trying to exclude.

The questions, accordingly, that we are going to address in this section of the chapter are whether his vertical-gradient results can even be formulated without recourse to molecular-kinetic theory and, if so, how precisely they need to be stated. So, only here, strictly speaking, will we be presenting what his vertical-gradient experiments established about Brownian motion in and of itself, independently of molecular-kinetic theory. To do this will force us into considering intricacies in the logic of theory-mediated measurement that usually remain tacit.

To begin with, the demand that the measured results in the experiments be well-behaved provides a test of a slightly weaker form of the assumption that bypasses any need to appeal to molecular theory. For the vertical variation of n, the surrogate for p_{gos}, was going to turn out to be logarithmic only if δp_g is at least proportional to, if not the same as, δp_{gos}. Weakening the assumption in this way, so that the experimental results themselves can support it, requires a corresponding weakening of the claim about what the measured values of W represent: a quantity at least proportional to, if not identical with, the mean translational kinetic energy of the granules. (A parallel weakening holds for the measured values of N as well.) That weakening, however, does not compromise the primary conclusion about Brownian motion that came out of Perrin's results, namely that *at any given temperature the mean translational kinetic energy of the granules is the same (within experimental accuracy) in all dilute emulsions.*

This evidence supporting the key assumption came from the well-behaved character of the measured values of an observed quantity, n. Well-behaved results for the target quantity in theory-mediated measurements yield evidence too. To show this, we need to introduce a couple of technical terms. By a *measure* for a quantity, we mean a formula relating it to quantities whose values can be independently obtained. The equation given at the end of the preceding section in which W_T appears is thus a measure for W_T. (Other Brownian motion measures for this quantity will emerge in subsequent sections of this chapter.) Most of the parameters in this equation—a, $(\rho_{gran} - \rho_{fluid})$, T, and h—are subject to experimental control and hence can be varied from one experimental trial to another; so too can parameters—like the shape and size of the container and, even, the geographical location of the experiment—that do not appear in the equation. Call a measure of a quantity (or a functional relationship between quantities) *stable* if it continues to yield the same result, to within experimental accuracy, independently of the values of these manipulable parameters. (Claims that measures are stable thus amount to inductive generalizations for which the warrant depends in part on the ranges of the parameters covered in the experiments.)

Correspondingly, then, Perrin was claiming that his vertical-gradient measure is stable when he said, "the values of the granular energy deduced from the preceding series are concordant within the limits of accuracy of the experiments, although the mass of the granules has varied as 1:40, the difference of density between the granules and the medium has varied as 1:4.7, and the rapidity of rarefaction as a function of height has varied as 1:30."[19] Let us grant him that he had adequate grounds for claiming that his measure for W, and for W_T as well, is stable. The question is, what more did that show about Brownian motion itself, beyond the primary conclusion already noted?

Perrin refers to the rapidity of rarefaction in the quote we just gave. Following him, let us call $H_{½}$ the height over which the volumetric concentration of the granules diminishes by a factor of 2. A further conclusion that the stability of the measure of W_T—that is, its constancy—showed was that, *ignoring variations in temperature, the rate of vertical rarefaction of the granules caused by gravity in any dilute emulsion is proportional to the buoyant weight of the granules.* This was the result referred to in Perrin's Nobel Prize citation as "his discovery of sedimentation equilibrium." It parallels the textbook result Maxwell noted in the quotation with which we began section 4.2, *ignoring variations in temperature, the rate of vertical rarefaction of each constituent of the atmosphere, or of any other gas, caused by uniform gravity is proportional to the molecular weight of the constituent.*[20] We will be discussing the relationship between these two in Chapter 5. For now, notice only that replacing *molecular weight* by *weight of the individual molecules* in the second still leaves the contrast between *weight* in it and *buoyant weight* in the other.

The logic of the evidence for Perrin's "law" of sedimentation equilibrium is worth spelling out fully, for it reveals a feature of evidence coming out of the well-behavedness of the results from theory-mediated measurements that is not always noticed. As we have emphasized, Perrin's vertical-gradient measure of W_T presupposes that the fraction of the vertical pressure increase in the fluid caused by the weight of the granules, δp_g, is identical with δp_{gos}, that is, with a vertical difference in the pressure exerted on the wall of the container from impact of the granules. While the logarithmic variation with height of the volumetric concentration of the granules cannot show the identity, it does provide clear evidence that δp_{gos} is at least proportional to δp_g. That evidence is independent of the question of whether the measure of W_T is stable. And the proportionality between δp_{gos} and δp_g is all that Perrin's "law" of sedimentation equilibrium, as we have stated it, requires. So, the overall logic of the theory-mediated evidence for that "law" has a "p→(q↔r)" form:

IF
δp_g is at least proportional to, if not the same as, δp_{gos},
THEN
the rate of rarefaction of the granules in dilute emulsions is strictly proportional to their buoyant weight and the temperature
IF AND ONLY IF
the vertical-gradient measure of W_T is stable.

Insofar as measurements are never exact, the full logic requires both clauses in the bi-conditional to begin with the phrase, "to the extent that."

The bi-conditional is the feature of the evidence coming out of the well-behavedness of theory-mediated measurements that is not always noticed. Because of it, the evidence is generally stronger than it would be if the experiments were merely testing a "law," or proportionality claim, taken as a hypothesis. We note again that the logarithmic variation of n provided evidence for the antecedent of the bi-conditional independently of the question of whether the measure of W_T is stable; the evidence for the latter then came from the indirectly measured values for W.[21]

The stability of the vertical-gradient measure provided evidence for a still further relationship or "law" for dilute emulsions beyond this one of equilibrium. It requires us to strengthen the previous antecedent slightly to obtain the corresponding bi-conditional. In our derivation of the measure for W_T we assumed that the pressure associated with the motion of the granules truly is an osmotic pressure, and hence a partial pressure in the fluid, in order to substitute δp_{gos} for the fraction of the pressure gradient in the fluid supporting the granules. Whether p_{gos}, the pressure associated with granular motion, *is* identical with the partial pressure in the fluid associated with their presence—that is, identical with the difference in the pressure in the fluid when the granules are present versus when they are absent—is not something that could be verified directly, for that difference was much too small to measure. The claim is unproblematic once one assumes that all the pressure in the fluid results from motion of discrete constituents comprising it; but this amounts to a molecular assumption that we have here barred.

The antecedent to the bi-conditional we now need is not so strong as to require this identity, but it still amounts to an assumption: *the fractional difference in the pressure in the fluid when the granules are present versus the pressure in their absence, p_g, is at least proportional to, if not identical with, the pressure produced by their impact on the walls of the container, p_{gos}.* This amounts to a strengthening of the antecedent in the previous bi-conditional since it follows from this one. Moreover, the logarithmic variation of the volumetric concentration of the granules with height provides evidence for this one as well as that one, for that variation confirms that p_{gos} itself varies in just the way it should if it is at least proportional to the partial pressure in the fluid added by the presence of the granules.

Now, as we noted during our derivation of the measure, as a matter of mechanics the pressure from the impact of the granules on the wall of the container is given by

$$p_{gos} = \frac{2}{3}\left(\frac{n}{V}\right)W$$

Our new assumption then yields

$$p_g \propto \frac{2}{3} W_T \left(\frac{n}{V} \right) T$$

So, under this assumption, *the partial pressure in the fluid added by the presence of the granules in any dilute emulsion is proportional to the product of their volumetric concentration and the absolute temperature* if and only if W_T *is a constant*, or, in other words, if and only if *the vertical-gradient measure for* W_T *is stable*. The evidence for the stability of the measure thus gave evidence for a "law" for dilute emulsions that parallels both van't Hoff's law for dilute solutions and the ideal gas law—in both of those cases, $p = (n/V)RT$, where (n/V) is the mole density. The remarkable implication in all three cases is that the pressure in question does *not* depend on the mass of the constituents, but only on their volumetric concentration: replace a constituent by another with a totally different mass, the pressure will remain the same so long as the volumetric concentration and temperature remain the same!

Perrin himself expresses this result for emulsions in a still stronger fashion: "we see the law of gases, already extended by Van't Hoff to dilute solutions, extended to uniform emulsions."[22] His basis for this stronger statement, however, has to involve molecular considerations. For historically the other remarkable feature of van't Hoff's result was that the constant of proportionality in it turned out, upon measurement, to be the same to at least three significant figures (according to Nernst) as the ideal gas constant R. Absent an appeal to molecular theory, no means was available to determine whether the constant of proportionality in the "law" for dilute emulsions is equivalent to R. One reason for this was that the pressure p_g, unlike the osmotic pressure in dilute solutions, was far too small to measure, and hence too the constant of proportionality for it was not amenable to measurement. Another reason was the absence of reliable chemical formulas for the granules, leaving no way to relate their volumetric concentration to a mole concentration except by taking the value for $N = (3R/2W_T)$ to be Avogadro's number N_0.[23] Even so, our weaker form still leaves intact the remarkable consequence of the dependency of pressure on volumetric concentration, not on mass concentration.

Because both the conclusions reached in this section about Perrin's vertical-gradient results and the approach to reaching them are new to the literature on Brownian motion, we should recapitulate them before turning to his further experiments. In order to replace molecular-kinetic assumptions in the design of his vertical-gradient experiments, we have relied on a demand of all theory-mediated measurement, namely that the measure—that is, the relationship between the target quantity and the accessible quantities authorizing the indirect measurement—be *stable*—that is, yield the same result for the target quantity as other quantities in the relationship are varied. We shall discuss the stability requirement further in the penultimate section of this chapter. In this section we have relied on one feature of the stability of a measure, its providing validating evidence, through a bi-conditional, for some, but not all, of the assumptions underlying the measure. To be specific, we have relied on it to distinguish the assumptions so validated from those that are not, so that we could replace the assumptions from molecular-kinetic theory by ones independent of that theory while maintaining the same logical structure of the evidence coming out of the measured results in Perrin's experiments.

Our argument, then, that the evidence supporting Perrin's three extraordinary conclusions concerning Brownian motion remained no less forceful without the molecular-kinetic assumptions proceeded in five steps:

(1) Recognition that his assumption, *the fraction of the vertical pressure increase in the fluid, δp_g, is identical with the vertical difference in the pressure exerted on the wall of the container from the impact of the granules, δp_{gos}*, presupposes a molecular-kinetic interpretation of van't Hoff's results for osmotic pressure in the absence of any way of measuring the granule osmotic pressure in the experiments.
(2) Replacement of this assumption by the weaker assumption, δp_g is at least proportional to, if not identical with, δp_{gos}, an assumption supported by the logarithmic vertical variation of granule concentration.
(3) Recognition that this change entails that W may no longer be a measure of the mean translational kinetic energy of the granules, but instead a quantity proportional to this energy—that is, differing from it only by a multiplicative constant.
(4) The conclusion that these changes have no effect on Perrin's result on the universality of the mean translational kinetic energy of the granules and still yield his "law" of sedimentation equilibrium.
(5) A slight strengthening of the new assumption in order to obtain a weaker version of Perrin's claim that van't Hoff's law for solutions extends to dilute emulsions as well, namely a version of the same form as van't Hoff's, but with the question left open of whether its constant of proportionality is the same as the universal gas constant R.

The upshot of the section, then, is worth repeating: Perrin's vertical-gradient experiments produced strong evidence, *independent of molecular theory*, for three extraordinary (at the time) conclusions concerning Brownian motion in uniform dilute emulsions: (1) the universality of the mean translational kinetic energy of the granules per degree Kelvin, regardless of their mass; (2) the "law" of sedimentation equilibrium; and (3) a formal analogue of van't Hoff's law for dilute solutions. We postpone the question of the implications all of this has for molecular-kinetic theory until we have considered Perrin's other experiments on Brownian motion.

This summary requires a further note. We are not claiming that anyone at the time announced the preceding steps showing how Perrin's conclusions about the vertical-gradient do not presuppose molecular-kinetic theory. We doubt anyone even expressly noted the lacuna in Perrin's analogic appeal to van't Hoff. By 1908 proponents of molecular-kinetic theory had long since adopted a molecular interpretation of van't Hoff's findings and would therefore not have paused to question whether the granules of Brownian motion have an osmotic pressure akin to van't Hoff's. But then too, by 1908 the proponents of the theory were so intent on finding ways to silence its critics that they would have scarcely paused to ask whether Perrin's vertical-gradient results indispensably presuppose it. Indeed, they were not especially interested in the Brownian motion results, save to the extent that they might offer grounds for ending questions about the existence of molecules.

Ostwald had championed van't Hoff's results from the time they were put forward, yet without adopting a molecular-kinetic interpretation or even viewing them as providing much in particular evidence for molecular-kinetic theory.[24] Our effort has been to assess Perrin's results from the perspective of someone writing a careful review article at the time. As such, we did not simply concede him his reasoning by analogy with van't Hoff, even though others, we grant, seem to have done so at the time. To the contrary, we have retained Ostwald's demand in the 1890s for gaining substantive experimental access to molecules and their actions before granting a molecular-kinetic interpretation of van't Hoff's results. If nothing else, this is allowing us to separate the question of the evidence for Perrin's vertical-gradient results for granules from the question of the evidence they provided for molecular-kinetic theory.[25]

4.4. Perrin's Granule-Displacement Experiments: The Two Measures

Perrin presents his granule-displacement experiments as tests of Einstein's theory of Brownian motion. Einstein, and by the time Perrin had learned of them others as well,[26] had derived from the theory of molecular action two formulas that provide measures of what the two of them called Avogadro's number, but that we have been calling measures of N, a proxy for the mean translational kinetic energy of the granules:

$$\xi^2 = \tau \frac{RT}{N} \frac{1}{3\pi a \zeta}$$

and

$$D = \frac{RT}{N} \frac{1}{6\pi a \zeta}$$

where ξ^2 is the mean of the square of the displacements of the granules over the time τ projected on an axis,[27] ζ is the viscosity of the liquid (which generally varies with temperature), D is the diffusion coefficient characterizing the rate of diffusion of the granules in the liquid, and a, as before, is the radius of the presumably spherical, uniform granules. The logic of the tests was to infer values for N by measuring the other terms in the formulas, with agreement of those values with one another and with values for N obtained independently of Einstein's theory providing confirmation of his theory.[28]

While Einstein and Perrin both took N to be the number of molecules in a mole, that is, Avogadro's number N_0, strictly speaking what the formula provides an indirect measure of is the number of granules required for their mean kinetic energy to correspond to the characteristic energy in a mole of gas, RT. Their authorization for their interpretation of N came out of the steps within molecular-kinetic theory that led to the formulas. We are here excluding considerations from that theory in describing the results of the experiments, and hence for us the values obtained for N represent not a count of molecules, but a count of granules. Alternatively, the identity between N and N_0 follows if the (universal) mean translational kinetic energy of molecules in gases at any given temperature is the same as the (universal) mean translational kinetic energy of granules in uniform dilute emulsions at that temperature. But that is a question we are postponing until Chapters 5 and 6.

Therefore, for us here again, N is a surrogate for the mean translational kinetic energy of the granules, W, and the formulas are best viewed as giving measures for W or, presuming it (subject to verification) to be a linear function of temperature, measures for W_T:

$$\xi^2 = \tau \frac{2W_T T}{9\pi a \zeta}$$

and

$$D = \frac{W_T T}{9\pi a \zeta}$$

Our question in this section, then, concerns what Perrin's evidence that the value for W_T obtained by means of these formulas is the same for all uniform dilute emulsions implied about Brownian motion itself, independently of molecular theory. We are thereby also postponing—until section 4.8—any discussion of the significance of the agreement between values for W_T obtained from these two measures with the values obtained from the vertical-gradient measure of the previous subsection.

Einstein published more than one derivation of the two formulas, all proceeding from "molecular-kinetic" theory, some of them tied to abstract features of Boltzmann's development of statistical mechanics.[29] Our need here is solely to identify the assumptions needed to obtain the formulas. For that we have relied on a textbook version of the 1908 derivation by Perrin's colleague, Paul Langevin,[30] and Einstein's 1908 derivation (for chemists) combined with points emphasized by Deborah Mayo.[31] The core idea is the same in all derivations: the motion of each granule involves continual random changes of velocity in both direction and magnitude, with the velocity at all times counteracted by a resistance force proportional to the viscosity of the liquid. The viscous resistance term is obtained from Stokes's law for spheres moving through a viscous medium under the action of a uniform force. So, one assumption is that the emulsions consist of uniform spherical granules that (1) are not so small that Stokes's law does not hold for them, and (2) that are not so large or moving so fast that the inertia of the liquid exerts a resistance force counteracting their motion as well. Perrin regarded his evidence that the radii of the granules can be indirectly determined from Stokes's law as sufficient to justify assuming that it holds for the granules he tested;[32] and his techniques of preparation assured reasonable uniformity and sphericity of the granules in his emulsions.

The assumptions concerning the continual rapid changes in velocity of the granules are the more significant ones for our purposes. Our ground rules in this section preclude deriving the formulas from statistical mechanics by assuming that the changes in velocity result from unbalanced impact forces of molecules on the granules. But they do not preclude our borrowing the mathematics employed in such a derivation and applying it not to hypothetical molecules, but to the observable granules—just as we relied earlier on the mechanics and mathematics employed in kinetic theory to justify the relation, $p_{gos}V = (2/3)nW$, for the pressure on the walls of a container generated by the impact of the granules. In that case, the relation tied an in-principle measurable quantity, *pressure*, to the mean translational kinetic energy of the granules. In the present case, we need a relation between an in-principle measurable quantity and a measure of the variability or spread of the velocities, and hence of the translational kinetic energies, of individual granules over time. For that, we need assumptions first about the statistical distribution of the translational kinetic energies of the granules over time, and second about the forces governing the changing magnitudes of that energy for individual granules.

Consider the latter first. We have already identified one force acting on each granule, the resistance force arising from the viscosity of the liquid. This force, under the assumption of Stokes's law, is given by $6\pi a\zeta v$, or for simplicity by Kv. Whatever the force F is that is producing the velocity that the viscous force resists, Newton's second law yields what has come to be known as Langevin's equation for Brownian motion:

$$m\frac{d^2x}{dt^2} = -K\frac{dx}{dt} + F$$

where, as above, the displacement x is the component of an individual granule in a single direction, say, the x-axis. Multiplying both sides of this equation by x and calling on some calculus identities transforms it into one involving x^2:

$$\frac{m}{2}\frac{d^2(x^2)}{dt^2} - m\left(\frac{dx}{dt}\right)^2 = -\frac{K}{2}\frac{d(x^2)}{dt} + Fx$$

This is Newton's law for a single granule, but under what amounts to an energy formulation. Now consider the "ensemble" average of the square of the displacements across all the granules; the term Fx drops out because, whatever the driving force may be, we know empirically that it does not add energy to the totality of the granules. Averaging across all the granules, then, and replacing the second term on the left by the x-component of the mean of their translational kinetic energies, that is, W, yields

$$\frac{m}{2}\frac{d^2(\overline{x^2})}{dt^2} - \frac{2}{3}W = -\frac{K}{2}\frac{d(\overline{x^2})}{dt}$$

Assuming W does not vary with time—an assumption we will shortly show how to justify—integration gives a solution for the rate of change of the ensemble mean of the squared displacements of the granules in a single direction versus time:

$$\frac{d\overline{x^2}}{dt} = \frac{4}{3}\frac{W}{K}\left[1 + e^{-\frac{K}{m}t}\right]$$

A historically important step introduced by Einstein has occurred in reaching this equation. Efforts before him to test the hypothesis that molecular agitation is the cause of Brownian motion had taken observed values of granule displacement over finite periods of time as a measure of their velocities. The conclusion drawn from these efforts was that the granule velocities are far too small versus the theoretical values of the molecular velocities. If, however, the number of changes in both the direction and magnitude of the velocity of the observed granule over the course of the time of observation is much too large for the eye to detect, the approach taken in these efforts could have radically misrepresented the granule velocities. Einstein's switch to the square of the displacements bypassed that worry, enabling the evidence Perrin subsequently obtained to countermand this previous evidence against the molecular agitation hypothesis.[33]

Although we did not note it in our discussion of Perrin's vertical-gradient experiments, this point about avoiding reliance on the apparent velocities was no less true of them. Perrin knew the masses of the granules, and hence in principle he could have determined values of their mean translational kinetic energy directly by squaring their apparent velocities. As he already knew at the time he was conceiving these experiments, doing so would most definitely not have resulted in measured values of granule mean translational kinetic energy of remotely the order of magnitude of that of the hypothetical molecules. Instead, he adopted an indirect, theory-mediated measure for this energy, and with it managed to reach conclusions about it and Brownian motion that could never have been reached so long as a direct approach to determining granule velocities continued to be employed.

To put the point in the case of Einstein slightly differently, anyone observing a granule in motion through a microscope is not seeing all the changes in its motion that have doubtlessly occurred, but only the net change in position over a period of duration covering a possibly huge number of changes in motion. In terms of the preceding equation, accordingly, in order to describe what is observed, we should replace the instantaneous derivative on the left-hand side by $\Delta x^2/\tau$; where τ is the duration over which the observer records a change in granule location; and in doing so, we are free to require those durations to be long enough that the exponential term can be ignored owing to its being smaller than the limits of the precision of the observed granule locations. This yields the first of the two measures for W_T stated at the beginning of this section:

$$\overline{\Delta x^2} = \xi^2 = \tau \frac{4}{3} \frac{W}{K} = \frac{2W_T T}{9\pi a \zeta}$$

The assumption of uniform spherical particles made in reaching it is a matter of experimental control. The assumption that W does not vary with time so long as T remains constant can be cross-checked by confirming that ξ^2 does indeed vary linearly with τ in each individual experiment. So, those assumptions need not be worrisome.

The step taken in the preceding paragraph involves more than may at first be evident. In requiring τ to be long, we have introduced for the second time a step of statistical aggregation, smoothing over the many changes in motion of all the granules for which the squares of their individual displacements are averaged to give ξ^2 versus τ. Have assumptions been tacitly made with this step beyond τ being long enough for the exponential term to become negligible? Not so far, as we have laid it out. But now consider the actual process of observation that has to be carried out in any experiment. The first step is always going to be to determine the displacements of individual granules over a sequence of time increments of duration τ; and the second step will be to take the average of the squares of those displacements of the different granules at the end of each time increment. In other words, in the experiments themselves the temporal-aggregation step has to occur first and the ensemble-aggregation step second—the reverse of the sequence in our derivation, where the ensemble mean was taken first.

So, the combination of our derivation of the ξ^2 measure and the way in which the measurements are made in practice assumes that the sequence of the statistical aggregation steps makes no difference. That assumption amounts to a special case of the ergodic hypothesis in statistical mechanics.[34] Because the ergodic hypothesis need not automatically hold, that in turn calls for us to be more specific about just what "random" means in our assumption that F and hence the changes of motion of individual granules over the course of any sufficiently large time increment are random. Although it is anachronistic to put it this way, what was assumed in the various derivations of the ξ^2 measure is that the displacement x of each granule amounts to a normally distributed random variable with mean 0 and variance ξ^2. Though he could not at the time have phrased it in this way, it is what Perrin meant when he stipulated of the granules that "their movement is *perfectly irregular* not only at right angles to the vertical (as under ordinary conditions) but in all senses."[35]

Our way of putting the matter we have taken from Deborah Mayo, who made it on the way to making an extremely important point that others in the philosophic literature on the subject had failed to make: *Perrin, in his granule-displacement experiments, developed strong evidence that the apparent randomness of Brownian motion does in fact stem from a Gaussian statistical distribution of granule displacements;*[36] *and furthermore that evidence was entirely independent of his measurements of mean translational kinetic energies in those experiments*. The evidence Perrin presented[37] consists of distributions of displacements at specified times in samples in comparison with the distributions theoretically predicted to hold in the limit as the number of granules approaches infinity. Mayo analyzes this evidence in the light of more recent developments in statistical hypothesis testing.[38] The conclusion she reaches is the point we emphasize here: Perrin provided strong evidence for a historic conclusion about Brownian motion itself: *the motion of the granules is, in a specifically defined sense, random, with a Gaussian distribution*. The corresponding claim had, since Maxwell, been a central hypothesis for the motion of molecules within kinetic theory. Perrin's confirmation of it for Brownian motion, however, was independent of molecular theory! We shall return to this confirmation in the next section.

Returning now to the ξ^2 measure of W and W_T, the evidence we have been discussing served to justify a key assumption presupposed by it. With that assumption now explicit and confirmed, reasoning parallel to that in statistical mechanics yields the other measure we listed earlier, the D measure. For that reasoning leads to identifying the diffusion coefficient D with $\xi^2/2\tau$ generally whenever the displacements accord with the Gaussian distribution.[39]

Insofar as D and $\xi^2/2\tau$ are identical under a Gaussian distribution of displacements, one might ask what this second D measure of W_T can have added. Phenomenologically, the diffusion coefficient is the constant of proportionality in the relation between the rate of change in the concentration of a substance and the spatial variation of the concentration at any point (at a given value of temperature), expressed by the diffusion equation (also called Fick's second law of diffusion):[40]

$$\frac{\partial v}{\partial t} = D\nabla^2 v$$

or, restricting ourselves to a single dimension,

$$\frac{\partial v}{\partial t} = D\frac{\partial^2 v}{\partial x^2}$$

where v is the (volumetric) concentration. Statistical mechanics explains this law in terms of the random motion of the molecules colliding with one another, with the magnitude of D dictated by the mean speed and the mean free path of the molecules. Einstein's derivation has the diffusion of granules in uniform dilute emulsions arising from their random motion, in the specific sense given earlier, constrained by resistance from the viscosity of the liquid, with the magnitude of D dictated by the mean translational kinetic energy of the (spherical) granules and the resistance force per unit speed, $6\pi a\zeta$, acting on them.

In Chapter 2 we noted that the analysis of diffusion within molecular-kinetic theory is inextricably related to the mean free path of molecules. The diffusion of granules in a dilute solution has nothing whatsoever to do with any mean free path of granules among themselves. The restriction of all the research on Brownian motion that we have been discussing to *dilute emulsions* was tantamount to assuring that *any* interaction of the granules with one another had negligible effects. In this respect, Brownian motion was anything but molecular motion made large enough to be visible.

Within molecular-kinetic theory, the claim that diffusion, a process that can involve a definite directional orientation, results from a random motion of molecules lacking any directional orientation offers a *hypothetical* explanation of a phenomenological "law." By contrast, any evidence in support of our second D measure, given the independent evidence for the Gaussian distribution of granule displacements, is going to be evidence that diffusion in uniform dilute emulsions does *in fact* result from such a random motion of granules that lacks any inherent directionality! It is also going to be evidence that diffusion in emulsions, a process that phenomenologically displays no statistical character, is nonetheless fundamentally statistical—regardless of whatever the cause of Brownian motion may be.

4.5. Perrin's Granule-Displacement Experiments: Displacement Results

Each of the two measures presented challenges when put to practice beyond those that Perrin had overcome with the vertical-gradient measure. That measure required granules to be counted in a state of equilibrium. So, while the stochastic character of Brownian motion introduced some variance from one count to another, reasonably stable values of the count did emerge from taking the mean of a few counts. The ξ^2 measure, by contrast, required individual granules to be followed over time while engaged in a highly stochastic process that was sure to yield wildly different values for the displacements versus time from one granule to another. This raised questions, as Mayo has stressed, about how many granules had to be included for the sample mean value for ξ^2 to be trusted as a measure of the population mean. To put the point in another way, in the case of the vertical-gradient measure, each count of the granules had *prima facie* claim to being representative; in the case of the ξ^2 measure, no sequence of displacements of any one granule over a finite period of time had any claim to being representative. Perrin's granule-displacement experiments thus fell into the realm of statistical experimental design, a realm that was then in the process of coming to be understood in the field of genetics, but was new to physics.

In the experiments that Perrin described in 1909, the location of an individual granule was marked at 0, 30, 60, 90, and 120 seconds, and then another granule, etc. Two series of mean values for 50 horizontal displacements at the end of each of these intervals in one experiment involving a single emulsion indicated at least a rough linear variation of ξ^2 with τ[41]—a test noted earlier of the assumption that W remains constant over the course of any one sequence of measurements.

Values for N (for us the surrogate for W) inferred at the end of the time intervals ranged from 59 to 89 × 10^{22} in the first series and from 62 to 71 × 10^{22} in the second. In a subsequent experiment involving a different kind of emulsion, with more granules followed and measurements extended to 240 seconds, the values at the end of the time intervals ranged from 57 to 70 × 10^{22} in one series and from 64 to 88 × 10^{22} in a second.[42] The mean values in the four series were 73, 68, 65, and 72 × 10^{22}, giving evidence that the departures from linearity across the time intervals within each of the four series were primarily from sampling variability.

Judgments of how well-behaved data are in any experiment should always take into consideration the fact that thoroughly novel experiments are like innovations in technology: however much development effort is required to achieve a prototype, much more in the way of further development is invariably needed for the innovation to begin fulfilling its full potential. The strongly positive response to J. J. Thomson's heralded cathode-ray experiments of 1897 was not because of the individual measurements of the mass-to-charge ratio of what subsequently came to be called the electron, for these results varied by more than a factor of 2. What so impressed others at the time were the uniform order of magnitude of the results (a factor of a thousand smaller than that of the hydrogen ion) and the absence of systematic variation with changes in the gas and the material of the electrodes.[43] Within a decade, the mass-to-charge ratio in experiments of the same basic design was being measured to four significant figures. We should not be surprised at Perrin's having felt no need to explain away the variability of his values within each series of his measurements, especially insofar as his measurements were being made in what we now call a stochastic process.

Anyway, the more important issue for us concerns the stability of the ξ^2 measure of N and W_T over a range of different types of emulsions with granules of different sizes and liquids of different viscosities. Referring to a table listing averaged values for N obtained with each of seven different types of emulsion, Perrin concluded in 1913,

> It may be seen from the table that the extreme values of the [granule] masses bear a ratio to one another of more than 15,000 to 1, and that the extreme values of the viscosities are in the ratio of 1 to 125. Nevertheless, whatever the nature of the intergranular liquid or of the grains, the quotient $N/10^{22}$ remains in the neighborhood of 70, as in the vertical distribution experiments. This remarkable agreement proves the rigorous accuracy of Einstein's formula and in a striking manner confirms the molecular theory.[44]

Let us ignore the last remark until Chapter 5, and the agreement with the vertical distribution experiments until section 4.8. The values for N that Perrin listed in the table ranged from 55 to 80 × 10^{22}. Only three of the seven cases involved more than 400 displacement recordings. For the three in which the values for N were based on 900, 1,000, and 1,500 displacement recordings, the values were respectively 69.5, 72.5, and 68.8 × 10^{22}.

To assess what this evidence for the stability of the ξ^2 measure showed about Brownian motion, we need first to lay out the logic of the theory-mediation in parallel with the way we did in the case of the vertical-gradient measure:

IF
the granule displacements over time are random, with a Gaussian distribution over time, and the resistance force from the viscosity of the liquid on the spherical granules is in accord with Stokes's law, and the duration τ covers a very large number of random changes in velocity
THEN
ξ^2/τ is strictly proportional to the temperature directly and the product of the uniform radius of the spherical granules and the viscosity of the liquid inversely
IF AND ONLY IF
the ξ^2 measure of W_T is stable.

As indicated earlier, Perrin presented evidence for each of the clauses of the antecedent to the bi-conditional that was independent of both the measure and molecular theory. Given this evidence, Perrin's evidence for the stability of the ξ^2 measure of W_T was accordingly evidence not only that the mean translational kinetic energy of the granules is the same at any given temperature, but also that the relation given in the left-hand side of the bi-conditional holds for Brownian motion in all uniform dilute emulsions.

Our offhand remark about Perrin having presented evidence of the Gaussian distribution of granule displacements glosses over what was arguably the single most important result historically to come out of Perrin's granule displacement measurements. The result was announced in a short note late in 1908 by Perrin's doctoral student Chaudesaigues, the main point of which was the confirmation of Einstein's ξ^2 formula together with a value of 64 × 10^{22} obtained in the process. The following paragraph ends the note:

> The perfect irregularity of Brownian motion allows of calculation, by application of the *law of chance errors*, the number of the displacements comprised between two fixed values. The agreement between the calculated numbers and the observed is absolutely striking, and all the better that the number of observations is very large. It is without doubt a most beautiful application of the calculus of probability to a natural phenomenon.[45]

Projections (in μ) comprised between	First Series.		Second Series.	
	n (found).	n (calc.).	n (found).	n (calc.).
0 and 1.7	38	48	48	44
1.7 ,, 3.4	44	43	38	40
3.4 ,, 5.1	33	40	36	35
5.1 ,, 6.8	33	30	29	28
6.8 ,, 8.5	35	23	16	21
8.5 ,, 10.2	11	16	15	15
10.2 ,, 11.9	14	11	8	10
11.9 ,, 13.6	6	6	7	5
13.6 ,, 15.3	5	4	4	4
15.3 ,, 17.0	2	2	4	2

Figure 4.1. Chaudesaigues's comparison of observed granule displacements after 30 sec with displacements he calculated on the basis of a Gaussian distribution.
Source: Perrin (1910), p. 65

We indicated earlier that displacements of individual granules had been recorded over a sequence of time intervals. Chaudesaigues made the calculation, "relatively to an arbitrary horizontal axis, for the displacements suffered in 30 seconds by the granules of *gamboge*" ($a = 0.212\ \mu$) within limits of multiples of 1.7 μ.[46] Figure 4.1, from Perrin's 1909 monograph, displays Chaudesaigues's comparison between the calculated and the observed distribution of the horizontal displacements.

Following Langevin's suggestion of "transporting parallel to themselves the observed displacements, in such a manner as to give them all a common origin," Perrin plotted 350 observations of granules of mastic ($a = 0.52\ \mu$) in order to display confirmation of the Gaussian distribution visually (Figure 4.2).[47]

There is more than one reason why this result was so much more important historically than its merely having confirmed an assumption underlying the ξ^2 measurement of W_T. The main reason was its being the first ever empirically confirmed instance of "random" motion in nature—that is, motion exhibiting a Gaussian statistical distribution. As we stressed in section 2.5, while Maxwell had proposed just such a distribution for molecular velocities, none of the evidence supporting molecular-kinetic theory, centered as it was on mean values, reached deeply enough to provide any support for this proposal. The question thus remained open whether any motion in nature, under strictly deterministic laws, could even in principle exhibit such a distribution. The observed distributions of displacements in Brownian motion in Perrin's granule-displacement experiments showed not only that the answer to that question is yes, but by implication showed that the velocities in Brownian motion have a

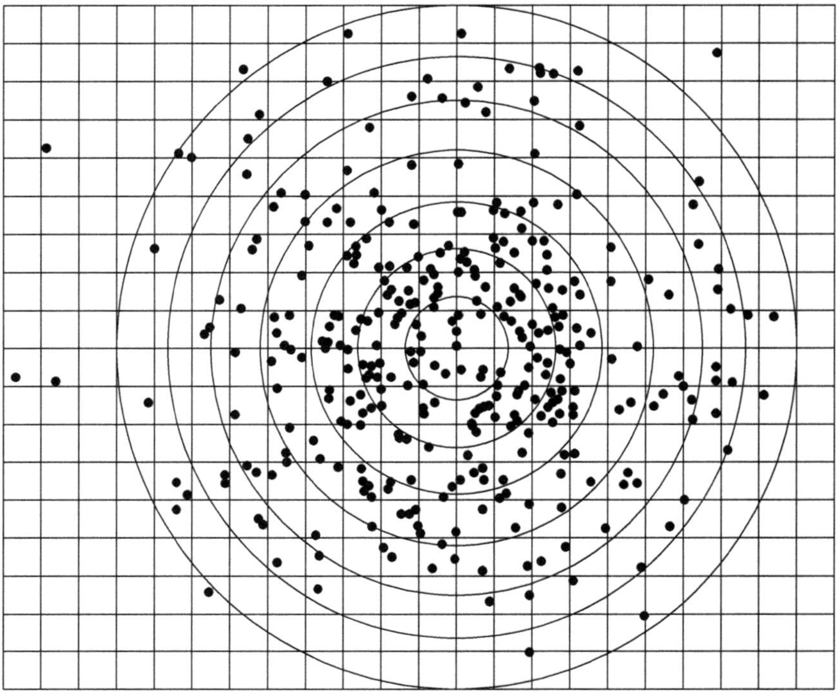

Figure 4.2. Graphical display of the observed displacements, in the manner of bullets fired at a target.
Source: Perrin (1910), p. 66

Gaussian and hence Maxwellian distribution. As we shall see in Chapter 5, none of Perrin's findings was more heralded at the time than this one. It was also why the appellation "Brownian motion" acquired a new referent, namely a subfield of mathematics.[48]

A second reason why this result was so important was that, unlike the other results coming out of research at the time on Brownian motion, it was established not only independently of molecular-kinetic theory, but as well without recourse to any form of theory-mediated measurement. The evidence for it consisted only of tabulated observations through a microscope, something that required extraordinary patience and effort, but no theoretical intermediary enabling the observations to serve as proxies for some other parameter. Put another way, the evidence for it was not indirect in the way that *any* evidence for Maxwell's claim about molecular velocities could not help but be indirect. As a consequence, this result had an epistemic standing more robust than Perrin's and Svedberg's determinations of what they claimed to be Avogadro's number.

A third reason why this result was so important will emerge in the next section. For now, let us recapitulate this section: Perrin's granule-displacement experiments produced strong evidence, *independent of molecular theory*, for three at the time extraordinary findings concerning Brownian motion in uniform dilute emulsions: (1) additional support, complementing that from the vertical-gradient experiments, for both the universality and the magnitude of the mean translational kinetic energy of the granules per degree Kelvin;[49] (2) confirmation of Einstein's equation for ξ^2/τ, the mean-square-displacement of granules in Brownian motion per unit time; and (3) the Gaussian, and hence Maxwellian, statistical distribution of granule displacements, and thus by implication, such a distribution as well for their velocities. Here again we postpone the question of the implications all of this has for molecular-kinetic theory until Chapter 5.

4.6. Perrin's Granule-Displacement Experiments: Diffusion Results

So much then for the ξ^2 measure of the mean granule translational kinetic energy. The parallel algebraic form of the D measure masks ways in which it amounts to a very different sort of measurement from that one. The diffusion coefficient is a bulk property of a solution that, other things remaining the same, should not vary with time or location; and measuring it involves neither the need to track individual granules over time nor the statistical sampling worries that arise in measuring ξ^2. In spite of this, Perrin and his associates at first saw no way of measuring the diffusion coefficient in emulsions. In his 1909 monograph, Perrin instead considers the known diffusion of what were taken to be very large molecules—for example, sugar—in solutions, with estimates of their "radius," to provide some evidence for Einstein's formula. In the years between the 1909 monograph and Perrin's 1911 Solvay Conference paper, however, the young Léon Brillouin[50] managed to turn an idea of Perrin's into a well-behaved measurement of the diffusion coefficient in one specific kind of emulsion.[51] The corresponding value of N inferred by means of the D measure was $69 \times 10^{22} \pm 3$ percent.

The only parameter that could be varied in Brillouin's measurements was the temperature, and thereby too the viscosity of the liquid. Claims about the stability of the D measure therefore require something of a reach. Nonetheless, evidence for its stability does amount to evidence that the diffusion coefficient in uniform dilute emulsions is strictly proportional to the temperature directly and the product of the radius of the spherical granules and the viscosity of the liquid inversely. To put the point more formally, keeping the antecedent the same as before, this relationship forms the other half of the bi-conditional together with the stability of the D measure.

To stop with this, however, is to miss a far more important implication of the evidence Perrin provided for the validity of the D measure. Let us call two (or more) stable theory-mediated measures of the same quantity *convergent* if they yield the same value within the experimental accuracy of the respective measurements. Keeping the antecedent the same as before, the two measures of N and hence W_T together yield a further bi-conditional:

$$\xi^2 / \tau = 2D$$

IF AND ONLY IF

the ξ^2 and D measures of W_T are stable and mutually convergent.

Therefore, Perrin's evidence for the agreement of values of N obtained by means of the two different measures was evidence for an identity between a *statistical property* of the motion of individual granules in Brownian motion and a *bulk property* of uniform dilute emulsions in which Brownian motion takes place.

A slight reformulation of this point is needed to bring out the full significance of what the evidence from the granule-displacement experiments was showing about Brownian motion. As we noted earlier, Perrin developed evidence that the displacements of individual granules in Brownian motion over time are random, with a Gaussian distribution that has a mean of 0 and a variance of ξ^2 over durations τ of sufficient length to encompass a large number of changes in motion and associated minute changes in location; and the random character of those displacements leaves them with no directional character at all—that is, no preference for any one direction over another. Moreover, that evidence was entirely independent of any question about the ξ^2 measure of N, W, or W_T. To quote Perrin, "Maxwell's law of irregularity is verified indisputably in its application to the displacement of the granules of an emulsion."[52] Therefore, the quantity ξ^2/τ represents a statistical aggregate itself, the rate of growth of the mean of the squares of the displacements versus time very long in comparison to the time over which individual granules undergo their changes of motion.

By contrast, the diffusion coefficient D is the characteristic parameter for the slow, gradual, and apparently continuous rate of change of a bulk quantity, the volumetric concentration of the granules; and the process for which it is the characteristic parameter has a clear directional character dictated by the spatial gradient of that concentration at any given time (in Brillouin's measurements, the gradient toward the walls of the container). *Nothing in the measurement of it, including Brillouin's specific measurements, presupposes or requires that it be a net consequence of a random or stochastic process.* Moreover, just as earlier, the results Brillouin obtained for D were entirely independent of any question about the D measure of N, W, or W_T.

The evidence for the convergence of the two measures, as we just noted, was evidence that $\xi^2/\tau = 2D$. Taking that evidence at face value, it implies that D in the case of uniform dilute emulsions is in truth itself a statistically aggregated quantity, even though nothing in the measurement of it requires it to be so. That constituted a major discovery about emulsions and the Brownian motion that occurs in them.

The point can be put in other ways. For example, the evidence implied that a seemingly continuous, bulk phenomenon, the diffusion of granules in emulsions in which they are not uniformly distributed and which has a distinct directional character, is *constituted* by a random or stochastic process that has no directional character at all. As we said earlier in anticipation of the conclusion we are reaching here, statistical mechanics was offering a parallel *hypothetical* claim about diffusion in gases and dilute solutions—"hypothetical" here meaning that the evidence for it at the time came entirely from the success that statistical mechanics had achieved in explaining a whole host of phenomena. The evidence Perrin and Brillouin developed for the claim about diffusion in dilute emulsions was not hypothetical in this sense!

Two points need to be made before we recapitulate what Perrin's granule-displacement experiments showed about Brownian motion. First, we must acknowledge that the point we are making about diffusion amounts to the same point Deborah Mayo contributed to the philosophic literature decades before us. The only notable difference is that she was looking at the matter from the point of view of testing Einstein's theory of the Brownian motion. We are looking at it from the point of view of theory-mediated measurements and the evidence they yield when their results are well-behaved.

Second, something said in section 4.5 should be said again about what the word "random" means here. It does *not* mean that the changes in motion of individual granules are a matter of chance. What it means is that (1) certain quantities—ξ^2 and ξ^2/τ, in particular—are statistical aggregates of a quantity of extremely high variability, namely the displacements of individual granules over time; and (2) the aggregation in question does not depend on any specific feature of those displacements, but only on their statistical distribution. To say it another way, the process of aggregation—physically in general, but especially during a measurement—glosses over the specifics of the changes in motion of the individual granules. The discovery that diffusion in dilute emulsions definitely arises out of a random or stochastic process means nothing more than that the measurable quantities employed in characterizing the process—the diffusion coefficient and volumetric concentrations—turn out to be statistical aggregates of quantities that amount to random variables, that is, quantities characterized not by their values, but by the statistical distribution of those values.

We have spoken throughout of statistical mechanics as explaining various phenomenological relationships among quantities. All that means is that statistical mechanics offers theoretical accounts of what is really being measured when those quantities are measured, and the answers to those questions then entail that the relationships in question among the quantities hold. In other words, classical statistical mechanics was first and foremost a theory about the nature of certain quantities and the physics of their measurement. Correspondingly, Perrin's experiments provided evidence that various quantities characterizing dilute emulsions—pressure, volumetric concentration, temperature, and diffusion rates, in particular—are all in fact statistical aggregates arising from a particular kind of statistical distribution of other quantities.

To recapitulate, then, Perrin's granule-displacement experiments provided evidence for several conclusions about Brownian motion that had been at best matters of conjecture before them. For one thing, they provided still further evidence that what we have labeled W_T—and correspondingly too the mean translational kinetic energy of the granules at any given temperature—remains the same in all dilute emulsions regardless of the granules and the liquid in which they are suspended. Second, they provided evidence for two strict proportionalities, one for ξ^2/τ and the other for D. These two proportionalities are different in one respect from the two supported by the evidence from the vertical-gradient experiments. Both of them had established phenomenological counterparts, while neither of the two here does. The reason for this difference lies in the fact that these two, unlike the earlier ones, include a quantity that has no phenomenological counterpart, the radius of the granules. The radius (indeed, cubed) enters the earlier ones, but only as a surrogate for their mass, which has atomic weight as a phenomenological counterpart.

Finally, and most importantly, the granule-displacement experiments provided evidence for a somewhat monumental conclusion about diffusion in dilute emulsions: this directional phenomenon is an aggregate consequence of the non-directional "random" motion of the granules, constrained by resistance arising from the viscosity of the liquid.[53] Like the conclusion about the "random" character of the motion, this one too has a hypothetical counterpart within statistical mechanics, but the significance of this we again postpone until Chapter 5.

4.7. Perrin's Granule-Rotation Experiments

Perrin presents, without derivation, a formula from Einstein that at first glance appears to provide a still further theory-mediated measure for his N:

$$\alpha^2 = \tau \frac{RT}{N} \frac{1}{4\pi\zeta a^3}$$

where α is the mean of the squares of the angles of rotation of the granules in time τ, ζ is the viscosity of the liquid, and a is the radius of the spherical granules.[54] Einstein obtained it as a variant of his formula for ξ^2 by considering the moment of the driving force—that is, the torque—acting on the surface of a granule resisted by the viscous action of the liquid.[55] But RT/N gets into either of these formulas only by substituting for the mean kinetic energy of the granules. In the case of this formula, unlike the one for ξ^2, the energy in question has to be the mean *rotational*, not *translational*, kinetic energy of the granules. And therefore N represents the number of granules required for their total *rotational* kinetic energy to match the characteristic energy RT of a mole of gas or dilute solution. Unless one assumes that the mean rotational kinetic energy of the granules matches their mean translational kinetic energy, as Einstein did in deriving the formula, N has changed meaning. But that indeed is the sole use to which Perrin puts this formula, to test that very assumption.

Because theory implied that the rotations would otherwise be too fast to measure, Perrin used spherical granules of diameter around 13 microns, with small visible inclusions on their surfaces that allowed change of angular position to be marked. (The large diameters, and hence moments of inertia, resulted in a rotational speed around 14° per minute—both as observed and as implied by the equipartition of energy assumption, given the mean translational energies obtained in the prior experiments.) The sole result he announced for N, based on 200 or so angular measurements, was 65×10^{22}, a value that matched, well within the margin for experimental error, the values obtained from his translational measurements. His conclusion is worth quoting, if only because of the controversy then surrounding the issue of equipartition of energy and the rotation of molecules:

> The agreement is remarkable, if one thinks of the difficulties of measurement and the complete uncertainty which *a priori* surrounded even the order of magnitude of the rotation. The granules utilized for these measurements were about 100,000 times heavier than the granules of gamboge first studied.
>
> So, the *equipartition of energy* is established throughout this great interval. Incidentally, its verification for the rotations is an experimental confirmation of the reasoning from the kinetic theory which has enabled the ratio C/c of the specific heats of gases to be predicted.[56]

The last sentence refers to fact that the measured values of the ratio of the specific heats of some (hydrogen, oxygen, and nitrogen), but not all (chlorine and iodine), diatomic gases at room temperature were very close to 1.4,[57] corresponding theoretically to three degrees of translational and two degrees of rotational freedom under equipartition of energy, assuming that the molecules are of a dumbbell shape.

Because Perrin examined only one emulsion, the question of evidence for the stability of his α^2 measure of N and thereby the mean rotational kinetic energy is moot.

The logic of the evidence that such stability would have yielded is still worth spelling out, if only for the sake of completeness:

IF
the granule angular displacements over time are random, with a Gaussian distribution over time, and the resistance force from the viscosity of the liquid on the surface of the spherical granules is in accord with Stokes's law
THEN
the kinetic energy of the granules in Brownian motion is equally distributed across translational and rotational degrees of freedom
IF AND ONLY IF
both the ξ^2 measure and the α^2 measure of N are stable and they yield the same value.

In subsequent works, Perrin continued to quote his rotational value for N as one among the many agreeing values for Avogadro's number obtained from a wide range of diverse phenomena. We consider this somewhat misleading, for the "characteristic energy" RT of a mole of gas corresponds only to translational kinetic energy. The issue of the ratio of the specific heats, from Clausius forward, concerned where the fraction of the total heat (that is, energy) added to a gas goes that does not show up as an increase in temperature. What Perrin's granule-rotation experiments showed was, *at least in one case*, the answer to the corresponding question for emulsions is, into rotation of the granules. That was historically important for just the reason Perrin says: it confirmed equipartition of energy for one case in a historical situation in which it had never been clearly confirmed before.

Perrin at the time showed no awareness of Einstein's proposal in his 1907 paper, "Planck's Theory of Radiation and the Theory of Specific Heats," that degrees of freedom may cease to absorb energy at low temperatures as a consequence of the quantized energy being too low to cross an associated threshold.[58] Already by 1912, Arnold Eucken, working in Nernst's laboratory, had shown that the presumed rotational kinetic energy of hydrogen molecules increasingly disappears, leaving only their translational energy, as the temperature drops below 100°K.[59] So, already by 1912 it was becoming clear that equipartition between translational and rotational degrees of freedom occurs only under certain conditions.

Inferring values for Avogadro's number from mean rotational energies was accordingly somewhat tenuous in comparison with the other approaches Perrin generally cited, though clearly from the numbers not invalid.

4.8. The Three Different Types of Experiment, Taken Together

So far we have been considering the evidence from each type of experiment separately from the others. We end our review of Perrin's experiments by considering the full body of evidence from them, taken together. The element they have in common, and hence the basis for doing so, is their separate determinations of what Perrin called Avogadro's number. Perrin himself singled out the evidence obtained from the four different determinations taken together by including a list in his 1911 and 1913 comprehensive presentations of his results.[60] Of course, he did so in large part to make his case that what he was measuring was Avogadro's number. That claim, however, presupposed the further assumption that his measured values for the mean translational and rotational kinetic energy of the granules represent as well the mean translational kinetic energy of the molecules forming the liquid in which they were moving. Because we are holding that assumption in abeyance in this chapter, the values he announced for Avogadro's number are here to be taken instead as values for our N, that is, our surrogate for W, the mean translational kinetic energy of the granules. So far as we have been able to determine, Perrin verified the linear variation of W with temperature only for his vertical-gradient experiments. If we grant him this linear variation in the others as well, our N can even more appropriately be taken as a surrogate for our W_T, the granule mean translational kinetic energy per degree Kelvin. Regardless, notwithstanding our listing his announced values for Avogadro's number in the following, what we will be considering here is the evidence from the four different measures of the mean kinetic energies of the granules.

A second cautionary note is needed before we turn to Perrin's announced values. Those values did not all remain the same from 1909 to 1911 and then 1913. In particular, noting that the granules ($a = 0.212\ \mu$) in what he had regarded as his preferred trial among the vertical-gradient experiment from earlier "did not appear to me sufficiently" uniform, he did the experiment anew with grains of "radius .367μ to within 1 per cent," obtaining a revised "probable mean value" for both his 1911 and 1913 accounts.[61] In the case of the ξ^2 measure, he announced seven different values in 1909 without singling out any one of them, yet in two places citing their mean, 71.5×10^{22}, as the result of the experiments.[62] We cited the three values with the most recorded observations in section 4.5. In 1911 and

1913, Perrin singled out the one of these three with the most observations. But then, in a later edition of *Les atomes* that was translated into English, he noted that René Constantin had convinced him that the motions of granules near the walls was too constrained to be representative and then added:

> Working at a sufficient distance from the walls with the grains I had used, he obtained the value $N = 64 \times 10^{22}$; unfortunately the number of observations (about 100) was too small. These measurements will be repeated.[63]

For reasons we will give in section 4.9, we applaud Perrin's revisiting earlier experiments in an effort to achieve greater precision. By 1913, however, both Millikan and Rutherford and his associates had obtained multiple values for Avogadro's number, invariably no greater than 62×10^{22}. Surely Perrin was feeling some pressure to reconsider his values in the light of the substantial difference between these results and his. We shall return to this difference in Chapter 6.

With these cautions in mind, Table 4.1 lists Perrin's announced values from his four measures in his 1909, 1911, and 1913 publications. We have included both the dependence of each measure on the radius a of the granules and the range of radii tested in preparation for our assessing the evidential import of the collective results.

By the time of the Solvay Conference in 1911, three of the values were sufficiently tightly clustered together to be considered equal to one another within

Table 4.1 Perrin's Quoted Values for Avogadro's Number, from 1909 through 1913

Phenomenon	1909	1911	1913	Dependence on Granule Radius	Range of Granule Radii for Which Measurements Made in Microns*
Vertical-gradient	70.5×10^{22}	68.2×10^{22}	68.3×10^{22}**	Cubed	0.212, 0.52, 0.367, 0.14, 0.29
ξ^2/τ Displacements	71.5×10^{22}	68.8×10^{22}	68.8×10^{22}***	Linear	0.52, 0.212, 0.367, 0.385, 5.5
Diffusion	—	69×10^{22}	69×10^{22}	Linear	0.52
Rotation	65×10^{22}	65×10^{22}	65×10^{22}	Cubed	6.5

* Granule radius for which results said by Perrin to be most reliable is listed first.

** Value listed at the end of the book; 68.2×10^{22} at the end of the chapter on vertical gradients.

*** 64×10^{22} in a later edition of *Les atomes*, in response to Constantin's proposal not to include granules near the wall.

experimental error, 68.2, 68.8, and 69, all times 10^{22}, with a mean of 68.67×10^{22}. Including the value from rotation shifts the mean only to 67.75×10^{22}. We are uncertain what, if any, significance should be attached to this fourth value falling outside the cluster of the other three. Keep in mind that the rotational energy was being obtained from *observed* magnitudes of rotational velocities, and those velocities may well have been changing far more often, as in the case of the apparent translational velocities, than the eye could catch.

The obvious point to draw from the listed values is that the vertical-gradient and granule-displacement experiments, and even the granule-rotation experiments with the qualification given earlier, taken together provided stronger evidence than any of them individually that ***the mean translational kinetic energy of the granules W in dilute emulsions is the same at the given temperatures regardless of the character of the granules or the liquid in which they are suspended.*** If we take the linearity of *W* with temperature to have been adequately established, the same can be said about the collective evidence for the claim, ***the mean translational kinetic energy of the granules per degree Kelvin, W_T, is a universal constant in dilute emulsions.*** Quite independently of the relation of the tabulated numbers to Avogadro's number, or for that matter their relation to molecular-kinetic theory, these were extraordinary findings for Brownian motion.

The reason the combined evidence for this claim is stronger is not merely that there is more of it. The evidence is coming out of the stability of different theory-mediated measures of *N*, and hence of *W*, employing different background assumptions. The convergence of those measures reduces the risk that the values obtained from any one of them are somehow an artifact of its background assumptions. *The greater the number of convergent stable measures of any theoretical quantity, the stronger the evidence is that a definite quantity is being measured.* That much is trivially obvious. If only we had convergent measures for, say, "intelligence quotient."

Less obvious perhaps is the evidence each convergent measure provides in support of the validity of the others. The antecedent to the bi-conditional characterizing the vertical-gradient measure of *W* differs fundamentally from the antecedent in the case of the two translational granule-displacement measures. The antecedent for the vertical-gradient measure concerns the pressure attendant to the motion of the granules; that, in turn, concerns the mean of the squares of the velocities of the granules. The antecedent for the two translational measures, by contrast, concerns the variability of the displacements of the granules over time. (In all cases, the antecedent we have stated highlights only the principal assumption; a fully fleshed out antecedent would include such further assumptions as the uniform sphericity of the granules.) In every case, Perrin offered evidence independent of the measures for the principal assumption behind them. The convergence of the measures adds to that evidence in the form of each measure providing further support for the validity of the principal assumption underlying the other ones.

(We choose the word "validity" here rather than "truth" because the force of this additional evidence is narrow. What the evidence for the assumption of one measure from its convergence with another supports is the claim that the assumption is *safe to adopt for purposes of theory-mediated measurement of the quantity in question at least over a yet to be fully determined range of circumstances*.)

In short, the evidence from Perrin's different types of experiment, taken together, provided further grounds—beyond any evidence for the stability of each measure and any independent evidence for its assumptions—that the measures, theory-mediated though they were, could be trusted. Notice that such grounds were precisely what was missing in the case of the attempts to infer Avogadro's number and molecular dimensions from the deviations of real gases from the ideal gas law by means of van der Waals's formula.[64]

So much for the gain in evidence in support of the measures of N from considering the different types of experiments together. What about the gain in evidence for the other conclusions reached about Brownian motion itself, independently of molecular-kinetic theory? Including the one about the measures of N, there were six such conclusions put forward in the preceding sections, here stated in some instances in a slightly different form:

(1) The observed "irregularity" or randomness of Brownian motion reflects the fact that the displacements of the granules over time, and at least by implication the velocities as well, have a Gaussian statistical distribution.

(2) The irregularity of the motion notwithstanding, the mean translational kinetic energy in Brownian motion at any given temperature is the same regardless of the size and mass of the granules and the properties of the liquid in which they are suspended; and this kinetic energy varies linearly with absolute temperature.

(3) The (partial) pressure resulting from impact of the granules on any solid surface varies from one case of Brownian motion to another linearly with their volumetric concentration and the absolute temperature of the liquid, and not at all with their size or mass.

(4) The height over which the volumetric concentration of the granules drops by a factor of 2 from the effects of gravity alone varies from one case of Brownian motion to another linearly with the buoyant weight of the granules and the absolute temperature of the liquid, and with nothing else, so long as the temperature remains uniform throughout the liquid.

(5) The rate of diffusion of spherical granules through the liquid, in response to any given initial spatial variation of their volumetric concentration, is characterized by a single quantity, the diffusion coefficient D; and D varies

from one case of Brownian motion to another linearly with the absolute temperature of the liquid, and inversely with the size of the granules and the viscosity of the liquid, and with nothing else.

(6) The mean rotational kinetic energy of the granules in at least one combination of granules and liquid is the same as their mean translational kinetic energy.

For what it is worth, though less monumental, we should add one more:

(7) The apparent velocities, determined by observation of the time required for a granule to change direction and the distance it covered before doing so, are totally misrepresentative of the true granule velocities and the frequencies with which they change in magnitude or direction.

Perrin's different types of experiments, taken together, provided evidence that (1) and (2)—and perhaps (6) as well[65]—together characterize Brownian motion *sufficiently to account for why* (3), (4), and (5) are true of it. The key feature of Brownian motion underlying all three of these is the universality of the mean translational kinetic energy of the granules and its linear dependence on the temperature of the liquid. Given that feature, (3) and (4) result from the consequent mean value of the square of the velocity and the fact that the mean value of the displacements of the granules is 0; and (5) results from the consequent variance of the displacements of the granules, given that the mean of the displacements is 0. Finally, (7) results directly from the magnitudes of the mean translational kinetic energies in (2), but then too from (1), (3), and (4), if not (5) as well.

Two claims are being made here. One concerns evidence about the physical mechanism responsible for (3), (4), and (5). The other concerns evidence that *no further details of Brownian motion make any difference* so far as the relations expressed by (3), (4), and (5) are concerned. That in turn means that the principal quantities characterized by these relations—the pressure the granular motion produces, the equilibrium concentration of granules under gravity, and their diffusion coefficient—are in fact, and not merely in theory, statistically aggregated quantities.

4.9. Some Remarks about "Well-Behaved" Theory-Mediated Measurements

The *practice* of indirect or theory-mediated measurement goes back at least as far as Ptolemy's *Almagest*, a large fraction of which describes how to obtain reliable values for his orbital elements through theory-mediated measures. It came to the forefront in physics in the mid-seventeenth century when Christiaan Huygens employed conical, cycloidal, and small-arc circular pendulums to measure the strength of surface gravity into the fourth significant figure. It has been central to physics ever since. We chose to write this monograph largely because Perrin's four measures of the mean kinetic energies of granules in Brownian motion—or, as he would have had it, of Avogadro's number—illustrate the practice so well. This chapter has examined his four measures as measures of granule kinetic energy, postponing the topic of his four measures as measures of Avogadro's number until Chapters 5 and especially 6. Partly as preparation for those chapters, but also as summary of the preceding sections, this section steps back to consider demands that are typically put on theory-mediated measurements, demands which Perrin took the trouble to meet to a greater extent than anyone else working on Brownian motion at the time.

Our saying that Perrin met these demands does not mean that he expressly identified them and announced he had met them. He simply followed elements of an established practice that, like many evidential practices in science, have never been explicitly codified. The lack of codification has led to a number of misconceptions about theory-mediated measurement that may have stood in the way of readers' understanding of some of the points we made in the preceding sections. This is a further reason why we step back here and remark on the practice of theory-mediated measurement generally.

One frequent misconception is that indirect or theory-mediated measurement is needed only when the target quantity is totally outside the range of human observation, even assisted observation. The dominant role that theory-mediated measurement has played in twentieth-century microphysics encourages this misconception, but there were prominent examples before the twentieth century, most notably electric current, where the magnitude was successfully measured, typically by means of Ampère's law, but not the direction in which electricity was actually flowing. Nevertheless, theory-mediated measurement can be needed for quantities within the range of human observation. The quantity Huygens measured for the strength of surface gravity *was the distance of fall from rest in the first second in the absence of air resistance.* Save for qualifications about controlling for the effects of air resistance, this was trivial to observe with the naked eye, but not with much precision. Huygens achieved four-significant-figure precision by turning to his pendulum measures. As we noted more than once in the preceding, the same sort of situation occurred with the velocities and mean kinetic energies in Brownian motion. The velocities apparent to the human eye aided by the ultramicroscope were thoroughly misleading, necessitating resort to indirect measures.

Another common misconception is that the theoretical relation linking the target quantity to the more accessible quantity serving as its proxy has to be true for a theory-mediated measure to yield valid results. Huygens's pendulum measures of surface gravity again provide a counterexample, for his relations linking the distance of fall in the first second to the length and period of pendulums hold only for uniform gravity acting along parallel lines, not for Newtonian gravity directed toward the center of the earth. Newton took the trouble to validate Huygens's measures by showing how their enabling relations hold in an asymptotic limit as the ratio of the height of the pendulum to the radius of the earth approaches zero.[66] His doing so illustrates what is required of the enabling relations in theory-mediated measurements generally, namely *that they hold to an appropriate level of approximation over the range of cases to which they are applied*. As we noted in passing in the preceding sections, Perrin did what he could to confirm that his measures of the mean kinetic energies met *this* requirement—for example, by his efforts to achieve uniform granule radii in each experiment—but he was not remotely in a position to establish that their enabling relations were definitely true.

Still another misconception—the one that stands in starkest contrast to what we have said in the preceding sections—is that the evidence supporting the theoretical relations involved in such measurement cannot be stronger than the evidence supporting the theory from which the relations are derived: the more the theory is open to challenge, the more so too is the measure. As far back as Huygens—arguably even Ptolemy—the practice of theory-mediated measurement has called for various tests or cross-checks to safeguard against the possibility of the numerical results creating an illusion of a physically meaningful measurement while amounting to nothing more than artifacts of the theoretical assumptions underlying it. From the intuitive perspective of those engaged in the practice, the aim of these tests can be described as assuring that the measure is yielding "well-behaved" results, where a failure to do so gives reasons for not trusting it. The aim of our terms "stability" and "convergence" in the preceding sections was to lend some specificity to the demand that the results be well-behaved.[67] A little more, however, needs to be said about these demands.

The demand that theory-mediated measurements be stable is a generalization of the demand made of all measurements that they be repeatable: *a theory-mediated measure must yield the same results, to within experimental precision, when the values of the manipulable parameters in its defining relation are varied.*[68] Huygens's cycloidal pendulum measure of surface gravity, for example, would have been shown invalid if it had yielded substantially different values as the length of the pendulum was varied. Correlatively, confirmation of the stability of his measure provided evidence for its validity. What we have added in the preceding is that the logical form this support takes is a bi-conditional by virtue of which the confirmation of stability provides direct evidence for some, but typically not all, of the assumptions underlying the measure. This divides these assumptions into two groups insofar as the bi-conditional is the consequent of a conditional in which the other assumptions form an antecedent:

IF $B_1, B_2, \ldots, B_n$,
THEN
$A_1, A_2, \ldots, A_m$ **IF AND ONLY IF** the measure is stable.

Think of the A's as assumptions integral to the relation linking the proxy quantity to the target quantity, and the B's as background assumptions, an example of which in Perrin's experiments is the assumed uniformity of the granules.

Evidence for the stability of a measure, accordingly, is evidence for the validity of the relation between the target quantity and the other quantities comprising it. Caution is needed, however, concerning the limitations of the latter evidence. As we noted earlier, it should be taken not as evidence for the truth of the relation, but rather that the relation holds over the limited range of values of its manipulable parameters for which its stability was tested to the level of precision revealed in those tests. More importantly, tests for the stability of a measure are generally ineffective in exposing systematic error built into the measure itself; the one occasional exception to this is when the range of values of the manipulable parameters over which stability is tested is very large.

Still more important is that the evidence transferred from stability through the bi-conditional to the validity of the measure is itself conditional on the background assumptions; for they can house sources that give rise to a misleading stability. This is why a standard part of the practice of theory-mediated measurement is to pursue evidence independent of the measure for its background assumptions. Examples of this in the case of Perrin are the efforts he put into assuring the uniformity of the granules and, of course, the Gaussian distribution of their displacements.

Perrin systematically tested the stability of two of his measures of mean granule kinetic energy, the vertical-gradient and ξ^2 displacement measures, but both the diffusion and rotational measures were too difficult to execute to permit any real variation of the values of the manipulable parameters involved in them. His tests of the stability of the vertical-gradient measure involved using granules of two different materials, gamboge and mastic, of quite different densities, and in the case of the former, of several different sizes.[69] The linear variation of the mean kinetic energy with temperature appears to have been tested for only one set of granules, with T at both −9°C and 60°C, as well as the usual 20°C.[70] The stability of the ξ^2 measure was tested again with the two kinds of granules over a range of sizes, with two different liquid substrates.[71] We leave to the reader's judgment the degree to which Perrin and his associates confirmed the stability of these two measures. Our point is that he made appropriate efforts to do so.

The demand for convergent complementary measures acts as a response to the three limitations of stability listed in the two paragraphs before the last, especially the latter two. Two theory-mediated measures of the same quantity are *complementary* to the extent that their governing relations and the background assumptions entering into them are logically independent of one another. Huygens's conical and cycloidal pendulum measures of surface gravity provide a good example, for not only are their governing relations different, but the former presupposes as a background assumption a special case of what later emerged as Newton's first two laws of motion, while the latter, being purely kinematic, does not. Two complementary measures of the same quantity are then *convergent* to the extent that their results agree with one another, preferably to within specifiable limits of precision of each of them. We say, "to the extent," because convergence is definitely a matter of degree and subject to judgment. Huygens's two measures agreed beyond their third significant figures; the results J. J. Thomson obtained for the mass-to-charge ratio of cathode rays did not quite overlap, though they were of the same order of magnitude, and that order of magnitude was sufficiently extraordinary to support his claim of convergence.[72] The question of degrees and judgments of convergence will be important in Chapter 6.

Failure to meet the demand that complementary measures of the same quantity converge has been the classic way to expose systematic error in individual measures. One of the more famous examples is the systematic 0.6 percent error in Millikan's measurements of e resulting from an incorrect value for the coefficient of viscosity in his oil-drop measurements.[73] Needless to say perhaps, a failure of convergence between complementary measures can expose only systematic error arising from assumptions that they do not have in common. A lack of sufficient convergence between Perrin's determinations of Avogadro's number from Brownian motion and various determinations from other phenomena, including Millikan's, will be considered in Chapter 6.

As we acknowledged in the case of the high degree of convergence of Perrin's three measures of mean granule kinetic energy, the convergence of complementary measures of a quantity constitutes evidence that the same definite physical quantity is being measured, especially when the measures in question have been shown to be stable. No less important is the evidence that the convergence of two measures provides for the acceptability, at least for purposes of measurement, of the assumptions that each of them has that the other does not. To pursue a second measure with some different assumptions amounts to a way of testing the assumptions peculiar to the first. To give a historically salient example of this, the combination of Newton's first two laws of motion provided a theory-mediated means for measuring forces from changes of motion. The strongest evidence in support of their doing so—virtually the only evidence when they first appeared together in the 1687 *Principia*—came from the fact that Huygens's conical pendulum measure of the distance of fall in the first second, a measure that presupposed a special case of them, gave the same value into the fourth significant figure as his cycloidal and small-arc circular pendulum measures, which did not.

We have already remarked on the convergence of Perrin's three measures of the mean translational kinetic energy of the granules. He also relied on converging measures in his determinations of the sizes of the granules. Of the accessible quantities entering into his kinetic energy measures, the most difficult to determine, owing to their very small sizes, was their radius *a*. Perrin pursued three indirect methods for doing so, concluding that "the three methods employed justify themselves by their concordance."[74] Further evidence for the reliability of the values of *a* he thus determined came from the good convergence of his three translational kinetic energy measures, in two of which it enters linearly as *a*, while in the third it enters as a^3.

In addition to stability and convergence, a third demand that is customarily put on indirect, theory-mediated measures is reasonably well captured by the phrase *amenability to increasing precision*. One factor that limits the precision of any theory-mediated measurement stems from any respects, recognized at the time or not, in which the mediating relation holds only to a certain level of approximation. A second factor stems from imprecision in the measurement of the accessible quantities entering into it, whether from insufficient control of, or corrections for, externalities, or simply from the direct limits of precision to which values for a quantity can be determined. *A theory-mediated measure is amenable to increasing precision if corrections introduced to compensate for externalities and known respects of approximation or improvements in direct measurement of accessible quantities yield more precise values of the target quantity.* The form the latter takes is typically a decrease in the error bands for the target quantity or an increase in the number of its justifiable significant figures. Huygens's small-arc circular pendulum measure of surface gravity here again provides a historical example. It was still the preferred means for measuring *g* in the 1950s, but over the intervening three centuries refined design of the pendulums and corrections for the small effects of air resistance and flexibility in the pendulum support had led to an increase in the number of significant figures to which values were justifiably being quoted from three to four in the time of Newton to seven in 1958.[75]

We mention this further demand here only because Perrin took some steps toward addressing it. In particular, we noted earlier that he had revisited his vertical-gradient measurement of kinetic energy between his monograph of 1909 and his Solvay Conference paper of 1911, taking increased pains to achieve granule uniformity. The upshot was a revision of his earlier preferred value for Avogadro's number from this measure of 70.5×10^{22} to 68.2×10^{22}, in closer agreement with his announced values of 68.8×10^{22} and $69 \times 10^{22} \pm 3$ percent from the two granule-displacement measures. We also noted that he was revisiting his ξ^2 measure sometime after 1914 once Constantin pointed out a possible source of systematic error owing to the effects of the walls on the displacements of granules near to them. These efforts notwithstanding, one has to wonder just how amenable to greater precision Perrin's measures were, given the enormous difficulties in making precise observations of Brownian motion. Perrin himself, nevertheless, expressed the view in 1909 that the vertical-gradient measure was "*capable of unlimited precision*," remarking,

> The preparation of a uniform emulsion and the determination of the magnitudes other than N which enter into the equation can in reality be pushed to whatever degree of perfection desired. It is simply a question of patience and time; nothing limits *a priori* the accuracy of the results, and the mass of the atom can be obtained, if desired, with the same precision as the mass of the earth.[76]

As we shall see in Chapter 6, by two years later enough alternative determinations of Avogadro's number differing from his by more than 10 percent had appeared to make one wonder whether he would still have been willing to say this. Regardless, Perrin saw the importance of what we are calling *amenability to increasing precision.*

If one thinks of the demand for stability of a measure as a response to the question of whether it is measuring any definite value for its target quantity at all, then the pursuit of increasing precision can be thought of as a response to the question of how near to exact the measure is—that is, how near to exact the functional relation is between its target quantity and its proxy. As noted earlier, the pursuit of convergent complementary measures also addresses the question of the exactness of a measure, first affirmatively through the extent of agreement, and second, negatively through shortfalls in agreement exposing systematic error. The demand for convergent complementary measures is also a response to the question of whether the values yielded by a stable measure truly are values for the theoretical quantity that they are intended to be, or in other words, whether the target quantity truly is what the derivation of the functional relation comprising the measure claims it to be. The extent of convergence of Perrin's two measures shown to be stable, complemented as well by the other two measures, thus provided strong grounds that what was being measured in all cases was indeed kinetic energy of granules in Brownian motion.

Perrin's further claim that what was being measured was Avogadro's number is *not* supported by the convergence of his four measures; for this claim required the further presupposition that the mean kinetic energies being measured are equal to the mean kinetic energies of the molecules in the liquid substrate; and this presupposition had to be added to all four of the measures. Hence, they are *not* complementary with respect to it. For a convergent measure to support this presupposition, it needs to be predicated on a relationship between a molecular magnitude like Avogadro's number and some measured quantity other than the mean kinetic energies of granules in Brownian motion. We shall consider such measures in Chapter 6.

The point of our excursion on theory-mediated measurement in this section has been to stress the extent to which Perrin's measurements of granule kinetic energy fell within the established practice of such measurement. Perrin and those working under him truly have claim to having measured the mean translational kinetic energy of granules to a reasonably high precision—a far stronger claim than anyone else at the time working on Brownian motion. This achievement has gone almost entirely unnoted. Perrin himself said little about it, focused as he was on claiming to have measured Avogadro's number; and the literature has largely followed him in this regard. The additional step Perrin introduced to bridge the gap between his measurement of granule kinetic energy and his claim of having measured Avogadro's number, as we shall see in Chapters 5 and 6—and for that matter in the final section of this chapter—was far more open to dispute than anything involved in his measurement of the granule energy. The latter, as we have said, was an indisputable achievement.

It is time we turn to his claim to having measured molecular magnitudes, considering first in Chapter 5 any evidence supporting this independently of presupposing the equality of the kinetic energies, and then in Chapter 6 any evidence supporting this equality from complementary measures of Avogadro's number. As a step that will help prepare for the next chapter, we conclude this one with a 1912 assessment of the state of Brownian motion as seen not from the perspective of microphysics, but from that of colloid chemistry.

4.10. A Critical Assessment of Perrin's Results at the Time

We said at the beginning of this chapter, and in Chapter 1 as well, that we were going to try to maintain the point of view of someone writing a review article on Brownian motion just before or soon after the first Solvay Conference—in particular, before the summer of 1913, when Bohr's three-part paper tying the structure of the atom to spectra, Mosley's two-part paper tying x-ray spectra to atomic number, and the Braggs' first papers on crystal structure appeared, totally transforming the approach that was being pursued to atomic and molecular structure. It goes without saying that we have violated that commitment in two respects. First, we have examined Perrin's experiments through the lens of a philosophical analysis of theory-mediated measurement and the evidence produced when measured results are well-behaved. No scientist writing a review article at the time would have done so. (We like to think that scientists at the time, presented with our analysis, would have said, "Of course, isn't that obvious—too obvious to need saying"; but nothing in the preceding offers any basis for this.) Second, we have considered only Perrin's experiments. Any review article at the time, even one stressing Perrin's results, would have considered them in the context of research by others on Brownian motion. Some of that research, most notably by Svedberg, paralleled Perrin's, but much of it, stretching back a quarter century, had yielded results in *prima facie* conflict with Perrin's.

So, how would a review article on Brownian motion itself, written between 1910 and 1913, have assessed Perrin's findings? Fortunately there was one, namely the final chapter of the third (German) edition of *Grundriss der Kolloidchemie* by Ostwald's son, Wolfgang Ostwald (1883–1943).[77] The chapter, which runs some 80 pages in the English edition, is the last part of three covering "Mechanical Properties of Colloids," the first two of which concern "relations of volume and mass in colloids," and "internal friction and surface tension of colloids"; entitled "Movement in Colloid Systems and Its Results," it opens with a section on Brownian motion, followed by sections on diffusion, dialysis, osmosis, and filtration that include further remarks on Brownian motion. Keep in mind that not all colloid systems exhibit Brownian motion, and hence not all the phenomena covered are necessarily tied to it. In particular, Perrin's experiments involved only dilute emulsions, that is, emulsions in which there is negligible interaction among the granules! Ostwald's chapter adopts no such restriction.

Because we have stressed results on Brownian motion that did not have to presuppose molecular-kinetic theory, the best place to start with young Ostwald's chapter is with a pair of long quotes on the relation to that theory, quotes that make important points to which we shall return in Chapter 5:

> These investigations, often of a most painstaking nature, show that the source of energy for Brownian movement lies within the disperse system itself and is obviously of a very general nature for it evidences its effects under the most varied external conditions. Brownian movement is, however, observed only in disperse systems, more particularly only in such as have a high degree of dispersion. The kinetic hypothesis according to which gases and liquids are regarded as conglomerates of rapidly moving molecularly dispersed particles, has recently been applied to Brownian movement. In fact, some have seen in this direct evidence for the correctness of the kinetic theory as applied, say to the movement of liquid particles. We return to this question on p. 205. While marveling at the successful applications that have been made of this kinetic hypothesis, it seems to me not impossible that future investigations may yield

another more universal and less hypothetical explanation of this spontaneous movement.[78]

Upon returning to the question on page 205, he reviews the evidence (1) that Svedberg and Henri had developed for the Einstein-Smoluchowski ξ^2 formula; (2) that Perrin had offered for the validity of Einstein's rotational formula; and (3) that Perrin had marshaled in establishing his vertical-gradient "law"—all of which, he notes, had "been deduced from considerations of the kinetic theory of gases."[79] He then turns to the values for "the constant N of the Einstein-Smoluchowski equation," noting that they "agree surprisingly well with those obtained by entirely different means," and adding that they may be yielding "the most exact figures of this fundamental value now obtainable."[80] This leads him to conclude:

> These brilliant results fill one with admiration for the remarkable fertility of the Einstein-Smoluchowski equation, especially when it is remembered how many still purely hypothetical factors enter into its composition. Nothing better illustrates the daring, we might say, of this train of thought than the remark of Perrin, to whom, with Svedberg, science owes the most in this field, about the theorem of the equality of the distribution of energy which is the nucleus of all kinetic deductions. "The word theorem should deceive no one, for it is full of hypotheses as is almost every theory of mathematical physics." It is safest, perhaps, to hold that the future will preserve but a *part* of our present kinetic notions to work over into a more general, less supposititious theory. As a matter of fact, several of the laws governing Brownian movement may be deduced even *without* recourse to kinetic assumptions, as for example the inverse proportionality of velocity to viscosity, from Stokes' law. Possibly this purely inductive method will some day discover these same laws; in fact, considerations of the methods of science demands it, but when the day will come must remain a matter of opinion.[81]

What restrictions young Ostwald wanted to place on assumptions entering into theory-mediated measurements in a "purely inductive method" he does not say. Clearly he was uncomfortable with at least some of the assumptions required by what we have called the ξ^2 measure of N. We will return to the issue of the hypothetical character of this and the other measures in Chapters 5 and 6.

Although he did not list it in the preceding quote, Ostwald's presentation of Perrin's vertical-gradient "law" earlier in the chapter makes clear that it too did not require recourse to kinetic assumptions. Specifically, his review of Perrin's data leads to the conclusion: "These measurements yielded an important law: *The concentration of the disperse phase increases in geometric progression with the algebraic decrease in the height of the level.*"[82] (Note that this statement does not include reference to the buoyant weight of the granules, the further factor needed to infer values for *N* from the rate of attenuation.) He concludes as well that observations had shown that "*the square of the path length is directly proportional to the ratio of absolute temperature and the viscosity*" in Brownian motion.[83]

Subsequent sections of the chapter emphasize departures from the law of diffusion and van't Hoff's law exhibited by colloids generally. Along the way, however, Ostwald concedes that the data from highly disperse colloids do support the Einstein-Smoluchowski equation for the diffusion coefficient *D*;[84] and he grants that the "Boyle-Gay-Lussac law" appears to hold for osmotic pressure "in greatly diluted colloid systems."[85] Accordingly, our (3), (4), and (5) at the end of section 4.8 are at least candidates for conclusions about Brownian motion that Ostwald would have agreed did not require recourse to molecular-kinetic theory.

The most important point, to our way of thinking, that we made earlier about (3), (4), and (5)—their tie to the conclusions (1) and (2) about the motion itself—was not in any way granted by the young Ostwald. He does not present the granules as engaged in an extraordinarily rapid random motion, but instead as "vibrating," each with a period corresponding to the length of time over which it is observed to change direction in an irregular fashion and an amplitude equal to the displacement over that time; and, in contrast to elastic vibration, the mean of the displacements, he stresses, is proportional to, and not independent of, that period.[86] Worse than that, he then treats the distance divided by the time as the velocity of the granules, ignoring the warnings of Einstein and Perrin that each granule has undergone a huge number of changes in velocity during the time in which we observe one change in direction to the next, and hence that their velocity at any time is not really accessible to measurement, much less observation. In keeping with this, he makes no mention anywhere in his book of a "random error"—that is, what we now call a Gaussian—distribution. And, while citing the 1910 German translation of Perrin's 1909 monograph, he totally ignores the evidence meticulously presented in it that the displacements of the granules have a Gaussian distribution. In short, for the young Ostwald in 1912, Brownian motion was not what we have come to mean by the term at all.

His most striking failure to appreciate the Maxwellian character of Brownian motion was in his account of granule diffusion:

> In the irregular, particularly in the forward, movements of small particles, as observed, for example, by Zsigmondy in colloid solutions, it is to be expected that an *accidental* wandering of the particles over into the pure dispersion means must take place. But such accidental migration cannot wholly explain all diffusion, the laws of which A. Fick formulated in 1855. In order that Brownian

> movement may lead to diffusion, it must become directive in character toward the pure dispersion means or toward the "more dilute" parts of any continuous system.
>
> . . . As noted above (p. 196), R. Zsigmondy observed *less* movement in dilute systems than in concentrated ones. Because of this, equilibrium cannot exist, so far as average velocities are concerned, in a system consisting, say, of a colloid solution covered by a layer of the pure dispersion medium. In places of greater concentration, the particles will be moving faster than in those of a lower one. The sources of energy for Brownian movement, whatever they may be, must therefore have different values in different parts of the system at the beginning of diffusion.[87]

Young Ostwald just does not see how the statistical character of Brownian motion can produce an effect in the direction from higher to lower concentration gradient. That is, he does not see that the *fraction* of granules whose displacement is large in any given time—for example, in excess of the standard deviation—can be the same everywhere, but that the *number* of such granules in the high concentration region will, in the aggregate, be proportionately greater than the number in the low concentration region. No mysterious force or gradient in the source of the energy is needed to explain the "directive character" of diffusion.

Does this mean that we have violated our intent of viewing the situation from the point of view of someone writing a review article on Brownian motion at the time? Not necessarily. Ostwald senior had emphasized the Maxwellian character of Brownian motion in the section devoted to the subject that he introduced for the first time in the 1909 edition of his *Grundriss*:

> These movements are in apparent contradiction to the Second Law. The velocity depends on the viscosity, the effect of which is to use up the energy of the moving particles. The movement continues in spite of this, unless the particles coagulate together (a suspension of vermillion had not lost its movement after two years), so that energy at rest appears to become energy in motion. As early as 1863 Chr. Wiener regarded the phenomenon as a consequence of the kinetic nature of heat, which he supposed to consist of the motion of the smallest particles of liquids or of their molecules, their impacts on the grains giving rise to the Brownian movements. This kinetic theory of heat, which was propounded in the early ages, has been extensively developed by Clausius, Maxwell, Boltzmann, and by many later scientists, especially in its application to gases, in spite of the very serious difficulties encountered. Still, until recently there was wanting direct proof of the necessity for such a representation, which accordingly remained an arbitrary though useful hypothesis.

> If the movements of a small particle suspended in a liquid are calculated on the basis of this hypothesis, *the agreement with the movements actually observed is so close that we are compelled to regard this agreement as a fairly satisfactory proof of the kinetic nature of heat.* A brief account of the kinetic theory of gases will be found farther on.[88] [italics added]

The evidence Ostwald goes on to present showing this agreement came purely from Svedberg's 1906 confirmation of the Einstein-Smoluchowski ξ^2 formula. Ostwald did not know of Perrin's far more definitive evidence that granule displacements have a Gaussian distribution at the time he wrote this.

We draw two lessons from this contrast between Ostwald *père* and Ostwald *fils*. First, the thoroughgoing statistical view of Brownian motion was difficult for many to grasp at the time. Second, whether a review article written at the time was going to emphasize Perrin's evidence for the Gaussian distribution of granule displacements and its implications for the phenomenon of granule diffusion depended on who was writing the article.

Notes

1. W. Ostwald (1907), p. 500f.
2. Two-thirds of the translational kinetic energy because the pressure exerted in each orthogonal direction by impact is $1/3mv^2$.
3. See Mayo (1996), where in Chapter 7 on Brownian Motion (pp. 214–250) she makes the same point that Perrin had to condense and organize his raw data to put his displacement formula to any kind of test. Nothing like a bull's-eye of the Gaussian distribution of granular displacement is actually observed in experiment, and "the actual data consists of 500 scratch marks" (p. 224).
4. We say, "not all that much" because Nernst (1916, p. 469) calls attention to Svedberg's having announced values for Avogadro's number (that is, *RT* divided by 3/2 of the mean kinetic energies of granules) for Brownian motion more than 10 percent lower than Perrin's values in experiments he conducted following Perrin's. See Svedberg (1912), p.136f, where he emphasizes the consistency of his value, in contrast to Perrin's, with those obtained from other phenomena by Regener, Rutherford, and Millikan. Even before the 1911 Solvay Conference, Millikan had outspokenly challenged Perrin's inferred value for the elementary charge, and hence too by implication of Avogadro's number, in the widely read *Philosophical Magazine* (Millikan 1910, especially p. 226f).
5. A natural thought, suggested to us by Mary Jo Nye, is to attribute the discrepancy to inaccuracy in Perrin's measurements of the sizes of his granules, a measurement he acknowledged was especially demanding. The trouble, however, is that Perrin's Avogadro values (as we shall see) vary inversely with the size of the granules in one of his three approaches (the displacements of granules with time), and inversely with

the cube of their size in the other two; the values from the three approaches display no pattern of such sensitivity to their size.

6. Halley (1686). Halley, who was writing at a time when he had received the manuscript of only Book 1 of Newton's *Principia*, assumed uniform gravity; Propositions 21 and 22 of Book 2 cover $1/r$ and $1/r^2$ gravity, with an acknowledgment of Halley. Halley correctly assigns the variation of barometric pressure to differences in the "weight" of the air, but his explanations for why that weight varies with the weather (as caused by the prevailing winds) have matters largely upside down.
7. Maxwell (1875), p. 434.
8. Perrin gives this relationship in a footnote from the earliest editions of *Les atomes* forward. See Perrin (1990) and (1916), p. 91, in English; in the first French edition, Perrin (1913), p. 130, or (2014), p. 188.
9. Perrin (1990) and (1916), pp. 89–94 and 36–39 in English; in the first French edition, Perrin (1913), pp. 128–131 and pp. 52–56, or (2014), pp. 186–190 and 136–139; see also Perrin (1910), p. 23.
10. Perrin (1910, p. 47) speaks of the osmotic pressure in his emulsions as not reaching a "thousand millionth of an atmosphere," adding:

    > This last figure shows to what point I was removed from the conditions under which it has been possible to reveal (Malfitano), and then to measure (J. Duclaux) the osmotic pressure of colloidal solutions with very fine granules closely crowded together. It may be that a generalization, more or less analogous to that of Van der Waals, will give, one day, by means of a reasoning from kinetic theory, the osmotic pressure of such solutions.

11. Ibid., pp. 41–44.
12. Einstein (1956b), p. 27; the paper itself, the second of Einstein's on Brownian motion, appeared in *Annalen der Physik* in 1906 (vol. 19, pp. 371–381). Perrin subsequently cites Smoluchowski as well in this regard.
13. Perrin (1910), p. 33.
14. Ibid., p. 40.
15. Perrin (1926), p. 9f; for more details of the efforts, see pp. 24–44 of Perrin (1910), or Perrin (1990) and (1916), pp. 94–102, in English; Perrin (1913), pp. 134–146, or (2014), pp. 191–198, in French.
16. Perrin (1910), p. 48.
17. Perrin (1990) and (1916), p. 106f; in French, Perrin (1913), p. 151f, or (2014) p. 202f; also, Perrin (1912a), p. 185f.
18. See Perrin (1910), p. 23f; and Perrin (1990) and (1916), pp. 89–93, in English; Perrin (1913), pp. 128–133, or (2014), pp. 186–189, in French.
19. Perrin (1910), p. 46.
20. The assumption of uniform gravity in this phenomenological "law" does not compromise it. For, even at an altitude of 10 miles or 50,000 feet above the surface of the Earth, the change in g under inverse-square gravity amounts to less than 10^{-7} percent. And, of course, both it and the parallel relationship for dilute emulsions can be modified to incorporate inverse-square gravity in the manner Newton showed.

21. Harper (2012) has argued that this sort of evidence is central to the approach to evidence that Newton took in his *Principia*.
22. Perrin (1910), p. 46.
23. In fact, from our retrospective vantage point, we know that the constant of proportionality in the law for dilute emulsions, had it been independently determinable, would not have amounted to the same thing as the gas constant, for Perrin's values for N were too high, and hence his values for W_T were too low, compared with our current values for N_0 and the mean translational kinetic energy of molecules at any given temperature. Whether this was because of experimental error in Perrin's experiments or a difference between macroscopic (relatively speaking) granules and molecules we are leaving as an open question.
24. See W. Ostwald (1890b), pp. 127–130.
25. We thank Jody Azzouni for calling our attention to the need for the remarks made in this and the preceding paragraphs.
26. Einstein (1956), pp. 1–18, originally published as Einstein (1905); ibid., pp. 19–35, originally published as Einstein (1906b); ibid., pp. 36–62, originally published as Einstein (1906a); ibid., pp. 63–67, originally published as Einstein (1907); and ibid., pp. 68–85, originally published as Einstein (1908); Smoluchowski (1906a); and Langevin (1908). Smoluchowski's formula differed from Einstein's by a factor; Langevin's derivation most limits premises from molecular-kinetic theory.
27. The observations amounted to projections of displacements of individual particles on a plane—for example, a horizontal plane—but only the components of those observed displacements in a single direction then entered into the formula; hence, their projection on an axis. A tacit assumption here is that the mean velocities and hence mean translational kinetic energies are distributed equally over the three directions, and hence the relevant portion of the mean translational kinetic energy, once we replace RT/N, is 1/3 of W.
28. See Perrin (1910), pp. 51–54.
29. See Renn (2005) for a thorough discussion of the background leading up to Einstein's derivations, and Brush (1968) and Fürth's "Notes" in Einstein (1956), pp. 86–119, for discussions of these and other derivations.
30. Specifically, Loeb (1934), pp. 399–405.
31. In particular, in Mayo (1996), pp. 219–231, or her (1988).
32. Perrin (1910), p. 53. H. Fletcher, working under Millikan, subsequently challenged the applicability of Stokes's law to emulsions in Fletcher (1911), but that need not concern us here insofar as the issue raised involves only the precise accuracy of the formulas.
33. For a more extended discussion of this point, see Perrin (1910), p. 22f and pp. 54–56.
34. The following statement of the ergodic hypothesis is from Hecht (1990), p. 9:

    > The time average of any thermodynamic variable ϕ in an actual system is equal to the ensemble average of ϕ in the limit as the number of replicas η in the ensemble goes to infinity, provided that the members of the ensemble copy precisely the thermodynamic state and environment of the actual system, that is, adhere to the specified macroscopic conditions.

35. Perrin (1910), p. 52; see also p. 20, where Perrin refers to "Maxwell's *law of distribution of velocities*."
36. The term we use here, "Gaussian," was not then available to Perrin (or Einstein); they spoke instead of randomness and, on occasion, of a "random error" distribution. For clarity, we are going to stick with "Gaussian," notwithstanding the anachronism of doing so.
37. Perrin (1910), pp. 64–68.
38. See note 31.
39. See Einstein (1956), pp. 74–81; Hecht (1990), pp. 346–348; and Atkins and de Paula (2010), pp. 766–774.
40. See Atkins and de Paula (2010).
41. Perrin (1910), p. 60f.
42. Ibid., p. 63.
43. See Smith (2001b).
44. Perrin (1990) and (1916), p. 123; in French, Perrin (1913) p. 174f, or (2014), p. 220f. At the Solvay Conference, Perrin made the same point (1912a, p. 206), in conjunction with a slight variant of the table.
45. Chaudesaigues (1908), p. 1046.
46. Perrin (1910), p. 65.
47. Ibid., p. 66. In both his 1911 Solvay Conference paper (1912a, p. 201) and his *Les atomes*, Perrin presents a similar figure, but for 500 observations of granules of radius $a = 0.367\ \mu$. See Perrin (1990) and (1916), p. 118; Perrin (1913), p. 169, or (2014), p. 216.
48. See, for example, Mazo (2002) and Chung (2002).
49. In stating this conclusion in terms of W_T, instead of *W at any given temperature*, we are giving Perrin the benefit of the doubt, for he did not present results at different temperatures. We shall return to this point later.
50. Léon Brillouin (1889–1969) subsequently became one of the major figures in the history of solid state physics, among other things identifying the "zones" named after him. He was barely 20 years old when he carried out the experiments under Perrin described in the text.
51. The idea exploited the fact that had originally been a nuisance, that the granules in a glycerine emulsion upon contact adhered to the walls, gradually reducing their concentration in the emulsion. Counting the number of granules adhering to a unit surface of the wall over time (in hours) yielded a well-defined value of their diffusion coefficient. For details, see Perrin (1990) and (1916), pp. 129–132, where he adds that the rate of diffusion measured was 140,000 times slower than that of sugar in water. (In the French original, Perrin (1913), pp. 184–188, or (2014), pp. 226–229.)
52. Perrin (1910), p. 68. He immediately adds, ". . . it is now difficult to doubt Maxwell's law of distribution of velocities, although the completely direct verification, realized here for the displacements, is still wanting for the velocities."
53. Somewhat ironically, as remarked earlier in our text, "Brownian motion" has come to refer not to a physical phenomenon in various dilute solutions, but to a general mathematically characterized "stochastic process" of a certain sort, "random walks," the

principal historical empirical instance of which is Brownian motion itself. See note 48 for examples, as well as S. Chandrasekhar's seminal (1943).

54. Perrin (1910), p. 73.
55. Einstein (1956), p. 32f; (1906b), p. 379; (1989a), p. 342; and (1989b), p. 188.
56. Perrin (1910), p. 74. The duration of the "great interval" referred to in the text is not specified, beyond remarking that "he has succeeded in fixing from minute to minute the orientation of spheres of mastic" (p. 73).
57. See Partington and Shilling (1924), p. 188ff, for a listing of measured values from 1813 to 1924 for a wide range of gases (not including iodine).
58. Einstein (1989b), pp. 214–228.
59. Partington and Shilling (1924), pp. 234–241, especially p. 238.
60. Specifically, p. 211 in Perrin (1912a), and p. 132 of Perrin (1990) and (1916); in French, Perrin (1913), p. 188f, or (2014), p. 229f.
61. Perrin (1912a), p. 188; (1990) and (1916), p. 107; (1913), p. 154, or (2014), p. 203.
62. Perrin (1910), pp. 63 and 90.
63. Perrin (1990), p. 124. He cites Constantin, who had died in 1916 in the War, in his Nobel Prize lecture as well, Perrin (1926, p. 13) again replacing his earlier value of Avogadro's number from displacements, 69×10^{22} with 64×10^{22}, but adds nothing more in the way of details. Constantin is not cited in either the fourth or fifth editions of 1914, nor consequently in the 1916 English edition.
64. In particular, as we stated in section 2.2 of Chapter 2, Meyer (1899b, pp. 317–332, especially p. 329) notes the failure of what we are calling convergence of two other measures of the sizes of molecules of different sorts with a measure of those sizes based on van der Waals's formula, with consequent implications (p. 332f) for the value of Loschmidt's and hence Avogadro's number.
65. The equipartition of kinetic energy among the different degrees of freedom of the granules would become significant if experiments were to show that the amount of heat needed to raise the temperature of dilute emulsions in which Brownian motion occurs is greater than the energy taken up in their translational motion.
66. See Newton (1999), pp. 552–556 (*viz.*, Book 1, Propositions 51 and 52).
67. The terms "stability" and "convergence" are taken from Smith (2001b).
68. We have used the phrase "same results," to allow the statement to apply not just to the measurement of putative constants, but to cases in which what is putatively being measured is a functional relationship between two variables. In the case of a putative linear variation, for example, the results must continue to yield a linear relationship as the values of other parameters are varied. Thus Boyle's measurements of pV being a constant, where pressure was being measured indirectly, would have been invalidated if the same relationship had not occurred when the diameter of his tubes, for example, was varied.
69. Perrin (1910), pp. 41–47.
70. Perrin (1990) and (1916), pp. 104 and 106; in the original in French, Perrin (1913), pp. 149f and 152f, or (2014), pp. 200f and 202f.
71. Ibid., p. 123; in the original in French, Perrin (1913), p. 175, or (2014), p. 220.
72. See Smith (2001b).

73. See Birge (1942), esp. pp. 116–119, and (1945). For a subsequent review, see, for example, Cohen, Crowe, and Dumond (1957), pp. 112–128.
74. Perrin (1910), p. 39. The sentence following the quotation is, "Further, this concordance brings out certain important consequences concerning the law of Stokes."
75. See Heiskanen and Vening Meinesz (1958), p. 86.
76. Perrin (1910), p. 48.
77. C. W. W. Ostwald (1912); English translation (1915). This book, the standard reference on colloid chemistry throughout the first half of the twentieth century, appeared in several subsequent editions, some of them translated into English.
78. C. W. W. Ostwald (1915), p. 192.
79. Ibid., p. 208.
80. Ibid. We have corrected a misprint in the English text, replacing "new" with "now" in the last phrase quoted.
81. Ibid., p. 209. This paragraph remained nearly word-for-word the same in the second English edition of 1919 (p. 214), translated from the third German edition again by Fischer, with the same publisher.
82. Ibid., p. 204.
83. Ibid., p. 199. Italics his.
84. Ibid., pp. 215–219.
85. Ibid., p. 261f.
86. Ibid., p. 195ff.
87. Ibid., p. 210. We have adopted the correction to the translation from the next edition, replacing the word "means" in "pure dispersion means" with the word "medium."
88. W. Ostwald (1912), p. 484f.

5
Implications for Molecular-Kinetic Theory

In his Presentation Speech at the Nobel ceremony in 1926, readers will recall, Oseen said, "The object of the researches of Professor Jean Perrin which have gained for him the Nobel Prize in Physics for 1926 was to put a definite end to the long struggle regarding the real existence of molecules." We have been considering these researches as if their aim were to establish conclusions about Brownian motion itself. But that was surely not their aim. And, judging from Ostwald's singling out Perrin's research, along with Thomson's on the electron, in the Preface to the 1909 edition of his *Grundriss der Physikalische Chemie*, Brownian motion surely did make a key contribution toward closing the question of the reality of molecules within the scientific community at the time. What precisely it was about Brownian motion that made a difference, however, is not so obvious. Ostwald, in granting "the experimental proof of the atomic nature of matter," cites "the agreement of the Brownian movements with the requirements of the kinetic hypothesis." But that does not tell us what those requirements were, or the specific respects in which the findings on Brownian motion fulfilled them. We must also keep in mind that the foremost historian of kinetic theory, Stephen Brush, looking back on the time 60 years later, found "the willingness of scientists to believe in the 'reality' of atoms after 1908, in contrast to previous insistence on their 'hypothetical' character, . . . quite amazing."

Three factors make it difficult to pinpoint what it was about Brownian motion that made a difference. First, so many conclusions emerged from the research, each of them a candidate to sway one individual or another. The verification that the *kind* of irregular motion hypothesized in kinetic theory actually exists in the case of Brownian motion may alone have been enough for some. Second, the experimental results on Brownian motion were published at a time when so much other research connected to the molecular point of view was making historic advances. The research by Thomson and others on the atomicity of electricity is one example; the research by Rutherford and his associates, the Curies, and others on radioactivity is another. Especially important in this regard was Einstein's reformulation of Planck's work in his 1907 paper proposing the quantization of energy in order to account for the specific heats of solids at low temperatures, for that paper, far more than anything on Brownian motion, seems to have spurred Nernst into initiating the first Solvay Conference. Perhaps Perrin's research merely amounted to a final straw.

Brownian Motion and Molecular Reality. George E. Smith and Raghav Seth, Oxford University Press (2020). © Oxford University Press. DOI: 10.1093/oso/9780190098025.001.0001.

Third, the closing of any long-standing question in science is always a complex sociological process. It is not a matter of a large number of individuals reaching a decision one by one. The scientific community is highly stratified. Within the physical chemistry community, in particular, the pronouncements by two of the long-established leading figures, Ostwald and Nernst, put any continuing outspoken holdout against atomicity under a threat of ostracism. In practice, the decision took the form of a refusal of leading journals to publish further papers opposing atomicity. A huge fraction of research in leading laboratories in chemistry, though less so in physics, could proceed independently of any stance on atomicity, but research at the forefront came to focus on questions that presupposed the reality of molecules without having to say anything about the grounds for this presupposition. Figuring out all the respects in which Brownian motion entered into the sociological process, and how it entered, is not something we are about to take on in this monograph—with one exception, namely what it meant within the community for the question to be closed, a topic of Chapter 7.

Rather, then, than trying to pinpoint just one or two aspects of Perrin's results as making some sort of crucial difference, this chapter and the next are going to examine five different ways in which the Brownian motion results were cited *at the time* as bearing on the molecular-kinetic hypothesis. We emphasize "at the time" to indicate why we are not instead proceeding through the list of proposals that have been put forward in the philosophy of science literature. One of those proposals —van Fraassen's "grounding of kinetic theory"—nevertheless occupies a pivotal position in the discussion that follows, first because it provides a minimal position that has *prima facie* claim to being correct historically, at least as far as it goes, and second because, apparently unbeknownst to van Fraassen, it fits so perfectly with the demands Ostwald made in spelling out (in his *Monist* article quoted in section 2.4 of Chapter 2) why in 1907 he found the kinetic theory inadequate. Section 5.3 will consider the grounding-proposal, with sections 5.4 and 5.5 then considering two candidates for ways in which the Brownian motion results had *more* of a bearing on the molecular-kinetic hypothesis beyond just grounding it.

Because our conclusions in this chapter run counter to so much of the "lore" attached to Perrin, we had best summarize them here to enable readers to anticipate them. *Other than through an appeal to some form of inference to the best explanation,* Perrin's results on Brownian motion in and of themselves had far more limited evidential implications for molecular theory than is generally thought without the further presupposition that the mean kinetic energies of the granules match the mean kinetic energies of molecules in the liquid substrate, and hence without presupposing the reality of molecules. The reason for this is that his results did not gain sufficient—indeed, any—experimental access to molecules themselves and their interaction with the granules.

Nevertheless, Perrin's results did have major evidential implications concerning the liquid substrate and incessant motion within it, and hence for the kinetic theory of heat.

We are delaying until Chapter 6 the evidential significance of the agreement—to the extent that there was agreement—between the Brownian motion determinations of Avogadro's number and determinations involving other phenomena. Perrin emphasized this agreement, to say the least, in all three of his more comprehensive publications on Brownian motion and molecular reality. We have two reasons for considering it in a separate chapter after considering what evidence his results on Brownian motion, taken unto themselves, provided for molecular reality. First, the majority of the cluster of agreeing values for Avogadro's number that emerged between 1907 and 1913 were from phenomena at some remove from Brownian motion, even though Perrin was the one who at the time made the strongest claims about what they showed. Second, and more important, questions about the evidential force of agreeing values for fundamental constants have become a central topic not only within philosophy of science, but within physics itself, where it has become a primary form of evidence within microphysics.[1] Simply put, we think the topic is too important not to merit a chapter unto itself.

A word of warning is needed about this chapter. We will be subjecting the proposed evidence from Brownian motion for molecular-kinetic theory to unusually detailed, rigorous scrutiny—indeed, to an extent that may try readers' patience in many places. Our doing so is in keeping with our goal of maintaining the perspective of a careful review article written at the time. More to the point, insofar as van Fraassen has challenged the force of this evidence, we will be endeavoring not to mistakenly grant it more force than careful logical analysis can justify. Ultimately, this will pay off with a quite substantial conclusion in section 5.6. For those whose patience is tried in the meantime, never forget the extent to which evidence in science so often lies in the most recondite of details.

Before turning to the topics of grounding and how the evidence from Brownian motion may have gone beyond merely grounding molecular-kinetic theory, we first need to consider ways in which the research on it could have been regarded as *testing* this theory, for some may take that to be what Ostwald *père* was intimating with his comparison of calculation and observation in the quotation at the end of the preceding chapter; and Perrin often words matters in this way in his 1909 monograph, even to the point of referring to an *experimentum crucis*, a cross-roads experiment "deciding for or against the molecular theory of the Brownian movement."[2] Sections 5.1 and 5.2 will accordingly consider whether Perrin's results can be construed as tests of the theory.

5.1. Testing Molecular-Kinetic Theory: The Kinetic Energy "Test"

Scientists tend to speak of "testing" a theory in a very broad way, including virtually any form of evidence for or against a theory. Ostwald, in the quotation just cited, referred more narrowly to the comparison of calculations of Brownian motion based on the molecular-kinetic hypothesis with observations of the motions. The comparisons he went on to cite were (1) between calculated values of the root mean square of granule displacements in a given time—our ξ^2—and values observed by Svedberg; and (2) between the predicted logarithmic relation of granule concentration versus height to their buoyant weight and Perrin's measured concentrations.[3] Here we are considering testing of a theory only in this narrow sense: *a "prediction" of specific values of a quantity or a specific functional relationship among quantities deduced from a theory (with the help, if need be, of auxiliary assumptions) that can be directly compared with independently measured values of that quantity or those quantities.*

Ostwald initially cites Einstein (1905) and Smoluchowski (1906) in referring to calculated values of ξ^2, but then lists values for different emulsions calculated by Svedberg from Einstein's and Smoluchowski's formulas.[4] Einstein did indeed calculate a value of 0.8 microns after one second for hypothetical granules in water, but only by assuming a specific value for N_0, namely 6×10^{23} "in accordance with the kinetic theory of gases."[5] As we saw earlier, however, the values for N_0 that had been obtained with the kinetic theory of gases ranged quite widely, so that Einstein's choice, for which he gave no further explanation, had to have been somewhat arbitrary. Moreover, Svedberg's measured values for platinum particles in four different liquids, as listed by Ostwald, were four to five times greater than the values he calculated from Einstein's formula. One way to interpret Svedberg's measurements, then, was that, rather than confirming kinetic theory, they were disconfirming the value for N_0 of 6×10^{23}. Ostwald chose a more charitable interpretation:

> The paths calculated from Einstein's formula are roughly one-fourth the observed paths, from Smoluchowski's formula one-third. As the theoretical value of the numerical factor is obviously not satisfactorily settled, the appropriate proportionality between the observed and calculated values (which remains unaffected by the numerical factor) may well be considered confirmation of the theory.[6]

(We include this quote in order to remind the reader of just how much the question of agreement between observation and calculation depends on the historical context in which the comparison is being made.)

An important point nevertheless emerges from this comparison. No specific value of any of the quantities pertaining to Brownian motion that we have been discussing can be derived from molecular-kinetic theory without (1) a specific value of Avogadro's number or its counterpart, the mean translational kinetic energy of the molecules, $3RT/2N_0$, and (2) the assumption that the mean translational kinetic energy of the granules is equal to the mean translational kinetic energy of molecules in gases and solutions at the same temperature. Because no supportable value of Avogadro's number was available at the time, (1) alone entails that Brownian motion could not then provide a test, in the narrow sense, of any claim derived from kinetic theory involving a specific value of a quantity! Beyond that, the assumption (2) was far from straightforward. For, granules in an emulsion, unlike the molecules of kinetic theory, are subject to such bulk effects as resistance forces from the viscosity of the liquid and buoyancy; and granules in a dilute emulsion do not constantly interact with one another in the manner of the molecules of kinetic theory.

The lack of good agreement between calculation and measurement for ξ^2 was why Ostwald then cited Perrin's vertical-gradient measurements, saying they afforded "more definite confirmation of the [kinetic] theory" than Svedberg's and provided "probably the most accurate value for the molecular constant obtained as yet."[7] How Ostwald saw the vertical-gradient results as a test of kinetic theory is unclear. On the one hand, he singles out the geometric progression of granule concentration, suggesting that the test was only of that functional relationship. On the other hand, he says that this progression depends on "the corresponding formula of the kinetic theory of gases," and he notes the extremely short distance over which the concentration drops by a factor of 2. That suggests that here too he was thinking in terms of a comparison of observed values of this distance with values calculated using the full equation and an already existing value for the mean translational kinetic energy of the molecules, which is tantamount to N_0. If the latter, then the limitations of this test are the same as those of the test of Einstein's and Smoluchowski's equation: no supportable value of N_0 was available, and the test presupposes the identity between granular and molecular kinetic energies.

Perrin's reference to an "*experimentum crucis*" construed the test in this second way, but as a comparison of his indirectly measured values of granular mean translational kinetic energy with ones calculated on the basis of kinetic theory:

> If it is to measure the magnitudes other than *W* which enter into this equation, one can see whether it is verified and whether the value it indicates for *W* is the same as that which has been approximately assigned to the molecular energy.

> In the event of an affirmative answer, the origin of Brownian movement will be established, and the law of gases, already extended by Van't Hoff to solutions, can be regarded as still valid even for emulsions.[8]

When he finally comes to compare magnitudes, however, it is not for *W*, but for its surrogate *N*, the number of granules required for their total translational kinetic energy to match $3R/2T$.[9]

> But, further, it is manifest that these values agree with that which we have foreseen for the molecular energy. The mean departure does not exceed 15 per cent., and the number given by the equation of Van der Waals does not allow of this degree of accuracy.
>
> I do not think this agreement can leave any doubt as to the origin of the Brownian movement.[10]

The specific comparison is between four values for *N* obtained from vertical-gradient measurements—75×10^{22}, 65×10^{22}, 72×10^{22}, and 70×10^{22}—with 60×10^{22}, "the number indicated by Van der Waals' equation," and the question the comparison is answering is whether the measured value for *N* "is placed in the neighborhood of that number."

Perrin's *experimentum crucis* accordingly did indeed test what we labeled earlier as an assumption, namely that the mean translational kinetic energy of the granules matches the mean translational kinetic energy of molecules, with the former measured in his vertical-gradient experiments and the latter inferred by means of van der Waals's equation from deviations of real gases from the ideal gas law. The outcome of this test Perrin construed as a Baconian fingerpost at the crossroads, so to speak, indicating that the energy displayed in the perpetual motion of the granules comes from exchanges of kinetic energy with the postulated molecules of the liquid.

The chief limitation of the test lies in the arbitrariness at the time of the 60×10^{22} number. That it is the same as the number Einstein had invoked in 1905 did not make it less arbitrary. Perrin himself, a page later, concedes as much:

> I scarcely need observe, on the other hand, that even perfect measurements of compressibility might not be able to prevent an uncertainty of perhaps 40 per cent. in the value of *N* deduced from the equation of Van der Waals, by means of hypotheses which we know are certainly not completely exact*.

The footnote to this quote then explains why even the 40 percent estimate is somewhat arbitrary:

> *Sphericity of the molecules, and various simplifications in the reasoning which lead to the expression for the mean free path, make it impossible to specify exactly what uncertainty exists in the numerical coefficients of the approximate equations which connect the viscosity, the mean free path, and the molecular diameter.[11]

Perrin makes this concession as a step in his argument for adopting the mean of the four values he has listed, 70.5×10^{22}, as the preferred value for Avogadro's number, from which preferred values for the mean translational kinetic energy of molecules at 273°K and the charge of the electron can then be obtained. (Ostwald was a little more cautious when he said that Perrin "has developed the method to such an extent that it is probably the most accurate value for the molecular constant obtained as yet."[12])

The whiff of circularity in basing a conclusion on a specific number, 60×10^{22}, which the conclusion then gives reason for replacing, need not detain us. The more interesting question is what Perrin's *experimentum crucis* showed independently of any limitations arising from the arbitrariness of that number. Before turning to that, however, we should quote what Perrin proposed that it showed:

> Thus the molecular theory of Brownian movement can be regarded as experimentally established, and, at the same time, *it becomes very difficult to deny the objective reality of molecules*. At the same time we see the law of gases, already extended by Van't Hoff to dilute solutions, extended to uniform emulsions. The Brownian movement offers us, on a different scale, the faithful picture of the movements possessed, for example, by the molecules of oxygen dissolved in the water of a lake, which, encountering one another only rarely, change their directions and speed by virtue of their impacts with the molecules of the solvent.[13] [italics in original]

The last sentence we will return to in section 5.4. The immediate question is why Perrin's *experimentum crucis* made it "very difficult to deny the objective reality of molecules."

Let's suppose for the sake of argument that Perrin's measured values for the mean translational kinetic energy of granules matched, without qualification, the values obtained via the kinetic theory of gases for the mean translational kinetic energy of molecules. What conclusion would have been licensed at the time? The mystery of Brownian motion was the source of energy sustaining the movement in spite of the resistance forces from the liquid—a *prima facie* violation of the second law of thermodynamics. External sources had largely been eliminated, and initial efforts to attribute it to molecular action had found the apparent granule velocities much too small. So, confirmation that the energies do nevertheless match would have made the molecular source the sole available candidate for explaining how the motion continues indefinitely! In that respect, therefore, a finding of matching energies would have amounted to an *experimentum crucis*, where the choice was between a successfully tested answer to the mystery and no answer at all. And denial of "the objective reality of molecules" would have been more difficult than before, if only because of the burden on anyone faced with having to propose some alternative answer.

Notice, however, that this would have been an *experimentum crucis* about the sustaining source of Brownian motion, not about molecular-kinetic theory. The mere matching of the energies by itself would have added no new information about molecules. A natural question accordingly is why this finding would have done more to establish molecular-kinetic theory than van't Hoff's finding two decades earlier that the gas law extends to dilute solutions, for that finding too had only a single candidate explanation, the action of molecules producing the osmotic pressure on semi-permeable membranes. Or, for that matter, consider Maxwell's own initial *experimentum crucis*, his verification of the counterintuitive consequence of kinetic theory that viscosity does not vary with density.[14] The point generalizes. Why should findings about the source of Brownian motion have been more telling evidence for molecular-kinetic theory than any of the prior findings bearing on the theory? The answer seems obvious: the findings on Brownian motion entailed values for molecular magnitudes in a way that none of them did. That, however, is precisely van Fraassen's point in claiming that what was special about Perrin's results was that they finally *grounded* molecular-kinetic theory—the topic of section 5.3.

Perhaps we are not being fair to Perrin in this response to his *experimentum crucis* remark. As we have noted repeatedly, before 1905 an empirical impediment to accepting molecular motion as the source of Brownian motion was the clear discrepancy between the apparent velocities of granules and the velocities of molecules implied by kinetic theory. Clausius had pointed out from the outset that the mean velocities of gases can be inferred. Meyer puts the point forcefully after deriving a kinetic-theory form of Boyle's law, $p = 1/3\rho G^2$:

> In this new form, however, it [Boyle's law] teaches us much more; it empowers us to draw a remarkable and very important conclusion. Since two of the magnitudes occurring in the formula, viz. the pressure p and the density ρ, are directly amenable to observation and measurement, the formula allows us to deduce from them the third, viz. the value of G, the mean speed of the molecules, *in absolute measure*. It was Joule who by this conception opened up to investigation a field which one would have been tempted to think was closed to human knowledge; and Clausius followed along the path thus trodden to explore an unseen world.[15]

Meyer goes on to list values of *G* at 0 deg C as a function of the square root of the specific gravity of the gas. These values were the ones deemed so much larger than the apparent speeds of granules in Brownian motion.

Now, one can indeed take this discrepancy as the outcome of a test of the claim that molecular motion is the source of Brownian motion, a test the claim had failed. Perrin says as much:

> As a matter of fact this was at first essayed, and values can be found in different papers which are always some microns per second for the mean speed of granules of the order of a micron, and which will assign to these granules a mean energy about 100,000 times less than the kinetic theory indicated for the molecules, and this would completely overthrow the theory of the equipartition of the energy.[16]

Perrin appears never to have numerically compared the mean speeds of granules implied by his values for their mean translational kinetic energy with the mean speeds of molecules obtained via Clausius's route. The nearest he comes is a highly qualified comparison of energies:

> it is manifest that these values agree with that which was foreseen for the molecular energy. The mean departure does not exceed 15 per cent., and the number given by the equation of Van der Waals does not allow of this degree of accuracy.[17]

He and his student Chaudesaigues did, however, emphasize that they had invalidated the apparent velocities.[18] Moreover, the mean granule kinetic energies Perrin obtained sufficed to eliminate any gross discrepancy between granule and molecular mean speeds. Even prior to Perrin, Einstein in 1907 had reacted to Svedberg's evidence supporting his 1905 equation for ξ^2/τ by insisting, contrary to Svedberg, that the actual velocities of the granules were not only far greater than the apparent velocities, but also beyond the scope of observation.[19]

Perhaps, then, the charitable way to construe Perrin's *experimentum crucis* remark is to take the prior experimental results to have amounted to an *experimentum crucis* against the claim that molecular motion is the source of Brownian motion; and then to take Perrin's results as having *removed* this counter-evidence, leaving molecular motion as the sole available explanation of Brownian motion. Construing it this way would make the comparison of the two energies a less stringent test of the claim that molecular motion is the source of Brownian motion.

Given the intellectual climate around 1905, however, this was all that many physicists needed to regard this claim as having been established once and for all, and with it the claim that Brownian motion had provided an *experimentum crucis* for molecular-kinetic theory. One should never underestimate the extent to which a hypothesis having the status of the sole available explanation for a phenomenon can suffice to convince a scientific community to accept it. Put more pointedly, one should never underestimate the force of the question "Whatever other explanation can there be?" when no other explanation is available. This may be the most instructive lesson to draw from the virtually universal acceptance of the ether in the first 80 years after Fresnel's wave theory of light.

We are therefore prepared to grant, as a matter of social history, that this weaker construal of Perrin's *experimentum crucis* claim may account for much of the positive reaction at the time to his results on Brownian motion, especially among those learning of them second hand. We cannot concede this, however, without calling attention to two of its ramifications. First, one should never lose sight of the fact that the phrase "inference to the best explanation" originally entered the philosophical literature as denoting a fallacy of reasoning.[20] So, at the very least, one should be asking for still further evidence. Second, to weaken the demand that granule kinetic energy be identical with molecular kinetic energy is to make the test arising from comparing the two a test merely of the compatibility of the respective mean velocities and hence a test merely of the possibility—to use Newton's term in this regard—of molecular motion being the cause of Brownian motion. And this in turn at least compromises any claim to the effect that the comparison of the two kinetic energies itself served to *ground* molecular-kinetic theory.

5.2. Testing Molecular-Kinetic Theory: Other "Tests"

So far we have been considering the evidence on the cause of Brownian motion provided solely by Perrin's conclusion, however tenuous, that the kinetic energy of granules matches that of the molecules of kinetic theory. Questions of causation will come up again in sections 5.4 and 5.5, after we have considered grounding. We are not quite finished yet with questions of testing, however. For what about Perrin's other results on Brownian motion? Didn't they too successfully test molecular-kinetic theory?

Without question, Perrin did successfully test a number of other "predictions" about Brownian motion in his experiments: (1) the logarithmic vertical-gradient in concentration; (2) the functional relation between that gradient and the buoyant weight of the granules; (3) the Gaussian distribution of granule displacements versus time; (4) the functional relation between the mean of the square of those displacements and granule size and the viscosity of the liquid; (5) the corresponding functional relation between the coefficient of diffusion and these parameters; and (6) the equipartition of energy between the translational and rotational degrees of freedom of granules. Still further, to the extent that he varied the manipulable parameters in the equations he employed to measure *N* and W_T (the mean translational kinetic energy of the granules per degree K), the stability of his measures tested the equations themselves; for that is a classic way to test equations among physical quantities, namely to test whether their constants of proportionality really are constant. Each of these "predictions" was derived, either by Einstein or by Perrin in concert with Langevin, from molecular-kinetic theory together with some auxiliary assumptions. So, didn't his successful testing of them constitute a test of at least some central claims of that theory?

Of course it did. There are nevertheless three reasons not to attach too much significance to those tests. First is the already alluded to complaint that Isaac Newton lodged against hypothesis testing as a form of evidence: "For if the possibility of hypotheses is to be the test of the truth and reality of things, I see not how certainty can be obtained in any science."[21] By not falsifying the "predictions," Perrin's tests confirmed their consistency with the considerations that entered into their derivation from kinetic theory, and that in turn left the possibility of the reality of molecules intact. But, again, why did that confirmation of conclusions derived from molecular-kinetic theory do any more than, for example, the derivations of van't Hoff's finding from kinetic theory toward settling the question of the reality of molecules and related questions?

Second is the worry raised in 1912 by Ostwald's son in the last section of Chapter 4: after granting that "experience confirms the molecule-kinetic deductions of the authors," he reminds readers of "how many purely hypothetical factors enter into" them.[22] These, of course, included the auxiliary assumptions as well as those specific claims from kinetic theory that entered into the deductions. So, what was tested, some specific claims within kinetic theory or some ad hoc assumptions needed to get from the invisible hypothetical to the visible? To give one particular instance of this sort of worry, Perrin reminds the reader more than once that his evidence was for the Gaussian distribution of granule *displacements*, not for the thesis at the heart of statistical mechanics, the Maxwell-Boltzmann distribution of molecular *velocities*.

Third, and most important, by paying attention to the specific contributions made by the auxiliary assumptions, we have shown in Chapter 4 that no hypotheses from molecular-kinetic theory were categorically required for Perrin's results on Brownian motion itself. Not only could he have started from hypotheses about Brownian motion, he in fact developed comparatively direct evidence for those hypotheses. This does not mean that his results did not test hypotheses from molecular-kinetic theory that he and Einstein employed in their derivations. It means only that the evidence from those tests was quite limited, showing at most the consistency of Perrin's results with those hypotheses. Those hypotheses were nonetheless eliminable from the logical design of the experiments without undercutting any of the results! We grant full well that tenets of kinetic-molecular theory had a historically indispensable role in motivating and guiding the logical designs; that is, as a matter of historical fact they were indispensable *heuristically*. But that is the mark of heuristic, in contrast to constitutive, elements in the design of an experiment. They can be eliminated and the results remain intact.

Recall that in the earlier editions of his *Grundriss*, Ostwald always stressed the enormous heuristic value of molecular and atomic theory. What he demanded beyond that, as he made clear in his 1907 *Monist* paper, were theory-mediated measurements of the quantities within that theory independently of one another, for otherwise meaningful tests of proposed relationships among those quantities could not be carried out. Under the premise that the mean translational kinetic energy of the granules matches that of the molecules of kinetic theory, Perrin's Brownian motion experiments did provide measurements of the latter energy and hence too of Avogadro's number; and that opened the way to more reliable theory-mediated measurements of other quantities within molecular-kinetic theory. But now once again, though this time more from Ostwald's point of view than van Fraassen's, we find ourselves concluding that the contribution made by Perrin's experiments was not one of testing molecular-kinetic theory, but of grounding it.

One last note before turning to grounding. Our list of "predictions" about Brownian motion that Perrin successfully tested did not include the *prima facie* historically most notable "prediction" of them all, $2D= \xi^2/\tau$, with the root-mean-square granule displacement ξ (as we would now say) a random variable with a Gaussian distribution. The equation itself can be obtained merely by integrating the diffusion equation—$\partial n/\partial t = D\nabla^2 n$—without regard to whether the displacements are of granules or molecules, so long as the (average) displacements ξ of localized elements or "particles" of the fluid are well defined.[23] Without question, however, the claim that the granule displacements are random, in the technical sense given earlier, with a Gaussian distribution, would never have been put forward in the absence of molecular-kinetic theory. Without question too, the Perrin-Brillouin evidence that granule diffusion is a statistical phenomenon had a major impact on attitudes toward molecular-kinetic theory—if only through showing that such a statistical phenomenon really can occur in the world, and not just in theory.

Clear evidence of this impact can be found in a remark by Paul and Tatiana Ehrenfest in their landmark 1912 defense of Boltzmann, *The Conceptual Foundations of the Statistical Approach in Mechanics*:

> The above statement [a "visible" state S_A at time t_A completely determines the "visible" state for any subsequent time t_B] is therefore a postulate, which goes considerably beyond the possibility of an experimental check. Our observations cannot tell us anything about what the sequence of state of an actual isolated gas quantum would be over a very long period of time and whether it would satisfy the principle of determinacy.
>
> As soon as we include a microscope among the instruments of observation of the "visible" state, we discover the Brownian motion. For such a more precise "visible" state the principle of determinacy does not seem to hold any more. On the other hand, it has been shown that this phenomenon is surprisingly well suited to a statistical interpretation.[24]

For Brownian motion to be a proven counterexample to the principle of determinacy in mechanics is nevertheless one thing; evidence pertaining to the indeterminacy of molecular motion is another.

Mayo is certainly correct in the emphasis she has put on Perrin's conclusion that displacements in Brownian motion are irregular, with a Gaussian distribution. And she is also correct in stressing how Perrin's evidence for this conclusion consisted of a series of tests of a statistical hypothesis—indeed, taken together, a quite stringent test.[25] The hypothesis that Perrin tested, however, pertained to the displacements of granules, not to the Maxwell-Boltzmann claim about the statistical distribution of the velocities of molecules. For all its heuristic value in Einstein's hands, the latter claim was not ineliminably presupposed in Perrin's test of the Einstein-Smoluchowski equation, and hence by definition it was not a constitutive element of that test. We shall return to the important relationship between the "observed" statistical character of Brownian motion and the postulated statistical character of molecular motion in section 5.4. Our point here is simply that it is a mistake to construe Perrin's test as a decisive or even particularly stringent test of molecular-kinetic theory.

5.3. Grounding Molecular-Kinetic Theory

Van Fraassen contends that the resistance to molecular-kinetic theory before Perrin's experiments came not from molecules being invisible, but from the theory not being "empirically grounded." To remind the reader, three requirements have to be met in order to ground a theory. First, the significant parameters of the theory must have definite values obtainable through measurement—usually indirect measurement involving theoretically posited connections between those parameters and parameters amenable to measurement that does not presuppose the theory, or at worst presupposes it only weakly. This requirement corresponds to what we called the demand for *stability* of any measure in Chapter 4. Second, measured values for any parameter obtained through different theoretically posited connections must agree with one another, at least to within the limits of experimental precision. This requirement corresponds to the one we called *convergence* of measures in Chapter 4. Third, the outcome of the measurements cannot be guaranteed in advance: there must be an alternative possible outcome for the same measurements to have refuted the theory on the basis of the same theoretically posited connections. This corresponds to our point that the stability of any measure must be established inductively and hence is always open to future counterexamples.

Accordingly, although our focus was on the issue of evidence deriving from well-behaved theory-mediated measurements, we see ourselves in full accord with van Fraassen's requirements. In a subsequent paper, van Fraassen has argued that "what counts as a measurement at all is theory relative"[26] insofar as the theory must itself validate that the measurement procedure is properly tied to a well-constituted parameter of the theory. We see no reason to disagree with him about this either. So, we think we are on common ground with him in posing two questions, both historical. Was the resistance within the scientific community to molecular-kinetic theory before Perrin's experiments at least primarily because of its lack of grounding? And was it the grounding of the theory provided by those experiments that to all intents and purposes ended the resistance, at least within the community of those engaging in research on the fundamental physics and chemistry of matter?

The answer to the first question is straightforward. Van Fraassen provided a sufficient sketch of the prior history to show that molecular-kinetic theory had not yet become empirically grounded at the end of the nineteenth century, notwithstanding the many efforts toward that end. We have supplemented van Fraassen's account by quoting, in section 2.4 of Chapter 2, Ostwald's clear pronouncement to the same effect in his 1907 *Monist* paper. Faced with the task of giving a philosophically precise statement of the requirements that molecular-kinetic theory had not yet met, Ostwald singled out the demand for proper independent measurement of the quantities in the functional relations of the theory in order for those proposed relations to be tested. Ostwald was one of the four seminal figures of the community of physical chemists, and hence what he said in 1907, especially given the care with which he phrased it, can be taken as a statement he viewed as reflecting the standards of at least that community.

The answer to the second question is not so straightforward. In granting the "atomic hypothesis" has been "raised to the position of scientifically well-founded theory," in the Preface to the 1909 edition of his *Grundriss*, Ostwald singles out "the isolation and counting of gas ions" and "the agreement of the

Brownian movements with the requirements of the kinetic hypothesis" as "experimental proof of the atomic nature of matter." The phrase "scientifically well-founded theory" is clearly reminiscent of his *Monist* paper of the preceding year, but how the agreement of Brownian motion with the kinetic hypothesis constitutes grounding is not immediately clear. The discussion of Brownian motion, added to the *Grundriss* for the first time in the 1909 edition, singles out first the order of magnitude agreement between Svedberg's observed granule displacements and those calculated from Einstein's and Smoluchowski's theories and then Perrin's determination of N_0, noting that it is "probably the most accurate value for the molecular constant obtained as yet."[27] The emphasis on N_0 supports van Fraassen.

One might accordingly expect Ostwald's chapter on kinetic theory a few pages later to show how the value for N_0 serves to determine values for other molecular parameters, opening the way to the kind of testing of molecular functional relations he had demanded of scientifically well-founded theories in his 1907 *Monist* paper. Other than moving the chapter from early in the book to late, Ostwald left it virtually word-for-word the same until near the end.[28] In particular, Ostwald reviews the estimates that had been made from Clausius forward of the speeds of molecules, their mean free paths, and their sizes ("supposed cubical in form"). The only number that changes is the size of a hydrogen molecule, from 6×10^{-8} to 1.6×10^{-8} cm, and that only because Ostwald has accepted van der Waals's insistence that the volume of a molecule derived from his formula represents 4 times its physical volume owing to the effects of its motion. Notice here the uncertainty that still remained not just about the value, but more tellingly about what physical quantity was being determined!

At this point Ostwald's chapter adds a new paragraph, followed by a new subsection. The new paragraph first notes that the sizes inferred for other molecules "are usually a little smaller, but are generally independent of the atomic weight and the complexity of the substance," adding that they are too uncertain to merit listing. The paragraph then explains why the size matters:

> Still, "the dimensions of molecules" thus found have a physical significance quite apart from kinetic theory. They are the dimensions below which the properties of the substances become different from those exhibited by the quantities we are accustomed to deal with.[29]

The subsection that follows concerns a question not raised in the earlier chapter, the number of molecules in a mole. Ostwald first derives a value of "$.85 \times 10^{24}$, or in round numbers 10^{24}" from the 1.6×10^{-8} cm side of the hydrogen cube; he then concludes:

> Calculations of this fundamental value of the kinetic theory, based on various other data, yield results which are a little higher or a little lower than this. The most probable value is $.71 \times 10^{24}$ (Perrin, 1908).[30]

This is the closest Ostwald comes in the book to indicating how Perrin's results served to ground molecular-kinetic theory or, as he said in the Preface, to turn it into "a scientifically well-founded theory." He adds nothing about whether Perrin's value for N_0 can help clarify molecular sizes and, with them, the micro-physical properties of substances that chemists had long sought.

As we noted at the very beginning of this monograph, Ostwald was not alone in citing Perrin's value for N_0 as a milestone for molecular-kinetic theory. Poincaré remarked that its first significant figure had been determined, completing "the triumph of atomism." Nernst authorized his translator to add Perrin's determination, 71.0×10^{22}, in the English edition of his 1909 German *Theoretische Chemie*,[31] and subsequent editions cited the values of N_0 as important "fresh proof of the correctness of the theory of thermal motion." So, van Fraassen is surely correct in claiming that Perrin's comparatively precise value had a major effect on the attitude toward molecular-kinetic theory. Our question, however, is whether van Fraassen is correct in claiming that it was the grounding of the theory provided by Perrin's value that ended the resistance to it. Our examination of Ostwald has raised a prior question: To what extent did that value actually ground the theory?

The place to start answering this question is with the list of molecular quantities that Perrin claimed in 1909 to have determined from his value of Avogadro's number, a list that Ostwald had seen in a prior version in Perrin's October 1908 paper at the time he was finishing the new edition of his *Grundriss*. Perrin's list does not stop with the quantities straightforwardly derivable from N_0, the *constant of molecular energy* (in cgs units, 1.77×10^{-16})[32], the *mean translational molecular kinetic energy* at 273°K, and the *masses* of such molecules and atoms as oxygen and hydrogen. He turns to the Clausius-Maxwell equation for mean free path,

$$L = \frac{1}{\pi\sqrt{2}} \frac{V_0}{N_0 s^2}$$

(where V_0 is the volume of a mole) to obtain values for the "diameters" in centimeters of the molecules of helium (1.7×10^{-8}), argon (2.7×10^{-8}), mercury (2.8×10^{-8}), hydrogen (2×10^{-8}), oxygen (2.6×10^{-8}), nitrogen (2.7×10^{-8}), chlorine (4×10^{-8}), and ether (6×10^{-8}). He ends the list with a cautionary note:

> It is clear that, in the case of polyatomic molecules, one can only deal with a badly defined diameter, of which the determination, although but little affected

> by variations in mass, cannot from its nature have the certainty possible for the masses.[33]

This cautionary note is the key to answering our question about the extent to which Perrin's value for N_0 grounded molecular-kinetic theory. He was using the same mean-free-path formula as Ostwald. The mean free paths of molecules were regarded as known to within experimental accuracy thanks to the agreeing values inferred from the viscosity, diffusion, and heat conduction in different gases.[34] The equation relating them to the sizes of molecules, however, presupposes that the molecules are spherical. So, even with the value of N_0 in hand, Perrin was still stuck with having to make an assumption about the shapes of molecules. He lists monoatomic molecules first because they are the only clear candidates for being spherical. Van Fraassen emphasizes the recognized arbitrariness of the sphere assumption in his argument that molecular-kinetic theory was not yet grounded before Perrin. Perrin's value for N_0 did not eliminate the need for that assumption. Hence there were limits to the extent to which it grounded the theory.

Perrin's value nevertheless did amount to a gain toward grounding the theory. Ostwald had turned to van der Waals's equation to derive his value for the size of the hydrogen molecule, and then derived from it a value for N_0. With his value for N_0 in hand, Perrin *had no need for van der Waals's equation, or any other such supplementary relationship, in arriving at his values for the sizes*. This is the one respect in which this fundamental parameter of the theory was more grounded as a result of Perrin's value for N_0 than it had been before. Was that enough to end the resistance to molecular theory? We are not sure.

The continuing worry over the sphere assumption was voiced at the time. A decade later, Nernst repeated Perrin's calculation for the molecules in the list (excepting ether), but using instead Millikan's value for N_0, 60.64×10^{22}, and calling the sizes obtained "the spheres of influence affecting the collision of two gas molecules." He prefaced his list by cautioning,

> The question of the sphere of influence of the molecular forces, on the contrary, is less definitely settled [than the diameter of the hydrogen nucleus]. The size of the sphere of influence was formerly considered to be identical with the atomic or molecular diameter. From the remarks on p. 508 it is evident that the sphere of influence is variable, but it is of interest at any rate to calculate its value under given conditions.[35]

The reference to page 508 concerns replacing the two-billiard-ball model of molecular impact with one in which, in the light of post-Bohr atomic models, "mutual forces come into play while the molecules are still some distance apart." So, not just the question of shape remained hanging after Perrin; so too did the question van der Waals had raised concerning just what *s* in the equation for the mean free path represents physically.

These remarks by Nernst were made in passing. The first book-length review article of the kinetic theory of gases (at least in English) after 1913, the 1916 edition of James Jeans's *The Dynamical Theory of Gases*, makes the point far more forcefully in its opening chapter:

> It is a matter of some difficulty to determine or even to define the size of a molecule. The trouble arises primarily from our ignorance of the shape and other properties of the molecule. If the molecules were known to be elastic spheres the question would be simple enough, and the size of the molecule would be measured by the diameter of the sphere. If, however, the molecules are *assumed* as a first approximation to be elastic spheres, experiment leads to discordant results for the diameters of these spheres, shewing that the original assumption is unjustifiable. The divergences arise not only from the fact that the shape of the molecules is not spherical, but also from the fact that the molecules are surrounded by fields of force, and in most experiments it is the extension of this field of force, rather than that of the molecules themselves, with which we are concerned.[36]

The Lennard-Jones potential that gives an approximate model of the field of force in question dates from 1924. The problem of calculating the field from the configuration of molecules, even for helium, remained a challenge well into the 1980s.[37]

Both Ostwald and van Fraassen demand empirical grounding of the parameters of a theory in order for testing of it to be possible at all. The doubts we are raising concern how much more effectively molecular-kinetic theory could be tested with Perrin's value for N_0 in hand than before. Van Fraassen ends his paper by concluding,

> To properly credential the theory, the procedures that count as tests and measurements in the eyes of the theory must provide an empirical grounding for all its significant parameters. The completion of this task, which made the kinetic theory into, at least in principle, a truly empirical theory, was Perrin's real achievement.[38]

Suppose we grant without qualifications Perrin's value for N_0—which, we shall see shortly, those at the time did not do. His results still did not provide an empirical grounding for *all* of the significant parameters of molecular-kinetic theory. It is not even obvious how they made the theory much more testable than it was before.

One exception needs to be noted to this last remark. Once a value for N_0 is available, it can be plugged into formulas like the one given earlier for the mean free path to reduce the unknowns to just *L* and *s*, under the assumption, of course, of spherical molecules. Jeans does this in his 1916 edition in order to compare the theoretical characterization of viscosity, heat conduction, and diffusion of gases with experimental values. The upshot of those comparisons can be read as clearly experimentally establishing that molecules are *not* spherical.[39] In this respect, therefore, having a value for N_0 did permit a claim to be more definitively tested than it had been before, though only with a negative outcome. The question of the specific features of different molecules and their relationship to the variability of viscosity and diffusion from one gas to another—emphasized in our section 2.7 of Chapter 2—did not, however, become more susceptible to experimental investigation.

Although less in keeping with van Fraassen's general viewpoint, there is another way of making our point about the limits to which Perrin's value for N_0 grounded the theory. Van Fraassen and Ostwald view the challenge posed by research into the microphysics of matter as one of achieving well-founded theory-mediated measurement of the parameters of a theory. That view can be turned on its head. The challenge—whether of research into microphysics or of research into the deep structure of the Earth—is one of *gaining empirical access* to the realm in question by means of theory-mediated measurements. From this viewpoint, Perrin's value for N_0, *taken at face value*, provided more access to the molecular realm than had been available before, in large part through providing a value for Boltzmann's constant $k = R/N_0$ that occurs ubiquitously in the form kT throughout statistical thermodynamics. This alone constrained the answers to other questions. Yet it did not by itself provide sufficient access for experiments to answer a wide range of questions central to microphysics and chemistry that Ostwald and Nernst had left hanging in their respective books. At most, therefore, it was merely a step forward.

We emphasized the phrase "taken at face value" in the preceding paragraph because gaining access through theory-mediated measurement always demands cross-checks on the measurements. Without them, the apparent gain in access may be illusory. This explains why both Ostwald and Nernst, while heralding Perrin's value, remained so cautious in their initial reaction. That a controversy, spearheaded by Millikan, ensued after 1911 over whether the value of N_0 is closer to 60×10^{22} than to 70×10^{22} was entirely appropriate. Perrin's celebration of the extent of the agreement among the diverse determinations of N_0 at the end of his 1909 monograph metamorphosed soon thereafter into a dispute about their disagreement.

A clear indication of the uncertainty about which value for N_0 should be adopted can be seen in the 1916 edition of Jeans's *The Dynamical Theory of*

Gases. Its opening chapter lists the available values for Loschmidt's number: 2.77×10^{19}, 2.70×10^{19}, and 2.68×10^{19} from the Rutherford-Geiger, Regener, and Millikan determinations of the fundamental charge *e*, respectively; 2.76×10^{19} from Planck's blackbody determination of Boltzmann's constant; and 3.06×10^{19}, 3.09×10^{19}, 2.92×10^{19}, and 3.10×10^{19} from Perrin's four Brownian motion determinations. Jeans then adopts 2.75×10^{19}, corresponding to a value of N_0 of 61.2×10^{22} "for the purposes of calculations in the present book."[40] By contrast, in the next edition of the book five years later, Jeans adopts 2.705×10^{19}, corresponding to 60.62×10^{22}, for these purposes, based on Millikan's 1913 oil-drop value for *e*, reissued with tighter bands of uncertainty in 1917. He then adds the following remark about Perrin's values, which he no longer lists:

> Another determination of N_0 can be made from observations on the Brownian Movements. The method was first developed by Perrin, who obtained values uniformly larger than Millikan's value given above. More recent observations by H. Fletcher give the value $N_0 = 6.03 \times 10^{23}$, with a probable error of about 2 per cent. of the whole.[41]

From the viewpoint we have been briefly considering, the dispute over the value of N_0 was raising questions about whether Perrin's measurements of the mean kinetic energy of granules in dilute emulsions was truly providing access to the mean kinetic energy of molecules. Van Fraassen refers to this as Perrin's "main hypothesis,"[42] and we have spoken of it as an assumption. Both Perrin and Ostwald, however, treat it as falling somewhere between a hypothesis and a result. For, they invoke it—in the form of the number of granules *N* required for their total kinetic energy to match *RT* for a mole of gas—in the context of the order of magnitude values for N_0 obtained from kinetic theory from the combination of mean free paths and van der Waals's equation. In other words, they present the value of *N* from Brownian motion, and hence the measured values of the mean kinetic energy of the granules, as more precise determinations of molecular quantities for which order of magnitude estimates were already available. And at least part of their grounds for doing so was that the Brownian motion values fell within the range of those estimates.

The question is whether this was grounds enough to insist that the mean kinetic energy of granules was truly providing access to the mean kinetic energy of molecules. Granules in Brownian motion not interacting with one another are one thing, a mind-boggling number of molecules in motion interacting with one another at a mind-boggling rate is another. Asking for more of a cross-check of the match between the two energies than merely order of magnitude agreement was not inappropriate. One cross-check, as Ostwald intimated in citing J. J. Thomson's results for the electron, could be supplied by comparison with values for N_0 derived from phenomena other than Brownian motion. We will consider how those values contributed to the change in attitude toward molecular theory in Chapter 6. Our question here is how much the value Perrin derived from Brownian motion, by itself, contributed to this change.

Among the most important features of Perrin's value was the cross-check his experiments had provided by obtaining values for N and hence granular energy in such good agreement with one another from three different phenomena—vertical-gradients, granule displacements, and diffusion—not to mention the nearby value from granule rotations. That agreement, however, provided a check only on whether the parameter N has a definite value and whether it represents what it is being said to represent, namely the number of granules required for their total energy to match the characteristic total energy RT of a mole of gas. It did not provide a cross-check on whether $N = N_0$, nor correspondingly whether the energies match. Ostwald thus had two complementary reasons for referring to Perrin's value as "*probably* the most accurate value of the molecular constant *as yet*." On the one hand, it had claim to being a determinate value; on the other hand, its being the value for Avogadro's number was still awaiting further confirmation.

We need to add a brief remark about how the "gaining access" viewpoint we have adopted in the last few paragraphs contrasts with van Fraassen's "grounding" viewpoint. The point of grounding a theory from his perspective is to provide values for its parameters sufficient to enable its "empirical adequacy" to be thoroughly tested, usually hypothetico-deductively—that is, by testing of empirically accessible claims derived from it. His focus is on theory as an end-product. From a "gaining access" viewpoint, theory is not an end-product so much as it is an instrument that enables questions about empirically less accessible matters to be answered by specific empirical observations and experiments. Evidence accrues to a theory being used in this way through the definiteness of the answers achieved, as illustrated by the evidence supporting the validity of Perrin's measures coming from his results for the mean kinetic energies of granules being well-behaved. Bohr's 1913 model and its subsequent refinements are best viewed from a "gaining access" viewpoint, for they enabled spectral data that had been gathering for three-quarters of a century to become evidence bearing on questions of atomic structure. So too is the theory of x-ray diffraction developed by the Braggs contemporaneously with Bohr's model, insofar as it enabled questions about the atomic and molecular structures of crystals to receive empirically dictated answers.

That said, we nonetheless shifted to the "gaining access" viewpoint not to draw a contrast with van Fraassen, but in an effort to clarify our point about the limits to which Perrin's value for Avogadro's number grounded molecular-kinetic theory. We now see that the grounding of the theory that it provided was also, if not tenuous, then at best tentative, pending further developments. All of this has been in response to the question, was it the grounding provided by Perrin's value—and this by itself—that ended the resistance to molecular-kinetic theory in the years between 1905 and 1913? Given the limited and tentative character of the grounding and the diversity of the community, this turns out to be a complex sociological question requiring resources beyond those we have at hand to answer. Nevertheless, whether phrased in terms of grounding molecular theory or gaining empirical access to the microphysical realm, the question has at its core an epistemic issue. And van Fraassen is surely not wrong in stressing the importance of the grounding of molecular-kinetic theory that Perrin's value provided in ending resistance to it. All we are saying is that reason remains to question whether more than just this element of grounding was involved in ending the resistance.

Our intent in using the phrase "end resistance" has been to delay (until Chapter 7) the question of precisely what the change in attitude toward molecular theory that occurred between 1905 and 1913 amounted to. Van Fraassen submits his answer to this question: Perrin's "work laid to rest the idea that it might be good for physics to opt for a different way of modeling nature, one that rivaled atomic theories of matter."[43] If this is all that the change in attitude amounted to, then perhaps it may have been Perrin's value for N_0 by itself that produced the change. But that still leaves us wondering just what it was about this value, given the others that had been put forward before it, that made the difference. Readers will recall Steven Brush's remark quoted in Chapter 1: "The fact that one could determine Avogadro's number and the charge on the electron by one more method seems hardly sufficient to justify such profound metaphysical conclusions." Is it any more clear at this juncture of our discussion of Perrin's efforts why it was sufficient to justify Ostwald's abandoning his "energetics" approach?

Van Fraassen's proposal for what the change in attitude amounted to is expressly intended to deny that it involved profound metaphysical conclusions. This is clear from the paragraph that immediately precedes the proposal quoted earlier:

> It is still possible, of course, to also read these results as providing evidence for the reality of molecules. But it is in retrospect rather a strange reading—however, much encouraged by Perrin's own prose and by commentaries on his work in the scientific and philosophical community. For Perrin's research was entirely in the framework of the classical kinetic theory in which atoms and molecules were mainly represented as hard but elastic spheres of definite diameter, position, and velocity. Moreover, it begins with the conviction on Perrin's part that there is no need at his late date to give evidence for the general belief

> in the particulate character of gases and fluids. On the contrary (as Achinstein saw) Perrin begins his theoretical work in a context where the postulate of atomic structure is taken for granted.[44]

Contrast this, however, with Ostwald's remark: "on the one hand . . . the brilliant researches of J. J. Thomson and, on the other, the agreement of the Brownian movements with the requirements of the kinetic hypothesis, established by many investigators and most conclusively by J. Perrin, justify the most cautious scientist in now speaking of *the experimental proof of the atomic nature of matter*."[45] The issue of the "reality of molecules," whether "metaphysical" or otherwise, cannot be dismissed quite so easily. Nor should it be, for after all the title of Perrin's 1909 monograph was "Mouvement brownien et réalité moléculaire."

The point van Fraassen makes about Perrin beginning his work in a context in which molecules were being taken for granted, in some respect or other, is beyond dispute. For, according to Perrin himself, the pivotal question of his work was whether the indirectly measured value for the mean kinetic energy of granules in Brownian motion "is the same as that which has been approximately assigned to the molecular energy."[46] But the question of the magnitude of molecular energy can be raised absent any conviction of there being no need at the time "to give evidence for the general belief in the particulate character of gases and fluids." In particular, it can be raised as one among a group of conditional questions—that is, questions under a supposition:

> Suppose gases and liquids consist of discrete elements, which for the moment we shall call "molecules." These elements should have mass and kinetic energy and occupy a distinct region of space, and there should be a definite number of them in any given quantity of gas or liquid. So, what is the mass of a molecule of, say, hydrogen, what is its size and its shape, what is its kinetic energy, and how many of them are there in a gram of hydrogen?

Thus formulated, these questions can be put forward as a challenge to molecular theory and hence not as ones already committed to it.

We fully grant van Fraassen that resistance to molecular-kinetic theory at the time came from the lack of empirical grounding of its significant parameters. Notice, however, that the demand for empirical grounding, as he lays it out, is a demand for *empirical* answers to questions of the sort our preceding list illustrates. Those questions have a referential presupposition—that is, a necessary condition for them to have answers at all: if gases and liquids do not consist of discrete elements—"molecules"—then the questions will have no answers. Suppose now that experiments are designed that yield robust answers to those questions. Then concomitantly the evidence that those answers are robust is evidence that the referential presupposition of those questions is true. And experiments that fail to yield robust answers—for example, *What is the velocity of the Earth through the ether?*—give reason to reject the referential presupposition. In other words, while one can ask whether the mean kinetic energy of granules is the same as the mean kinetic energy of molecules without being committed to molecules, nevertheless any experiment giving evidence that the energies are the same is concomitantly evidence for the "reality" of molecules. The scare quotes are added here because what the word "molecule" refers to was so underspecified. The whole point of "question-answering" experiments that would ground the theory was to reveal what, if anything, the word refers to.

Both Perrin and Ostwald were eminently aware of an example of just such an approach to establishing the "reality" of hypothesized elements, namely the efforts of J. J. Thomson over the preceding decade in establishing the particulate character of electronic charge. He initiated his 1897 experiments on cathode rays in the context of a dispute over whether they consist of discrete negatively charged particles or are instead some further sort of electromagnetic propagation. His experiments were designed to measure the mass-to-charge ratio of the putative particles. Those experiments did not yield a definite value for that ratio, but they did show it (1) to be of an extraordinary order of magnitude (three orders of magnitude less than the mass-to-charge ratio of the hydrogen ion known from electrolysis), and (2) to have no systematic variation from one cathode ray to the next (that is, with variations in the gas and the electrode material in the cathode ray tube). This led Thomson to conclude that cathode rays do consist of discrete particle-like elements, "matter in a new state" theretofore unknown. Subsequent experiments revealed elements of the same order of magnitude mass-to-charge ratio showing up in other electrical discharges, leading him to the conclusion with which he opened his Nobel Prize lecture in 1906:

> The carriers of negative electricity are bodies, which I have called corpuscles, having a mass very much smaller than that of the atom of any known element, and are of the same character from whatever source the negative electricity may be derived.[47]

By that year their mass-to-charge ratio had been determined to at least three significant figures, and, even though many questions about the electron remained open, including its precise value of charge, the question of the particulate nature of electricity—or, in other words, the question of the "reality" of electrons—had come to be regarded as closed.[48]

In citing them together, was Ostwald saying that Perrin's experiments had done the same thing for the particulate character of matter that Thomson's had done for the particulate character of electronic charge when he spoke of "the experimental proof of the atomic nature of matter"? Perhaps, but we suspect not. Perrin did extract answers to several questions about molecules from his results for Brownian motion. Those answers, however, presupposed that granular mean kinetic energy is the same as molecular mean kinetic energy. That presupposition, we have argued, was still calling for further confirmation. As a consequence, Perrin's answers to questions about molecules, we claim, were at the time only promising *candidates* for answers. They stood in marked contrast to the robust answers—some of them historically monumental—his experiments had provided to questions about Brownian motion itself.

In particular, we do not see how Perrin's answer to the question of the mean kinetic energy of granules in Brownian motion by itself had yet clearly bridged the gap between the visible realm of granules and the microphysical realm remotely to the extent that the research on the electron had. But this is just another way of saying what we said earlier in opposition to van Fraassen: the extent to which Perrin's value for the granular energy grounded molecular-kinetic theory was at most limited and tentative. If that alone was enough for Ostwald, then he must have been far more favorably predisposed toward molecular-kinetic theory than his 1907 *Monist* paper indicates.

The contrast between experiments designed to answer questions about granules in Brownian motion and ones designed to answer questions about molecules leads to a final point on which we may be disagreeing with van Fraassen. He sees the "enrichment" of molecular-kinetic theory "through additional theoretical hypotheses" over the course of the nineteenth century as an effort to make empirical grounding of its parameters possible. We do not disagree with this, nor do we disagree with extending this claim to other cases in which theories have been enriched by adding further hypotheses over time. We do, however, disagree with the intimation—van Fraassen never explicitly says this—that the theoretical efforts on Brownian motion leading to Perrin's experiments were an enrichment or extension of molecular-kinetic theory. We grant that the theory of Brownian motion that emerged was heuristically guided by molecular-kinetic theory, and that it was developed with the express purpose of obtaining decisive evidence for that theory. Nevertheless, as we have shown in Chapter 4, the theory of Brownian motion could be formulated and tested entirely independently of molecular-kinetic theory. Indeed, it was in fact formulated by Perrin and Langevin and tested by Perrin independently of that theory. To intimate otherwise is to understate, if not misrepresent, the force of the evidence Perrin developed for his conclusions about Brownian motion itself.

Consider, in particular, two conclusions about Brownian motion that Perrin emphasizes, yet van Fraassen scarcely mentions: (1) the displacements versus time of the granules are random, with a Gaussian distribution that governs their diffusion; and (2) the rate of vertical rarefaction of the granules varies linearly with their buoyant weight and temperature, that is, the conclusion the Nobel Committee labeled Perrin's law of sedimentation. Each of these, of course, has a parallel counterpart in kinetic theory. Perrin's evidence for them, however, came from measurements involving only the granules themselves. His evidence for (1) supported the claim made by Einstein and others that the granules were undergoing changes in velocity more rapidly than the ultramicroscope-aided eye can detect, a conclusion that was supported by the measured values of the mean translational kinetic energy of the granules. And the evidence for (2) had to include evidence that their mean translational kinetic energy varies only with temperature, and not with any characteristic of the granules themselves. None of this evidence presupposed molecular theory; nor, for that matter, did Perrin's evidence for the conclusion of equipartition of energy between translational and rotational degrees of freedom of granules in Brownian motion.

Perrin, needless to say, emphasized the parallels between these three conclusions and their counterparts within molecular-kinetic theory in his arguments in both the 1909 monograph and the 1911 paper at the Solvay Conference, and then too in his *Les atomes*, for the "reality of molecules." More important for us, in the section on Brownian motion that Ostwald added in the 1909 edition of his *Grundriss*, he cites the parallel between (2) and its counterpart in the atmosphere; and even with only Svedberg's initial evidence in hand, he says that the agreement between calculated and observed granule displacements "is so close that we are compelled to regard this agreement as fairly satisfactory proof of the kinetic nature of heat"—this notwithstanding "the very serious difficulties encountered" by that theory since Clausius, Maxwell, and Boltzmann had developed it.[49]

Perrin also constantly calls attention in all three of his extended presentations to the thought that granules in Brownian motion amount to nothing more than molecules in the large, insofar as the next step beyond the smallest granules he tested is to exceptionally large molecules. Strikingly, Ostwald expresses the same thought at the end of his chapter on disperse systems, the chapter containing his new section on Brownian motion:

> The kinetic theory places the size of gas molecules at about 10^{-8} cm. There is thus unbroken continuity. The limit of spatial discontinuity which is physically observable is still lower, for electrical phenomena take us at least a further power of ten.[50]

Perhaps the *parallels* between Brownian motion and the theoretical motions of molecules and the *continuity* between the visible realm of the granules and their invisible realm were factors, together with Perrin's value for N_0 being "probably the most accurate obtained as yet" that led him to acknowledge "the experimental proof of the atomic nature of matter." It is to these two we turn in sections 5.4 and 5.5, postponing until Chapters 6 and 7 questions about how Ostwald might have seen J. J. Thomson's estimated value for Avogadro's number to have contributed.

We should not wait until section 5.5, however, to point out how misleading the suggestion was that Brownian motion is continuous with molecular motion, and hence that Brownian motion can be thought of as molecular motion writ large and its granules as outsized molecules. The minor respect in which this is misleading is that Brownian motion occurs only in sufficiently dilute emulsions for the granules not to interact with one another. The more serious respect lies in the picture of a granule interacting with an individual molecule, or even a handful of molecules, at any instant. The smallest granules for which Perrin published results in which he expressed confidence were 0.212 μ in radius, that is, 0.212×10^{-4} cm. Consider now, as we did in Chapter 1, the spherical shell of thickness equal to that radius encapsulating any such granule, that is, a shell of radius 0.424×10^{-4} cm. The liquid within this shell has *prima facie* claim to being the portion of the substrate with which the granule is interacting at any time. But by Perrin's announced numbers for Avogadro's number, which imply a Loschmidt number around 30×10^{18}, the number of molecules in that amount of liquid would be greater than 8 million were it a gas at standard conditions. It was not a gas, however, but in the case of the 0.212×10^{-4} cm radii granules in question, it was water consisting of more than 33×10^{21} molecules per cubic centimeter, so that the number of molecules within one radius of each granule was around 9 billion. The interaction of granules in Brownian motion with their liquid substrate, therefore, cannot remotely resemble what a billiard-ball picture conceives it to be. Brownian motion is anything but molecular motion writ large.

Although not essential to the narrow point we are making, a broad-brush picture of how we are disagreeing with van Fraassen here may help readers at this point. In saying that measurement and evidence are relative to theory—a claim on which we do not necessarily disagree—he seems to be leaning toward a holistic view of evidence for theories; that is, theories as a whole are and are not "empirically adequate." Perrin's results, on this view, served to ground kinetic theory, but their doing so was only a step that finally made the theory properly testable. The issue then became the overall empirical adequacy of this theory. This view in turn can lead naturally into the view that modern molecular-kinetic theory, for all its empirical adequacy, is but one among many possible empirically adequate representations of the phenomena it covers.

While we will return to van Fraassen's general view in Chapter 7, we are not, as such, contesting it here. We are, however, insisting that taking Perrin's results as presupposing kinetic theory and hence as merely opening the way to assessing its overall empirical adequacy grossly misrepresents the force of the evidence they provided concerning Brownian motion itself. Precisely because those results did not presuppose molecular-kinetic theory, they established some extraordinary conclusions about Brownian motion. Those conclusions did not in any way have to await further assessments of the empirical adequacy of kinetic theory.

Insofar as van Fraassen's account of Perrin is what gave rise to our study in the first place, we should summarize our assessment of it before we go on. We grant van Fraassen that resistance within the research community at the time to molecular-kinetic theory, as typified by Ostwald, was from insufficient empirical grounding of its parameters. We also grant that the most important contribution Perrin's results made to ending the resistance at the time was his measured value of the kinetic energy constant for granules, our W_T, for it provided the most promising candidate to date for grounding its molecular counterpart, Boltzmann's constant k, and with it Avogadro's number, N_0. We claim, however, that the step from W_T to k was nevertheless too tentative, and the consequent grounding of molecular-kinetic theory too limited, for Perrin's measured value of W_T by itself to have ended resistance to the theory, again as typified by Ostwald, so abruptly. The robustness of Perrin's measured value for W_T was not by itself enough to assure that the mean kinetic energy of granules in Brownian motion is the same as that of hypothesized molecules of the liquid surrounding them. So, where van Fraassen stops, we are instead turning to other conclusions Perrin reached about Brownian motion to see how they might have given added grounds for taking the kinetic energies to be the same, or have otherwise provided evidence for molecular-kinetic theory.

5.4. Parallels with Molecular-Kinetic Theory

In addition to the two parallels between Brownian and molecular motion we cited in the preceding paragraphs, one involving diffusion and the other, vertical rarefaction, Perrin cites a third immediately following his sentence (in italics), "*it becomes very difficult to deny the objective reality of molecules*":

> At the same time we see the law of gases, already extended by Van't Hoff to dilute solutions, extended to uniform emulsions. The Brownian movement offers us, on a different scale, the faithful picture of the movements possessed, for example, by the molecules of oxygen dissolved in the water of a lake, which, encountering one another rarely, change their direction and speed by virtue of their impacts with the molecules of the solvent.[51]

In the opening paragraphs of our monograph we quoted others citing the visible motion of granules as analogous to the invisible motion of molecules. Recall that, even before Perrin had published his 1909 monograph, Nernst had said, "In the face of these *ocular* confirmations of the kinetic theory of the molecular world, we may well acknowledge that the theory begins to lose its hypothetical character."[52] Poincaré in 1912 may have had Perrin's remark in mind in saying, in the long quotation we gave in Chapter 1,

> Besides, the theory of solutions leads us very naturally to that of the Brownian movement, in which it is impossible to consider the thermal disturbance as a mere figment of the imagination, since it can be seen directly under the microscope.[53]

And in that same year the Ehrenfests, as we noted near the end of section 5.1, cited the established randomness of visible Brownian motion in support of Boltzmann's thoroughgoing statistical formulation of thermodynamics.

Such remarks at first glance appear to be responding to worries that had been raised, especially by Ernst Mach and Pierre Duhem, about the invisibility of atoms and molecules. Duhem, for example, had objected in 1906 to both molecular theory and the theory of the luminiferous ether by contrasting them with the vibrational theory of sound, to which he had no objections: "the explanation which acoustic theories give of experimental laws governing sound claims to give us certainty; it can in a great many cases make us see with our own eyes the motions to which it attributes these phenomena, and feel them with our own fingers." A theory, on the other hand, that "cannot render accessible to the senses the reality it proclaims as residing underneath those appearances" must content itself "with proving that all our perceptions are produced *as if* the reality were what it asserts."[54] Ostwald's view of molecular theory as merely a useful way of picturing matter in the earlier editions of his *Grundriss* and Nernst's insisting on the purely hypothetical status of molecular theory over the same period were undoubtedly taken by many to be expressing the same objection as Duhem. And Perrin's remark about the motion of oxygen molecules dissolved in water was at least reminiscent of Duhem's demand, of which he was undoubtedly aware.

The visibility of Brownian motion under a microscope was surely crucial in some respect or other to the contribution Perrin's experiments made to ending the resistance to molecular theory. Whatever importance those at the time attached to the parallels between Brownian motion and theoretical molecular motion, it surely stemmed from the visibility of the former. This visibility, however, definitely did not "render accessible to the senses the reality" of molecules. At most, Brownian motion provided a visible analogue of the hypothesized motion of molecules. And even that has to be qualified, first because the granules of Brownian motion do not visibly collide with one another in the manner claimed for molecules, and second because what is visible is not really Brownian motion itself, for the eye cannot catch the rapid changes of velocity that the measured irregularity of the granule displacements implies are occurring. That both Duhem and Mach did not follow Ostwald's lead in ending their resistance to molecules should therefore not be surprising, for, unlike him, they seem to have been insisting on experiments that render the reality of molecules accessible to the senses.[55]

Here too, therefore, we agree with van Fraassen that the invisibility of molecules was not at the time the source of resistance to molecular theory within the community of those actively engaged in research in physical chemistry. The empirical access Ostwald was demanding was not one of experiments that would render molecules visible, but of experiments that would derive from observable phenomena determinate values for such among their hypothesized characteristics as their number per unit volume, their kinetic energies, their masses, and at least the effective volumes they individually occupy, if not their specific sizes and shapes. Ostwald said in 1909 that Perrin's value for the first of these, and hence the values implied by it for most of the others, was "probably the most accurate . . . obtained as yet." As Steven Brush noted, however, others had obtained values of the same order of magnitude as Perrin's; and in the case of the electron, Ostwald in 1909 was content with order of magnitude values for charge and mass.[56] Even more to the point, in speaking of "the experimental proof of the atomic nature of matter" in his Preface dated November 1908, Ostwald cited not Perrin's values, but "the agreement of the Brownian movements with the requirements of the kinetic hypothesis." In what, however, did the agreement he had in mind consist, and what, if anything, might it have had to do with granules being visible in an ultramicroscope?

The one clue Ostwald gives is the pair of results he emphasizes in his review of Brownian motion, "the agreement [of the kinetic theory of heat] with the movements actually observed" by Svedberg and Perrin's count of the granules arranging "themselves under the action of gravity in the same manner as the atmosphere is arranged over the earth's surface," where "the constants [of the geometric series] depend on the size and the density, and also on the corresponding formula of the kinetic theory of gases."[57] (The formula referred to has to be $pV = (2/3)nW$, where W is the mean translational kinetic energy of the granules or molecules and n is their number in the volume V.) In each of these two, the agreement is between an observed characteristic of the visibly accessible granules in dilute emulsions and a theoretical characteristic of the hypothetical molecules in gases: (1) between the pattern of displacements over time of the granules and the pattern of displacements kinetic theory requires of molecules; and (2) between the rate of decrease in the number of granules counted in layers at increasing heights in the test apparatus and the rate of decrease in the number of molecules of, say, oxygen that kinetic theory requires in layers at increasing heights above the surface of the Earth.

This gives us as well an answer to the question about the visibility of granules. It allows one side of each of the two to be established entirely independently of the other. That, of course, is the point we have been insisting on all along: thanks to the visibility of the granules under ultramicroscopes, Perrin's results for Brownian motion itself did not depend in any way on molecular-kinetic theory.

But so what? Why is the reasoning here from the visible realm of Brownian motion to the invisible microphysical realm anything but reasoning by analogy? And wasn't the resistance to molecular-kinetic theory all along at least in part from distrust of such reasoning by analogy from visible hard spheres in impact with one another to hypothetical molecules of unknown shape, size, and hardness interacting with one another? Why should the analogy Ostwald seems to be invoking have been of any greater evidential weight than those invoked by Dalton, Clausius, and their like, decades earlier?

Our answer to this question, which not surprisingly centers on theory-mediated measurement, derives from the second of Newton's proposed rules for inductive reasoning in science: *To natural effects of the same kind the same causes are to be assigned, so far as possible.*[58] Newton most prominently employs this rule for licensing the conclusion, from a common constant of proportionality, that terrestrial gravity extends to the Moon, and then the conclusion, from inverse-square centripetal forces holding bodies in orbit around the Sun, Jupiter, and Saturn, that those forces too are one-in-kind with terrestrial gravity. As the phrasing of this and his other rules indicates, these are provisional conclusions, subject to reconsideration as further evidence emerges. The rule also serves to authorize measurements that are otherwise inaccessible. In particular, it provides Newton the basis for inferring how much more any object on the surface of the Earth would weigh were it on the surface of the Sun, Jupiter, or Saturn.

Huygens, as much as he objected to Newtonian gravity in other respects, accepted not only the conclusion that the centripetal forces holding bodies in orbit in our planetary system are one-in-kind with terrestrial gravity, but also the claim to be able to infer what objects would weigh on the surfaces of the Sun, Jupiter, and Saturn:

> There still remains something for me to remark on his [Newton's] System, something which has pleased me greatly. It is that he finds means, supposing the distance from here to the Sun to be known, to define what gravity the inhabitants of Jupiter and Saturn would feel, compared to what we feel here on the Earth, and also what its measure is at the surface of the Sun. These are things which previously seemed quite removed from our knowledge, and which are nevertheless consequences of the principles I reported a little earlier.[59]

Huygens went on to point out that Newton's values were surely wrong because he had used too small of a distance "from here to the Sun," and he then supplied corrected values based on a more reliable estimate of that distance. From those corrected values he obtained the speed with which the particles in the vortices that produce gravity, according to his theory of the mechanism producing it, must be moving at the surface of the Sun, namely "49 times greater than what we have found near the Earth, which was already 17 times greater than the velocity of the Earth's surface at a point on the equator."[60]

Two points about the Newton-Huygens example need to be made before we return to Brownian motion. First, Huygens's view of Newton's proposed measurements was in no way contingent on the empirical adequacy of Newton's theory of gravity, or for that matter on his own theory of gravity. Huygens simply granted the evidence that the forces retaining the bodies in orbit around Jupiter and the Sun vary in an inverse-square ratio with their mean distances from their principals, and the evidence that terrestrial surface gravity, as he had measured it, is in an inverse-square ratio with the force retaining the Moon in orbit around the Earth. Those two sufficed to conclude, at least provisionally, that the centripetal forces retaining the bodies in orbit around Jupiter and the Sun also hold in an inverse-square ratio right down to their respective surfaces. All the rest of Newton's theory of gravity, and Huygens's as well, was beside the point.

Second, what Newton's same-effect-same-cause rule of analogical reasoning provided the proposed measures of gravity at the surfaces of Jupiter and Saturn was a legitimacy that they would otherwise have lacked. Robert Hooke, Edmund Halley, and Christopher Wren had all proposed, on the basis of Huygens's published solution for uniform circular motion, that the forces retaining bodies in orbit around Jupiter and Saturn vary in an inverse-square ratio with the mean radii of those orbits. Absent any evidence that terrestrial gravity varies in an inverse-square ratio, any further proposal about what terrestrial bodies would weigh on the surface of Jupiter and the Sun would have amounted to nothing but conjecture. Newton identified the same effect, orbits requiring inverse-square forces, and extended it to the Moon. His confirmation that the centripetal tendency of the Moon is in an inverse-square ratio with the tendency of bodies to fall at the surface of the Earth implied a cause, an inverse-square tendency throughout the space around the Earth to fall toward it. His rule of reasoning then authorized extending that cause to Jupiter and the Sun as well, licensing the inference from the tendency in orbit to a tendency on their respective surfaces. The values of the tendency inferred from the orbits could thereby provide a measure of the tendency at their surfaces. It was through this line of reasoning that the proposed measure of weight at the surfaces of Jupiter and the Sun acquired legitimacy when otherwise it would have been mere conjecture.

5.4.1. The Vertical-Gradient Parallel

To see how Newton's rule can be viewed as doing the same thing in the case of the parallels between Brownian motion and molecular theory that Ostwald cited, consider first the vertical-gradients in concentration of species under gravity. As we pointed out earlier, Maxwell had cited the faster diminution of oxygen than nitrogen in the atmosphere as we ascend in proposing that "the theorem of Boltzmann may be applied not only to determine the distribution of velocity among the molecules, but to determine the distribution of the molecules themselves in a region in which they are acted on by external forces."[61] The formula from kinetic theory giving the vertical distribution of molecules of any constituent of the atmosphere at uniform temperature is,

$$\ln\frac{\rho_0}{\rho} = \frac{3}{2}\frac{(mg)_{molecule}}{W_{molecule}}h$$

where ρ is the density of the constituent, W is the mean translational kinetic energy of its molecules, and mg is the weight per molecule, ignoring variation of g with altitude. The formula provides a determination of the distribution of the molecules vertically insofar as the ratio of the densities of any constituent at constant temperature, according to molecular theory, just is the ratio of the numbers of molecules.

This theoretical formula, *together with the claim that the mean translational kinetic energy of the molecules of all constituents of a gas is the same at any given temperature*, offered an explanation of a phenomenon we noted earlier: the rates at which the different constituents of the atmosphere diminish with altitude, once allowances are made for variations in temperature, vary strictly with their molecular weights. The explanation it offered, however, was in Duhem's sense purely hypothetical; it is "as if" gases consist of different species of molecules in motion, with each species having its own characteristic weight and the mean translational kinetic energy of the molecules of all the species the same at any given temperature. So long as the weight of the individual molecules and the mean of their individual mean translational kinetic energies could not be determined, this formula could not be tested or used to measure one quantity indirectly in terms of another, even when stated in terms of densities, much less when stated in terms of numbers of molecules per unit volume.

The molecular formula should look familiar, for it has the same form as Perrin's "law of sedimentation equilibrium" that we gave in section 4.2:

$$\ln\frac{n_0}{n} = \frac{3}{2}\frac{m_{gran}g_{eff}}{W}h$$

The difference in the case of Brownian motion was that Perrin managed to determine values of the variables in it except for *W*, verifying in the process that the distribution of *n* is logarithmic. The formula thereby provided a measure of *W*. Moreover, by varying the masses of the granules and the ratio of their density to that of the liquid, he showed that, while the rate at which the concentration diminishes changes, *W* remains the same once allowances are made for any differences in temperature. Further confirmation that *W* is the mean translational kinetic energy of the granules came from his subsequent granule displacement experiments. Consequently, in the case of his law of sedimentation equilibrium for dilute emulsions, Perrin had compelling evidence that the reason why the rates at which granule concentrations diminish at any given temperature vary strictly with the effective weights of the granules is that the mean translational granular kinetic energy is the same at a given temperature, regardless of the species of granules.

Now apply Newton's same-effect-same-cause reasoning: the reason why the rates at which the concentrations of the different constituents of the atmosphere diminish with altitude vary strictly with constituent molecular weight is that each constituent consists of "molecules" in motion with a characteristic weight, where the mean translational kinetic energy of these "molecules" is the same in all constituents. Here the term "molecule" refers only to not yet further specified discrete elements, the mass of which depends on the number of them required to comprise a gram-molecule of the substance. With the uniformity of the molecular kinetic energy inferred through this reasoning, the molecular formula has gained notably more license than it had before to be used, as Maxwell said, "to determine the distributions of the molecules themselves" of, say, oxygen in our atmosphere. In other words, the molecular formula had more legitimacy to provide theory-mediated measurements of any quantity in it from the other quantities than it had before, when the uniformity of mean molecular kinetic energy was merely a hypothesis. Indeed, just as Newton used still uncertain quantities to infer weights on the surfaces of Jupiter and the Sun, Perrin's "probably most accurate as yet obtained" values for this energy and molecular weights gave license to infer estimates of the number of oxygen molecules per unit volume at higher altitudes.

Comparison with Newton's use of same-effect-same-cause reasoning offers a way of interpreting such reasoning when applied to the atmosphere that does not require a direct counterpart within molecular theory. Let *r* be the radius of an orbit and *P* be the time required to complete an orbit. Phenomenologically, the r/P^2 values for the four satellites of Jupiter, which are in virtually circular orbits, are proportional to the respective $1/r^2$ values. Insofar as this proportionality might be just a numerical accident, a natural question is whether the constant of proportionality in the relationship has any physical significance.

Newton's same-effect-same-cause reasoning licensed the inference from the case of the Earth to the conclusion that the constant of proportionality for Jupiter is a measure of the strength of the inverse-square centripetal tendency experienced by all bodies in the space surrounding it. And, of course, a parallel inference can be drawn for the constant of proportionality for the Sun.

The phenomenological analogue in the case of the atmosphere has the height over which the density of any constituent diminishes by a factor of 2 proportional to the molecular weight of the constituent. The question in this case concerns the physical significance of the constant of proportionality in this relation. In the case of Brownian motion, Perrin's law of sedimentation expresses a parallel proportionality between the rate of rarefaction and the buoyant weight of the granules. In that case, however, Perrin provided clear evidence that its constant of proportionality *is* a measure of the mean translational kinetic energy of the granules, which at any given temperature is the same regardless of the mass of the granules. Same-effect-same-cause reasoning therefore analogously licensed the inference that the constant of proportionality in the case of the atmosphere is a measure of the mean translational kinetic energy of discrete elements composing it, which at any given temperature is the same regardless of the mass of those elements. This inference was merely provisional, but the license for it became far stronger after Perrin's results than it had been before them.

5.4.2. The Diffusion Parallel

Turning now to the other parallel between Brownian motion and molecular theory that Ostwald cited—the agreement between the theoretical and the observed motions—the application of same-effect-same-cause-reasoning is less straightforward. Such reasoning requires a parallel between two phenomena in which the cause of one has been determined. Ostwald refers to the kinetic theory of heat, for which the phenomenon is the proportionality of pressure in a gas at constant mole density, or of the osmotic pressure in a dilute solution at constant volumetric concentration, to absolute temperature. Perrin, however, never measured the very small osmotic pressure of granules in dilute emulsions, much less its variation with temperature. The "pressure" that he showed to vary with absolute temperature amounted only to the momentum of the impact of the granules on the walls of the container inferred from their translational kinetic energy and masses. To complete the parallel with van't Hoff, he would have had to show that the osmotic pressure of the granules matches this momentum. So, we need some other phenomenon in parallel in order for same-effect-same-cause reasoning to apply.[62]

Insofar as Ostwald cited Einstein and Smoluchowski in conjunction with Brownian motion confirming the kinetic theory of heat, the obvious phenomenon to turn to is diffusion. Here, however, a different problem arises: there was no equation within classic kinetic theory parallel to Einstein's equation for the diffusion coefficient in dilute emulsions. To appreciate why this was, we need to consider the history of the theory of diffusion within kinetic theory. That history will also bring out more clearly why grounding kinetic theory remained a challenge even after Perrin's results.

The standard text in kinetic theory during the years we have been considering was O. E. Meyer's *The Kinetic Theory of Gases*.[63] We shall follow its account here. No sooner had Clausius proposed the kinetic theory and concluded that mean molecular speeds are of the order of 400 meters per second than the challenge was posed to reconcile these speeds with the long time diffusion requires in gases. Clausius's reply was the concept of *mean free path*. Let λ^3 be the volume per molecule, that is, the total volume divided by the number of molecules. One can think of this as the cube within which, on average, a single molecule is located. Let s then be the smallest distance between the centers of two molecules at which they interact, with a transfer of momentum between them and hence a change in their velocity. Clausius obtained the following formula for the mean free path under the assumption that the speed of all the molecules is always the same, but all directions are equally probable:[64]

$$L = \frac{3}{4}\frac{\lambda^3}{\pi s^2}$$

The randomness of the molecular motion on this view came only from the randomness of the angles to which molecules are deflected when they interact with one another. From this Clausius concluded that, on the one hand, the distance on average a molecule covers between interactions is large compared with the minimal distance between them, but, on the other, the frequency of interaction whenever their number per unit volume is large explains why none of them goes that far in any one direction before being redirected.

Maxwell augmented this analysis with his proposed velocity distribution, yielding a small modification of Clausius's formula:

$$L = \frac{\lambda^3}{\pi s^2 \sqrt{2}}$$

Maxwell also called attention to the implication of Regnault's measured real gas deviations from Boyle's law that molecules repel one another at close distances. This possibility led Meyer to speak of s as the radius of the "*sphere of action*, meaning thereby that the mean point or centre of gravity of another molecule cannot penetrate into it."[65] (The formula we gave for the mean free path in section 5.3, in which the numerator is 1.0 and the denominator includes Avogadro's number as a factor, is obtained by taking the unit volume to be that of a mole of gas and then multiplying both numerator and denominator by N_0.)

One thing to notice here is the extent to which the formula bypasses the need to take into consideration any details at all of what happens when two molecules interact and their velocities change. Another is the absence of any way of moving from L to the parameter central to the Einstein-Smoluchowski equation, ξ^2/τ, the mean value of the square of the displacements of molecules in any one direction over times τ long in comparison with the times between interactions. Neither of these, however, stood in the way of arriving at Maxwell's equation for the viscosity of a uniform gas in terms of the mean free path:

$$\zeta = 0.30976\rho L\Omega$$

where ρ is the density of the gas and Ω is the arithmetical mean of the speeds in Maxwell's distribution.[66] (The famous conclusion that the viscosity is independent of the density of the gas follows from the fact that L is inversely proportional to the density.) Insofar as Ω can be inferred from the overall energy level of the gas, this formula permitted values for L to be inferred from measured values of the viscosity for a wide range of uniform gases.[67] The precise physical significance of those inferred values, however, was open to question because of assumptions made in deriving the equation and uncertainty about the physical meaning of s. Viscosity, moreover, concerns only the lateral transfer of momentum, not the gradual displacements of molecules over time involved in diffusion.

Reasoning similar to that used in deriving the equation for viscosity yielded an equation for the coefficient of diffusion of one species of gas through another in terms of mean free paths:

$$D = \frac{\pi}{8N}\left(N_1\Lambda_2\Omega_2 + N_2\Lambda_1\Omega_1\right)$$

where the symbol representing the mean free paths is changed to Λ in recognition of its representing the mean of paths of one type of molecule in a gas composed of two types of molecules.[68] A host of questions arise with this formula. For example, how should the sphere of action governing the mean free paths be determined? The usual proposal was to define it as the mean of the spheres of action for each of the gases alone, but this served only to bypass the question of how molecules of different types encounter one another. A further question concerned whether molecules of both types limit the diffusion of those of each type, or only the molecules of the other type. This led to competing theories of diffusion, one by Stefan and Maxwell adopting the second stance, and one by Meyer adopting the first. We could go on.[69] Comparisons with measured values of the coefficient of diffusion with values derived from the preceding equation, using values for mean free paths from viscosity, showed at best rough agreement, with no clear pattern to the discrepancies between the two.[70]

Commenting on the formula Meyer gives for the mean free path, Ostwald still saw the theory of diffusion within kinetic theory as unsettled in 1908, even after he had ended his resistance to molecular-kinetic theory:

> The theory of these [diffusion] processes, as well as of the nearly related phenomena of gaseous friction and conduction of heat, is still by no means completely worked out, in spite of the many efforts in this direction; yet we have got so far that the values of the free paths determined in different ways agree pretty well with each other.[71]

This statement occurs in the book after Ostwald had acknowledged confirmation of the kinetic theory of heat from the data supporting the Einstein-Smoluchowski relation. But then, Perrin conceded as much in a footnote in his 1909 monograph:

> Sphericity of the molecules, and various simplifications in the reasoning which lead to the expression for the mean free path, make it impossible to specify exactly what uncertainty exists in the numerical coefficients of the approximate equations which connect the viscosity, the mean free path, and the molecular diameter.[72]

In short, the analysis of diffusion within classical kinetic theory had not reached the point of providing an equation for the diffusion coefficient for one gas diffusing through another that could meaningfully serve as a counterpart to Einstein's equation for diffusion in dilute emulsions. Worse, the latter concerns diffusion within a liquid solvent, not diffusion within a gas, and the kinetic theory of liquids was in an embryonic stage compared with that of gases. Insofar as the smallest granules Perrin tested were, however, orders of magnitude larger than the molecules of the liquid, it is not clear even how to make sense of their having a mean free path. Mean-free-path analysis was based on the frequency of the encounters of two molecules of comparable size moving on average at comparable speeds, not the encounters of a very large molecule diffusing through a host of smaller molecules moving on average at much greater speeds. Finding a way to relate Einstein's displacements of granules to mean free paths of molecules thus seemed hopeless. (This is likely what Perrin meant when he remarked, "Without further troubling about the infinitely tangled trajectory which the granule describes in a given time, Einstein considered simply its displacement during this time, the displacement being defined as the length of the rectilinear segment which separates the point of departure from the point of arrival."[73])

Still, diffusion in dilute emulsions does have its counterpart, not only in diffusion within dilute solutions, but also within gases in which the concentration of the diffusing species is small. The great virtue of Einstein's theory came from his not deriving it from within the kinetic theory of diffusion! For to do so, as Perrin's remark indicates, he would have had to introduce assumptions about the interaction of granules with molecules of just the sort that the younger Ostwald complained about, assumptions that would have stood in the way of Brownian motion grounding the two key parameters, the mean free path and the sphere of action. Instead, the central equation within Einstein's theory presupposed nothing within kinetic theory at all, but only that the motion of the granules is totally irregular, with a constant mean translational kinetic energy over time:

$$D = \frac{1}{2}\frac{\xi^2}{\tau}$$

Perrin provided compelling evidence for that assumption by showing that the displacements accord with a Gaussian distribution and their mean translational kinetic energy is not only constant at any given temperature, but the same regardless of the type of granule; and Brillouin managed to confirm the equation itself for granules of gamboge, totally independently of any equation for the term on the right. So, while issues can be raised about the extent of the evidence, the experimental results confirmed this equation and the assumption behind it for diffusion in dilute emulsions.

But then same-effect-same-cause reasoning yields the conclusion that, *in dilute solutions and gases with a small concentration of one species, diffusion results from an irregular motion of discrete elements, yielding a Gaussian distribution of their displacements over comparatively long times τ; and the value of the diffusion coefficient is dictated by the mean of the squares of those displacements versus τ.* Nothing specific is said about these elements, other than that they are discrete. In particular, no assumption has been made about their shape, how often they interact, or what happens when they do, beyond assuming that their mean translational kinetic energy persists over time. Granted, though the conclusion reached via this same-effect-same-cause reasoning was only provisional, it nevertheless licensed inferences from the measured values of the diffusion coefficient in such solutions and gases to values of ξ^2/τ for the invisible diffusing discrete elements. This inference, one should note, fully parallels that of Newton's inference from observed orbiting bodies to the otherwise inaccessible weights of objects on the surfaces of Jupiter and the Sun.

Neither Einstein nor Perrin stopped with this bare equation for the diffusion coefficient. Einstein assumed Stokes's law in order to specify a force impeding the motions of the granules over short times, yielding the equation,

$$\frac{\xi^2}{\tau} = \frac{RT}{N}\frac{1}{3\pi a\zeta}$$

This equation presupposes that the mean translational kinetic energy of the granules is the same as that of the molecules of the liquid. As we indicated in section 4.3 of Chapter 4, however, this assumption can be dropped by restating the equation in terms of the mean kinetic energy at any given temperature of the granules themselves, independently of any claim about the liquid. The equation then becomes

$$\frac{\xi^2}{\tau} = \frac{2W_T T}{9\pi a\zeta}$$

Again, Perrin's results confirmed this equation, including the universality of W_T across granules of different types and the applicability of Stokes's law to spherical granules. It holds independently of molecular-kinetic theory.[74]

Both Einstein and Perrin elected to apply the formula preceding this one to sugar molecules in solution, extracting values for Avogadro's number and, in Einstein's words, "the hydrodynamically-effective radius of the molecule." (Perrin remarked, "It may be hoped that this formula still applies roughly to the case of molecules as small as sugar or phenol."[75]) Notice in this case, however, that same-effect-same-cause reasoning is open to challenge, for not one, but two inferences from effect to cause are involved, one concerning diffusion and the other, the short-term action of the liquid on the velocity of the diffusing molecule. Einstein subsequently provided a way of applying same-effect-same-cause reasoning for solutions and gases by substituting a generic parameter in place of the term in the denominator obtained from Stokes's law:

$$\frac{\xi^2}{\tau} = \frac{RT}{N}\frac{1}{Z}$$

where Z he called the frictional resistance of the liquid (or the gas) on the diffusing molecule, which "cannot in general be deduced theoretically."[76] (The inverse of Z subsequently became called the mobility μ of the molecule.[77]) Perrin's results can be viewed as confirming the analogue of this formula for dilute emulsions by measuring ξ^2/τ, Z, and $W_T T$ in place of (RT/N).

Same-effect-same-cause reasoning can now be applied in two steps, first to obtain values for ξ^2/τ from measured values of the diffusion coefficient in dilute solutions (or gases in which the diffusing species is at low concentration) and then, from these values, values of NZ under the assumption that the mean translational energy of the diffusing molecules at any given temperature is the same as that of granules in dilute emulsions. Adopting Perrin's provisional value for Avogadro's number then allows values to be inferred for Z and its inverse μ, the mobility of the diffusing molecule within the fluid surrounding it. We argued earlier that the physical significance of the mean free path and the radius of the sphere of action in molecular encounters remained unclear at the time. The same is true for Z and μ. Now, however, we have formulas for the diffusion coefficient expressed in terms of mean free paths and spheres of action, and a provisional formula for it expressed in terms of mobility, for which values can be obtained. This opened at least the possibility of gaining more information about mean free paths and spheres of action from the inferred values of the mobility.

Perrin included a brief section on the mobility of ions in liquids under the assumption that the fluid resistance to their motion is given by Stokes's law, but he found the results disappointing.[78] Ostwald, later in his book, called attention to the "velocity of migration" of gas ions in connection with his discussion of the research of J. J. Thomson "and his pupils," concluding that a mean-free-path calculation based on those velocities yields a number of ions per mole that agrees with Perrin's value.[79] Ostwald did not provide a source, but he could well have been referring to measurements made by John Townsend in 1900. Employing measured values of the mobility of ions that had been obtained by Thomson and Rutherford, Townsend's measurements of diffusion coefficients had shown that the total charge *Ne* per unit volume of gas ions produced by Röntgen-rays in a variety of gases corresponds to within roughly 1 percent of the total charge *Ne* Faraday had obtained in electrolysis:

> Since N is a constant, we conclude that the total charges produced by Röntgen-rays in air, oxygen, carbonic acid, and hydrogen are all the same, and equal to the charge on the hydrogen ion in a liquid electrolyte.[80]

Although Ostwald did not mention Townsend's research in his 1909 edition, Perrin did in his 1909 monograph, calling it "the first exact demonstration of the invariability of the atomic charge."[81] Ostwald did, however, remark in the new chapter, "Conduction of Electricity in Gases and Radioactivity," added to the 1909 edition of his *Grundriss* that it was important to the developing theories of the electron "that the charge on an electron was found to be the same as that on an electrolytic ion, i.e. 10^{-19} coulombs"[82]—a conclusion for which the principal evidence came from Townsend.

Once Townsend had his result for *Ne*, he drew a number of further conclusions, including an estimated value for *N* (45×10^{22}) based on the value for *e* (6×10^{-10} esu) that Thomson was at the time proposing. More important here, Townsend also turned to Maxwell's mean-free-path theory of diffusion to obtain values for *s* for the ions of different gases (8.2×10^{-8} cm for hydrogen and 9.2×10^{-8} cm for oxygen). Perrin ignored these values in his discussion of Townsend, but he did emphasize that Townsend's equation for the migration of ions in his experiment was equivalent to Einstein's equation for diffusion stated in terms of mobility.[83] Once this connection was made, the way was open for systematically exploring the relation between the rates of migration per unit time implied by Townsend's measurements and Maxwell's mean-free-path equation for diffusion. For example, one could take the values of *s* that Townsend obtained for the radius of the sphere of action of ions, calculate the resistance to their motion implied by Stokes's law, and compare this with the ion mobilities that had been measured by Thomson and Rutherford. This illustrates what we mean when we say that, even though the Brownian motion results did not do much to directly ground such parameters as the mean free path and the radius of the sphere of activity, they did open the way to bringing new research to bear toward grounding those parameters than had been possible before.

* * *

Throughout this discussion of same-effect-same-cause reasoning we have stressed the provisional status of the measurements licensed. Moreover, while we have at times used the term "molecule" when referring to results of the licensed measurements, the most that the reasoning licensed was for this term to be referring to invisible discrete elements, with other questions about these elements remaining entirely open. The parallels between phenomena within them notwithstanding, the gap remained between the visible realm of activity in dilute emulsions, on the one hand, and the invisible realms of activity in dilute solutions and gases, on the other. Perrin's results had validated a set of theory-mediated measurements in the former. Their legitimacy was no longer in question. *What same-effect-same-cause reasoning in conjunction with his results added was to provide more legitimacy to the corresponding formulas for theory-mediated measurements in the invisible realms than those formulas had had before his results.*

We have no reason to think that those at the time appreciated the extent to which Perrin's Brownian motion results themselves were independent of molecular and molecular-kinetic theory. As we remarked at the end of section 4.3 in Chapter 4, so many of them were more intent on silencing critics than on finally establishing specifics about molecules. Beyond this, Perrin had presented his results as measures of Avogadro's number without calling attention to the possibility of the mean translational kinetic energy of the granules in Brownian motion not being identical with the mean translational kinetic energy of the molecules in the liquid. One has to wonder, therefore, how many at the time appreciated how much any claim Perrin was making to having gained experimental access to the molecular realm turned on his postulating this identity. Once the identity was granted, the natural conclusion was not, as noted earlier, that Perrin's results had provided more legitimacy to the corresponding formulas for the molecular realm than they had had before; the natural conclusion was that his results had gone a long way toward establishing those formulas for the molecular realm. Our italicized conclusion at the end of the preceding paragraph is to be read as not having granted this identity.[84]

We concluded in the preceding section on grounding that, contrary to van Fraassen, the value Perrin had obtained for Avogadro's number was too tentative, and the grounding it provided for the parameters of kinetic theory was too limited, for it alone to have ended Ostwald's resistance to molecular-kinetic theory. While Ostwald mentioned that value at the end of his newly added discussion of Brownian motion, he expressly regarded it as still tentative; and in the revised discussion of kinetic theory later in the book he drew no conclusions at all about such key parameters as the mean free paths and the sizes of molecules from Perrin's results, citing instead values that Meyer had published before Perrin.[85] Yet, as we have seen, Ostwald's complaint against molecular-kinetic theory before Perrin was a lack of grounding of its parameters, and he did cite results on Brownian motion in the Preface to the new edition of his book as answering that complaint. The sole textual evidence of what he saw in the Brownian motion results that did this were the two parallels, one between the irregular motion of the granules confirmed with the confirmation of the Einstein equation and the irregular motion hypothesized in molecular-kinetic theory, and the other between the confirmed rate of reduction of the number of granules with height under gravity and the explanation within kinetic theory for the parallel phenomenon in the atmosphere. The question is how Ostwald viewed these parallels as providing any more grounding of molecular-kinetic theory than Perrin's proposed, yet tentative, value for Avogadro's number provided.

The answer, we submit, is that he thought that the formulas within molecular-kinetic theory that had a potential for providing theory-mediated measurements of some of its parameters had gained more legitimacy from the two parallels with Brownian motion than they had had before. Granted, this was a matter of promise, but as we shall argue in Chapter 7, the new promise for grounding molecular-kinetic theory seems to have been what led Ostwald to end his resistance to it. Ostwald did not make explicit how the *parallels* with Brownian motion—or the analogies, if you prefer—provided any such added legitimacy. It may well have remained at an intuitive level for him, as it often does for most scientists. We are proposing Newton's same-effect-same-cause reasoning as the logical basis for making it explicit. At the very least, the parallels provided a promise for grounding kinetic theory beyond that deriving from Perrin's tentative value for Avogadro's number. We propose that this further promise for grounding was a major part of what persuaded Ostwald to change his attitude about the "scientific well-foundedness" of molecular-kinetic theory.

5.5. Continuity with Molecular-Kinetic Theory

We have been considering what Ostwald saw in the Brownian motion results that gave him grounds for concluding that "the atomic hypothesis is thus raised to the position of a scientifically well-founded theory" within a year after he had denied it this status in his 1907 *Monist* paper. We have been ignoring, however, the precise statement in support of which he cited J. J. Thomson's researches and Perrin's results:

> I am now convinced that we have recently become possessed of experimental evidence of the discrete or grained nature of matter, which the atomic hypothesis sought in vain for hundreds and thousands of years.[86]

This evidence, he went on to say, consisted in "the isolation and counting of gas ions" and "the agreement of the Brownian movements with the requirements of the kinetic hypothesis." It is easy enough to see why Ostwald would have regarded the isolation and counting of gas ions as evidence for the discrete nature of matter. It is not so easy to see why he would have regarded the results for Brownian motion as evidence for it. Yet he was not alone in doing so, for remember that Perrin's Nobel Prize citation noted "his work on the discontinuous structure of matter."

One of van Fraassen's central points is to deny that the results on Brownian motion provided any evidence at all for the discrete nature of matter. In section 5.2 we agreed with him that Perrin's value for Avogadro's number presupposed that his measured values for the mean translational kinetic energy of granules were identical with the hypothetical mean translational kinetic energy of molecules, and that in turn presupposed discrete molecules. Consequently, the value Perrin obtained for Avogadro's number by itself could scarcely have constituted evidence for discrete molecules. Section 5.4 led us to the same conclusion. For while the same-effect-same-cause reasoning provided added legitimacy to formulas within kinetic theory supporting theory-mediated measurements, those formulas presuppose the discrete nature of matter. In other words, if one takes the molecular realm for granted, the Brownian motion results altered the status of various theoretical claims about discrete elements forming matter. By its presupposing such discrete elements, however, the reasoning we examined in both sections 5.2 and 5.4 cannot have provided evidence for them except through Brownian motion providing some sort of test of that presupposition. In section

5.1 we concluded that the results on Brownian motion themselves were logically independent of molecular-kinetic theory and hence could not have been testing it beyond tests of its consistency with Brownian motion. So, how, if at all, might those results have provided evidence for the discrete nature of matter?

The sizes of the granules in Perrin's experiments ranged in order of magnitude from 10^{-3} to 10^{-5} cm. One natural thought, correspondingly, is that the granules provide an intermediate yet continuous sequence bridging the gap between bodies obeying the laws of everyday mechanics and the molecules of chemistry, a microscopic intermediary that fills the gap between the macrophysical and the microphysical realms. Perrin himself suggested such a line of thought in the opening portion of his 1909 monograph:

> These sugar molecules contain 35 atoms; the molecules of sulphate of quinine contain more than 100 atoms, and the most complicated and heaviest molecules to which the laws of Van't Hoff (or of Raoult which are deduced from them) can be extended may be cited. The mass of the molecule appears absolutely unlimited.
>
> Let us now consider a particle a little larger, still, itself formed of several molecules, in a word, a *dust*. Will it proceed to react towards the impact of the molecules encompassing it according to a new law? Will it not comport itself simply as a very large molecule, in the sense that its mean energy has still the same value as that of an isolated molecule? This cannot be averred without hesitation, but the hypothesis at least is sufficiently plausible to make it worth while to discuss its consequences.[87]

The principal law to which he was referring in this passage was, "*At the same temperature all the molecules of all fluids have the same mean kinetic energy, which is proportional to the absolute temperature.*"[88] And, indeed, he did confirm that the counterpart of this law holds for granules in Brownian motion as well.

This passage is nevertheless quite misleading, for the largest molecule for which Perrin or Einstein carried out calculations, sugar, they reckoned to be two and a half orders of magnitude, that is, roughly 500 times, smaller than the smallest granules in any of the experiments. Townsend's inferred values for the radii of the spheres of action of hydrogen and oxygen ions were a little less than 10^{-7} cm; the radius of the smallest granule Perrin tested was a little more than 10^{-5} cm. So, while Perrin tested granules ranging over two orders of magnitude in size, the gap from them to molecules was still at least two more orders of magnitude. That is quite a gap. Thinking of the realm of the granules tested as continuous with that of molecules in such a way that the motions of the former can be extended to those of the latter was therefore at most of heuristic value. That is why we entitled this section "Continuity with Molecular-Kinetic Theory," not "Continuity with the Molecular Realm."

Even continuity between the theory of Brownian motion and the theory of molecular motion is doubtful on its face. In Perrin's experiments the concentration of granules in the emulsions was kept low enough to minimize interaction among them. That very interaction was hypothesized to be the source of the randomness in molecular motion and hence at the heart of the theory of it. (The low concentration, of course, was why Perrin's remark about Brownian motion offering "on a different scale, the faithful picture of the movements" concerned oxygen dissolved in the water of a lake.[89]) More importantly, not only was frictional resistance to granule motion, in accord with Stokes's law, a confirmed factor in the Brownian motion of granules; Perrin had confirmed that the law holds for his spherical granules independently of Brownian motion. By contrast, no sort of bulk resistance akin to friction impeding the motions of molecules was hypothesized in the theory of their motion; and the sphericity of molecules and the applicability of Stokes's law to them had no basis other than to allow first-pass estimates to be made of parameters of kinetic theory. Even continuity between the pressure exerted by impact of granules on the walls of the container and van't Hoff's osmotic pressure remained open to question in the absence of measurements of the pressure exerted on semi-permeable membranes by the granules in dilute emulsions. One might conclude, therefore, that any proposed continuity between the theories of Brownian and molecular motion at the time we are considering amounted to not much more than wishful thinking.

There was, nevertheless, one proposition that the theories of the two kinds of motion had in common (rephrasing the quotation from Perrin only slightly): *At the same temperature the discrete elements—whether in dilute emulsions, on the one hand, or in gases and in dilute solutions, on the other—in each realm have the same mean kinetic energy, which is proportional to the absolute temperature and does not vary with the masses of the elements.* This was not a hypothesis in the kinetic theories of gases and dilute solutions. It was a derived consequence, within the initial fragment of the kinetic theory of heat, of the empirically established universality of both the ideal gas law and van't Hoff's parallel law for non-electrolytic dilute solutions. Nor was it a hypothesis in either of the theories of Brownian motion we have been considering, Perrin's theory of granule concentration-gradients at equilibrium under gravity, and Einstein's of displacements of granules over time. Rather, it was an empirical result that emerged repeatedly from Perrin's experiments, a discovery that at the time bordered on the remarkable.

Here, then, we have a specific respect in which the empirically confirmed theory of Brownian motion was continuous with the hypothetical theory of molecular motion—a respect much stressed by Perrin. What conclusions pertaining to the discrete nature of matter could at the time have legitimately been drawn from it? And were there other results from Perrin's experiments that, together with it, could have yielded such conclusions? These questions will take up the rest of this section.

In his 1909 monograph, Perrin first confirmed that the mean kinetic energy of the granules at given temperatures is the same regardless of the size of the granules (corresponding to a value for Avogadro's number of 70×10^{22}), and next that it matches the kinetic energy inferred for molecules to within 15 percent, remarking that "the number given by the equation of Van der Waals does not allow this degree of accuracy."[90] The paragraph that follows opens with "I do not think this agreement leaves any doubt as to the origin of the Brownian movement" and then notes that his vertical-gradient experiments could in principle have yielded values for granule kinetic energy anywhere from zero to infinity in underscoring "how striking the agreement is." From this he drew the conclusion we have quoted before:

> Thus the molecular theory of the Brownian movement can be regarded as experimentally established, and, at the same time, *it becomes very difficult to deny the objective reality of molecules*.[91]

Needless to say, this was not so much an argument as a challenge to those holding out against molecular-kinetic theory to come up with an alternative explanation for the universality of granular kinetic energy and its agreement with an estimate (from argon alone) of molecular kinetic energy.

Perrin was slightly less polemical in his second September 1908 paper, "L'origine du movement brownien," the announced goal of which was to present results supplementing his prior paper, which had proposed "to judge by experiment the hypothesis which makes molecular agitation the cause of the movement."[92] After indicating that the announced result in his prior paper had a mistake, he compared an estimate (40×10^{-15}) of the mean kinetic energy per molecule in dilute solutions (from presupposing that Avogadro's number is 60×10^{22}) with values for granule mean kinetic energy obtained from vertical-gradient experiments employing three different size granules: 42.5×10^{-15}, 40.4×10^{-15}, and 44×10^{-15} (corresponding to values for N of 57×10^{22}, 60×10^{22}, and 54×10^{22}, respectively). Here too he called the agreement "very striking" ("*extrêment frappant*"), noting how extreme the inferred values for the energy could in principle have turned out to be, ranging from a case of no gradient at all (and hence a "practically infinite" value for the energy) to all the granules settling on the bottom. He concludes, "This agreement can hardly leave any doubt as to the rigorous exactitude of the kinetic theory of Brownian movement."[93] Nothing, however, was said about "the objective reality of molecules."

With this one exception, therefore, the reasoning concerning the origin of Brownian motion was basically the same in both Perrin's second September 1908 paper and his more argumentative 1909 monograph. His earlier May 1908 paper, "L'agitation moléculaire et le mouvement brownien," which contained the error mentioned earlier, was notably less rhetorical. The more modest conclusion it drew from the general agreement between the value of mean granule kinetic energy (corresponding to a value of *N* of 67×10^{22}) and an estimate for its molecular counterpart was that "the kinetic theory of fluids seems a bit stronger, and the molecules a bit more tangible."[94]

Three points need to be made about Perrin's argument. The first concerns the context in which these two 1908 papers were written.[95] Perrin's May 1908 paper was written in response to experimental results challenging Einstein's theory of Brownian motion. In July, Duclaux had objected as well to Perrin's paper, especially its assumption, and Einstein's too, of the applicability of Stokes's law to granules in emulsions.[96] Perrin published two papers in September in response to this, the first, "La loi de Stokes et le mouvement brownien," providing the evidence, repeated in his 1909 monograph, that Stokes's law does apply.[97] Little wonder, then, that Perrin's second September 1908 paper and his monograph were more polemical. He was once and for all trying to end questions about whether Brownian motion is at least consistent with a molecular origin.

Second, little wonder too that the rhetorical thrust of Perrin's argument was not so much to prove the reality of molecules as to put the burden of proof forcefully on those opposing the molecular origin of Brownian motion, to whom he was responding. Our point here goes beyond our earlier concession to van Fraassen that Perrin cannot compare the kinetic energy of granules to estimated values of the kinetic energy of molecules without having presupposed that "molecule" refers to something. Perrin was not so much offering a step-by-step argument for the molecular origin of Brownian motion as he was saying to those who question it, "Your turn, you offer some other explanation of the granule kinetic energy results from the vertical-gradient experiments." Maybe this challenge was part of what Ostwald had in mind in citing Perrin in the 1909 edition of his *Grundriss*. Yet it is not immediately obvious why Ostwald would have found this challenge any more forceful than challenges he had been hearing for years to the same effect: "Your turn, you offer some other explanation of the laws for gases and dilute solutions." If the explanatory successes of molecular theory had never convinced him before, why should this one have done so? For it in no way offered a resolution to "the very serious difficulties" he had spoken of the kinetic theory as having encountered.

Third, Ostwald's two "Perrin (1908)" citations in the new edition were not to either the May or the September papers, but to a paper that appeared in early October 1908 entitled "Grandeur des molécules et charge de l'électron."[98] The main point of this short paper was a claim that his vertical-gradient experiments were providing a method for determining the value of Avogadro's number that is "direct and susceptible to unlimited precision." The claim was based on counting 13,000 granules with 16,000 readings in three different experiments "with granules of very different sizes" that "have led to substantially the same value for *N*." The paper then gave that value as 71×10^{22}, inferring from it values for the charge of the electron (4.1×10^{-10} esu), the kinetic energy per molecule, the masses of the oxygen molecule, the hydrogen atom, and the electron, and the diameters of molecules of oxygen and helium. Given Ostwald's demand for grounding of the parameters of molecular-kinetic theory, we can see why he would have cited this paper. Again, however, it is not obvious why he would have taken this paper to be giving evidence for the "discrete or grained nature of matter" unless all he had ever been asking for in the way of such evidence was well-behaved theory-mediated measurement of molecular parameters.

Judging from what he said in the section on Brownian motion added to the 1909 edition, Ostwald also saw the recent experimental results as confirming the proposal that this motion is a "consequence of the kinetic nature of heat, which he [Chr. Wiener] supposed to consist of the motion of the smallest particles of liquids or of their molecules, their impacts on the microscopic grains giving rise to the Brownian movements."[99] The evidence he then cited was the rough agreement between (differing) calculations of granule movements made by Einstein and Smoluchowski, on the one hand, and Svedberg's measurements. It would appear, therefore, that Ostwald was looking as well for evidence of the thoroughgoing irregularity of motions within the liquid substrate consistent with the kinetic theory of heat, evidence that he found in Svedberg's results.

As we shall discuss shortly, however, Einstein had strongly criticized the conclusions Svedberg had drawn from his results in a journal Ostwald knew well. Moreover, at the end of November 1908, shortly after Ostwald's book went to press, Chaudesaigues, working under Perrin, published an article providing significantly stronger evidence for Einstein's equation than Svedberg's.[100] After contrasting the value for Avogadro's number inferred from his results (64×10^{22}) with Perrin's value, Chaudesaigues concluded that his results had confirmed the perfect irregularity of Brownian motion calculated in accord with "the errors of chance," remarking that "this is without doubt a very beautiful application of the calculus of probabilities to a natural phenomenon." A few months later, Perrin drove this point home by publishing in his 1909 monograph far more extensive evidence of the Gaussian distribution of the granule displacements.

By the time the rest of the world was reading the new edition of Ostwald's *Grundriss*, therefore, the evidence that Ostwald was citing had been replaced by significantly stronger evidence. Precisely what Ostwald had in mind when he cited Svedberg and Perrin (1908) therefore ceased to be relevant, insofar as everyone, including him, would by then be citing Perrin's 1909 monograph. That still leaves us with the question we posed earlier, now reformulated to reflect the later date: What conclusions pertaining to the discrete nature of matter could at the time have legitimately been drawn from Perrin's evidence for the thoroughgoing irregularity of Brownian motion and the universal proportionality of the mean kinetic energy of the granules to temperature alone?

To answer this, we can still start from Ostwald's statement of the challenge posed by Brownian motion in his book:

> Such particles do not remain at rest, but are in a state of continuous, irregular motion, the paths being longer or shorter straight lines, which follow each other in a zigzag fashion. The movements are not due to diffusion equalising accidental differences of concentration, for suspensions of indifferent substances have been kept for years without any appreciable diminution of the movements. External causes appear to have been completely eliminated.
>
> The smaller the particles and the lower the viscosity of the liquid, the more rapid are the movements. Spontaneous suspension, which occurs when very fine powders are covered with water or other liquids, appears to be directly due to this effect. As this process rapidly falls off with increase in size of the particles, while sedimentation under the action of gravity increases, it is possible to draw conclusions regarding the size of suspended particles by observing whether sedimentation occurs or not.
>
> These movements are in apparent contradiction to the Second Law. The velocity depends on the viscosity, the effect of which is to use up the energy of the moving particles. The movement continues in spite of this, unless the particles coagulate together (a suspension of vermilion had not lost its movement after two years), so that energy at rest becomes energy in motion.[101]

This statement of the problem has to be amended in one important respect, as a consequence of Perrin's values for the mean translational kinetic energies of the granules. Those values confirmed Einstein's insistence that the mean velocity of the granules is much greater than it appears to the eye to be, that is, much larger than values based on the time any single granule appears to be moving in a straight line of an observed length. Svedberg, in the paper which Ostwald cited as support for Einstein and Smoluchowski, had failed to appreciate this. This had triggered Einstein's short note in 1907 in *Zeitschrift für Elektrochemie* criticizing Svedberg.[102]

The announced purpose of Einstein's paper was "to facilitate for physicists who handle the subject experimentally the interpretation of their observations as well as the comparison of the latter with the theory." In the first of three enumerated points, he gave an estimate of the "mean value of the instantaneous velocity which a particle may have at the absolute temperature T"—namely 8.6 cm/sec for the platinum particles Svedberg had employed. In the second, he deduced from this velocity that the friction of the fluid on Svedberg's particles would deprive them of 90 percent of that velocity within 3.3×10^{-7} seconds, leading him to conclude:

> But, at the same time, we must assume that the particle gets new impulses to movement during this time by some process that is the inverse of viscosity, so that it retains a velocity which on an average is equal to $\sqrt{\text{mean } v^2}$. But since we must imagine that direction and magnitude of these impulses are (approximately) independent of the original direction of motion and velocity of the particle, we must conclude that the velocity and direction of motion of the particle will be already very greatly altered in the extraordinarily short time θ, and, indeed, in a totally irregular manner.[103]

The final point contrasted the short time θ with the observed changes in position over times τ "substantially greater than θ" in order to conclude:

> Since an observer operating with definite means of observation in a definite manner can never perceive the actual path traversed in an arbitrarily small time, a certain mean velocity will always appear to him as an instantaneous velocity. But it is clear that the velocity ascertained thus corresponds to no objective property of the motion under investigation—at least, if the theory corresponds to the facts.[104]

We have quoted Einstein's note at length because it drives home how radically Ostwald's statement of the problem posed by Brownian motion has to be amended. To begin with, if Einstein's theory "corresponds to the facts," what was being observed under the microscope was not even remotely the actual motions of individual granules. The time-scale on which they lose energy to and then regain it from the liquid is miniscule compared with the time-scale of the changes in their location and direction of movement that were being observed and recorded. The spatial-scale of any straight line paths of the individual granules before they change direction is correspondingly miniscule compared with the "paths" that were being observed. And, still more important, the irregularity of the granule motion observed is merely a statistical aggregate, and hence a smoothing, of the far more extreme irregularity that is actually taking place. Considering all of this, just what was the theoretical challenge posed by Brownian motion, (1) the challenge posed by the observed motions, or (2) the challenge posed by the motions implied by Einstein's theory?

The issue posed by Ostwald was whether Brownian motion, as he had characterized it, is a "consequence of the kinetic nature of heat," consisting of "the motions of the smallest particles of liquids or of their molecules, their impacts on the microscopic grains giving rise to the Brownian movements."[105] Strictly speaking, Einstein's theory did not presuppose that impacts of the smallest particles or molecules on the granules are what give rise to their motions. The assumptions it made concerned the motions of the granules themselves, including both the irregularity of that motion and the proportionality of their mean kinetic energy to the absolute temperature of the liquid. In other words, it was not a theory of the cause of Brownian motion. Rather, it was a theory about how the true motions of the granules yield their apparent motions. Of course he intended it to eliminate any other cause of Brownian motion except molecular agitation. The theory for which Perrin provided such strong evidence, however, concerned only the motions of the granules themselves. Two steps were thus involved, the first from the observed motions to the motions claimed by Einstein's theory, and then from those motions to their being caused by the motions of discrete elements of the liquid. Perrin's evidence pertained to only the first of these two.

5.6. Continuity from Granule to Liquid Substrate

Perrin's evidence nevertheless went well beyond merely confirming Einstein's equation for ξ^2/τ, the mean of the squared granule displacements over a sequence of 30-second time increments. As we noted earlier, Perrin provided clear evidence, independently of Einstein's theory, that

(1) At any given temperature, granules in Brownian motion in uniform dilute emulsions have the same mean kinetic energy, proportional to the absolute temperature, regardless of their magnitude.

And, as we also said, his measured values of the kinetic energy confirmed the order of magnitude of Einstein's claim about the mean instantaneous velocities of the granules being so much larger than their apparent velocities. This was correspondingly evidence as well that the granules must be undergoing far more frequent changes in both speed and direction than they appear to be under a microscope.

Perrin's evidence went still further, however, as Deborah Mayo has so correctly emphasized. Again independently of Einstein's theory, Perrin's laboratory provided strong evidence that the statistical distribution of the displacements of granules over observable time intervals is Gaussian. This evidence left open questions about the precise time-scales on which granules change speed and direction and the statistical distribution of their velocities, either at any one time across all the granules or over time for any one granule. The evidence for the Gaussian distribution nevertheless was evidence for one claim about the motion below the level of observability:

(2) The observable irregularity of the granules in Brownian motion arises from a thoroughgoing irregularity in the changes in speed and direction that they individually undergo at time-scales much shorter than can be observed.

Put in more modern terms, it was evidence that Brownian motion arises from a stochastic process at very short time-scales. And hence it was evidence, too, that the changes in motion of the granules, while conserving the mean of their translational kinetic energy, are uncorrelated with one another.

Regardless of what Ostwald intended with his statement of the problem we quoted earlier, we have reason to take the post-Perrin challenge posed by Brownian motion to be one of identifying an origin for it that was consistent not only with the seeming eternality of the motion that he stressed, but with (1) and (2) as well. We grant that (1) and (2) reach beyond observation and hence involve inferences. Those inferences, however, do not presuppose molecular theory, nor any other view of the nature of the matter in the liquid substrates of the emulsions. Our post-Perrin statement of the challenge posed by Brownian motion therefore does not beg any questions about its origin. We also grant that, even fully aware of what Einstein and Perrin had said in print, those at the time would have required effort to grasp this post-Perrin description of Brownian motion—especially so if they were to see that it does not presuppose molecular-kinetic theory and hence were thereby to avoid merging it into that theory.

Let us return, then, to our question concerning what conclusions, if any, could have legitimately been drawn at the time about the discrete nature of matter from Perrin's results on Brownian motion, but now starting from this post-Perrin description of the motion. Each granule is constantly, and randomly, changing speeds, losing kinetic energy to the liquid as a consequence of the resistance to the motion arising from the viscosity of the liquid, yet gaining kinetic energy as a consequence of the action of the liquid on its surface. The viscous resistance to the motion does not change its direction; but the other action of the fluid constantly, and randomly, changes the direction in which the granule is moving, with no direction at any time favored over any other. These changes in speed and direction are occurring, if not strictly continuously, then thousands, if not millions, of times per second. Yet, through all of this, the mean translational kinetic energy of each granule over a long enough period of time remains the same, and the mean of those energies over all the granules remains the same, where both of those means vary linearly with the absolute temperature of the liquid. Furthermore, in uniform dilute emulsions at equilibrium, the temperature of the liquid remains the same over the course of this myriad of energy exchanges between it and the granules, and their motion, in seeming defiance of the second law of thermodynamics, persists for months, even years.

What does all this require of the liquid? In order for the action of the liquid to be changing the direction of motion of a granule, regardless of how that action changes its speed, the pressure on the surface of the granule must be nonuniform. In other words, pressure-gradients must be present in the liquid even though it is in thermodynamic equilibrium. Moreover, in order for these pressure-gradients to be changing the motion of the granules thousands if not millions of times per second, the pressure-gradients have to be highly local, on a spatial scale smaller than the diameter of the smallest of the granules Perrin tested. While any such local pressure-gradient is acting, the granule on which it is acting is transferring energy to the liquid as a consequence of the liquid's viscous resistance to the granule's motion. Consequently, the local pressure in the liquid must constantly be changing, again at a rate at any one location of thousands if not millions of times per second.[106] The local pressure-gradients must therefore be associated with and hence arising in conjunction with highly localized, extraordinarily rapid pressure fluctuations occurring continually throughout the liquid. Insofar as the changes of motion of the granules are totally irregular, with no correlation from one granule to the next, the magnitudes, if not the periods, of the pressure-fluctuations must be irregular. The hydrodynamic upshot of (2), accordingly, is that at every point in the liquid the pressure must be fluctuating irregularly at extremely high rapidity.

All of this is occurring at a scale and magnitude below our capacity to detect, not just by means of a microscope, but with any instrument we have for measuring local pressure gradients. Our access to the fluctuations in the liquid is coming purely from the effects they must be having on granules in Brownian motion.

Hydrodynamics further requires any pressure fluctuation to be accompanied by motion within the liquid. Attendant to the velocity of this local motion is kinetic energy, for a mass, however small, of the fluid moves during the course of each fluctuation. If the maximum pressure-gradient in the pressure fluctuations at any point varies with time, so must the levels of kinetic energy of the associated fluid at each point. In other words, just as the kinetic energy of each granule varies irregularly over times that are very short, the kinetic energy at each point in the liquid must vary irregularly over times that are very short. The answer to Perrin's question of the "origin of Brownian motion" lies in the constant exchanges between the kinetic energy of each granule and the kinetic energy of the liquid surrounding it. Moreover, because the mean kinetic energy of the granules is the same at any given temperature regardless of the mass of the granules or the liquid, the pressure fluctuations within the liquid must be present independently of the granules.

Now consider the mean, at any instant over the whole liquid or at any point over extended periods of time, of these locally varying levels of kinetic energy within the liquid, for mean values should exist regardless of what they are taken over. Because the mean kinetic energy of the granules varies linearly with the absolute temperature of the liquid, so too must the mean kinetic energy of the local pressure fluctuations within the liquid. And because the mean kinetic energy of the granules persists indefinitely over time, so too must the mean kinetic energy of the local pressure fluctuations within the liquid. So, just as Ostwald said, we see *that Perrin's results for Brownian motion provided evidence for the kinetic theory of heat in fluids generally*—i.e., *for heat giving rise to highly localized motions in fluids—without any need to invoke chemical molecules themselves.* Moreover, with the exchanges of kinetic energy persisting indefinitely and the energy of the granules arising from the energy of the pressure fluctuations, the mean kinetic energy of the granules, if not identical with the mean kinetic energy of the pressure fluctuations, must at least be proportional to it.

Our conclusions concerning the liquid have so far resulted from hydrodynamic reasoning and hence from considerations that lie within continuum mechanics. Those conclusions, however, are not really consistent with the liquid behaving hydrodynamically, as if it formed a continuum, for any local pressure fluctuation gives rise to a pressure wave propagating through the fluid at the speed of sound; but pressure waves in a viscous fluid, unlike our local pressure fluctuations, gradually die down as a consequence of the internal friction of the fluid. For that matter, any local pressure fluctuation in a viscous fluid dies down as a consequence of the internal friction of the fluid. So, our local pressure fluctuations do not accord with the laws governing pressure and velocity in continuous fluid media, the Navier-Stokes equations. In order for the mean value of the kinetic energy of the pressure fluctuations to persist over time, yet their local kinetic energies to constantly vary, kinetic energy has to transfer from one fluctuation to another. But the process of that transfer cannot be hydrodynamic.

There is a still stronger reason to conclude that the liquid cannot be acting as a continuum mechanical fluid, a feature implied by the post-Perrin description of Brownian motion that we have not yet taken into consideration. According to that description, the motions of the individual granules are in no way correlated with one another. To say it more cautiously, Perrin's evidence that the displacements of granules over extended periods of time have a Gaussian distribution was evidence as well that the motions of the individual granules are not correlated with one another. But these motions are arising from the local pressure fluctuations in the liquid. Therefore, Perrin's evidence was also evidence that these local pressure fluctuations are not correlated with one another. Even though kinetic energy is being exchanged among them, the irregularly varying magnitude and timing of any one pressure fluctuation must accordingly be uncorrelated with and hence in that regard independent of the varying magnitudes and timings of the adjacent pressure fluctuations. Because this cannot occur within a fluid continuum, the natural conclusion is that the liquid consists of highly localized cells that can transfer kinetic energy among one another at their boundaries, but within each of which the matter forming the liquid is constantly moving in an extremely irregular fashion, creating the irregular local pressure gradients, the effects of which then become visible in Brownian motion.

Accordingly, the post-Perrin description of Brownian motion, with emphasis on its characterization of the irregularity of that motion, did provide evidence that the matter forming the liquid in which the motion takes place is not continuous. Earlier we quoted Ostwald's statement of the problem posed by Brownian motion, which we then amended on the basis of Perrin's evidence pertaining to the degree of the irregularity of motion called for in Einstein's theory. Let us now quote the rest of Ostwald's paragraph and the assertion immediately following it:

> As early as 1863 Chr. Wiener regarded the phenomenon as a consequence of the kinetic nature of heat, which he supposed to consist of the motion of the smallest particles of liquid or their molecules, their impacts on the microscopic grains giving rise to the Brownian movements. This kinetic theory of heat, which was propounded in the early ages, has been extensively developed by Clausius (1857), Maxwell, Boltzmann, and by many later scientists, especially in its application to gases, in spite of the various serious difficulties encountered. Still, until recently there was wanting direct experimental proof of the necessity for such a representation, which accordingly remained an arbitrary though useful hypothesis.
>
> If the movements of a small particle suspended in a liquid are calculated on the basis of this hypothesis, the agreement with the movements actually observed is so close that we are compelled to regard this agreement as a fairly satisfactory proof of the kinetic theory of heat.[107]

Assuming Ostwald knew and took to heart what Einstein had said about the degree of irregularity of the motion of the granules in his 1907 note criticizing Svedberg,[108] and therefore was thinking in terms of our (1) and (2), the only complaint that can be lodged against his reasoning here is that he was relying on evidence that was weak compared to the evidence that Perrin presented a year later in his 1909 monograph.

So, we have an answer to our question concerning how the results on Brownian motion could have contributed to Ostwald's statement in the 1908 Preface of the new edition of his *Grundriss*, "I am now convinced that we have recently become possessed of experimental evidence of the discrete or grained nature of matter, which the atomic hypothesis sought in vain for hundreds and thousands of years."[109] His phrasing, "the necessity for such a [discrete or grained] representation," fits the reasoning we have presented, for its main conclusion is that treating the liquid as a true continuum is inconsistent with what its action must be to maintain Brownian motion. Phrases like Ostwald's "smallest particles of liquid" and our "highly localized cells"—in the same vein as "discontinuous structure of matter" in Perrin's Nobel Prize citation—point to how little in the way of specifics Brownian motion can give us about the nature, or for that matter even the number per unit volume, of the discrete elements forming the liquid. To pursue empirically grounded answers to questions about the actual size and activity of the discrete elements, scientists at the time had to turn to phenomena other than Brownian motion. And until such answers began to emerge, the conclusion about the existence of these elements—or, if you prefer, "the necessity of such a representation"—had to remain provisional.

We have pointed out more than once that, in order to reach his conclusion about the molecular origin of Brownian motion, Perrin consistently compared his Brownian motion values for Avogadro's number with ones inferred using van der Waals's theory. The reasoning we have offered for the conclusion about the discrete nature of the liquid stops short of giving any tie to "molecules"—even the molecules of Maxwell, which had no characteristics beyond their proposed dynamic activity, much less the molecules of chemistry. From the perspective of the reasoning we have offered, therefore, Perrin required such a further line of evidence, beyond his results for Brownian motion, in order to conclude that molecular agitation is the origin of the motion. Even the paragraph we quoted back in section 3.5 of Chapter 3 from the 1916 edition of Jeans's *The Dynamical Theory of Gases* notes the gap which we are here emphasizing:

> The second hypothesis, the identification of heat with molecular motion, is that with which the Kinetic Theory of Matter is especially concerned. This hypothesis was for long regarded as pure conjecture, incapable of direct proof, and probable just in proportion to the number of phenomena which could be explained with its help. In recent years, however, the study of the Brownian movements has provided brilliant visual demonstrations of the truth of this conjecture, and the actual heat-motion of molecules—*or at least of particles which play a role exactly similar to that of molecules*—may now be observed by anyone who can use a microscope.[110] [emphasis added]

This quotation remained word-for-word intact on page 1 of each of the two subsequent editions of Jeans's book.

Ostwald equally had reasons for limiting his conclusion about the origin to discrete elements moving in general accord with the kinetic theory of heat. He knew well the final chapter of Meyer's 1899 *Die kinetische Theorie der Gase*, which made clear how open to question any conclusions about Avogadro's number were from van der Waals's theory; and, anyway, his ultimate interest was with the molecules of chemistry. Ostwald must have seen the evidence from Brownian motion as at most a step, with a multitude of questions still demanding empirically based answers in order to complete the bridge between the visible and micro-chemistry.

This section was prompted by the intuition, which we today still share with those at the time, that even without full continuity between the motion of granules and the motion of molecules, Perrin's results on Brownian motion verified continuity between the theory of it and molecular-kinetic theory. Even though the reasoning we have offered falls far short of reaching all the way to *molecular*-kinetic theory, it should still be regarded as continuity reasoning. For it has proceeded from confirmed extraordinary features of Brownian motion to the next level down, to features required of the liquid in order for it to support such motion; and these features correspond at least to the rudiments of the kinetic theory of heat. So, just as we proposed about same-effect-same-cause reasoning at the end of section 5.4, we end this section with the proposal that the continuity reasoning we have offered makes explicit and thereby displays the legitimacy of a line of thought among scientists at the time that remained largely intuitive. *Independently of any assumptions about molecules, Perrin's results really did provide evidence for the discontinuous structure of matter.*

We can make this point in a slightly different way that may perhaps state both aspects of the conclusion we have reached more precisely. On the one hand, Brownian motion, especially its confirmed random character, provides sufficient experimental access to kinetic activity in the liquid substrate to show that this activity does not occur in the manner of motion in a continuous medium, and hence to show the discontinuous localized character of this motion. On the other hand, Brownian motion does not provide sufficient access to the kinetic activity in the liquid substrate to give any information whatever on details of the interaction between a granule and the liquid surrounding it. This would require gaining access to a far smaller scale of activity than the experimental investigation of granule motion itself can provide. Absent such more discriminating access, Brownian motion does not even show that there are any individual countable items forming the fluid, for the highly localized pressure fluctuations that it does reveal need not persist through time as distinct, countable individuals, nor does it show that they must be composed of such individuals. Consequently, all that Perrin's determinations of the number of granules required for their mean translational kinetic energy to match $3RT/2$ amount to are counts of the number of granules required for their mean translational kinetic energy to match this macroscopic quantity.

5.7. A Brief Recap

The principal question in this chapter has been, what evidence in support of both the kinetic theory of heat in general and the molecular-kinetic theory in particular did Perrin's Brownian motion results, *by themselves*, provide? Subordinate to that has been the question of whether van Fraassen is correct in claiming that those results "grounded," in his sense, molecular-kinetic theory, thereby for the first time turning it into (in Ostwald's words) a scientifically well-founded theory. Our answer to the latter question is a qualified no. We grant him that Perrin's values for Avogadro's number provided a more promising candidate than any of those to be found in Meyer's *The Kinetic Theory of Gases*, and hence too more promising candidates for values of some further parameters of molecular-kinetic theory, including, in particular, molecular masses. Perrin's determination of Avogadro's number was too tentative, however, depending as it did on an assumed equality of the measured mean kinetic energy of Brownian motion granules and the mean kinetic energy of hypothesized molecules; and the span of the parameters fundamental to molecular-kinetic theory was too limited to meet van Fraassen's requirement of measured values for all, or even most, of them.

Van Fraassen also argued that Perrin's results for Avogadro's number cannot have amounted to a test for the reality of molecules because those results presupposed this equality of kinetic energies, and hence the existence of molecules. We have reached an even stronger conclusion about Perrin's Brownian motion results not amounting to a test of molecular-kinetic theory. We grant that Einstein, and to an extent Perrin too, derived the equations underlying the impressive theory-mediated measurements achieved in the latter's experiments on Brownian motion from molecular-kinetic theory. As any careful review article at the time would have noted, those equations could have been derived from assumptions far weaker than the corresponding claims of molecular-kinetic theory. Langevin's derivation of the Einstein equations for granule displacements and diffusion show that no specific claims from molecular-kinetic theory are required.[111] What the successful measurements therefore really tested was not molecular-kinetic theory, but the weaker assumptions about Brownian motion itself presupposed by the equations in question. Indeed, for most of these assumptions Perrin provided direct evidence independently of his indirect measurements of granule mean translational kinetic energy. We grant that the derivation of the governing equations from molecular-kinetic theory did provide a test of that theory, but only a very weak one, confirming merely its consistency with the findings on Brownian motion.

By virtue of same-effect-same-cause reasoning, a slightly stronger conclusion than mere consistency came out of two of Perrin's findings on Brownian motion, the exponential vertical-distribution of granules under gravity, and granule displacements versus time under fluid resistance during diffusion. Both of these had macroscopic counterparts in gases, and molecular-kinetic theory had supplied equations paralleling those confirmed for Brownian motion to explain those counterparts. We grant that those molecular equations had gained more legitimacy than they had had before from Perrin's clear confirmation of the parallel equations for Brownian motion. What we mean by this is that inferences from the macroscopic counterparts to conclusions about the vertical distribution of molecules under gravity and the displacements of molecules versus time during diffusion gained more legitimacy than they had had before.

Here again we must add that our assessment is meant to represent what a careful review article would have concluded at the time, not what most of those in the community did in fact conclude. They took Perrin to be giving what appeared to be a reliable value for Avogadro's number; doing so was tantamount to their taking the mean translational kinetic energy of the granules to be identical with the mean translational kinetic energy of the molecules in the liquid. As van Fraassen said, however, this in turn was tantamount to taking the existence of molecules for granted and viewing Perrin's results as experimentally determining specifics about them that theretofore had been elusive. We thus agree with van Fraassen on this point.

That said, by far the strongest test that Perrin's results provided, on our view, was for the kinetic theory of heat, construed broadly as the claim that heat takes the form of unseen local motions within gases, liquids, and solids. The combination of (1) the seeming eternity of Brownian motion, (2) the universal mean translational energy of the granules per degree Kelvin, as measured indirectly by Perrin, and (3) the Gaussian, or random-walk, statistical distribution of granule displacements over time entails the existence of extraordinarily rapid, highly localized pressure and velocity fluctuations within the fluid, uncorrelated with one another. Again, that is entirely consistent with those fluctuations in fact consisting of molecules in motion with a Maxwellian-Boltzmann statistical distribution of velocities. The Brownian motion results, in and of themselves, nevertheless do not entail that the pressure fluctuations in the fluid are composed of molecules in motion. There is still a gap, not between the realms of granules and molecules, but between the unseen but implied action within the fluid and the molecules of molecular-kinetic theory.

In sum, the proponents of molecular-kinetic theory had every reason to celebrate Perrin's Brownian motion results, for they left their opponents with an enormously increased burden, namely to offer an alternative to molecular-kinetic theory that would explain those results along with all the other phenomena for which that theory was offering explanations. Insofar as the kinetic theory of heat had been strongly verified to hold for granules in Brownian motion, so too must the kinetic theory of heat hold for the liquid substrate as well, whatever its discontinuous elements in motion may be. Any careful review article at the time, nevertheless, would have noted how many questions about molecules the Brownian motion results left without even the beginnings of an answer, including the problem of reconciling equipartition of energy with the measured values of the specific heats of gases. With those questions still hanging, however much the Brownian motion results in and of themselves increased the attractiveness of molecular-kinetic theory, the question of the reality of the molecules posited by that theory—or, in Nernst's words, to whatever extent molecular-kinetic theory had "begun to lose its hypothetical character"—was not yet experimentally settled. On this, too, we agree with van Fraassen.

This point can be put in another way. In the last section of Chapter 2 we emphasized the extent to which nineteenth-century research had failed to gain experimental access to specific features of atoms, molecules, and their motions. Perrin's Brownian motion experiments had not done so either. Consider the four equations he had used to determine the mean kinetic energies of the granules and then to infer values of Avogadro's number under the assumption that these mean energies match those of the molecules in the underlying substrate. None of the equations relates any specific feature of Brownian motion to any specific feature of those atoms and molecules. The governing parameters in the equations are all macrophysical: the radius of the granules, their buoyant density, and the viscosity of the liquid. Hence, the only Brownian motion parameter Perrin's equations linked to any microphysical parameter was *mean kinetic energy* of the granules to *mean kinetic energy* of molecules, and this they did only through an assumption that the respective energies match.

Even the Gaussian character of the granule displacements gives evidence at most for some form of microphysical Gaussian motion of discontinuous elements in the substrate, and not for any specifics of that motion and how those specifics affect the motion of the granules. Absent further evidence that the energies match, therefore, the extent to which Perrin's efforts gained experimental access to specific features of atoms, molecules, and their motions was at best marginal. That evidence, moreover, could come only from measurements of Avogadro's number independent of Brownian motion, for only they could provide evidence bearing on the question of whether the energies match.

In the years following his initial publications on Brownian motion in 1908, Perrin himself came increasingly to emphasize the extent of agreement between his values for Avogadro's number and values obtained from phenomena that had no apparent relationship to Brownian motion. A shortcoming of this chapter has been its considering his Brownian motion results in isolation from the rest of what was occurring. Indeed, Ostwald himself did not cite Brownian motion alone. To other developments that were taking place during the same period we now turn, still asking our question about the evidence for molecular-kinetic theory in general and for the reality of molecules in particular.

We would be remiss, however, to end this chapter without acknowledging the rhetorical force of putting Brownian motion on display. According to Charlotte Bigg, Perrin initially used a camera lucida to project the motion live on screens in his lectures, but then, starting in 1911, he turned to cinematographers for this purpose.[112] He displayed such a film during his Nobel Prize lecture of 1926, saying, "you will see for yourselves the equilibrium distribution of an emulsion formed from the spherules which are agitated by the Brownian movement," followed by the remark, "the observations and the countings which this film summarizes for you prove that the laws of gases apply to dilute solutions."[113] The film did indeed enable the audience to appreciate the visually accessible evidence for the vertical gradient of the granules. One has to wonder, however, how many in the audience took the film itself "to *prove* that the laws of gases apply to dilute solutions." Still, in fairness to Perrin, we must grant that the film went beyond rhetoric in supporting a key point he had made earlier in the lecture—a point entirely in keeping with the conclusion we reached in section 5.6:

> We obtain from this an essential property of what is called a liquid in equilibrium: *its repose is only an illusion due to the imperfection of our senses, and what we call equilibrium is a certain well-defined permanent system of a perfectly irregular agitation.* This is an experimental fact in which no hypothesis plays any part.[114]

A glance at the internet reveals several films of Brownian motion now available to all. (Among these, according to Bigg, is a 1923 film featuring Perrin himself.[115]) With regard to the use of these in classrooms, we wonder here too whether those observing them are made aware of the gap between what they are seeing and the true motion of the granules, or the much larger gap between it and the actual motions of molecules in a gas, much less in a liquid. Regardless, visual displays of Brownian motion surely do have great rhetorical force in a classroom. Unfortunately, the main effect of that force may be to strengthen all the more the sense students acquire in classrooms that the predominant form of evidence in the physical sciences is inference to the best (available) explanation.[116]

Notes

1. See, for example, Taylor, Parker, and Langenberg (1969a). This monograph presents a program for developing evidence for quantum electrodynamics from agreeing measurements that depend in different ways, and to different degrees, on QED; that program has been continued ever since within CODATA.
2. Perrin (1910), p. 24 footnote, in reference to the logarithmic vertical gradient. We are here taking Perrin to be using the phrase "*experimentum crucis*"—literally, "experiment of the cross"—in its classic Baconian sense, an experiment that selects among alternatives posed in the manner of fingerposts at a crossroads.
3. W. Ostwald (1912), p. 485f.
4. Ibid.
5. Einstein (1956), p. 18; (1905), p. 559; (1989a), p. 234; (1989b), p. 133f.
6. Ostwald, (1912), p. 485f.
7. Ibid.
8. Perrin (1910), p. 24.
9. In his (1908c), Perrin listed granule kinetic energies, though there too he gave corresponding values for *N*.
10. Perrin (1910), p. 46.
11. Ibid., p. 48.
12. W. Ostwald (1912), p. 486.
13. Perrin (1910), p. 46.
14. See Brush, Everitt, and Garber (1986), especially pp. 300, 351f, 354–358, and 360–386.
15. Meyer (1899b), p. 24; italics in the original.
16. Perrin (1910), p. 22. In a footnote Perrin cites Wiener, Ramsey, Exner, Zsigmondy, and himself as those who published such results. Of course, he immediately goes on to deny that the apparent velocities have any relation to the actual ones.
17. Ibid., p. 46.
18. See Chaudesaigues (1908) and Perrin (1990) and (1916), p. 121; in the original French, Perrin (1913), p. 173, or (2014), p. 219.
19. Einstein (1907). Ostwald does not cite Einstein's paper in his discussion of Svedberg's results, but he is quick to refer to Perrin as having provided "still more definite confirmation" than Svedberg in 1908.
20. Sidgwick (1884). Sidgwick's discussion of "best explanation" is within a chapter entitled, "The Employment of Guesswork, Continued."
21. Cohen and Schofield (1978), p. 106.
22. C. W. W. Ostwald (1915), pp. 207 and 209.
23. See Atkins and de Paula (2010), p. 772f.
24. P. Ehrenfest and T. Ehrenfest (1959), p. 37.
25. Mayo (1996), pp. 222–231.
26. van Fraassen (2012), p. 781.
27. Actually, what Ostwald says (1912, p. 486) is that Perrin "obtained a value for *R*, the gas constant, in close agreement with the generally accepted value, but has developed the method to such an extent that it is probably the most accurate value for

the molecular constant obtained as yet." We assume that the reference to R is just a careless slip, first because nowhere does Perrin propose a new value for R, and second because Ostwald himself gives a value for R to four significant figures (1912, p. 163), which far exceeds the precision of the value of any parameter that Perrin obtains from Brownian motion.

28. In the second (1895) English edition, the chapter entitled "The Kinetic Theory of Gases" occupies pages 61–67; in the third (1912) edition, it occupies pages 503–510. The only change over the first seven of the pages is dropping two paragraphs, the first commenting on the significance of the speeds of molecules for chemical reactions (1895, p. 65), and the second, on the a/V^2 term in van der Waals's equation (1895, p. 67).
29. W. Ostwald (1912), p. 509.
30. Ibid., p. 510.
31. Nernst (1911), p. 437; in fact, Nernst's translator attributes a value of 3.17×10^{16} molecules per cubic millimeter at 0°C and atmospheric pressure to Perrin, citing Perrin (1910); multiplying this number by 24,000,000 yields 71×10^{22}.
32. This is our W_T, 3/2 of Boltzmann's constant.
33. Perrin (1910), p. 51.
34. See Nernst (1923), p. 219. Also, Meyer (1899b), pp. 273–276 and pp. 291–296.
35. Nernst (1923), p. 515.
36. Jeans (1916), p. 9. This book had appeared, in a very different form, in a first edition in 1904, and it subsequently appeared in two further editions, in 1921 and 1925, the last of which is available in Dover Press, 1954.
37. See Anderson, Boghosian, and Traynor (1993), especially pp. 345–348.
38. van Fraassen (2009), p. 23.
39. See Jeans (1916), pp. 266–347.
40. Ibid., p. 8.
41. Jeans (1921), p. 8. We have replaced the symbol for Avogadro's number used by Jeans with the one we are using in this book.
42. van Fraassen (2009), p. 21.
43. Ibid., p. 23
44. Ibid., p. 22f.
45. W. Ostwald in the Preface to the 1909 edition of his *Grundriss*, (1912), p. vi; emphasis added.
46. Perrin (1910), p. 24. Recall that he goes on to say, "In the event of an affirmative answer, the origin of Brownian movement will be established, and the laws of gases, already extended by Van't Hoff to solutions, can be regarded as still valid even for emulsions with visible granules."
47. Thomson (1906).
48. For a detailed account of this construal of Thomson's efforts on the electron, see Smith (2001b).
49. W. Ostwald (1912), p. 485.
50. Ibid., p. 491.
51. Perrin (1910), p. 46.

52. Nernst (1911), p. 207, italics his; for the original German, see Nernst (1909), p. 212.
53. Poincaré (1963), p. 90.
54. Duhem (1991), p. 8.
55. Duhem at one point (1991, p. 98) refers to J. J. Thomson's "curious experiments." By the time of the second edition of *The Aim and Structure of Physical Theory* in 1914, values for the charge and mass of the electron were being quoted to four significant figures. He seems to have been at a loss at what to say about the particulate character of electronic charge.
56. W. Ostwald (1912), p. 538f.
57. Ibid., p. 485. Readers should keep in mind that Ostwald was citing the first of these two before Perrin's results on the statistical distribution of the displacements of the granules were published.
58. Cohen and Koyré (1972), p. 550 (p. 387 in the original Latin, p. 795 in Newton, 1999). The translation given here follows the Latin text of the second and third editions literally, in particular, the verb *assignandae*. In the first edition the phrasing is not "are to be assigned," but simply "are."
59. Huygens (1690), p. 167 of the French original, p. 477 in *Oeuvres Complètes*, vol. 21.
60. Ibid., p. 168 of the French original, p. 477f in *Oeuvres Complètes*, vol. 21.
61. Maxwell (1875), p. 433.
62. We elaborated on the underlying point made in this paragraph in section 4.3.
63. Meyer (1899b). The book presents the theory in terms of its historical development, relegating much of the mathematics to a series of six appendices.
64. Ibid., p. 161.
65. Ibid., p. 155.
66. Ibid., p. 189.
67. Ibid., pp. 188–200.
68. Ibid., pp. 256–260. Meyer replaces the symbol L with a script-L not available to us. How to treat diffusion in terms of a mean-free-path analysis remained contentious in the years following Perrin's experiments in spite of Chapman and Enskog on non-uniform gases. See, for example, the 1925 edition of *The Dynamical Theory of Gases*, Jeans (1954), pp. 307–323.
69. See Meyer (1899b), Chapter VIII.
70. Ibid., p. 275.
71. Ostwald (1912), p. 508.
72. Perrin (1910), p. 48.
73. Ibid., p. 51.
74. We are being generous here in granting that Perrin showed W_T is constant in his granule displacement results, for (so far as we can tell) the temperature was systematically varied only in his vertical-gradient experiments (Perrin, 1990 and 1916, p. 104; Perrin 1912a, pp. 185–187). In all three of these works, however, Perrin's summaries (1990 and 1916, p. 132; 1912a, p. 216) of all his results from Brownian motion include in their list the variation of temperature from –9 to +58°C. Insofar as such a change in temperature has a notable effect on the viscosity of the liquid substrate, and viscosity is a major factor in granule displacement, but not in vertical distribution, one would

naturally have expected the displacement experiments to have included such a systematic variation in temperature.

75. Einstein (1956), p. 60f; Perrin (1910), p. 75f.
76. Einstein (1956), p. 73; (1908), p. 236; (1989a), p. 498; (1989b), p. 320f. Einstein used a different symbol from our Z.
77. See, for example, Hecht (1990), p. 348.
78. Perrin (1910), pp. 76–78.
79. W. Ostwald (1912), p. 535.
80. Townsend (1900), p. 153.
81. Perrin (1910), p. 81.
82. W. Ostwald (1912), p. 539.
83. Perrin (1910), p. 81.
84. We thank Jody Azzouni for calling our attention to the need to remind readers that those at the time tended to accept Perrin's conclusions about the molecular realm from his results without much hesitancy, clearly less hesitancy than we are claiming was appropriate. As we shall see in Chapter 6, and then again at the end of Chapter 7, the realization that Perrin's values for Avogadro's number, and hence too his equating the mean translational kinetic energy of granules with that of the molecules in the liquid, are not so reliable as first thought had begun to emerge at more or less the same time as the first few editions of his *Les atomes* were published.
85. W. Ostwald (1912), p. 508ff.
86. Ibid., p. vi.
87. Perrin (1910), p. 20.
88. Ibid., p. 19. Italics in the original.
89. Ibid., p. 46.
90. Ibid.
91. Ibid., italics in the original.
92. Perrin (1908c), p. 530.
93. Ibid., p. 532. The final paragraph of the article proposes to employ further experiments of the same type to "determine, with a precision I believe hitherto unknown, the universal constant N and various constants depending on it, including the charge of the atom or corpuscle of electricity."
94. Perrin (1908a), p. 970.
95. For more details on this context, see Maiocchi (1990).
96. Duclaux (1908).
97. Perrin (1908b).
98. Perrin (1908d).
99. W. Ostwald (1912), p. 484f.
100. Chaudesaigues (1908), p. 1046. Note, in particular, the very close agreement between calculated and observed values for ξ^2 over four consecutive 30-second intervals.
101. Ostwald (1912), p. 484.
102. Einstein (1907). For details on how far removed from Einstein's account of Brownian motion Svedberg was in the paper that Ostwald took to be confirming it, see Maiocchi (1990), pp. 270–272, and Kerker (1976), especially pp. 197–201.

103. Einstein (1956), p. 65f; (1907), p. 41; (1989a), p. 399; (1989b), p. 230.
104. Einstein (1956), p. 67; (1907), p. 42; (1989a), p. 400; (1989b), p. 231.
105. W. Ostwald (1912), p. 484f.
106. We are stating this from the perspective of the period 1905–1913. S. Chandrasekhar, writing in 1943, makes our point far more forcefully: "Under normal conditions, in a liquid, a Brownian particle will suffer about 10^{21} collisions per second and this is so frequent that we cannot really speak of separate collisions. Also, since each collision produces a kink in the path of the particle, it follows that we cannot hope to follow the path in any detail—indeed, to our senses the details of the path are impossibly fine" (Chandrasekhar, 1943, p. 20).
107. Ostwald (1912), p. 484f.
108. We have found no way to confirm that Ostwald had read Einstein's paper when it appeared in 1907, but the journal was the official organ of an organization he had founded in 1895, the Deutschen Elektrochemie Gesellschaft, and he was still publishing in it in Volume 14, the one immediately after the one containing Einstein's note.
109. W. Ostwald (1912), p. vi.
110. Jeans (1916), p. 1.
111. See Nye (1972), p. 121, and Langevin (1908).
112. Bigg (2014), pp. 195–198; see also Bigg (2008), (2011).
113. Perrin (1926), p. 10.
114. Ibid., p. 5.
115. Bigg (2014), p. 197.
116. We thank an anonymous reviewer for insisting that we include some acknowledgment of the rhetorical force of witnessing Brownian motion.

6

Converging Values for Avogadro's Number

Perrin's Comparisons

Those familiar with the philosophical literature on Perrin might be wondering why we have postponed to last his appeals to agreement among several different theory-mediated ways of measuring Avogadro's number, both in his 1909 monograph and his 1911 Solvay Conference paper. Perrin became a subject in this literature only after Wesley Salmon, in his 1984 *Scientific Explanation and the Causal Structure of the World*, argued that Perrin's appeal, at the end of a later edition of his 1913 book, to different measurements of N_0 represents an instance of common-origin reasoning.[1] That appeal consisted of a list of 16 different theory-mediated determinations of Avogadro's number, all but one of them within the range from 60×10^{22} to 69×10^{22}, followed by two sentences representing the primary conclusion of the whole book:

> Our wonder is aroused at the very remarkable agreement found between values derived from the consideration of such widely different phenomena. Seeing that not only is the same magnitude obtained by each method when the conditions under which it is applied are varied as much as possible, but that the numbers thus established also agree among themselves, without discrepancy, for all the methods employed, gives to the real existence of the molecule a probability bordering on certainty.[2]

But if this is Perrin's principal argument, why haven't we treated it as such, focusing our attention on it?

There are, moreover, good reasons independently of Perrin's appeal to devote philosophical effort to the nature of the evidence arising from agreeing theory-mediated measurements of constants. Over the course of the twentieth century, this form of evidence has become increasingly important within microphysics. Consider, for example, the following conclusion drawn in the 2006 CODATA Report:

> The good agreement of the highly accurate values of [the fine-structure constant] α inferred from $h/m(^{133}Cs)$ and $h/m(^{87}Rb)$, which are only weakly dependent on QED theory, with the values of α inferred from [the electron

Brownian Motion and Molecular Reality. George E. Smith and Raghav Seth, Oxford University Press (2020). © Oxford University Press. DOI: 10.1093/oso/9780190098025.001.0001.

magnetic moment anomaly] a_e, muonium transition frequencies, and H and D transition frequencies provide support for the QED theory of a_e as well as the bound-state QED theory of muonium and H and D. In particular, the weighted mean of the two values of α inferred from $h/m(^{133}Cs)$ and $h/m(^{87}Rb)$, $\alpha^{-1} = 137.03599934(69)[5.0 \times 10^{-9}]$, and the weighted mean of the two values of α inferred from the two experimental values of a_e, $\alpha^{-1} = 137.035999680(94)[6.9 \times 10^{-10}]$, differ by only $0.5u_{diff}$, with $u_{diff} = 5.1 \times 10^{-9}$. This is a truly impressive confirmation of QED theory.[3]

If we believe that agreeing values from diverse theory-mediated measurements provide such important evidence, then why have we postponed discussing the evidence from Perrin's appeal to such agreeing measurements to the very end? We shall answer that question in the next two sections before turning to Perrin's actual comparisons. This will prepare us for the several different perspectives from which his comparisons need to be considered.

6.1. Some Historical Background

One reason for delaying until now Perrin's focus on multiple values for N_0 comes from the long history of appeals to agreeing values in diverse theory-mediated measurements and the often very different kinds of evidential reasoning drawn from them. Complementary theory-mediated determinations of the values of key parameters go back at least as far as Ptolemy's *Almagest*, initiating a practice in mathematical astronomy that has nowhere been employed to greater effect than in Kepler's *Astronomia Nova*. In that tradition, however, complementary determinations served only to safeguard against being misled by values that depended on potentially erroneous theoretical assumptions. In other words, they served only as cross-checks on one another, not as evidence for their theoretical presuppositions.

The most spectacular instance of complementary theory-mediated measures in the seventeenth century was Huygens's two four-significant-figure determinations of the acceleration of surface gravity, one employing pendulums and the other conical pendulums.[4] Huygens too seems to have viewed these two different approaches as merely cross-checking one another in a historical context in which the value for the acceleration of gravity he obtained in the two ways was notably greater than any direct determination of it that had been made before him.[5] A case can nevertheless be made that Newton saw the measurements as providing the strongest evidence then available for his first two laws of motion insofar as Huygens's conical pendulum measure presupposed a special case of those laws, while his pendulum measure presupposed only Galilean kinematic principles.[6] If this is correct, then Huygens's two measures may have been the prototype for taking agreeing theory-mediated measurements to be providing evidence for the theoretical claims presupposed in them.

Regardless, in the aftermath of Newton, agreeing theory-mediated measurements did come to be viewed as providing evidence beyond mere cross-checks on one another. A celebrated, though now comparatively less known, example became the principal test of Newton's universal—i.e., particle-to-particle—gravity during the second half of the eighteenth and first half of the nineteenth centuries. The oblateness of the Earth could be measured directly. It could also be inferred from the variation of surface gravity with latitude by means of Clairaut's equation, which presupposed universal gravity and hydrostatic equilibrium of the Earth. Decades of effort went into resolving the failure of the values of oblateness from these two to agree with one another, culminating finally in the realization that the Earth is not in hydrostatic equilibrium.[7] Unsuccessful though it was, this was clearly an example of the demand for agreeing measurements serving as a test of a theory by virtue of that theory being presupposed in one method of measurement, but not the other.

Maxwell in 1864 provided the first truly spectacular instance of evidence for a fundamental—indeed, "ontological"—theoretical conclusion coming from agreement between comparatively direct measurements of the speed of light and the inferred speed of electromagnetic propagation implied by his equations. Looking to show that light consists of electromagnetic waves propagating through "an elastic medium," Maxwell derived an equation in his famous monograph on the electromagnetic field for the velocity of propagation of those transverse waves:

$$v = \sqrt{\frac{k}{4\pi\mu}}$$

where k is "the ratio of the electromotive force to the electric displacement" and μ is "the ratio of the magnetic induction in a given medium to that in air under an equal magnetizing force." Using experimental results from Weber and Kohlrausch from the 1850s, Maxwell concluded that the velocity of propagation in air is 310740 km/sec. He compared this with speeds of light in air measured using interferometers by Fizeau, 314,859 km/sec, and Foucault, 298,000 km/sec, and the speed in the space surrounding the Earth inferred from the aberration of light, 308,000 km/sec. From this he concluded:

> Hence the velocity of light deduced from experiment agrees sufficiently well with the value of v deduced from the only set of experiments we as yet possess. . . . The agreement of the results seems to show that light

> and magnetism are affections of the same substance, and that light is an electromagnetic disturbance propagated through the field according to electromagnetic laws.[8]

A good deal of further experimental effort was needed before this proposal became established, and by the time of Perrin's Brownian motion experiments, Maxwell's "elastic medium" had begun falling by the wayside. That made his comparison of the speeds no less a watershed, however, for it did open the way to the science of optics becoming subsumed under electromagnetic theory. Indeed, a preferred value for the speed of light subsequently came to be obtained from Maxwell's equation, not from any direct measurement.

The other famous example, just prior to Perrin, of a fundamental theoretical—again "ontological"—conclusion coming from agreeing measurements involved the determination by J. J. Thomson and then others of the mass-to-charge ratio m/e of what was coming after 1900 to be called the electron. Initially, Thomson had developed a second, complementary way of measuring m/e for cathode rays during the summer of 1897, employing crossed electric and magnetic fields, as a cross-check on his first way, in the process confirming, contrary to Hertz's earlier claim, that the rays are displaced by electrostatic forces. There was no overlap at all between the sets of values obtained for m/e that Thomson reported from the two ways, but strikingly all the values were three orders of magnitude smaller than the value of m/e for hydrogen inferred from electrolysis; and the values showed no systematic variation with changes in the gas and the material of the electrodes in the cathode ray tubes. From this, Thomson concluded that cathode rays all consist of the same negatively charged discrete elements with an m/e of a distinctive order of magnitude.[9]

Thomson's far more monumental conclusion from his m/e measurements was announced in the December 1899 issue of *Philosophical Magazine*. Using a crossed-field method related to the earlier one, he confirmed that m/e for both the photoelectric and the thermionic discharges is of the same distinctive order of magnitude as for cathode rays. Same-distinctive-order-of-magnitude-same-particle reasoning led to the conclusion, subsequently supported by measurements of m/e for other electrical phenomena by others, of the ubiquity of this carrier of negative electricity. On that basis, Thomson made the following watershed proposal:

> It thus possesses the characteristics of being a fundamental conception of electricity; and it seems desirable to adopt some view of electrical action which brings this conception into prominence. These considerations have led me to take as a working hypothesis the following method of regarding the electrification of a gas, or indeed matter in any state.[10]

What follows his proposal, over the next three pages, is the twentieth-century conception of electrical action. At the time Thomson proposed it, he was still relying on merely order of magnitude agreement of values for *m*/*e* in different phenomena. By 1906, however, the value was being measured to three significant figures.

Although not likely known to Perrin, one other example of agreeing measurements from the end of the nineteenth century should be noted. Starting in the 1840s, a good deal of effort went into refining Cavendish's famous 1798 torsion-bar experiment measuring the mean density of the Earth. Even in the 1890s, many of the results from such experiments were still being stated in terms of the mean density of the Earth, but by then it had become at least as common to quote them instead in terms of the fundamental constant of gravity *G*.[11] *G*, however, is just the constant of proportionality in Newton's law of gravity. The agreeing measurements for *G* were thus serving to confirm that law—that is, they were evidence that the three variable quantities entering into it and no others affect the magnitude of gravitational forces. This corresponded to the agreeing measurements of the gas constant *R* earlier in the century, again a constant of proportionality, but for the ideal gas law. As we noted in Chapter 2, the most impressive aspect of van't Hoff's results for dilute solutions was the emergence of this same value *R* for his constant of proportionality. The agreeing measurements for his constant of proportionality thus did more than just confirm van't Hoff's law; as Perrin himself emphasized, and Poincaré as well, they linked dilute solutions to gases.

The main point of this review of historically salient appeals to agreeing measurements was to lay out one reason why we have adopted a cautious attitude toward what to make of Perrin's appeal to alternative determinations of Avogadro's number: *historically, different kinds of evidential reasoning were being drawn from agreeing measurements*. Therefore we should not be too quick to assign one form of evidential reasoning to all the places where Perrin compares his values from Brownian motion with values derived from other phenomena. We had best look at the comparisons one by one, asking in each case what evidential conclusions each can support, before turning to his appeal to the entire collection of them.

As a matter of fact, Perrin's aim in the first of his articles on Brownian motion that compared his value for Avogadro's number with values obtained in other ways was not one of establishing the reality of molecules. The first paragraph of that short October 1908 article is worth quoting in full:[12]

> The number of molecules N that the total gram-molecule contains, the charge *e* of the electron and the quotient α of the mean energy of a molecule by its absolute temperature T are the *universal constants* which are all known as soon

as any one of them is known. In effect, in the electrolysis of a monovalent salt, 96550 coulombs are transported by N atoms, which gives

$$Ne = 3 \times 10^9 \times 96550 = 29 \times 10^{13} \text{ (electrostatic units, C.G.S.)}$$

and, on the other hand, according to the kinetic theory, the product 3RT measures the translational kinetic energy (*force vive de translation*) that any N molecules in a gram-molecule possess, so that

$$2N\alpha = 3R = 3 \times 83.2 \times 10^6.$$

Perrin went on then, in fine print, to give a range for Avogadro's number obtained from packing molecules into a volume and a sphere of equivalent dielectric capacity, 40×10^{22} to 100×10^{22}; a value from van der Waals's reasoning, 60×10^{22}, with an error as high as 40 parts in 100; a range from the measurements from J. J. Thomson's laboratory, 43×10^{22} to 96×10^{22}; the Planck value, 61×10^{22} (obtained from the 2.02×10^{-16} value of α measured for blackbody radiation by Kurlbaum), which Perrin immediately offsets with a value Lorentz "found from a different theory of radiation," 77×10^{22}; and, in a footnote attached to the paragraph, a range from the "beautiful" Rutherford-Geiger article a month earlier, 62×10^{22} to 72×10^{22}. Returning to normal print, Perrin states that he has a method that has the advantages of being "*direct and susceptible to unlimited precision*," namely from his vertical-gradient results, and gives as values 71×10^{22} for Avogadro's number, 4.1×10^{-10} for e, and 1.7×10^{-16} for α. Perrin's point in making the comparisons, accordingly, was not to call attention to the extent of the agreement between the various values, but to propose that his value should be regarded as more reliable than the others.

(Perrin's presentation of the values of others in this October 1908 article initiated an unfortunate practice that he continued at least through the eleventh edition of *Les atomes* in 1921, namely presenting them in a way that undercut any legitimate claims they had to being as well-founded as his own. The egregious example in 1908 was his presentation of the Rutherford-Geiger findings. They had announced a robust measure of $2e$, 9.3×10^{-10}, from collecting α-particles, arguing that it was to be preferred to other directly obtained values for e in spite of its departure from them. From this value they obtained values of 4.65×10^{-10} for "the charge carried by a hydrogen atom," 62×10^{22} for Avogadro's number, 1.61×10^{-24} grams for the mass of "the hydrogen atom," and 2.72×10^{19} for Loschmidt's number.[13] A footnote to this list reads, "It is of interest to note that Planck deduced a value of 4.69×10^{-10} E.S. unit from a general optical theory of the nature of temperature-radiation." A few pages earlier they had reviewed two cross-checks to their value for e, the second of which (by Boltwood) had given a value of 4.1×10^{-10} based on a half-life of radium of 2,000 years, versus their own value of 1,760 years. Immediately before the preceding list of values, they expressly rejected this value for *e* as too small owing to its assumed value for the half-life, and they never derived values for any further atomic parameters from it.[14] The second value Perrin attributed to them, 72×10^{22}, was calculated by him on the basis of this 2,000-year half-life, with nothing in their article in any way authorizing it.

Though not so egregious, Perrin's listing of the value implied by Kurlbaum's measurement of the constant in Planck's blackbody formula, side-by-side with a value Lorentz had "found from a different theory of radiation,"[15] undercut the well-founded claim to precision for the measurement of the constants in the blackbody formula attainable from measuring them at several different temperatures. And Perrin never notes the close agreement between the Planck and Rutherford values. To be blunt about it, one needs always to check the original source of any value for Avogadro's number besides his own that Perrin quotes. We shall return to this point later, for it has had some unfortunate historical consequences, and not just in the recent philosophical literature.)

6.2. What *Agreeing* Measurements?

We have a second reason for being cautious about Perrin's comparisons, a reason arising from the spread among the numbers being compared and concerns over criteria for *agreement* between measurements. Perrin himself kept revising his values for Avogadro's number. In particular, his announced preferred value from his vertical-gradient experiments was 6.7×10^{23} in May 1908 (subsequently corrected to 5.7×10^{23}), 6×10^{23} in September, 71×10^{22} in October, 70.5×10^{22} in his 1909 monograph, 68.3×10^{22} in the table at the end of his 1911 Solvay Conference paper and his *Les atomes* in 1913. From the granule displacement experiments, Chaudesaigues gave a value of 64×10^{22} in November 1908, while Perrin's preferred value was 71.5×10^{22} in his 1909 monograph, the mean of results from a series of multiple trials. A table in his 1911 paper and 1913 book lists results of the trials, with three entries involving more than 500 displacements recorded, yielding values of 69.5×10^{22}, 72.5×10^{22}, and 68.8×10^{22}; in both 1911 and 1913 he chose the last of these values, for which the number of recorded displacements was greatest, as the value he subsequently compared to others.

We are not questioning Perrin's judgment as an experimentalist in preferring later to earlier values, for the difficult experiments were undergoing continuing development over the years in question. Perrin himself offered a summary of his results from his vertical-gradient measure in 1913 that seems to us a reasonable statement of the overall situation for all his results: "In spite of all these variations [in the range of emulsions investigated], the value found for Avogadro's number N remains approximately constant, varying irregularly between 65×10^{22} and 72×10^{22}."[16] This, however, still raises a question about criteria of "agreement" and whether the values can be viewed as in agreement from one comparison to another. Perrin may have said that all the values agree with one another within the respective margins of experimental error; but that does not mean that everyone at the time should have agreed.

Perrin's spread was tight enough not only to deter questions about his values until 1911, but to prompt Ostwald's 1908 remark about Perrin having provided "probably the most accurate value for the molecular constant as yet obtained," 71×10^{22}.[17] By contrast, the spread in values for Avogadro's number obtained in other ways was immense, raising questions about the standard that has to be met in order for different values to agree with one another. Taking the difficulty of many of the experiments and the associated measurement uncertainty into consideration, Wesley Salmon remarked, "Let us say that values lying between 4×10^{23} and 8×10^{23} are acceptable."[18] Salmon was endeavoring to give a philosophical analysis of the evidential force of what he took scientists to be granting were agreeing measurements for N_0. Prior to the issue of the philosophic basis of any such evidence, however, is the issue of whether any, and if so what, claims could legitimately have been drawn at the time from comparisons among the different theory-mediated measurements of N_0. Rather than settling the issue of the proper range for agreement, therefore, Salmon's proposal simply raises the question, on what principled basis should the range for agreement be decided? In the case of J. J. Thomson's *m/e* measurements, taking the range of agreement to be an order of magnitude was justified because all his results were three orders of magnitude less than any prior value for *m/e*. In other words, the range of his measured values was appropriately assessed relative to the smallest prior value. Nothing like that was available, however, in the case of N_0. As Table 2.1 in Chapter 2 showed, the outstanding feature of the prior proposed values for N_0 from kinetic theory was their lack of agreement with one another.

This worry about what properly constituted agreement is our other reason for having postponed until last a discussion of the evidence from the comparisons Perrin made between the values for N_0 derived from Brownian motion and the values derived from other phenomena. It also gives a further reason to consider each such comparison individually, asking what to make of it, rather than turning immediately to the question of what to make of the claimed agreement among so many different ways of measuring it.

We remarked at the beginning of the paragraph before last that Perrin's values went largely unquestioned before 1911. Three individuals ended this in 1911. Millikan published his first oil-drop measurements of electric charge, announcing a value for *e* of 4.891×10^{-10} esu and for N_0 of 59.22×10^{22}.[19] And, working under Millikan, Harvey Fletcher published results on the Brownian motion of oil-drops in air, announcing a value for N_0 of 57.5×10^{22}, with a consequent value for e from eN_0 of 5.01×10^{-10} esu, within 2.5 percent of Millikan's value.[20] While Fletcher, like Perrin, obtained his value for N_0 by taking the mean kinetic energy of his drops to match that of molecules, his approach, he emphasized, had a special virtue:

> This shows that *all the distributions spoken of in Part I., and also the average value of the Brownian displacement in one second can be calculated without knowing the density of the drop, the viscosity or pressure of the gas, or the law of motion of a sphere through a viscous medium. The only assumption involved is that the velocity of steady motion is proportional to the moving force,* an assumption that is fully justified both here and elsewhere.[21]

And finally, published in the midst of the 1911 papers by Millikan and Fletcher, Boltwood and Rutherford had announced a value for Avogadro's number of 60.3×10^{22} by collecting a measured number of α-particles sufficient to form a gas and then determining the volume they occupy at standard conditions.[22] Even though none of these authors expressly challenged Perrin's values in 1911, the claims made for precision by Millikan and Fletcher and the comparative directness of the measure by Boltwood and Rutherford did raise questions about whether Perrin's values might be too large by more than 10 percent.

These questions gained added force in 1912 when Svedberg published results from new Brownian motion experiments on diffusion and displacement of gold hydrosol particles, announcing values for N_0 of 58×10^{22} and 62×10^{22}, respectively—both falling below Perrin's range.[23] Then in 1913 Millikan published his legendary oil-drop results, announcing values for e and N_0 of $4.774 \pm .009 \times 10^{-10}$ esu and $6.062 \pm .012 \times 10^{23}$, and adding, "So far as I am aware, there is at present no determination of e or N by any other method which does not involve an uncertainty at least 15 times as great as that represented in the above measurements."[24] The question of what Perrin's comparisons of his values for N_0 with those obtained in other ways yielded in the way of evidence was accordingly undergoing a metamorphosis over the course of the years between his first published result and our self-imposed historical boundary of the emergence of the Bohr model. This question continued to gain force beyond that boundary, for example when Fletcher in 1914 published a new value, $60.3 \pm 1.2 \times 10^{22}$, based on refined measurements for Brownian motion of oil-drops in the air.[25]

By the way, viewed retrospectively from the other side of our boundary, Millikan's error-bands were mistaken, but not his insistence that his value was more reliable than Perrin's. Perrin's range of values extended from 8 to 20 percent higher than our current (2018) value of $60.2214076 \times 10^{22}$. Subsequent textbooks have attributed the discrepancy to imprecision in his determinations of the diameters of his granules.[26] This, however, does not make much sense. His values obtained from his vertical-gradient and granule rotation experiments depend on granule diameter cubed, while his values obtained from granule displacement and diffusion depend on it to the first power; yet there is no corresponding variation in Perrin's values. Taken together with the viscous action of the liquid on his granules, Perrin's high values for Avogadro's number seem to us more likely to have resulted from his measured values of their mean kinetic energy not being equal to, but instead consistently a little less than, the mean molecular kinetic energy at the same temperatures. It goes without saying that Perrin's published values for mean granule kinetic energy, and hence for Boltzmann's constant, turned out to be less than subsequent molecular values.

In the light of these considerations and those in section 6.1, we are going to examine Perrin's comparisons with those of others in three steps: first, looking at the evidence each of the individual comparisons he made offered in support of the validities of the respective measures; then reviewing his downplaying of the seeming lack of agreement in those individual comparisons; and finally, in section 6.5, considering his claim that the overall set of values, taken together, provided decisive evidence for the reality of molecules, that is, for the salient presupposition that all the different measures had in common.

As an aid to the reader, Table 6.1 lists the principal published values for Loschmidt's number (the number of molecules in a cubic centimeter of ideal gas at standard conditions) and Avogadro's number (the number of molecules per mole)[27] from between Perrin's first published value in 1908 and the 1914 publication of the fourth revised edition of his 1913 book, along with two benchmark values obtained by Planck in 1900 from blackbody radiation and by J. J. Thomson in 1903 from water-drop measurements that were the forerunners of Millikan's oil-drop values. By way of clarification, to say that values are obtained from k means that they are obtained from a measured value of k in conjunction with the equation kN_0 = the gas constant R; and that they are obtained from e means that they are obtained from a measured value of e in conjunction with the equation eN_0 = Faraday's constant (the quantity of electricity required to liberate one mole of a monovalent element in electrolysis). During the period in question there were no "official" recommended values for either the gas constant or Faraday's constant, and hence the quoted numbers depend on the choices for these constants made by the respective authors. (We too often forget what a landmark effort Raymond Birge's 1929 paper, "Probable Values of the General Physical Constants," truly was.[28])

We shall be discussing most of these individual values in the following sections. Two points about them are appropriate to end this section. On the one hand, the spread of the values from 1908 to 1913 was remarkably small compared

Table 6.1 Principal Published Values for Loschmidt's and Avogadro's Numbers, 1900–1914

2.76×10^{19}	61.75×10^{22}	Planck	1900	From blackbody radiation $k = 1.347 \times 10^{-16}$
3.9×10^{19}	$[87 \times 10^{22}]$	Wilson, J. J. Thomson	1903	From $e = 3.1 \times 10^{-10}$ (water drop)
$[31.6 \times 10^{18}]$	$[70.8 \times 10^{22}]$	Millikan, Begeman	1908	From $e = 4.06 \times 10^{-10}$ (water drop)
$[3.0 \times 10^{19}]$	6.7×10^{23}	Perrin	1908	From Brownian motion: vertical gradient
2.72×10^{19}	62.0×10^{22}	Rutherford, Geiger	1908	From α-particle $2e = 9.3 \times 10^{-10}$
$[32 \times 10^{18}]$	71×10^{22}	Perrin	1908	From Brownian motion: vertical gradient
$[29 \times 10^{18}]$	64×10^{22}	Chaudesaigues	1908	From Brownian motion: displacements
$[28 \times 10^{18}]$	62×10^{22}	Perrin	1909	From van der Waals, applied to argon
$[31 \times 10^{18}]$	70.5×10^{22}	Perrin	1909	From Brownian motion: vertical gradient
$[31.9 \times 10^{18}]$	71.5×10^{22}	Perrin	1909	From Brownian motion: displacements
$[29 \times 10^{18}]$	65×10^{22}	Perrin	1909	From Brownian motion: rotation
2.70×10^{19}	$[60.4 \times 10^{22}]$	Regener	1909	From α-particle $2e = 9.58 \times 10^{-10}$
$[2.76 \times 10^{19}]$	$[61.9 \times 10^{22}]$	Begeman	1910	From $e = 4.669 \times 10^{-10}$ (water drop.)
2.72×10^{19}	$[61.0 \times 10^{22}]$	Millikan	1910	From $e = 4.65 \times 10^{-10}$ (water drop.)
2.644×10^{19}	59.22×10^{22}	Millikan	1911	From $e = 4.891 \times 10^{-10}$ (oil drop.)
2.69×10^{19}	$[60.3 \times 10^{22}]$	Boltwood, Rutherford	1911	From collecting α-particles to form a gas
$[2.57 \times 10^{19}]$	57.5×10^{22}	Fletcher	1911	From Brownian motion: oil droplets in air
$[2.75 \times 10^{19}]$	$[61.7 \times 10^{22}]$	Warburg	1911	From blackbody radiation
$[30.4 \times 10^{18}]$	68.2×10^{22}	Perrin	1911	From Brownian motion: vertical gradient

Table 6.1 *Continued*

[30.7×10^{18}]	68.8×10^{22}	Perrin	1911	From Brownian motion: displacements
[30.8×10^{18}]	$69 \times 10^{22} \pm 3\%$	Perrin, Brillouin	1911	From Brownian motion: diffusion
[2.6×10^{19}]	58×10^{22}	Svedberg	1912	From Brownian motion: diffusion
[2.8×10^{19}]	62×10^{22}	Svedberg	1912	From Brownian motion: displacements
2.77×10^{19}	62.0×10^{22}	Planck	1913	From blackbody radiation $k = 1.34 \times 10^{-16}$
2.705×10^{19}	$(60.62 \pm 0.12) \times 10^{22}$	Millikan	1913	From $e = (4.774 \pm 0.009) \times 10^{-10}$ (oil drop)
[2.69×10^{19}]	$(60.3 \pm 1.2) \times 10^{22}$	Fletcher	1914	From Brownian motion: oil droplets in air

As of CODATA 2014, a century later (but prior to the new SI units of 2019 in which values for these became exact), Loschmidt's number = $2.68665664 \times 10^{19}$, Avogadro's = $60.22140857 \times 10^{22}$, and $e = 4.803204673 \times 10^{-10}$ esu. The earliest quasi-official values, listed in Birge (1929), were $e = (4.770 \pm 0.005) \times 10^{-10}$ esu, from which he derived a value of $(60.64 \pm 0.06) \times 10^{22}$ for Avogadro's number (corresponding to a value of 2.705×10^{19} for Loschmidt's number). Numbers enclosed in brackets were not stated in the publications cited, but were instead calculated here from values stated in them. The numbers in the table are not entirely consistent with one another owing to different authors not employing the same values for the gas constant or Faraday's constant.

with that for the values obtained from kinetic theory during the 35 years before Planck's, as listed in Table 2.1. On the other hand, what value was to be preferred among those from 1908 to 1913 was surely not clear. Ostwald in 1908 chose 71×10^{22} from Perrin as "the most probable value."[29] In a letter to Perrin in late 1909, Einstein was already focusing more on the differences than on the agreement:

> A precise determination of the size of the molecule appears to me of the highest importance, moreover, because the radiation formula of Planck can be more sharply proved through this than through measurements of radiations. The Planck theory of radiation leads to a determination of the absolute size of atoms, with *exact* validity claimed. Since the difference
>
> 61×10^{22} (Planck)
> 70.6×10^{22} (Perrin)
>
> between the Planck value and your value appears critical to me, it must be said at this time that the Planck formula is apparently only theoretically

grounded. Why the question of whether Planck's formula is correct appears so important to me, you can see from the papers which I am sending you with this.[30]

Four years after this letter, Bohr, faced with having to select a value for *e* in order to derive a value for Rydberg's constant for hydrogen from his model of the hydrogen atom, chose 4.7×10^{-10} esu, corresponding to a value of 61.5×10^{22} for Avogadro's number, that is, a value at some remove from Perrin's.[31] The formula for the Rydberg frequency that Bohr employed in this calculation was:[32]

$$\frac{2\pi^2 me^4}{h^3}$$

Insofar as the only means for obtaining a value for *m*, the mass of the electron, was from the known value of *e*/*m* (for which Bohr employed 5.31×10^{17}), in effect the value for *e* entered into this formula to the fifth power. The value for the Rydberg constant that Bohr published in 1913 was 3.1×10^{15} Hz, within roughly 6 percent of the value measured at the time, 3.29×10^{15}. Had Bohr employed the value for *e*, 4.23×10^{-10} esu, implied by a Perrin value of 68.5×10^{22} for N_0, his calculated value for the Rydberg constant would have been 1.8×10^{15}, a discrepancy of 55 percent from the measured value. While we have not done the historical research to confirm it, Bohr's calculation appears soon thereafter to have put an end to anyone at the cutting edge of physics after 1913 taking Perrin's value for Avogadro's number to be as well-founded as those no greater than 62×10^{22}.[33]

6.3. Perrin's Comparisons, Individually

The comparison Perrin always made, beginning with his initial May 1908 paper, was with values of Avogadro's number obtained from kinetic theory, typically from the combination of the measured viscosity of gases and van der Waals's equation. The value from kinetic theory Perrin gave varied from 7×10^{23} in that first paper to 6×10^{23} in September 1908 (Einstein's choice in 1905 and 1906) and 60×10^{22} in October, and then (after stating bounds of 45×10^{22} and 200×10^{22} based on different gases) to 62×10^{22} based on argon in his 1909 monograph, the value he continued to cite in 1911 and 1913. As we said in section 5.6 and at the end of section 5.5 in Chapter 5, Perrin needed this comparison in order to link his results for Brownian motion to the molecules of kinetic theory, for this was the only way he had for bridging the gap between the two realms. In terms of the evidence deriving from the comparison, therefore, it provided grounds for equating the mean translational kinetic energies of the granules to the mean translational kinetic energies of molecules in gases. This was the sole empirical evidence that

Perrin had for *equating* the two! All the other values for Avogadro's number with which he compared his were not connected in any way to kinetic theory. Absent this one, therefore, all he had was a question—how does the granular kinetic energy compare with the kinetic energy attributed to molecules?—to which he had no other recourse for answering except stipulation. This is surely why he came to prefer the value from argon, and continued to do so after 1911, even though it fell closer to the values proposed by Planck, Millikan, and Rutherford and his coauthors than to his. It offered definite support for his equating the energies.

In one respect, nevertheless, we are understating the evidential value of this primary comparison. All of Perrin's Brownian motion values for Avogadro's number were based on a formula at the heart of kinetic theory:

$$\frac{2}{3}NW = RT$$

where W is the mean translational kinetic energy of the *hypothesized* discrete elements, that is, the molecules of Clausius and Maxwell. W, R, and T all represent numbers that are not integers; the only basis for saying that N represents an integer is its being stipulated to do so. Everywhere Perrin turned to this formula to obtain a value for N, the question he was answering was, *given the measured value of the mean translational kinetic energy of the granules, how many granules are required to reach the characteristic energy of a mole*? Since granules are discrete, the answer to this question has to be, strictly speaking, an integer. In other words, Perrin's reasoning indisputably involves a count of discrete items. The same cannot be said of the value for Avogadro's number obtained from the viscosity of gases and van der Waals's equation. Indeed, it cannot be said of all but a few of the other approaches to obtaining values for N_0 we are about to consider. In this respect, then, Perrin's comparison of his values with the kinetic theory value he cited for argon was lending at least a little support for the latter's also numbering discrete items.

The first place Perrin compared his value for N_0 with values not obtained from kinetic theory was in his October 1908 paper, "Grandeur des molécules et charge de l'électron," the paper Ostwald expressly cited. As we said earlier, Perrin there noted both the values J. J. Thomson and his colleagues were obtaining from their efforts to pin down the fundamental charge and the values obtained from radiation via the contrasting theories of Planck (61×10^{22}) and Lorentz (77×10^{22}), contending that his value should be preferred to those.[34] In a footnote he also noted the range of values (between 62×10^{22} and 72×10^{22}) from measurements involving radioactivity reported by Rutherford in his "beautiful memoir" of August 1908.[35] We shall look at these three comparisons in the order Perrin considers them in his 1909 monograph, where radioactivity precedes blackbody radiation.

6.3.1. Comparison with Values from Ions

The first question Perrin had to answer concerned which values of Avogadro's number put forward by Thomson and others he should compare with his values. In October 1908, Perrin cited four values for *e*—3×10^{-10} esu from Townsend; 3.4×10^{-10} and 6.8×10^{-10} from Thomson before 1900; and 3.1×10^{-10} from Harold Wilson working under Thomson—giving a range for N_0 from 43×10^{22} to 96×10^{22}. In his 1909 monograph, he cited slightly different values of *e*, giving a range 45×10^{22} to 85×10^{22}, which he remarked "is the same degree of precision of the determination of van der Waals."[36] Then, after commenting on "the large sources of error" in determining *e*, he cited the values of 4.05×10^{-10} and 72×10^{22} from new experiments by Millikan and Begeman (1908, employing suspended water droplets) that "appear to have permitted more accuracy."[37] But then Perrin immediately turned to experiments for measuring the charge, carried out independently by Ehrenhaft and de Broglie, that had avoided relying on the condensation of water droplets around ions, yet still had a significant source of uncertainty from the dust-particles they had used varying in size and perhaps shape. Perrin's 1909 conclusion:

> So it is found that the exact value of *e* cannot differ much from 4.64×10^{-10} (Ehrenhaft) or 4.5×10^{-10} (de Broglie), which makes N 64.10^{22}. In spite of the uncertainty pointed out, I am inclined to consider this method more precise and more easy to perfect than that which depends upon the condensation of drops of water by expansion.[38]

By the time of the first Solvay Conference at the end of October 1911, Millikan and Begeman had not only published refined values for water-drops, but Millikan had published his first oil-drop experiments and his value of 59.22×10^{22}, without error-bands, for Avogadro's number. Perrin's paper "Les preuves de la réalité moléculaire" for the conference went on at some length on the determination of *e* and the recent work by Millikan and Fletcher. It concluded that "Millikan's experiments demonstrate in a decisive manner the existence of an atom of electricity equal to the charge carried by the hydrogen ion in electrolysis."[39] Perrin repeated this sentence, and much of the discussion preceding it, in his *Les atomes*, where he cited Millikan's 1911 values, 4.89×10^{-10} and 59×10^{22}, saying of the latter that it "is, after all, in remarkable agreement with the value 68×10^{22} given above."[40] In the table at the end of his 1911 Conference paper, the value he gives for Avogadro's number based on "charged spheres (in a gas)" is 64×10^{22},[41] yet in the first five editions of *Les atomes*, this number becomes 68×10^{22}, still further removed from Millikan's.[42] By the 1921 edition of the book, however, the designation has become "charge as microscopic particles," and the value,

61×10^{22} followed by a question mark.[43] None of these variations in the listed number is explained in the text. Why he never lists Millikan's actual published values in his concluding table we leave to others.

By 1913, Perrin was in the ambivalent position of wanting to stress agreement between the value of Avogadro's number obtained from Brownian motion and the value from *e*, yet at the same time wanting to insist that his value was to be preferred to Millikan's. Perhaps we should, as earlier, just rely on Ostwald's assessment of the comparison in his 1909 book. There, without citing anyone, he referred to a method of calculating the number of molecules from the mean cross-section of gas ions, claiming it agrees with the value proposed earlier (71×10^{22}), with a corresponding value for the fundamental charge in agreement with the unit charge of an electrolytic ion, "about 10^{-19} coulomb" (3.3×10^{-10} esu).[44]

One reason for turning to Ostwald in this regard is his having singled out "the isolation and counting of gas ions, on the one hand, which have crowned with success the long and brilliant researches of J. J. Thomson" in conjunction with Perrin's results on Brownian motion as justifying "the most cautious scientist in now speaking of the experimental proof of the atomic nature of matter."[45] When we turn to the chapter "Conduction in Gases and Radioactivity" that Ostwald added in the new edition of his *Grundriss*,[46] however, we find little or no significance attached to comparing values for Avogadro's number in the two ways. Instead, Ostwald emphasizes three lessons from the work by Thomson and his associates: (1) "gas ions behave from the point of view of the kinetic theory like molecules of ordinary gases, and each ion carries the same charge as a univalent electrolytic ion";[47] (2) the difference between gas ions and the negatively charged particles "a thousand times smaller than the mass of a hydrogen atom" known as electrons, "the individual existence of which has been proved experimentally"—"the last product of analysis in this region";[48] and (3)

> the possibility of individualizing and counting the number of electrical charges and of ions. For, in contrast with most other effects, in which the sum of all the particles present is obtained, and which do not afford any means of *distinguishing between separate particles and a continuous system*, every particle here acts individually and summation occurs later, if at all.[49] [emphasis added]

Putting words in Ostwald's mouth, then, what he found so important in Thomson's research were the atomicity of electric charge, the discreteness of gas ions, and the similarity in diffusion between those ions and the molecules of kinetic theory. These three thus provided "experimental evidence of the discrete or grained nature of matter" quite independently of the results from Perrin and others on Brownian motion. All three Perrin stressed in his 1909 monograph and thereafter, and hence, although he also stressed the "remarkable agreement" in the values for N_0 more than Ostwald did, he and Ostwald did not otherwise differ on the significance of the research on the "atom of charge."

That still leaves us with our primary question: taking the agreement between the values for Avogadro's number for granted, what evidence did it, in and of itself, provide? The values from Brownian motion were always answers to the question, *given the mean translational kinetic energy* $W_T T$ *of the granules, how many of them are needed to match the characteristic energy RT of a mole of gas or dilute solution?* The values from the atom of charge, at least on their face, were answers to the question, *given the fundamental unit of charge e, how many of those units are needed to match the total electricity required in electrolysis to produce a mole of a monovalent gas like hydrogen—that is, to match Faraday's constant?* Obviously, something more was needed for the answers to either of these questions to be about the same thing, namely "molecules."

We have repeatedly noted the two-part additional premise serving that purpose in the case of Brownian motion: *the characteristic energy RT of a gas or dilute solution consists of kinetic energy distributed across discrete elements forming a gas or dilute solution—call them "Maxwellian molecules"; and the mean translational kinetic energy of those elements is the same at any given temperature as that of the granules.* Correspondingly, the two-part additional premise in the case of the atom of charge was: *a monovalent gas like hydrogen consists of discrete, chemically stable elements—call them "chemical molecules," like* H_2*; and one fundamental unit of charge is needed to produce each such element during electrolysis.* Framed this way, the evidence from agreement between the two measures of Avogadro's number can be thought of as amounting to evidence that "Maxwellian molecules," the molecules of the kinetic theory of heat, and "chemical molecules," the molecules of chemistry, are one and the same.

This way of thinking about the matter, however, fails to capture the historical situation. As Perrin invariably emphasized, Townsend's experiment had provided independent evidence that the *Ne* for a mole of a monovalent gas, like hydrogen, in electrolysis matches the *Ne* of a mole of ions in an ionized gas, like hydrogen subjected to x-rays. This result had been confirmed by others, most notably by Fletcher in 1911, who obtained a value for *Ne* for gaseous ionization of 2.88×10^{14} esu versus 2.896×10^{14} esu for electrolysis.[50] What's more, J. J. Thomson and his associates and Fletcher under different assumptions had provided evidence that ions in a gas are discrete and hence countable—Thomson under the assumption of one gaseous ion per condensed water droplet, and Fletcher, one per oil-drop. (Recall that Ostwald had emphasized that ions could be counted in the quote from him given earlier.) And finally, Townsend's experiment had centered on the diffusion and hence the mobility of gaseous ions in an electric field, so that the ions involved in his determination of *Ne* amounted to the Maxwellian discrete elements of kinetic theory.

Accordingly, the situation in which the two sets of values for Avogadro's number were being compared was one in which there was independent evidence that the chemically characterized discrete elements in electrolysis and ionization were the same as the dynamically characterized discrete elements of kinetic theory. But then the main thrust of the evidence from agreement between these values of Avogadro's number was not just to add more evidence for that conclusion. Rather, the situation became like that with Huygens's two ways of measuring surface gravity. The agreement was providing mutual support for the premises entering into each of the two kinds of theory-mediated measurement, one derived from the universality of the gas constant and the other from the universality of Faraday's constant. These two kinds of measurement had a key assumption in common: stated most generally (and using Ostwald's words), "the discrete or grained nature of matter." So, Perrin was fully justified in claiming that the agreement was evidence for the "reality of molecules"—*at least as an assumption licensing theory-mediated measurements*—for what the convergence of the values for Avogadro's number between the two approaches was doing was to simultaneously support the validity of the complementary measures.

How much to make of this evidence is another question—surely not as much as with the nine significant-figure agreement we presently have in the case of the fine-structure constant, or even the four significant-figure agreement that Huygens obtained for the strength of surface gravity in the middle of the seventeenth century. The obvious place to question the Brownian motion measurements was the premise that the mean translational kinetic energy of the granules matches that of the molecules in the surrounding liquid. The no less obvious place to question the atom of charge measurements was the premise of one fundamental unit of charge per molecule in electrolysis and per ion in gaseous ionization, with the consequence of one or more units of charge per water (or oil) droplet. The agreement to within 15 or so percent of the values for Avogadro's number obtained over the period from 1908 to 1911 from the two different phenomena reduced the grounds for challenging these premises. And, in the process, it gave grounds for concluding that phenomena of electricity in liquids and gases, on the one hand, and dynamical phenomena in emulsions, on the other, have something fundamental in common. Let this be enough of an answer for the moment. We shall return to the question later, after we have considered the other agreeing measurements that Perrin cited.

A final comment is in order about the agreement between the measurements of N_0 we have been considering so far. Playing complementary theory-mediated measurements off against one another can turn into a continuing process. Once Millikan published values for Avogadro's number with bands of uncertainty undercutting Perrin's values, the situation changed. Two questions then became central: Which value is preferred? And what are the sources of uncertainty? Tight bands of uncertainty are typically the basis for answering the first question, as they seemed to have been in many people's eyes in this case. How much effort to devote to pinning down sources of uncertainty or, for that matter, sources of systematic error, varies historically from one case to another. What makes the difference, it would appear, is how dependent ongoing research at the time is on the evidence from the challenged measurements. When Millikan's value for *e* came under dispute in the early 1930s, a great deal of effort was put into pinpointing the source of systematic error in his measurements.[51] Little effort, by comparison, was put into identifying specific sources for the deviation of Perrin's values from Millikan's, either at the time or in the ensuing years. Whether the mean kinetic energy of granules is exactly the same as that of molecules, or only nearly the same, seems at the time not to have mattered all that much to most of those other than Perrin.

6.3.2. Comparison with Values from α-Particles

Turning now to the phenomenon of radioactivity, while research on it had been going on for more than a decade, it was in June 1908, and hence virtually contemporaneous with Perrin's publishing his values from Brownian motion, that Rutherford and Geiger announced their values for Loschmidt's and Avogadro's numbers from α-particle radiation, publishing them two months later. The key step in both this determination and the one by Rutherford and Boltwood three years later was an ingenious experimental design that enabled each α-particle entering a narrow tube to be individually recognized, by an abrupt "throw" of an electrometer needle, and hence each of them to be truly counted one by one, at a rate typically around 4 per minute. From this, the rate of release of α-particles from one gram of radium could be inferred with only minimal assumptions concerning the particles themselves. That rate, given in the first of the two Rutherford-Geiger papers of 1908, was 3.4×10^{10} per second from a gram of radium in equilibrium.

(One of the main questions answered in that paper was whether "the number of scintillations observed on a properly-prepared screen of zinc sulphide is, within experimental error, equal to the number of α-particles falling on it, as counted by the [new] electrical method."[52] The answer was yes, and by virtue of the agreement between the two ways of counting, the count from scintillations provided a cross-check of the count from the new electrical method.)

This new electrical method thus yielded strong evidence that α-particles are discrete (and hence countable). It gave evidence as well that they all carry the same charge. Prior to it, the evidence for discreteness had come from the discrete character of the visible scintillations and, even more so, from extensive prior measurements of the charge-to-mass ratio of α-particles in crossed electric and magnetic fields that had yielded a *prima facie* universal value of 5.07×10^3 emu/gm (versus 9.63×10^3 measured for hydrogen); just as with the electron, the well-behaved character of α-particles in crossed fields was providing the principal evidence that they all have the same charge-to-mass ratio.

Given the rate of release of α-particles from radium, the charge per particle could then be determined from the total charge over fixed times accumulating on a collector. The values announced by Rutherford and Geiger in their 1908 paper accompanying the one noted earlier were 9.2×10^{-10} esu from a first experiment and 9.4×10^{-10} from a second, yielding the mean of 9.3×10^{-10} esu, their announced value.[53] This value fell between two and three times most of the values for *e* that had been announced before, in particular the value announced by Wilson in 1903. That initiated the next key step in their reasoning:

> On the general view that the charge *e* carried by an hydrogen atom is the fundamental unit of electricity, we conclude that the charge carried by an α-particle is an integral multiple of *e* and may be either 2*e* or 3*e*.
>
> We shall now consider some evidence based on radio-active data, which indicates that the α-particle carries a charge 2*e* and that the ordinarily accepted values of *e* are somewhat too small.[54]

They turned to two cross-checks (the second of which, based on work by Boltwood, Perrin treated as an independent determination of Avogadro's number, quoting a value from it of 70.6×10^{22} in his 1909 monograph). The upshot was a value for *e* of 4.65×10^{-10} esu, "a somewhat higher value than those found in the measurements of J. J. Thomson, H. A. Wilson, and Millikan."[55] Rutherford and Geiger then singled out the likelihood of evaporation in the water-drop experiments of each of these three, arguing that their new value for *e* was more reliable; in a footnote, they cited as well Planck's deduced value of 4.69×10^{-10} esu.

From this value and Faraday's constant they then derived 2.72×10^{19} as their value for Loschmidt's number and 62×10^{22} for Avogadro's. From the combination of their value for *e* and the prior value for *e*/*m* of the α-particle, they obtained 3.84 as an atomic weight for the particle. From this they concluded,

> The atomic weight of the helium atom is 3.96. Taking into account probable experimental errors in the estimates of the value of E/M for the α-particle, we may conclude that *an α-particle is a helium atom*, or, to be more precise, *the α-particle, after it has lost its positive charge, is a helium atom*.[56]

This last remark should remind readers that at the time the question of the structure of an atom—meaning, the question of how it has neutral charge in spite of apparently containing negatively charged electrons—was entirely unresolved. Rutherford did not propose his planetary model (on the basis of α- and β-particle scattering results obtained in experiments with Geiger and Marsden) until 1911, and Bohr did not provide an explanation for how such a planetary configuration could survive without releasing continuing radiation and hence energy until 1913.

(We have gone into such detail on the Rutherford-Geiger 1908 value for Avogadro's number because Perrin—in his 1909 monograph, his 1911 Solvay Conference paper, and his *Les atomes*—portrayed it somewhat differently from the way they had. *The Rutherford-Geiger value, on their view, was just a variant on the atom of charge value discussed earlier, but with a new, better founded value for the atom of charge.* Perrin in 1909 chose to present it as giving two complementary methods for obtaining Avogadro's number from phenomena of radioactivity, one of which Rutherford had regarded as a less reliable cross-check of the method by which he and Geiger had reached their values for *e* and hence N_0.)

This still leaves us with the question of the force of the evidence resulting from the "agreement" between Perrin's Brownian motion values for N_0 and the 1908 value Rutherford and Geiger had obtained from α-particle radiation. Notice first that this evidence hinged on an assumption stated by Rutherford and Geiger that there is such a thing as "the fundamental unit of electricity." For, on its face, the fact that their value for the charge of an α-particle lay between 2*e* and 3*e*, given the values for *e* that had been obtained from ionization, raised the possibility that there is no such thing as a single universal discrete unit of charge. That worry was not addressed until Millikan published his revised water-drop paper in 1910 and his first oil-drop paper in 1911,[57] both confirming the claims Rutherford and Geiger had made in 1908 that the value of *e* had been underestimated before them and that the positive charge of the α-particle is *2e*. Perrin was thus on sound grounds in saying, both at the Solvay Conference of 1911 and in his 1913 book, "Millikan's experiments demonstrate in a decisive manner the existence of an atom of electricity equal to the charge carried by the hydrogen ion in electrolysis."[58]

Accordingly, the question of the evidence from the "agreement" between Perrin's and the Rutherford-Geiger values for Avogadro's number is best considered as of 1911 rather than at the time of Perrin's initial appeal to it in 1908. By 1911, however, the significance to attach to the Rutherford-Geiger value had changed dramatically because of three further results. First, Erich Regener in Berlin had carried out a version of the Rutherford-Geiger experiments, publishing in 1909 a value of 9.58×10^{-10} esu for the charge of α-particles and hence of 4.79×10^{-10} esu for the elementary charge.[59] Second, Rutherford and Royds in that same year had provided "conclusive proof that the α particle after losing its charge is an atom of helium" by verifying that the fingerprint of helium, its spectrum, is observed after a few days of collecting the discharge from radium.[60]

Third, and most striking of all, Rutherford and Boltwood, proceeding from results by Sir James Dewar in 1908 and continuing through 1911 with further results by him and by Rutherford and Boltwood, had obtained increasingly reliable values for the rate of production of helium by radium, as measured by the volume occupied at standard conditions.[61] Coupled with the rate of α-particle production measured by Rutherford and Geiger, this volumetric rate yielded, in Perrin's words, "the number of monatomic molecules N of helium that occupy 22,400 cubic centimeters, and which therefore make up a gramme molecule."[62] The number per cubic centimeter at standard conditions announced by Boltwood and Rutherford in 1911 was 2.69×10^{19}, which translates into a value of 60.3×10^{22}, for Avogadro's number. (Perrin gave instead 64×10^{22} as their value for this at the Solvay Conference in late 1911.)

By 1911, therefore, α-particle radiation had yielded two values for Avogadro's number within 1.3 percent of one another, the first derived from the charge of the particles, and the second from the volume at standard conditions they come to occupy when collected over a period of time. The principal assumption these two had in common was the number of α-particles released per second from one gram of radium at equilibrium (among its isotopes), namely 3.4×10^{10}, as measured by Rutherford and Geiger in 1908, supported by the rate of visible scintillations observed when α-particles strike zinc sulphide, and then measured more exactingly by Boltwood and Rutherford. Both of these methods, and Millikan's oil-drop method as well, involved no assumptions at all from the kinetic theory of heat. The three values—60.3×10^{22}, 62.0×10^{22}, and Millikan's 59.22×10^{22} in 1911 (revised to 60.62×10^{22} in 1913)—all fell well below Perrin's values from Brownian motion, which by 1913 he was listing as 68.3×10^{22} from vertical-gradients, 68.8×10^{22} from granule displacements[63], 65×10^{22} from granule rotations, and 69×10^{22} from diffusion rates. (In the table at the end of his 1911 Conference paper, for which Rutherford was in the audience, and as well at the end of the 1913 and 1914 editions of the book, Perrin gives 62.5×10^{22} for the value derived from the charge of α-particles, said to be a mean of those from Rutherford-Geiger and Regener.)

While the discrete elements that were being counted in each of these three methods were not visible in the way that Perrin's granules were, each of the three was supplemented by strong evidence that discrete elements of one sort or another were in fact being counted. What seems less and less straightforward as we consider all these numbers in detail is what *evidence* could have been gained purely from comparing Perrin's values with these others. The most obvious thought is evidence *against* his assumption that the mean translational kinetic energy of his granules is identical with that of molecules at the same temperature.

Stopping with that, however, is a mistake. The agreement with Perrin's values was nevertheless close enough to provide evidence that, even if granular kinetic energy is not identical with molecular kinetic energy, it is close enough to support his claim that thoroughly random agitation by discrete elements in the liquid is the source of Brownian motion. That in turn, just as Ostwald concluded, provided the first empirically grounded evidence at least for the kinetic theory of heat, if not for the molecular-kinetic theory of heat. The comparison with these other values, in other words, *did* strengthen the evidence Perrin's Brownian motion experiments provided for that theory. Adding still further to that evidence were Fletcher's 1911 Brownian motion experiments involving diffusion in an electric field and hence kinetic theory, which, joined with the Millikan and Rutherford-Geiger values for e, gave a value for Avogadro's number around 61×10^{22}. The comparison of values could therefore have legitimately been viewed at the time as giving less reason to quibble with Perrin's kinetic energy values and more reason to take his results for what they were, the most compelling empirical evidence that had yet been obtained for the kinetic theory of heat.

Notice here, however, that the evidence in the comparison was not reciprocal. Perrin gained support from Millikan and Rutherford-Geiger, while they gained little or no support from Perrin. But then they scarcely needed it, for their results stood on their own, owing especially to the tight agreement among their values for e and N_0, the mutually corroborating values for the latter from the two different measures based on radioactivity, and the tight uncertainty bands Millikan was achieving. From an evidential point of view, the only limitation of the atom-of-charge and α-particle determinations of N_0 was their lack of bearing on kinetic theory.

One other aspect of those determinations, largely absent from Perrin's, should not go unnoticed. In our remarks about the initial atom-of-charge determinations, we contrasted the molecules of kinetic theory with the molecules of chemical theory. Consider now the *questions* prompted by the different determinations of N_0. The Brownian motion determinations did not so much prompt new questions as they underscored long-standing questions about the sensitivity of collision-cross-section and mean-free-path calculations to assumptions about the shapes and sizes of the molecules of kinetic theory. By contrast, the atom-of-charge and α-particle determinations raised a host of questions about the molecules of chemistry. Six different items were being counted in those determinations: (1) elementary units of charge e, (2) α-particles, (3) atoms of radium transmuted, (4) gaseous ions, (5) "monoatomic molecules" of helium, and (6) molecules of gases, like H_2, released in electrolysis. The questions concerned the relationships among these six. Questions about the structure and unity of molecules, if not about the structure of atoms, had been around throughout most of the nineteenth century. In introducing a variety of elements of specificity that had been absent before, the common count of the number of "molecules" per mole from these six that had emerged by 1911 put those questions into an entirely new form.

6.3.3. Comparison with Values from Blackbody Radiation

The one further individual comparison Perrin made of his values for Avogadro's number and an independently determined value was with Planck's from 1900. This determination of N_0 from blackbody radiation raised questions of a very different sort, for it was not altogether clear, especially to Perrin, how and why the value quoted by Planck in 1900, 61.75×10^{22}, had to be a count of the number of molecules in a mole. The number itself came from measurements of blackbody radiation determining the values of the two constants, h and k, in Planck's equation:

$$E_\nu = \frac{8\pi h\nu^3}{c^3}\frac{1}{\exp(h\nu/kT)-1}$$

In deriving this formula Planck had assumed (1) a discrete rather than continuous distribution of frequencies, and (2) that Boltzmann's probabilistic analysis of the second law of thermodynamics applied to individual monochromatic resonators radiating and absorbing electromagnetic waves. The constant k enters as a parameter in the expression for the entropy of such a system. Planck reformulated Boltzmann's definition of entropy to show that, for a gas in thermal equilibrium with radiation, the total entropy could be the sum of the entropy of the parts only if $k = R/N_0$.[64] Planck's value for Avogadro's number then followed from the value of k obtained from the blackbody radiation measurements, 1.346×10^{-6} erg/deg K.[65] That the equation successfully captured the distribution of energy across the spectrum as a function of temperature was beyond dispute. So too was the robustness of the measured value of k, relative at least to the precision of the measurements, for both h and k could be re-evaluated at many different temperatures.

The questions centered on Planck's derivation of his equation and its *prima facie* departure from classical physics. The questions were multiplied when Einstein in 1907 re-derived Planck's equation—under the more radical assumption that all energy takes the quantized form of integral multiples of $h\nu$—in proposing a new approach to the specific heats of solids at low temperatures. The quantization of energy, which was in keeping with the quantization of light proposed in his 1905 paper on the photoelectric emission, entailed that individual atoms cease to vibrate, and hence cease to contribute to the heat capacity, at temperatures below the requisite *hν* threshold. By 1912 Peter Debye had extended Einstein's approach to multiple degrees of freedom of atom vibration and had shown that, with this extension, it matches low temperature specific heat data for solids to high accuracy.[66] On the one hand, then, were two major theoretical results agreeing with otherwise recalcitrant experimental findings, both of them tied to the two constants h and k. On the other hand were their radical departures from and hence *prima facie* incompatibility with classical physics.

Perrin cited Planck's determination of N_0 in his October 1908 paper and his 1909 monograph, but only side-by-side with a classical theory of the infrared spectrum that had been put forward by Lorentz and had yielded, as the value of Avogadro's number, 77×10^{22}, in contrast to Planck's 61.75×10^{22}.[67] Perrin's paper for the 1911 Solvay Conference—befitting the preoccupation of the Conference with the Planck and Einstein theories—summarized Planck's determination of N_0 without mentioning Lorentz's, or for that matter Einstein's theory.[68] *Les atomes* devotes a full chapter to "Light and Quanta," though again only summarizing Planck's result without fully deriving it. Perrin's most telling remarks on the comparison with Planck's value came in a much later edition of the book:

> The agreement between this value and those already found is indeed marvelous. And at the same time we have acquired yet another means for determining accurately the molecular magnitudes.
>
> In spite of the importance of these results, we cannot conceal the fact that Planck's theory presents great difficulties. It will be (and already has been) profoundly modified, and we shall certainly see Planck's postulates replaced by others more comprehensible and in more accurate agreement with experiment (Bohr).[69]

Insofar as Perrin had provided no hint of any discrepancy between Planck's theory and experiment, it is unclear what he had in mind, unless a value for Avogadro's number more in keeping with his own.

What then are we to make of the evidence deriving from the comparison of Perrin's values with Planck's? In his book Perrin first presents Planck's theory of the spectrum, cites its agreement with experiment, and then, in turning to Planck's value, says, "But a still more striking verification lies in the agreement found between the values already obtained from Avogadro's number and the value that can be deduced from Planck's equation."[70] This view of the evidence deriving from the comparison is in full accord with the view Planck expressed in the final paragraph of his 1900 paper, which we repeat here:

> If the theory is at all correct, all these relations should be not approximately, but absolutely, valid. The accuracy of the calculated value is thus essentially

> the same as that of the relatively worst known, the radiation constant k, and is thus much better than all determinations up to now. To test it by more direct methods should be both an important and a necessary task for future research.[71]

The evidence deriving from the comparison in this case therefore went only one way, from other values for N_0 to support for Planck's theory, and not reciprocally. Do notice, however, that by 1911 Planck should surely have been stressing the values for N_0 coming from gaseous ionization and radioactivity, not the values coming from Brownian motion.[72]

* * *

To recapitulate, the four individual comparisons Perrin made at the time with values of Avogadro's number determined from other phenomena each yielded some evidence supporting the validity of at least one of the respective measures—especially if the various values were taken to agree with one another to within the margins of uncertainty of the respective determinations. The key unsupported presupposition in Perrin's determinations of Avogadro's number is that the mean translational kinetic energy of the granules is identical with that of the molecules in the substrate. His comparison of his values with the one he states for argon gave independent support for this presupposition by countering the claim that the granule velocities, based on direct observation, are too small for the kinetic energies to match.

So too did the values of Avogadro's number from determinations of the electric charge, from the Rutherford-Geiger (1908) and Regener (1909) values through those from Millikan and his associates from 1908 to 1913. The key open-to-challenge presupposition in the determinations of Avogadro's number from the electric charge was that the number of basic charges needed to liberate a mole of gas in electrolysis is identical with the number of molecules of the gas. Perrin's Avogadro's values gave independent support for this presupposition; and the values from these two together gave evidence as well that the molecules of kinetic theory are the same as the molecules formed in electrolysis, that is, the molecules of chemistry. The only comparisons in which the evidence went only one way were (1) with Planck's blackbody value of Avogadro's number as a consequence of novel aspects of his derivation of his blackbody equation, and (2) with the Boltwood-Rutherford determination, which, unlike the other determinations of Avogadro's number at the time, seemed to involve no salient open-to-challenge presuppositions calling for independent support other than that α-particles remain discrete when they lose their positive charge.

All of this evidence, however, is predicated on taking the various values to agree with one another to within the margins of uncertainty of the respective determinations. One worry, of course, was that those margins of uncertainty were anything but clear at the time, if only because so many distinct, *prima facie* well-founded novel determinations of Avogadro's number had emerged in the brief five-year period we (like Perrin) are considering. A more serious worry was that *all* the values of Avogadro's number with which Perrin compared his own were no greater than 62×10^{22}, the value he had stated for argon—that is, roughly 10 percent below the lowest of the values, 68.2×10^{22}, that he had quoted on the basis of his determinations of the mean translational kinetic energies of the granules. The central question of this chapter concerns what evidence the level of agreement among all the values in question provided for the reality of molecules. The systematic difference between the two sets of values threatened to downgrade the force of this evidence. We would therefore be remiss to move on to the central question of the chapter without at least reviewing how Perrin himself responded to this gap between his values and all these others.

6.4. Perrin on the N_0 Values $\leq 62 \times 10^{22}$

We claim to be adopting the viewpoint of someone writing a review article on Brownian motion and molecular reality at the time, say in the aftermath of the 1911 Solvay Conference. Readers might well ask why we need to be doing this when Jean Perrin wrote not one, but at least three review articles at the time, his 1909 monograph, his *Les atomes* in its five editions during 1913 and 1914, and his 1911 paper "Les preuves de la réalité moléculaire" for the Conference. What need is there for a review article beyond his? Part of our answer is that the 1909 paper and the book are too polemical to be considered review articles; their aim, often expressed in rhetorical questions, was to lay out the burden of argument on anyone still prepared to challenge the reality of molecules. The paper at the Solvay Conference, however, does throughout have the tone of a review article—quite appropriately given its extraordinary immediate audience, to whom it was circulated before its presentation at the Conference itself. Why shouldn't it be *the* review article on the subject at the time?

We grant that Perrin's 1911 paper was a review article, though the topic under review was just what its title indicates, not the state of the molecular hypothesis, but solely the body of evidence in support of it. Perrin does not mention the problem of the specific heats of gases and the questions it was raising about the equipartition of energy, nor does he note the extent to which questions about the structure of atoms and molecules remained intractable. (In the exceptionally brief discussion following his presentation, Lorentz did remark on J. J. Thomson's "spheres of positive electricity" in relation to "the hypothesis of perfect spheres."[73])

Still, the serious shortcoming of Perrin's 1911 paper as a review article, even on this topic, was its failure even to note the *systematic* difference between his values for Avogadro's number from Brownian motion and all the comparably well-founded values with which he directly compared his in the paper. Instead, he compares them one by one, never raising the somewhat obvious question, once the systematic difference is noted, whether it might be pointing to some systematic error in his results.[74] Nor did anyone in the recorded discussion raise this question, not even Rutherford or Warburg. Our concern in this section—in keeping with our general concern with theory-mediated measurement—is, primarily, with how he managed to avoid this question and the consequences of his doing so, and, secondarily, with why no one raised the question in the discussion.

We saw at the end of the preceding section how Perrin in his book undercut the significance of the value of Avogadro's number coming out of the law of blackbody radiation by questioning Planck's theoretical derivation of the law. As we saw earlier, Perrin had grounds for not endorsing Planck's derivation, if only from the letter he had received from Einstein in 1909 which had invoked the value for N_0 from Brownian motion as a challenge to that derivation. One can argue that a derivation of the blackbody radiation law that was well founded did not emerge until after Compton's evidence for light quanta from the effect named after him and Bose's light-quanta based derivation of Planck's law in 1924 in which the distinction between what came to be known as Bose-Einstein and Maxwell-Boltzmann statistics emerged. That said, however, by 1911 the constants in Planck's formula had become well established, with results over a range of temperatures. At Solvay, Warburg reviewed such results, choosing to express them in terms of their implied value for *e*, for which he gave a value of 4.69×10^{-10}, corresponding to a value for N_0 of 61.7×10^{22}.[75] Perrin listed Planck's value from 1901 (quoted as 61.6×10^{22}) and then noted Warburg's paper at Solvay with praise; yet when it came to adopting a value from blackbody radiation, he chose 64×10^{22} without explanation, thereby glossing over the large discrepancy with his own values, to which Einstein had called his attention.[76]

The systematic difference with his own values was more difficult to avoid in the case of the ones for N_0 obtained from the values of *e*. As Table 6.1 shows, by 1911, five careful "direct" determinations of *e* had been published, based on three different approaches to measuring it: two from α-particles (4.65 and 4.79 times 10^{-10} esu), two from water droplets (4.669 and 4.65 times 10^{-10}), and one earlier that year from oil droplets (4.891×10^{-10}). (Warburg listed all of these except the ones involving water-droplets in his Solvay Conference paper, along with Perrin's value (4.23×10^{-10}); he added as well a blackbody value, 4.32×10^{-10}, implied by a measurement of Kurlbaum's, but ended his paper by challenging its validity.[77] Those at the Conference thus had this list of values in front of them prior to Perrin's presentation.)

The mean of the values we listed is 4.73×10^{-10}, implying a corresponding mean value for N_0 between 60 and 61.2 times 10^{22}, depending in part on the value adopted for Faraday's constant. The convergence of these values for e thus lent as much support to them and their mean as the convergence of Perrin's three determinations from the mean translational kinetic energy of the granules W he announced in 1911 had lent to his—that is, 68.2, 68.8, and 69 times 10^{22}, the mean of which is 68.7×10^{22} (implying a corresponding mean value for e of 4.21×10^{-10}). Given the respective degrees of convergence of the results for e and W, this difference in the values for N_0 well in excess of 10 percent between the two approaches implies a systematic error either in the step from e or in the step from W to N_0, or in both. Faced with the audience he had at the Solvay Conference, along with his insistence on using the results for e to support the atomicity of electricity, Perrin could scarcely have avoided this worry in his 1911 paper.

Rather than consider the values of N_0 obtained from e in the aggregate, however, he reviews the ones from α-particle radiation entirely separately from the others, and he reviews the latter historically, starting with the efforts by Townsend, J. J. Thomson, and Wilson before 1905 and postponing the new oil-drop experiments to a separate section that begins with a review of the problematic Ehrenhaft results.[78] This has the effect of drawing attention to the uncertainties of the measurements of e—for example, those arising from questions about the evaporation of water droplets—instead of the trend after 1908 toward a more definite value coming out of Millikan's laboratory. Only in a short final paragraph at the end of the first of these sections does he note the more recent results for e, beginning with those by Przibram using alcohol-droplets ($e = 3.8 \times 10^{-10}$), followed by the water-droplet results, first by Millikan and Begeman in 1908 (citing a value of 4.6×10^{-10} that includes corrections Begeman added in 1910 instead of the originally published value of 4.06×10^{-10}) and then Begeman's new 1910 value (4.7×10^{-10}). Perrin then dismisses all of these values at the end of the paragraph with the remark, "But there is no need to discuss these [numbers] at length, much greater precision becoming possible by the study of individual charged particles."[79]

After then dismissing Ehrenhaft's results that challenged the existence of a fundamental atom of charge,[80] Perrin reviews in some detail Millikan's results in his first oil-drop paper, "The Isolation of an Ion, A Precise Measurement of Its Charge, and the Correction of Stokes's Law." (Insofar as Millikan was an American and his paper had been published in the American journal *The Physical Review*, and then only six months earlier, the detail may have been needed, for several of those at the Conference were surely hearing about it for the first time.) The claim of a precise measurement in the title appears not only in the four significant figures Millikan gives for the value, but in his stated assessment of it:

> The probable error in the final mean value 4.891×10^{-10}, computed least squares from the numbers in the last column, is four hundredths of one percent. . . . Since, however, the coefficient of viscosity of air is involved in the formula, the accuracy with which *e* is known is limited by that which has been attained in the measurement of this constant. There is no other factor involved in this work which has not been measured with an accuracy at least as great as .2 percent.[81]

The paper is nevertheless less famous for its claiming a precise value for *e* than it is for having shown that the charge on a large number of oil droplets was always an integral multiple of a single charge, independently of the absolute magnitude of that charge.

Perrin's account of the paper begins with this last point, laying out some of the results for individual droplets and citing Fletcher's related paper that had announced a "direct" value for *Ne* for gaseous ions, independently of the values of *N* or *e*, within 0.5 percent of Faraday's constant for electrolysis, the strongest evidence yet that *Ne* is the same for both.[82] Perrin drew an unqualified conclusion, contra Ehrenhaft, from these results: "In short, the experiments of Millikan demonstrate in a decisive manner the existence of an atom of electricity, equal to the charge carried by a hydrogen ion during electrolysis."[83]

Perrin then turns to Millikan's claimed precise value for *e*, opening a new section with the sentence, "But these beautiful experiments do not yet appear to me to give a reliable value for this elementary charge which they have made manifest."[84] The doubt, he goes on to say, stems from Millikan's treatment of the frictional action of the air on oil-droplets, that is, the version of Stokes's law for air that Millikan had employed. He then invokes efforts by Roux to repeat Millikan's experiments in his own lab, but in liquid as well as air, saying that they had yielded a value for Avogadro's number around 67.5×10^{22}, versus the value 59×10^{22}—not the more precise 59.22×10^{22}—that Millikan had announced on the basis of his value for *e*. Perrin then concludes by adopting a value for Avogadro's number from the charge on ions of 64×10^{22}, a mean between Millikan's 59×10^{22} and 69×10^{22} from Brownian motion. Perrin's audience, whether at the Conference itself or readers of the subsequent publication in French and German of its proceedings, were left with no reason to raise the question of whether Millikan's precise value for *e* might be exposing a systematic error in Perrin's Brownian motion values for N_0.[85] Indeed, they would have had to turn to Millikan's paper even to realize the level of precision he was claiming.

(Perrin was justified in citing Fletcher's value of 28.8 for *Ne* while ignoring the values for N_0 (57.5×10^{22}) and *e* (5.01×10^{-10}) given in the same paper, for Fletcher himself cited them only to claim that the theory of Brownian motion in air he had developed in the paper "gives a fairly accurate means of determining molecular magnitudes."[86] The situation changed radically a short time later, however, after Millikan had published his new value for *e* in 1913, reflecting a revised account of the friction of air on oil droplets. That revision gave Fletcher all he needed to obtain a "direct" value for N_0 through observing the displacements of individual oil drops, and improvements in the apparatus had made it possible to obtain from 1,000 to 6,000 observations. Fletcher's paper, "A Determination of Avogadro's Constant *N* from Measurements of the Brownian Movements of Small Oil Drops Suspended in Air," gave values of N_0 obtained from 12 individual droplets ranging from a low of 57.9×10^{22} to a high of 64.8×10^{22}, with a corrected weighted mean of $60.3 \times 10^{22} \pm 1.2 \times 10^{22}$.[87] Insofar as Fletcher's value came from Brownian motion and differed systematically from Perrin's, it surely should have provoked questions about the possibility of systematic error in Perrin's determinations of the granular kinetic energy were it not for the fact that Bohr's 1913 paper had focused attention away from N_0 and onto the need for a precise value of *e* in order to draw conclusions from spectra. As we noted earlier, Bohr himself had adopted a value of 4.7×10^{-10} (implying a value below 62×10^{22} for N_0) in his theoretical calculation of Rydberg's constant. A few years later, Sommerfeld adopted a value of 4.77×10^{-10}.[88] We know of no one doing research on atomic structure after Bohr who preferred a value for *e* consistent with Perrin's translational values for N_0, that is, a value no greater than 4.25×10^{-10}. All of this, however, falls outside the historical time-window to which we are confining ourselves here.)

In his subsequent section on "genesis and destruction of atoms"—that is, phenomena of radioactivity—at the Solvay Conference, Perrin quotes the values for *e* announced by Rutherford and Geiger, 4.65×10^{-10}, without any comparison to Millikan's value, and then gives 62×10^{22} as the value it implies for N_0, "with a maximum error of not more than 10 parts in 100." He goes on then to acknowledge Regener's greater value for *e*, 4.79×10^{-10} from α-particles and its corresponding value of 60×10^{22} for N_0, but argues that they are less reliable than the Rutherford-Geiger value. In the table comparing the values for N_0 from the various phenomena, he lists 62.5×10^{22} for the value obtained from the charge of α-particles, immediately preceded by a value of 64×10^{22} from "the charge of spheres in gases."[89] No one looking at this table would see any sign of a systematic difference between his values from Brownian motion and those obtained from measurements of the elementary charge.

Finally, in the case of the 1911 Boltwood-Rutherford determination from the volume of the gas occupied at standard conditions by a known quantity of α-particles, Perrin elects not to quote their published value for Loschmidt's number, 2.69×10^{19}, which corresponds to 60.3×10^{22} for N_0. Instead, he quotes a range of values from which he obtains a range for N_0 of 62.4×10^{22} to 64×10^{22}. He then notes the corresponding experiments carried out by Mme. Curie using polonium, the volume obtained from them by Debierne, and the value 65×10^{22} implied for N_0. In the table at the end, he gives 64×10^{22} as the value obtained from this approach. Whether Perrin consciously suppressed the 60.3×10^{22} implied by Boltwood and Rutherford's stated value for Loschmidt's number is an interesting question. Rutherford, who was an attendee at Solvay, is not recorded as saying anything in the discussion of Perrin's paper. Perhaps Rutherford himself felt there was enough uncertainty in their results, side-by-side with the Curie-Debierne results, to withhold any objection to the 64×10^{22} value.[90]

Perrin's account of the Boltwood-Rutherford results was unfortunate historically. Because his 1911 paper and subsequent book have been taken—universally, it would seem—as *the* authoritative review of the value of Avogadro's number at the time, the extraordinary character of their experiment has been largely ignored.[91] Of all the different determinations of N_0 at the time, this one involved the least tenuous assumption linking their observations to the molecular constant, namely merely that the helium gas resulting from the collection of α-particles consisted of the same number of monoatomic molecules as the measured number of α-particles. Boltwood and Rutherford say as much quite forcefully:

> There can be no doubt that the α particle during its flight consists of a helium atom carrying two unit positive charges, and that helium itself is monatomic. The agreement between calculation and theory verifies in a remarkable way the essential correctness of the atomic theory of matter. The number of helium atoms expelled per second by one gram of radium have been directly counted, and the corresponding volume of helium has been experimentally measured. From these two experimental observations, it is possible to deduce with a minimum of assumption the number of atoms of helium in one cubic centimetre of that gas at standard pressure and temperature. That number is 2.69×10^{19}. By Avogadro's hypothesis this gives the number of molecules in a cubic centimetre of any gas at standard conditions.[92]

One should probably not dismiss as a mere accident that their value for Avogadro's number came closer to the post-1940 value than any of the others we have been considering.

Perrin was not alone in contributing to the Boltwood-Rutherford results not receiving the attention they deserve in subsequent histories. Svedberg's *Die existenz der moleküle: experimentelle studien* of 1912 makes no reference to Boltwood and Rutherford at all. Insofar as the Proceedings of the Solvay Conference did not appear in print until 1912 in French and still later in German, Svedberg almost certainly had not read Perrin's Conference paper and its discussion of Boltwood and Rutherford at the time his book went to the printer.

While Svedberg's main title suggests that his book is a review of the evidence for molecules, akin to Perrin's 1911 paper, it is instead, as its subtitle indicates, a review of experiments from Svedberg's laboratory, with extensive lists of numerical results, on diffusion in emulsions and Brownian motion. It gives values of 58×10^{22} for N_0 and 5.0×10^{-10} for e derived from the diffusion experiments, noting the agreement between them and the values quoted by Millikan and Fletcher versus the difference with Perrin's values (cited as 68.8×10^{22} and 4.17×10^{-10})[93]; and from the Brownian motion experiments, 62×10^{22} and 4.7×10^{-10}, this time noting the agreement with Millikan, Regener, and Rutherford versus Perrin (cited as 69×10^{22}).[94] Perrin does not comment on the latter, but in a footnote in *Les atomes* he cites "the high degree of uncertainty involved in the measurement (and even the definition) of the radii of *invisible* granules" in dismissing the former.[95]

Svedberg thus called attention in 1912 to the difference in values that would seem to prompt questions about the possibility of systematic error in Perrin's results, yet so far as we have been able to determine, no one at the time pursued this question much further. In the 1921 edition of his *Theoretical Chemistry from the Standpoint of Avogadro's Rule and Thermodynamics*, Nernst cites both Perrin's and Svedberg's values of N_0 from diffusion and Brownian motion and then concludes:

> The relatively good agreement of the values of N_0 obtained by these various methods from experiments with suspensions is an important fresh proof of the correctness of the theory of thermal motion; the accuracy of the results obtained by Perrin and Svedberg in the work mentioned above is very remarkable, since it would appear very difficult to obtain precise measurements of N_0 by such methods. It is noticeable that Perrin obtained appreciably higher values throughout than Svedberg.[96]

Four pages later, when it comes to listing his preferred values, Nernst chooses Millikan's values of 4.774×10^{-10} and 60.64×10^{22} for e and N_0.[97] So far as we know, any question of systematic error raised by the difference between Perrin's values and the others simply became moot as Millikan's values came to be preferred.

We draw three lessons in concluding this section. First, Perrin was not the person to be writing a truly careful review article in 1911 or 1913. We should not automatically be taken to be criticizing him in saying this. High-quality measurements were coming out of three laboratories, directed by three extraordinary experimentalists, Perrin, Rutherford, and Millikan. Each of them knew the care taken in obtaining the numbers they were quoting from their own laboratory, but each of these three had little way to make such an assessment of the numbers coming out of the other two laboratories. Perrin thus had good reasons for trusting his numbers more than the others, and he should consequently not be viewed solely as protecting his numbers in raising questions about the others. The lesson to be learned is that truly careful review articles are best written by someone able to address all the results being reviewed from a truly neutral perspective.

(Our matter-of-fact statements of the results coming out of these three laboratories do not do justice to the exceptional ingenuity and extended effort that went into the design and development of the experiments that produced them. Millikan's experiments became most famous for this, yet the two-part Rutherford-Geiger paper of 1908 displays a more elaborate design in response to a notably more challenging sequence of questions, and the answers they achieved were no less monumental than Millikan's; and one has only to read any one of Perrin's monographs of 1909 and 1911 or his *Les atomes* to come to appreciate the extraordinary demands of his experiments.)

Second, partly because Perrin's paper and book did become the authoritative review article at the time, but also because so little time had passed between the first well-founded values of Avogadro's number in 1908 and 1911 (or for that matter, 1913), no one was in a position at the time to select a preferred value for it. Planck was referring to the value for *e* in the preface to his 1913 edition of *The Theory of Heat Radiation*, but the phrasing he used there, after citing Rutherford, Regener, Perrin, Millikan, and Svedberg, applies equally well to the value for N_0, "between the values of Perrin and Millikan"[98]—that is, between 59 $\times 10^{22}$ and 69 $\times 10^{22}$, or in other words, the one significant figure claimed by Poincaré. It was entirely reasonable for anyone outside of these three laboratories, at least as of 1911, to take all the values between these two as agreeing with one another.

Third, one needs to appreciate how radically the capacity to assess theory-mediated measurements of microphysical constants changed between 1911 and 1929, the year of Birge's "Probable Values of the General Physical Constants." The change came not from more precise or better experimentation, but from so many more constants being linked to one another, starting with Bohr's linking of *h*, *e*, *e*/*m*, and Rydberg's constant in 1913, but rapidly proliferating thereafter. The only effective way to detect systematic error is to play different measures of the same constant or quantity off against one another. If nothing more, this section has brought out how limited they were in effecting this approach before Bohr. Our monograph should accordingly be viewed as a study in theory-mediated measurement in the years leading into an extraordinary increase in the capacity for drawing evidence out of such measurements.

6.5. Perrin's Comparisons, Collectively

The four comparisons we have examined individually—with values for N_0 obtained from the viscosity of gases, from gaseous ionization, from radioactivity, and from blackbody radiation—are not the only comparisons Perrin came to cite. Starting with his 1909 monograph and then continuing with his 1911 Solvay Conference paper and his 1913 book, he also cited comparisons with values derived from such phenomena as the blueness of the sky, diffusion of dissolved substances, the mobility of ions in water, and, in 1911 and 1913, density fluctuations in concentrated emulsions and critical opalescence. One reason for our not considering these individually is that the uncertainty bands associated with them were far larger than with the values obtained from ionization, radioactivity, and blackbody radiation. Another reason, besides wishing not to try readers' patience more than we already have, is that Perrin himself focused on individual comparisons only when defending his value for N_0 against others. Perrin's preoccupation was instead with the evidence for the existence of molecules deriving from the *aggregate* general agreement among so many different ways of obtaining values for Avogadro's number.

A lesson we intend to be learned from our discussion of the individual comparisons is that any comparison in the aggregate threatens to gloss over details that, when taken into consideration, may limit, if not undercut, the evidence that can legitimately be drawn from the whole. That granted, the many different, highly diverse determinations of Avogadro's number did agree with one another to a remarkable extent, at least in comparison with the uncertainty attaching to all such determinations before 1900. Poincaré's 1912 summary of the situation that we quoted at the very beginning of this monograph is the statement from which we are now going to proceed:

> The brilliant determinations of the number of atoms computed by Mr. Perrin have completed the triumph of atomism. What makes it all the more convincing are the multiple correspondences between results obtained by entirely different processes. Not too long ago, we would have considered ourselves fortunate if the numbers thus derived had contained the same number of digits. We would not even have required that the first significant figure be the same; this first figure is now determined; and what is remarkable is that the most diverse properties of the atom have been considered.[99]

Our question is, what evidence could appropriately have been drawn from this? Keep in mind, as we proceed, Poincaré's further remark, "each new discovery in physics reveals a new complexity of the atom."[100]

A place to start is with the elements of logical structure that all of the determinations of Avogadro's number had in common. Each case involved a theory-mediated measurement of some primary quantity from which the value of N_0 was then inferred as a further step. For example, the primary quantity in all of Perrin's Brownian motion measurements was the mean kinetic energy of the granules—translational in three of the measures and rotational in the fourth. The primary quantities in the other determinations included the fundamental unit of charge e, the number of α-particles emerging from a gram of radium per second, the volume occupied at standard conditions by a gas formed by a specific number of α-particles, and the two constants in Planck's equation, h and k, with the latter initially representing a mean energy per resonator. In each case the integrity or, if you prefer, the validity of the theory-mediated measure of the primary quantity could be confirmed by testing what we called its *stability* in Chapter 4—that is, not merely its repeatability in the same conditions, but obtaining the same value for the quantity as the values of the other parameters entering into the measure (or in other words, the conditions) are varied. The stability of any theory-mediated measure provides the first layer, so to speak, of evidence in support of the premises entering into it. When the measure is not all that stable, the premises underlying it are subject to question.

Measures of all the primary quantities we have just listed had been confirmed to be stable by 1912. The same cannot be said for the measures involved in determinations of N_0 based on the viscosity of gases, the blueness of the sky, diffusion of dissolved substances, the mobility of ions in water, density fluctuations in concentrated emulsions, and critical opalescence. Until those measures were shown to be stable, the most they could contribute to the evidence coming out of the many determinations of Avogadro's number had to amount to nothing beyond ancillary support, specifically in the form of a *prima facie* promise for still more cases of stable measures yielding agreeing values for N_0.

Continuing with technical terms introduced in Chapter 4, when two different stable theory-mediated measures of the same quantity yield, to within experimental precision, the same values, we follow Perrin in calling them *convergent*. Thus he developed three distinct convergent measures of the mean translational kinetic energy of granules, one based on vertical-gradients, a second on granule displacements, and a third on diffusion coefficients. What made these distinct was a difference in the premises constitutive to each of them. As we said in the case of Huygens's two measures of surface gravity, convergent measures of the same quantity provide mutual support for the differing premises constitutive to the measures—specifically, stronger evidence for the reliability of their premises for purposes of measurement than provided by stability alone. Among the other primary quantities with stable measures listed earlier, only e had convergent measures as of 1912, principally from the oil-drop measure of Millikan and the cumulative charge measure of Rutherford and Geiger, independently confirmed by Regener. In principle, however, the Einstein-Debye equation for the variation of the specific heat of solids with temperature could have been turned into a convergent measure for *h* and, of course, many other measures of *h* were soon to emerge.

In conjunction with these considerations, we are going to interpret Poincaré's one-significant-figure claim in the manner of Planck's 1913 comment in regard to *e*, namely as N_0 falling within the bounds defined by the two convergent sets of stable measurements: 59×10^{22}, corresponding to Millikan's 1911 value on the low end; and 69×10^{22}, corresponding to Perrin's translational kinetic energy values from the same year on the high end.

A second feature of the primary quantity measures was that their derivations all included reference to something discrete and hence amenable to being counted. What this was varied from measure to measure, as did the strength of the evidence that it truly is discrete. Granules are visibly discrete, and Rutherford and Geiger were counting anywhere from one to five distinct "jerks" of their electrometer needle in each minute, which markedly strengthened the evidence from scintillations that the radiation coming off of radium consisted of discrete items. Millikan always insisted that he had been the one to finally show that electricity consists of discrete unit charges, and at the very least, as Perrin himself contended, Millikan's oil-drop experiments strengthened the evidence that had come from counting droplets presumed to form around ions and measurements of the common charge-to-mass ratio of various discharges of negative electricity. Planck's discrete resonators and the discrete packets of energy associated with them had the least independent evidence as of 1912, and hence were most controversial, in spite of the great stability of the measures of *h* and *k* that presupposed them.[101] The variability of the evidence for the claim of discreteness is not so important, however, as the indispensability of such a claim to all the measures. For, while the discrete elements entering into each measure differed, they were each crucial to the further claim that the value for N_0 derived from their measured primary quantities amounted to a count of discrete items in its own regard.

A third feature that the logical structure of the determinations of N_0 had in common was a premise linking the primary quantity measured in each to what was said to be Avogadro's number. As we have seen in the case of the Brownian motion determinations, this premise—*that the mean translational kinetic energy of the granules W is the same as the mean translational kinetic energy of the molecules of the liquid surrounding them*—had a very different status from the premises underlying the measures of the primary quantity. Not only was there no independent evidence for it, but nothing akin to the stability and convergence of the measures for the energy of the granules provided any support for it. The stability and convergence of those measures provided support for the premises underlying them, but so far as those measures were concerned, the premise relating W to N_0 was something further tacked on and hence something that went beyond them. This is why we have distinguished between the primary quantity being measured in each case and N_0. *The determinations of the latter in every case represented a different kind of theory-mediated measurement because of their relying on a further premise tacked on in order to link the primary quantity measured to what was said to be Avogadro's number.*

A more precise way of making this point is that the premise linking the primary quantity measured to N_0 was in no way *constitutive* of the measure of the primary quantity in the sense of being indispensable to it; this premise was constitutive only to the further inference from the primary quantity to N_0. As a consequence, this premise was open to question regardless of how stable, convergent, or otherwise well-founded the measures of the primary quantities happened to be. The least open to question was the premise linking the number of α-particles that had entered a volume and the number of monatomic helium molecules in the volume. That premise requires little beyond α-particles remaining discrete and otherwise intact when, in Rutherford's words, they lose their positive charge. Still, absent more empirically grounded detail about what it is for an α-particle to lose its charge, a gap did remain between counting α-particles and having a count of molecules of helium. Similarly, it was one thing to infer the number of fundamental unit charges required to match Faraday's constant and another to claim that one unit charge in electrolysis yields one H_2 molecule of hydrogen. Why not instead one atom of hydrogen or one unit weight of hydrogen? Tightening the uncertainty bands on the value of the fundamental unit of charge was not ever in itself going to close this question.

We should take stock, then, before turning to the question of the evidence deriving from the extent of the agreement among these different determinations of Avogadro's number. On the one hand, there were clear grounds for concluding that theory-mediated measurements had yielded reasonably precise values of the following quantities by late 1911:

(1) The mean translational kinetic energy W of granules in Brownian motion in dilute emulsions at a given temperature, and with it, though slightly more provisionally, the mean translational kinetic energy per degree Kelvin, W_T, in Brownian motion.
(2) The value of the fundamental discrete unit of charge *e*, whether negative or positive, in ions and in such discharges as cathode rays, the photoelectric effect, and both α-particle and β-particle radioactivity.
(3) The volume occupied at standard conditions by the electrically neutral helium gas formed by the number of α-particles released and then collected over any given period of time.

(4) The mean energy per degree Kelvin k per discrete absorber-emitter of radiation in the thermodynamically equilibrium state of blackbody radiation, and with it the value of the new constant h, subsequently named "Planck's constant."

While the values determined for each of these "primary" quantities were theory-mediated in one way or another, in no case did the theory-mediation include, constitutively, any assumption from molecular-kinetic theory, and hence any assumption requiring the reality of molecules.

Consequently, on the other hand, the following further assumptions were correspondingly respectively needed to obtain values of Avogadro's number N_0 from these quantities:

(1) The mean translational kinetic energy of granules in Brownian motion at a given temperature is equal to the mean translational kinetic energy of the molecules in the substrate in which the motion is occurring.
(2) The number of unit charges required to produce a mole of a monovalent gas in electrolysis is equal to the number of molecules in a mole.
(3) The number of α-particles collected in a given time to form a volume of helium gas is equal to the number of monoatomic molecules of helium forming that gas, which in turn is the same as the number of molecules forming any other gas in the same volume at the same temperature and pressure.
(4) The mean energy per degree Kelvin per discrete absorber-emitter in blackbody radiation is equal to the mean energy per degree Kelvin of the molecules forming an ideal gas.

Each of these further assumptions can legitimately be regarded as the "weak-link" in the respective determinations of Avogadro's number—this, for two reasons. First, there was no evidence tied specifically to any one of the determinations to support the further assumption in question. Second, as a corollary to this, each assumption, taken in isolation, was presupposing, and hence on its face begging the question of, the reality of molecules.

But now the answer to our question about the evidence from the aggregate of the agreeing measurements of N_0 is obvious: the aggregate provided support, indeed virtually the only empirically grounded support, for these further premises linking each primary measured quantity to N_0. All those premises had one element in common, the number of "molecules" in a mole. What was thrown into question whenever the value of N_0 obtained in any one approach having a robust measure of its primary quantity failed to be in agreement with the others was not the measurement of this primary quantity, but the premise yielding the number of "molecules" in a mole. We saw such a question arise earlier in the case of the emerging discrepancy between the Perrin values and, for instance, the Millikan and Rutherford-Geiger values. Were Perrin's measured mean translational energies of the granules really equal to the mean translational energies of the molecules of the surrounding liquid?

A good way always to pin down the positive thrust of any evidence is to ask what a failure of the results to meet expectations would have been evidence of. If lack of agreement raised doubts about the linking premise, then, correspondingly, agreement gave support to it—the closer, the stronger. Accordingly, agreement among the values obtained from diverse phenomena for N_0 was the principal evidence at the time supporting the individual inferences from the primary quantities measured in each realm of phenomena to what was claimed to be the value of Avogadro's number.

Another way to see this is to consider what happened between 1908 and 1912. Attempts to derive a value for Avogadro's (or Loschmidt's) number went back at least as far as 1865. As Meyer made clear in the last chapter of his 1899 edition of *The Kinetic Theory of Gases*, none of the approaches to doing so had managed to achieve well-behaved measurements of the primary quantities from which, the hope was, a value for Avogadro's number could be derived. Their weak links lay in their measures of their primary quantities. Between 1908 and 1912, stable measures for at least four distinct primary quantities emerged: (1) the mean kinetic energies of granules in Brownian motion; (2) the fundamental unit of charge; (3) the number of α-particles per unit volume at standard conditions; and (4) the mean energy k per degree Kelvin governing in part how the spectrum varies with temperature. Moreover, at least with the first two of these, convergence among complementary stable measures had been achieved. As a consequence, the weak link in deriving a value for Avogadro's number from each of these measured quantities lay in their respective premises relating them to it. Agreement among the different derived values for Avogadro's number provided evidence substantiating that remaining weak link. Order of magnitude agreement with still other approaches that continued to have their own weak links and hence greater degrees of uncertainty could then serve to reinforce that evidence.

From the point of view of theory-mediated measurement, this was the principal evidence supplied by Perrin's aggregate of comparisons. But that was not the only evidence coming out of them. For example, the agreement among the measures was evidence that they really were measuring a single definite quantity, nominally the number of "molecules" in a mole. That was important because the primary count in each of the four cases was not of the number of molecules, but of four different items:

(1) Given their mean kinetic energy *WT*, the number of granules required to match the characteristic energy *RT* of a mole of gas or dilute solution;
(2) Given the magnitude of the fundamental unit of charge *e*, the number of these units required to match the total electricity *F* needed to yield a mole of a monovalent gas in electrolysis or, as had been shown to be equivalent to this, the total electrical charge in a mole of gaseous ions;
(3) Given the rate of production of α-particles from a gram of radium, the number of these positively charged particles required to yield a standard volume, 22,400 cubic centimeters, of helium gas at standard conditions.
(4) Given the mean energy per degree Kelvin *k* per radiator-absorber of electromagnetic radiation, the number of radiator-absorbers required to maintain thermodynamic equilibrium with the energy *RT* of a mole of gas.

Agreement among these four numbers was evidence that they are each counting a different *aspect*, in the classical philosophical sense of the term, of something beyond them that they have in common. Failure of agreement—again the tell-tale indicator—would have been clear evidence that they could not be counts of one and the same thing.

Viewing agreement among these four in this way explains the scare quotes and adverb in our peculiarly guarded phrasing, "nominally the number of 'molecules' in a mole." These four at the time were raising at least as many questions about what molecules are as they were answering—all specific variants of the generic question, *how do these amount to aspects of one and the same thing?* For those at the time, what "molecule" was supposed to refer to was not a matter of dispute, namely stable chemical unities combining fixed proportions by weight of elements—like NO, NO_2, H_2O, and He—that, at least in gases, move as integral wholes, interacting with one another in accord with the laws governing bodies in motion. Some questions raised by this characterization had received nothing but conjectural answers for a long time. For example, what holds the parts of molecules together to form a unity, and how is it related to the release or uptake of energy during chemical processes? What are the sizes and shapes of molecules, and how do they interact, through force-fields surrounding them, as proposed by Maxwell,[102] or through impact? The aspects of molecules in the four ways of counting them shed little or no light on these questions, and added new more perplexing ones. How do molecules absorb and emit electromagnetic waves in a spectrum that varies with temperature? Do they contain separate charges that nullify one another, and if so how? What happens chemically to a molecule that includes radium when an α-particle is emitted? As Poincaré said, each new discovery during the period from 1897 to 1912 had revealed a new complexity.

At the same time, possibilities for designing novel experiments to attack these questions had increased because of the many aspects of molecules that had newly been measured and hence could be *taken* as known in further experiments, and also because of ties among diverse phenomena revealed by those measurements. The theoretical connection among the gas constant R, Avogadro's number N_0, and Boltzmann's constant k had been experimentally confirmed as a consequence of each of them being measured independently of the other two. In the process, the three of them had become linked not only to the fundamental unit of charge e, Faraday's constant and the total charge in a mole of ionized gas, but also to the new "quantum" constant h governing the variation with temperature of both the blackbody radiation spectrum and, with Debye's (1912) refinement of Einstein's theory, the measured specific heats of solids. Measurements of the charge-to-mass ratio of the electron, the hydrogen ion, and the α-particle, combined with the value of e, had determined their masses, and the agreement between the masses of the latter two and the masses of the hydrogen and the helium atom inferred from Avogadro's number had established, at least to reasonable approximation, not only the masses of hydrogen and helium molecules, but by virtue of the long-standing system in chemistry of atomic and molecular weights—or, as some had called them, "equivalent" weights[103]—the masses of molecules of substances generally for which chemical formulas were known. Finally, with these masses in hand, the number of molecules in any quantity of liquid of known chemical composition—whether uniform or in given relative concentrations of different components—could be determined.

Granted, the residual uncertainties in the values determined for many of these quantities limited the precision to which they could be considered known to one or two significant figures. Contrast that, however, with the situation in 1899, as summarized in the last chapter of Meyer's *The Kinetic Theory of Gases* and J. J. Thomson's paper proposing his new conception of electricity, where the attempts to assign values to many of these quantities were still falling so far short that it made no sense to estimate residual uncertainties.

Moreover, still further experimental prospects for getting at questions about molecules were continuing to emerge at the time of Poincaré's 1912 remark. Sommerfeld announced that summer that Max von Laue, then at Munich but until 1909 Planck's assistant, had shown that crystals diffract x-rays—a discovery that won him the Nobel Prize a mere two years later. Within months, the Braggs had extended von Laue's discovery to determine the spectra of x-rays and the structure and lattice dimensions of various crystals—which won them the Nobel Prize a year after von Laue. In 1911 Rutherford, in a two-page abstract that went largely ignored at the time, had proposed on the basis of α- and β-particle scattering "a type of atom which consists of a central electric charge concentrated at a point and surrounded by a uniform spherical distribution of opposite electricity equal in amount."[104] In the summer of 1913, Bohr published his model of the hydrogen atom, in the process linking *h*, *e*, and *e/m* for electrons to the Rydberg constant.[105] And later that year, in the same volume of *Philosophical Magazine*, Henry Moseley, also at Manchester alongside Rutherford and Bohr, took advantage of the work by the Braggs to establish a relationship between measured spectra of x-rays emitted by different chemical elements and their atomic number, linking both again to *h*, *e*, and *e/m* for electrons; this led to atomic number replacing atomic weight as the more fundamental parameter and in the process effectively established Rutherford's model once and for all.[106]

A key point to appreciate about these landmark developments in 1913 was the extent to which they presupposed and built off of the measured values of the various quantities considered in the paragraph preceding the last two. In taking some exception to van Fraassen on the subject of grounding in section 5.3 of Chapter 5, we pointed out how little Perrin's Brownian motion results, taken by themselves, contributed toward empirically grounding molecular-kinetic theory. We now see, however, that the aggregate of the various measurements of Avogadro's number had by 1912 or so done an enormous amount toward grounding the theory. Recall that Ostwald's demand for what van Fraassen calls grounding was for (1) experimentally measured values of sufficient key parameters of the theory to enable true testing of it, and, with that, (2) step-by-step experimentally driven further development of it, to become possible. In the same vein, van Fraassen says, "a theory *needs to be informative enough to make testing possible at all*."[107] Even though many of the "difficulties" with molecular-kinetic theory that Ostwald had stressed remained unresolved in 1912, not to mention new difficulties that were emerging, and many questions about molecules remained unanswered, his demand had clearly been met. And, of course, van Fraassen's requirement had been met as well, though not by Perrin's results alone, but by the whole group of determinations in aggregate with one another.

6.6. "Les preuves de la réalité moléculaire"?

So far we have singled out three respects in which Perrin's comparisons among the different determinations of Avogadro's number, taken collectively, yielded evidence in their own right: (1) evidence for the "weak-link" premises licensing each inference from its measured primary quantity to a value for N_0; (2) evidence that the diverse determinations, each with its own "weak-link" premise, were nevertheless measuring—that is, yielding counts of—a single quantity;

and (3) evidence showing that a host of separate quantities—some of them presumed to be constants and others variable—are systematically related to one another not just in theory, but in measured numerical fact, thereby providing empirical grounding for molecular-kinetic theory beyond any it had ever had before. Perrin was accordingly not just engaging in rhetoric when he ended his 1909 monograph, his 1911 Solvay Conference paper, and his *Les atomes* by citing more than a dozen different determinations of Avogadro's number and insisting that the "convergence" among the values provided the most important evidence of all.

That, however, still leaves us with the question of whether Perrin was overreaching in saying at the end of his book that, *by virtue of this convergence among the values*, "the real existence of the molecule is given a probability bordering on certainty."[108] Van Fraassen's dismissal of this proposal occurs in two steps. In the first he argues that the convergence of values is indeed important, but that importance lies in its fulfilling a key part of the requirements for empirically grounding a theory. His statement of those requirements, we remind the reader, is adapted from Weyl:

- *Determinability*: any theoretically significant parameter must be such that there are conditions under which its value can be determined on the basis of measurement.
- *Concordance*, which has two aspects:
  - Theory-Relativity: this determination can, may, and generally must be on the basis of the theoretically posited connections.
  - Uniqueness: the quantities must be "uniquely coordinated," there needs to be concordance in the values thus determined by different means.[109]

On first thought, therefore, van Fraassen's construal of the convergence of values for N_0 appears to be the same as ours, save for our having spelled out in more detail the evidence deriving from meeting the uniqueness part of his concordance requirement. But this thought is wrong. Central to van Fraassen's account of how molecular-kinetic theory became grounded is that the "theoretically posited connections" enabling the measurement of N_0 and related parameters were extensions to—his phrasing is "enrichments" of—that theory. On his account, the historical core of molecular-kinetic theory was not "informative enough to make testing possible at all."[110] "The appropriate, and typical, response in that case is to start *enriching* the theory so that it becomes more informative, informative enough to allow the design of experiments in which this empirical determination of the values [of its parameters] does become possible."[111] But then, once grounding is achieved, it is the *whole* theory that becomes tested, with claims about the "reality" of its posits really amounting to nothing more than statements about the empirical adequacy of models lending specificity to the otherwise abstract theory. Therefore, in particular, the convergence of values Perrin cited was not providing evidence, in van Fraassen's view, for the "reality of molecules" if for no other reason than that it was just putting molecular-kinetic theory in a position where it could finally be properly tested.

Needless to say, we agree with van Fraassen that the determinations of N_0 involved theory-mediated measurements, and hence theoretical relations between it and experimentally more accessible quantities. And we agree as well that those theory-mediated measurements had to meet requirements in order not to be question-begging and consequently vacuous. We, however, have gone to quite some pains to show *how little* the theory-mediated measurements of the four primary quantities in the determinations of N_0 we have examined in detail depended on molecular-kinetic theory. Rather than re-arguing the point here, we refer the reader to Chapter 4 where we showed how each of Perrin's theory-mediated measurements of the kinetic energy of granules did not require any claim from molecular-kinetic theory *constitutively*. It is because the measurements of the primary quantities were independent of molecular-kinetic theory that a further premise was needed in each of the four approaches to obtain a value for N_0 from the measured value of the primary quantity.

Departing still farther from van Fraassen, notice how these further premises—the "weak links," as we have called them—offered remarkably little information about "molecules." Because of this, they were "weak" in a second sense—*they presupposed as little about the realm of molecules as they could and still yield a value for N_0*. But that, on our view, was the great virtue of the four determinations of N_0 we have discussed in detail. They left so many questions about the microphysical realm without answers, instead opening the way for subsequent experimental research to answer them, even one by one. Van Fraassen regards the empirical grounding of molecular-kinetic theory achieved during the years we have been considering as making possible experimental testing of that theory as a whole. We regard it as having made possible a systematic experimentally based further development of that theory, which it so badly needed given the unresolved difficulties that had encumbered it for decades and the new ones Poincaré referred to that were then emerging.

The remark about presupposing as little as they could and still yield a value merits further comment insofar as it is central to theory-mediated measurement in general. The key presupposition in any such measurement is the one linking the less accessible target quantity to its more accessible proxy. Having it be as limited in its theoretical commitments as possible while still yielding a stable measured value of the target quantity has the virtue of making the measurement more robust—that is, secure from future theory change—than it would be otherwise. The stability of the measure, in the technical sense we have employed in this book, can provide evidence supporting this theoretical link between the quantities, at least for purposes of measurement over the range for which measured values are obtained. This gives stable theory-mediated measures an integrity in their own right independently of the evidence supporting the theory from which they are derived.

The difficulty during the late nineteenth century in obtaining values of Avogadro's number from such proxies as the coefficients of viscosity and diffusion came from having to assume so much about molecules and their interaction, such as their sphericity, or even weaker, their having a definite effective size *s* in terms of which mean free paths could be determined. By contrast, Perrin's determinations of Avogadro's number did not require any assumption about the millions of molecules in the immediate vicinity of each granule and how their interactions were creating local, extreme high-frequency pressure fluctuations that over time were restoring kinetic energy to the granules lost in their viscous interaction with the surrounding fluid. This does not mean that more general theory is irrelevant to theory-mediated measurement. Its role, in addition to the derivation of the link between the two quantities in the first place, is to address questions about why the measure is stable, what its envelope of applicability might be, what sources of potential systematic error it might incorporate, and ultimately what is truly being measured. These questions, one should note, are subject to continuing inquiry as theory develops further, especially as other measures of the same target-quantity emerge, often involving quite different general theory.

Because this difference between us and van Fraassen concerning the enrichment of kinetic theory is at the core of our disagreement with him, we had best spell out a specific instance of it before we turn to his second step in dismissing Perrin's claim about the reality of molecules. We agree with him that two major enrichments to molecular-kinetic theory occurred with the advent of statistical mechanics when Avogadro's hypothesis took the form of a universal constant N_0 tied to the gas constant R and Boltzmann then tied the constant k not only to these other two, but to entropy. So, we are not disagreeing with van Fraassen about either the character or the importance of the theoretically posited connections that occurred with the enrichment of molecular-kinetic theory made with the advent of statistical mechanics. Our point, developed in section 6.3, is that the actual determinations of N_0 and k cited by Perrin managed to bypass presupposing the relation $N_0 = R/k$. They did so through e providing values for N_0 independent of that relation, and Planck's blackbody relation doing so as well for k. As a consequence, each of the three quantities was measured independently of the theoretical relation among them, and consequently the values so obtained then provided evidence supporting that theoretical relationship—indeed, supporting it "locally," so to speak, without having to import more of the complex apparatus of statistical mechanics! And in doing so they introduced an empirical constraint on the further development of statistical mechanics at the time.

We will allow van Fraassen to state the second step in his dismissal of Perrin's proposal in his own words:

> It is still possible, of course, to also read these results as providing evidence for the reality of molecules. But it is in retrospect rather a strange reading—however, much encouraged by Perrin's own prose and by the commentaries on his work in the scientific and philosophical community. For Perrin's research was entirely in the framework of the classical kinetic theory in which atoms and molecules were mainly represented as hard but elastic spheres of definite diameter, position, and velocity. Moreover, it begins with the conviction on Perrin's part that there is no need at his late date to give evidence for the general belief in the particulate character of gases and fluids. On the contrary (as Achinstein saw) Perrin begins his theoretical work in a context where the postulate of atomic structure is taken for granted.[112]

Van Fraassen goes on to propose an alternative to the usual reality-of-molecules interpretation of what Perrin accomplished with which we have some sympathy; but let us postpone considering that until Chapter 7.

We have already shown in detail how Perrin's theory-mediated measurements did not presuppose classical kinetic theory, much less a hard elastic sphere model of atoms and molecules. The sole place where that model enters the discussion was when he compared his values for Avogadro's number with the far more widely varying values obtained from the measured viscosities of gases. Save for the more recent value of N_0 derived from argon cited by Perrin, the values from the viscosities of gases he considered corresponded to those in the last chapter of Meyer's 1899 *Kinetic Theory of Gases*. As we have seen, the more telling comparisons Perrin made were with the values obtained by Millikan and Fletcher, by Rutherford and Geiger, by Boltwood and Rutherford, and by Planck. None of those values presupposed the hard elastic sphere model of molecules. Indeed, as we remarked near the end of section 5.4 in Chapter 5, those values were regarded by some at the time as a starting point for reconsidering and perhaps refining that model, at least as it was being deployed in the analysis of viscosity and diffusion. From that point of view, the virtue of the values of N_0 we have been emphasizing was their very independence from classical kinetic theory!

The main point we want to address in the quotation from van Fraassen is his "rather a strange reading" remark. Before turning to it, however, we should again lay out a specific instance in which the values for Avogadro's number, rather than presupposing kinetic theory, were providing a basis for reconsidering it. Of the different measures we have been emphasizing, the one that would seem most to have presupposed statistical thermodynamics—more specifically, Boltzmann's probabilistic theory of entropy (which of course had nothing to do with hard elastic spheres)—was Planck's equation, whether obtained via his derivation of it or Einstein's. During the decade after its publication, Planck's *equation* had been repeatedly confirmed experimentally. E. Warburg and H. Rubens each gave reports on its verification at the 1911 Solvay Conference, immediately preceding Planck's own presentation. More than just the equation had been confirmed experimentally, however. That the coefficient of temperature k in the equation stands for Boltzmann's constant had also been confirmed through establishing that the value for it measured from blackbody radiation spectra agrees with values obtained for the mean energy per degree Kelvin from other phenomena. The most direct comparison yielding this confirmation was with the mean energy per degree Kelvin of Perrin's granules. Less direct was a comparison of the value for Avogadro's number inferred from R divided by the measured value of the coefficient k—the very point Planck was making when he derived a value for N_0 in his original paper.

Indeed, as we noted earlier, Warburg ended his report on the verification of Planck's equation at the Solvay Conference by comparing values for e derived from blackbody radiation with values derived from the other measurements we have been emphasizing: in esu, 4.69×10^{-10} (along with his challenged 4.32×10^{-10}) from blackbody radiation; 4.65×10^{-10} and 4.79×10^{-10} from α-particle radiation; 4.89×10^{-10} from Millikan; and 4.23×10^{-10} from Perrin's results for Brownian motion.[113] The agreement among these numbers not only confirmed that the coefficient k in Planck's equation represents Boltzmann's constant; *in doing so it also rendered that assertion independent of any theoretical derivation of this equation*! To put the point more generally, the experimental verification of Planck's equation rendered even its validity independent of statistical thermodynamics.

The importance of this should not be underestimated. One of the principal issues at the Solvay Conference was the proper theoretical derivation of Planck's equation. Einstein's derivation had quantized energy universally, Planck's had not, and some of the attendees, as we saw earlier in the quote from Perrin's book about his attitude toward these derivations, would have preferred a derivation requiring no quantization at all.[114] (In a letter following the Conference, Einstein remarked, "I was able to convince Planck to a large extent about my views, now that he has resisted them for years."[115]) The experimental verification of Planck's equation and the signification of the constant k in it had freed the equation from all derivations of it, in the process turning it into a constraint on further theorizing. To put the point slightly differently, the equation and the signification of k in it had become an established fact; the question then concerned what it was telling us about the physical world. Notice, however, that what had done this was not the measurement of k in blackbody radiation by itself, but the comparison with other measurements, whether of Avogadro's number or constants related to it like the fundamental unit of charge.

Notice as well the seeming element of circular reasoning between this argument for k being Boltzmann's constant independently of the derivation of Planck's equation and our earlier argument that the relationship among R, N_0, and k was being established empirically independently of its derivation within kinetic theory. There we relied on k being Boltzmann's constant to use the value obtained for it from blackbody radiation as grounds for saying that, numerically, $k = R/N_0$. But here we have appealed to the relation among these three along with the not exceptionally good match between the measured number and the mean energy per granule in Brownian motion as grounds for saying that, numerically, the value of k obtained from blackbody radiation matches, to within the precision of the experiments, the value determined independently of blackbody radiation for Boltzmann's constant.

This was not a careless mistake on our part. We intentionally chose the two examples in order for the seeming circularity between them to illustrate how the converging values for N_0 and the quantities related to it from diverse theory-mediated measures were able to contribute to ongoing research at the time. Metaphorically speaking, the values from the different measures could be *played off against one another*. Thus, when the issue under consideration concerned the derivation of Planck's equation, the other determinations of k allowed the equation to stand independently of any derivation of it, leaving a free range for considering alternative derivations. If, instead, the issue was something like the equipartition of energy in classical kinetic theory, the determinations of R, k, and N_0 independently of one another allowed the equation $k = R/N_0$ to stand independently of its derivation within kinetic theory. The overall collection of converging determinations did not make them all independent of any theoretical derivations of the equations constitutive of their theory-mediated measurement. Rather, it lessened the dependency of all of those constitutive equations on the theoretical derivations of them, thereby allowing them to be played off against one another when considering changes to the theories by means of which they were originally derived.

This was one way of playing off the converging theory-mediated measurements of the quantities in question against one another. A second, illustrated earlier in section 6.2, was to use differences among the values obtained from the quantities to expose sources of systematic error in the measurements. Thus, for example, we saw in that section a clustering of values for N_0 between 59×10^{22} and 62×10^{22} that had been obtained by Rutherford and Geiger, Boltwood and Rutherford, Millikan, and Planck by 1911—all markedly less than the values

Perrin had obtained, yet greater than Fletcher's 57.5×10^{22}—from Brownian motion. That raised questions not only about which values should be preferred, but also about what Perrin's consistently large values might be indicating about his measurements. What saying that converging measurements can be played off against one another in more than one way really means is that they can provide diverse forms of evidence in ongoing research, depending on the nature of the questions on which that research is focusing.

Returning finally to van Fraassen's "strange reading" remark about Perrin's results being evidence for the reality of molecules, we agree that it would be strange—not just in retrospect, but even at the time, long before Heisenberg and Schrödinger—to take any of the various determinations of Avogadro's number as evidence for the reality of molecules "as hard but elastic spheres of definite diameter, position, and velocity." We take ourselves to have shown that, regardless of the extent to which such a picture of molecules remained in the minds of physicists at the time,[116] that picture in no way entered constitutively into the determinations of N_0 or the quantities related to it that we (and Warburg) have emphasized. Indeed, judging from Jeans's 1916 edition of his *The Dynamical Theory of Gases*, the determinations of N_0 led to clear experimental evidence that molecules are *not* "hard but elastic spheres of definite diameter."

In dismissing van Fraassen's argument against taking those determinations to have been evidence for the reality of molecules, however, we have not explained how and why it was appropriate at the time to take them to be evidence for it. We turn now to that task, focusing on Perrin's paper at the Solvay Conference, "Les preuves de la réalité moléculaire."

Something that does appear strange in retrospect is why Perrin was invited to give *this* paper, more than twice as long, at 98 pages, as any other paper at the Conference save for Sommerfeld's 60 pages, and occupying 22 percent of the Proceedings volume, yet prompting the least discussion of any. The letter of invitation to attend the Conference, part of which we quoted in Chapter 1, made clear that the reality of atoms and molecules was not what was going to be at issue:

> It appears that we find ourselves at present in the midst of a new evolution of the principles on which the classical molecular theory and the kinetic theory of matter have been based.
>
> On the one hand, this theory leads to a logical development of a radiation formula whose validity is contradicted by the results of experiments; on the other, there follows from the same theory certain results on the specific heat (constancy of the specific heat of a gas with the variation of temperature, the validity of Dulong and Petit's law up to the lowest temperature), which are equally refuted by many measurements.

As Planck and Einstein in particular have shown, these contradictions disappear if one places certain limits (doctrine of energy quanta) on the motion of electrons and atoms in the case of their oscillations around a position of rest. But this interpretation, in turn, is so far-removed from the equations of motion of material points employed until now, that its acceptance would necessarily and incontestably lead to a vast reformation of our current fundamental theories.

. . . A great step in the direction of the orderly continuous development of atomistics would have already been taken if one could clearly establish as to which of our molecular and kinetic interpretations are in accord with observations and which others should undergo a complete transformation.

. . . The subjects of the conference treatises will be the following:

1 —The derivation of the Rayleigh formula for radiation.
2 —Comparison of the kinetic theory of ideal gases with the results of experiments.
3 —Application of the kinetic theory to emulsions.
4 —The kinetic theory of specific heats due to Clausius, Maxwell, and Boltzmann.
5 —The formula for radiation and the theory of quanta of energy ("Quantenhypothese").
6 —Specific heat and the theory of quanta.
7 —Application of the theory of quanta to a series of problems in physics.
8 —Application of the theory of quanta to a series of problems in physical chemistry and chemistry.

For each of these questions we are asking a particularly competent member to write a preliminary paper. These papers, written in French, in German, or in English, will be printed and distributed to the different members, if possible by the end of September; later they will be joined in one volume together with accounts of the discussions prompted by them.[117]

While Perrin was on the list of invitees from the outset, the one topic not in the original list proposed by Nernst, but added by Solvay or one of his advisors, was the "application of the kinetic theory to emulsions."

In keeping with the invitation, Perrin's subtitle, in parentheses, was "(*Études spéciale des émulsions*)"; and slightly more than half his paper concerns Brownian motion results that he and his colleagues had produced, including Léon Brillouin's results on granule diffusion not yet available at the time of Perrin's 1909 monograph. The one other slight departure from that monograph involved the range of values determined for Avogadro's number: 68.3×10^{22} (versus 70.5×10^{22}) from vertical-rarefaction and 68.8×10^{22} (versus 71.5×10^{22}) from granule displacement, along with Brillouin's 69×10^{22} from granule diffusion. The only question during the brief discussion of Perrin's paper concerning his results raised the possibility of his numbers being affected by radiational pressure acting on the granules, a possibility that he immediately dismissed. (Virtually all the rest of the discussion concerned the proper magnitude of the value of e, with Einstein calling attention to recent results that had shown Ehrenhaft's much smaller measured values of e to be a consequence of experimental error, and Madame Curie citing the agreeing values published by Rutherford and Regener.)

Insofar as Perrin's 1909 monograph had been translated into German and English in 1910 and was surely known to the attendees, one can ask what this part of his paper contributed to the Conference. So far as we can see, the answer seems to have been (1) the strong evidence for the thoroughgoing irregularity of Brownian motion and hence too of the pressure fluctuations driving it, and (2) the consequent evidence for the granular character of matter, along the lines we made explicit in Section 5.6 of Chapter 5. This was not a small contribution insofar as one of the topics at issue at the Conference was Boltzmann's probabilistic interpretation of entropy and the second law of thermodynamics. Thus we find Einstein opening the discussion of his own paper, at the end of the conference, insisting that "another principle whose validity we must retain unconditionally is Boltzmann's definition of entropy through probability."[118] The sole experimental evidence that Einstein cited in arguing for this proposal was the Gaussian distribution that Perrin had obtained in his experiments on granule displacements![119]

The remaining 35 pages of Perrin's paper present, in some analytical detail, values for Avogadro's number obtained from other phenomena. These included the recent oil-drop results by Millikan (who was not present), up-to-date results from α-particle radiation by Rutherford (who was present) and his colleagues, and results from blackbody radiation. As in Perrin's 1909 monograph and subsequent *Les atomes*, this discussion ends with proposed values for N_0, 67×10^{22}, and e, 4.3×10^{-10} esu, and a section entitled "*Réalité des molecules*" that includes a table[120] listing the converging values for N_0 followed by a pronouncement:

> We are struck with admiration before the miracle of such precise agreement from the basis of such different phenomena. That we not only find the same magnitude obtained by each method while varying experimental conditions as much as possible, but also that the numbers so defined without ambiguity by so many methods coincide: this gives molecular reality as much certainty as the principles of Thermodynamics can have.[121]

Notice that the final phrase in this conclusion, expressed to other leading figures in the field, differs a little from the one in the corresponding passage in his *Les atomes* that we quoted at the opening of the chapter—"gives to the real existence of the molecule a probability bordering on certainty."

In keeping with the plural "proofs" in the title of Perrin's Solvay Conference paper, the others present doubtlessly took each of the several agreeing determinations of Avogadro's number to have provided a proof of some sort or other. The passage we just quoted, however, suggests that his definitive proof of the reality of molecules, overriding any shortcomings in any of the prior ones, was the convergence of the diverse determinations. This raises two questions. How might the others at the Conference have construed this convergence as providing the strongest evidence? And, given that most of them, if not all,[122] had already accepted the existence of molecules, what might they have thought Perrin's proofs were contributing to the Conference?

Perrin's presentation of each of the determinations was careful and thorough, and several of those contributing to them—Einstein, Langevin, Rutherford, and Planck—were present. So, the conferees should have had little trouble seeing that the measurements of what we have called the primary quantities in the determinations we have emphasized were at most only weakly dependent on molecular-kinetic theory, if not entirely independent of it. That would have highlighted the premises linking those primary quantities to the claimed count of molecules in a mole: (1) that the kinetic energy per degree Kelvin of granules in Brownian motion matches that of molecules; (2) that the ions in a gas are just atoms, or molecules, with one or more unit charges; (3) that one fundamental unit of charge yields one molecule of a monovalent substance in electrolysis; (4) that α-particles retain their discreteness, and hence become molecules, when they lose their positive charge to form helium gas; and (5) that the coefficient of temperature in Planck's equation is identical with the Boltzmann constant of statistical mechanics, giving the mean energy per molecule per degree Kelvin. But then they would have viewed the convergence of the corresponding values for N_0, just as we said in section 6.5, as responding to any remaining uncertainty in any of these five considered separately from the others. And with that the evidence for the reality of molecules would have been seen as derivative from the strong evidence for the discreteness of the unit charge, the discreteness of α-particles, the discreteness of *hν* quanta of energy in radiation, and the discreteness required of localized elements in the fluid in order for their thoroughly random, mutually uncorrelated motions to sustain Brownian motion.

In section 6.5 we put scare quotes around the word "molecules" in essentially the same construal of the evidence for their existence. We did this because, construed in this way, it offered so little evidence pertaining to features of molecules that had been postulated within molecular-kinetic theory at one time or another in conjunction with proposed explanations of phenomena. As we said there, the evidence left so many questions unanswered precisely because the determinations of N_0 singled out were so minimally tied to molecular-kinetic theory.

Perrin included three determinations in his Conference paper beyond the ones we have been emphasizing that depended more strongly on that theory, namely those derived from viscosity, critical opalescence, and the blueness of the sky, with both of the latter tied to fluctuations in fluids, that is, to "regimes of permanently varying inequalities in the properties of the microscopic portions of matter in equilibrium."[123] From viscosity Perrin listed two values in his table, ca. 45×10^{22} in parentheses based on the volume of monoatomic mercury and 62×10^{22} based on van der Waals's equation and argon. No sooner did he give those values in the body of his paper, however, than he indicated that errors in the values of mean free paths, owing to the spherical-molecule assumption in the formula tying them to viscosity, rendered these values for N_0 uncertain.[124]

The values from both critical opalescence and the blueness of the sky (as a consequence of the diffraction of light) were both based on an equation derived by Keesom and later independently by Einstein for the intensity of light diffused by one cubic centimeter of a fluid.[125] The value Perrin listed in his table for the former was 75×10^{22}, while for the latter, 60×10^{22} with a question mark attached. In the derivation of the first of these two, Perrin had spoken of "approximation within 15 percent."[126] So, these two values, like the two obtained from viscosity, were likely viewed by others at the Conference as being rather uncertain compared to those from the determinations we have been emphasizing. They offered a promise of tying the latter to determinations more dependent on molecular-kinetic theory and hence giving more information about molecules themselves, but further research was going to be required for that promise to be realized.[127]

That the clearest evidence for the existence of "molecules" left so many questions about them unanswered was likely regarded at the Conference as not all that surprising and in many respects a virtue—for consider the topics at issue. The equipartition of energy principle at the heart of the classical kinetic theory of specific heats was under challenge from Einstein's quantum theory for solids, in which the vibration of individual atoms disappears at low temperatures from the absence of sufficient energy to exceed an $h\nu$ quantum threshold for those statistically at low energies, as well as from Nernst's proposal that rotational degrees of freedom of non-spherical diatomic molecules in gases drop out at low temperatures for the same reason.[128] Needless to say, the challenge here was not merely to the equipartition principle, but to the classical mechanics of motion as well, at least so far as atoms and molecules were concerned.

The equipartition principle deployed in the classical statistical thermodynamics of radiation was also under challenge by the failure of the Rayleigh-Jeans blackbody equation and the success of Planck's equation. Jeans's effort at the Conference to save the principle by deriving Planck's equation instead from hypothetical proposals concerning the process of energy exchange between matter and radiation also put at issue how neutrally charged atoms and molecules nevertheless release (and absorb) electromagnetic radiation, presumably through some sort of interaction with the ether or motions of charges, in particular electrons, within them. In rejecting such proposals in his paper, "On the Application to Radiation of the Theorem of the Equipartition of Energy," that opened the Conference, Lorentz concluded that a satisfactory formula for radiation and explanation of Planck's constant *h* were going to require "considerations of an entirely different order" from those of classical theory.[129] Reinforcing the need to look for new directions were Kamerlingh Onnes's short paper on the newly discovered phenomenon of superconductivity and papers considering possible further applications of the quantum principle, like Langevin's on the kinetic theory of para- and ferro-magnetism.[130]

In short, what was at issue at the Conference concerned how to move away from classical molecular-kinetic theory and statistical thermodynamics without abandoning them entirely. In such a circumstance it was a virtue, not a shortcoming, of the convergent determinations of Avogadro's number emphasized at the end of Perrin's paper that they were so minimally tied to classical theory; then the values for N_0 and the various other microphysical quantities tied to it could provide an empirical basis for further theorizing without imposing constraints that might have turned into obstacles to be overcome. Theory-mediated though they were, the determinations of these quantities did not presuppose stances on the theoretical points at issue at the Conference and hence were not going to require revision as these issues came to be resolved. The values, at least to the level of precision then achieved, could be regarded as established facts. It was, moreover, the convergence of the values from diverse phenomena that gave them their factual status, for because of what we have called the "weak-link" premises, any one determination by itself would have been regarded by those at the Conference as precarious until it was supported by others.

The mention of constraints on further theorizing points to the answer to our other question about the attitudes toward Perrin's paper at the Conference: What might the others have seen Perrin's "proofs of the reality of molecules" as contributing to it? It also points to why it was appropriate for Perrin's going to such length in his paper, covering not only his determinations of N_0 from Brownian motion, but also those from other phenomena, laying out in considerable detail the considerations in every case leading to the equations mediating the determinations. Not just the values of the quantities, but the equations used in deriving them from measured phenomena, were going to be serving as at least provisional constraints on further theorizing.

We have already cited an example of this with Einstein's invoking Perrin's finding of thoroughgoing randomness in his defense of Boltzmann's probabilistic account of entropy at the Conference. The detail to which Perrin went made clear to the other attendees just how robust were the various equations and values going into what we have called the measurement of the primary quantities in each determination of N_0, and equally how limited were the further claims made by the additional "weak-link" premises licensing the values for N_0 obtained from those quantities. In doing so, Perrin's paper, though not the only one at the Conference to focus on experimental results, made clear to all several important, by then more firmly established, empirical results pertaining to microphysics.

One of these results, tied at most loosely to classic molecular-kinetic theory, was that matter consists of chemically stable discrete elements—"molecules," as it were—numbering $(64 \pm 5) \times 10^{22}$ per mole. Moreover, a *mole* is constituted not just by pneumatic conditions of temperature and pressure, but also by chemical composition. Thus the mole density of a gas composed of NO_2 begins to decrease when conditions reach a point where N_2O_4, nitrogen peroxide, begins to form. But then the number of chemically stable discrete elements forming that gas must begin to decrease as well. The step from "molecules" to the molecules of chemistry was thus quite small as a consequence of the converging values for Avogadro's number, even without invoking the monoatomic molecules of chemically inert helium in the Boltwood-Rutherford measure. To put the matter only slightly differently, the converging values of N_0 had licensed the ideal gas law to be written not just as $p = (n/V)RT$, where (n/V) is *mole-density*, but also as $p = (N/V)kT$, where (N/V) is the number of chemically defined molecules per unit volume.

Furthermore, while a great many questions were left open about these molecules, the convergent determinations of N_0 had provided strong evidence for the links between them and such phenomena as the thoroughly random Brownian motion; ionization, electrolysis, and the fundamental unit of charge; α-particle radioactive decay and the transformation of radium into other elements; and blackbody radiation. The general question left for further theorizing was how to preserve those results and links among parameters while moving away from classical molecular-kinetic theory. All this, needless to say, underscores just how far removed the actual historical situation was from van Fraassen's claim that "Perrin's research was entirely in the framework of the classical kinetic theory in which atoms and molecules were mainly represented as hard but elastic spheres of definite diameter, position, and velocity."

Such spherical molecules continued to appear in works presenting *mathematical* statistical mechanics in the years after 1912, but this was because the actual shapes of molecules and the force-fields involved in their interactions remained unknown, and impact of elastic spheres was mathematically tractable; and, thanks in part to the successful determination of N_0, comparison of the dynamical theory of elastic spherical molecules with experimental results increasingly underscored how limited this theory was for linking any specific quantitative feature of phenomena like viscosity to specific features of chemically different molecules.[131]

To summarize, those at the Conference would scarcely have found Perrin's "proofs" strange at all. Equally, however, they would not have regarded them as addressing the question, Do molecules *really* exist at all? For no one at the Conference was questioning this. Rather they would have seen the "proofs" as laying out what about the molecular realm could be regarded as having been experimentally established during the preceding decade—or, perhaps better put, sufficiently established to be taken as constraints in ongoing research on the topic of the Conference, "the theory of radiation and the quanta." Such constraints were needed, even somewhat desperately, given how *prima facie* irreconcilable quantization seemed to be with classical physics in 1911. Perrin's paper was not the only one at the Conference to be presenting such constraints, but it did the most toward doing so. Its exceptional length was accordingly not inappropriate.

6.7. On Eliminating All Reference to the Invisible

As a final step in his argument that the empirical content of Perrin's results extended only to the grounding of molecular-kinetic theory, van Fraassen cites Perrin's remark near the end of his 1909 monograph, "we ought always to be in a position to express all the visible realities without making any appeal to elements still invisible."[132] Perrin followed the remark first by noting how "the constant N is simply a completely known numerical factor, figuring in the enunciation" of each equation employed in its determination, and then showing how to drop the term by using any one of these equations to substitute for N_0 in all the others, yielding equations in which "*only evident realities occur*."[133] Perrin made the same proposal at the end of both his 1913 book and his 1911 Solvay Conference paper. What are we to make of it?

As we have been doing, let us focus on this proposal at the Solvay Conference and how its attendees might have taken it. Perrin's specific phrasing there, which immediately followed the remark we quoted earlier about molecular reality having as much certainty as the principles of thermodynamics can have, was:

> Nevertheless, and as greatly necessary as the existence of molecules or atoms may be, we should always be in a position to express visible reality without making appeals to as yet invisible elements. And indeed this is very easy. It suffices to eliminate the constant N among the p equations that helped us determine it, in order to obtain $(p-1)$ relations in which only sensible properties figure and which express profound connections among phenomena at first

sight as completely independent as the viscosity of gases, Brownian motion, the blue of the sky, the blackbody spectrum, or radioactivity.[134]

Perrin proceeded to illustrate this by combining Einstein's equation for granule diffusion in dilute emulsions with Planck's equation:

$$\xi(\lambda, T) = \frac{8\pi ch}{\lambda^5} \frac{1}{e^{\frac{ch}{\lambda T}\frac{6\pi a\zeta D}{\theta}} - 1}$$

where c is the speed of light and θ is the temperature at which the value for the diffusion coefficient D is measured. Perrin then added, "This ensures that we can control, by means of spectrophotometric measures, points pertaining to emulsions!"[135]

Surely most of those at the Conference would have regarded this equation and even more so the remark following it as parody, ridiculing those like Duhem and Mach who were still continuing to deny the admissibility of molecules and atoms in physical theory. Just consider for a moment what sort of research would have revealed this relation between granule diffusion and blackbody radiation as a robust empirical result independent of any thought of molecules and atoms. Or consider the challenge to theoreticians to offer an explanation of it without appealing to microphysics.

That Perrin himself may well have intended it to be parody, at least at the Conference, is reinforced by his closing paragraph, immediately following the exclamation point quoted in the preceding:

> But we will not have the tactlessness, under the pretext of rigor, to rid our equations of the elementary magnitudes that enabled us to obtain them. This would not amount to taking away the scaffolding rendered useless to the completed building, but to concealing the pillars that make up its structure and its beauty.[136]

Although we have not managed to establish the direct connection, we find it difficult to believe that Perrin's choice of words here was not drawn from Ostwald, as reflected in a passage from his 1907 *Monist* paper that we chose not to quote earlier (that had been published in German earlier that year in the opening number of *Rivista di Scienza*):

> In the scientific language of to-day these two fundamentally different kinds of hypotheses are called by the same name, "hypotheses." I am in favor of leaving the name hypothesis for the unverifiable assumptions, since indeed most hypotheses of the science of to-day are of this character. The other assumptions

> which, like scaffolding, serve the purpose of the particular investigation on hand, and in the course of the work are replaced once or oftener according to our need for new usable assumptions until the condition sought for is actually found,—these assumptions made for the purpose of positive work, I call prototheses. A protothesis, therefore, is set up at the beginning of an investigation and disappears at its close if the work is successful; while a hypothesis is established when the work can be carried no farther. For this reason it is customary that in the production of scientific work the different prototheses which the investigator has employed are for the most part not mentioned at all, for it has been usual to communicate only those assumptions which the investigation has finally proved to be correct or at least approximate. Silence is kept with regard to the unsuccessful prototheses just as the scaffolding is removed after a building is completed. Only very rarely, as for instance in Kepler's reports of his astronomical researches, do we learn a little about unsuccessful prototheses. On the other hand the hypotheses in the narrower sense occupy a large place in literature. Because inasmuch as they have reference to scientifically inaccessible things, they can be neither proved nor refuted, an endless pro and con is the usual result; further, because among the problems there are also pseudo-problems which do not refer to demonstrable things, these problems are unsolvable and are dragged through science as unanswered questions which can not be disposed of, but not until they are recognized to be merely pseudo-problems.[137]

Ostwald himself, who had not been invited to the Conference, must have taken Perrin's words to be responding to his when he finally read the Conference proceedings, whether in the original French or the official German translation.

Parody though they may well have been, Perrin's closing remarks contained two important further points not stressed during his review of the different determinations of N_0. First, his reference to "elementary magnitudes" reminded his audience that not just Avogadro's number, but other terms as well appearing in the equations, involved "invisible elements." Even if the unit charge was not being identified with the electron, counts of invisible ions associated with water and oil droplets entered into some of the determinations, as did counts of α-particles. Getting rid of the latter, for example, would have required replacement of the rate of α-particle production in the equation by the number of "throws" per minute of an electrometer in an extraordinarily recondite experimental setup. But then too the number of electromagnetic cycles per second of light of any specific color was scarcely visible either. The challenge posed by the invisibility of molecules and the like in 1911 was not one of inaccessibility, for enough experimental access had been gained to answer *some* questions about them. The challenge was one of designing further experiments to gain still greater access. We should not be surprised that Laue and the Braggs were awarded Nobel Prizes, each within two years of their providing such further access by means of x-ray diffraction over the next two years.

Second, the emphasis Perrin placed on the wide range of "at first sight completely independent" phenomena yielding agreeing values for N_0 amounted to more than just an argument that the values were well founded. The more ubiquitously N_0 shows up in equations describing different phenomena, the stronger are the grounds that it represents a *fundamental* constant of nature.

This last point, however, can be turned on its head. Ostwald had introduced the word "mole" only a couple of decades earlier to stand for a fundamental unit in chemistry. What made this unit appropriate was the realization that the chemical makeup of a gas can change with temperature and pressure—for example, the formation of nitrogen peroxide in a gas composed initially of NO_2 as the temperature is lowered. To accommodate this, the gas law needed to be reformulated as a proportionality among pressure, absolute temperature, and, rather than density, mole density or its equivalent. The mole nevertheless had a somewhat artificial character as a unit, tied as it was to the system of atomic (or "equivalent") weights, and its somewhat arbitrary choice for the unit weight—not to mention the still unfolding system of chemical formulas. (This artificiality came even more into question two years later with Mosley's evidence that atomic number, not atomic weight, is the fundamental chemical parameter.) The diverse multiple determinations of N_0 served historically to remove worries about the artificiality of the mole. In particular, the two robust determinations from radioactivity, one (involving volume) dependent on R and the other (involving charge) not dependent on R, but on Faraday's constant, gave evidence that N_0 provides a physical characterization of something universal to a mole, namely its containing, at least so far as precision of measurement then allowed, a specific number of chemically stable independent discrete elements—molecules, as it were. The more diverse were the phenomena that yield increasingly convergent values, the greater was the evidence that N_0 represents a fundamental constant that provides such a count, and hence the stronger were the grounds that the mole is a well-constituted physical unit.

Both of these final two points that we have teased out of Perrin's closing remarks reinforce his claim that quantities like N_0 pertaining to invisible elements in the equations he had presented did not comprise mere "scaffolding" to be discarded in the future. They were going to remain as "pillars," regardless of how the reformulation of classical molecular-kinetic theory and statistical thermodynamics might unfold, not to mention molecular theory within chemistry.

6.8. A Parting Comment, Regarding *réalité moléculaire*

Our conclusion, therefore, is that the converging measures of Avogadro's number, though still at the time less convergent than one might want, provided much stronger evidence as of early 1913 for the "reality of molecules" than Perrin's results for Brownian motion alone did. While the Gaussian distribution of displacements in Brownian motion provided strong evidence for the discontinuous or "granular" character of the liquid substrate, a clear lacuna remained between this conclusion and the claim that the discontinuous character in question is from chemical molecules, or even Maxwell's "molecules." We grant that Perrin's results left molecular-kinetic theory as the only available explanation for both the perpetual and stochastic character of Brownian motion. To infer the "reality of molecules" from this, however, is to engage in inference to the best explanation. Regardless of what one thinks of inference to the best explanation, to invoke molecular-kinetic theory as an explanation of Brownian motion is to presuppose the reality of the molecules being invoked and then take the explanation they provide as added evidence for them—which is more or less what van Fraassen claimed about Perrin's reasoning.

By contrast, neither the "reality of molecules" nor the molecular-kinetic theory of heat has to be presupposed if the evidence provided by the converging measures of Avogadro's number is construed as a form of same-effect-same-cause reasoning—specifically as same-magnitude-same-quantity-being-measured reasoning. Consider the three least controversial measures that were yielding, to within reasonable precision, the same magnitudes: the number of unit charges, that is, electrons, required to yield a mole of hydrogen, or any other monovalent element, in electrolysis; the number of granules in Brownian motion required to match the total energy at any given temperature of a mole of gas (or liquid solution); and the number of α-particles required to form a mole of helium. Insofar as all three yield a numerical value characteristic of a *mole*, the proposed inference plus inductive generalization yields the conclusion of the existence of some distinctive quantity—a certain number of units of something—characteristic of any mole of any substance. Suppose we call this quantity *m-units*, with the three measures taken to be indirect measures of it.

As we noted earlier, the ideal gas law implies that this quantity has a systematic relationship to chemical formulas. Suppose, for example, we take a mole of nitrogen dioxide, NO_2, and lower its temperature to the point where it consists of half (by volume) nitrogen dioxide and half nitrogen peroxide, N_2O_4. Then the pressure-volume relationship of the mixture is going to entail a mole density of two-thirds of what it was when the gas was pure nitrogen-dioxide. But this in turn is going to entail a reduction in the number of *m-units* to two-thirds of what it was originally—half of which is a measure of the number of *m-units* of NO_2, and the other half, of the number of *m-units* of N_2O_4. This point generalizes: the *molar density* of any gas, pure or mixed, is going to entail a corresponding value for *m-unit density*.

Nothing in this reasoning so far requires the *m-unit* quantity to be a count of individual, discrete items. Each of the quantities taken to be indirect measures of it, however, are counts of individual discrete items: the fundamental negative charge, which has a definite measured (rest) mass associated with it and hence can no less appropriately be called an electron; granules in Brownian motion; and the α-particle, which again has a definite measured mass associated with it. Extending the sort of same-cause-same-effect reasoning we have been developing, the fact that these three are counts of individual discrete items gives grounds for concluding that the *m-unit* quantity too is a count of individual discrete items; or put more conventionally, it gives grounds for taking Avogadro's number to represent a count of the number of these chemically stable individual, discrete items in a mole, each with its characteristic chemical formula. The value of Avogadro's number then suffices to determine the mass of each such item from its chemical formula and the system of equivalent weights that had been worked out over the course of the nineteenth century. Moreover, we have reached these conclusions without invoking the evidence for the discontinuity or "granular" character of matter provided by Brownian motion, and hence that evidence adds further support for taking Avogadro's number to represent a count of something.

Insofar as no one had any claim to knowing anything more about the hypothetical entities called "molecules" than what we have listed in the preceding, and before 1908 the absence of a credible value for Avogadro's number had left them with no legitimate claim to knowing some of what we have listed, the reasoning we have offered provides grounds for calling the items counted by *m-units* "molecules." This reasoning required no assumptions whatever about the sizes and shapes of such molecules, nor about how they interact with one another. In particular, the molecular-kinetic theory of heat did not enter into the reasoning at all, nor, contrary to van Fraassen, did any assumption to the effect that what we are now calling molecules are "hard but elastic spheres of definite diameter, position, and velocity." Nor did the reasoning in any way conflict with the claim Nernst made so forcefully in the 1893 edition of his text that we quoted more than once in Chapter 2, a claim that he repeated word-for-word in the 1908 edition: "At present scarcely anything definite is known regarding either the nature of the forces which bind the atoms together in a molecule and which hinder them from flying apart in consequence of the heat motion, or regarding their laws of action." In the 1921 edition, Nernst again repeated this sentence, but appending to it a further clause: "... in this case also the application of the quantum theory is helping to throw light on the problem."[138]

Construed as we have been proposing, therefore, the converging measures of Avogadro's number gave clear grounds for *taking* molecules to be real while still leaving open the many questions about their specific features that we argued in section 2.7 of Chapter 2 are needed for molecular theory to explain any specific features of macroscopic phenomena. In the process these questions had gained more legitimacy than they had had before, and the related results on the fundamental unit of charge, ionization, and α- and β-particle radiation had added new constraints on answers to these questions, as well as some promising leads. In fact, the questions had gained more than just added legitimacy. They had gained an added demand to find experimental answers to them, for a failure to obtain such answers would prompt revisiting of the reasoning we have proposed, in particular the steps from the converging measures to their being indirect measures of a further quantity, *m-units*, and from their being counts of their respective individual discrete items to taking *m-units* to be a count of individual discrete items.

In particular, this last step to *m-units* being a count of individual discrete items involves a reach beyond the available evidence. Nothing in the evidence we have examined *requires* a mole to consist of $(64 \pm 5) \times 10^{22}$ *individuals* with sufficient integrity to persist in and of themselves over time. Even granting that atoms exist as individuals and chemical compounds consist of atoms in combination with one another, nothing in the evidence precludes those combinations from existing evanescently, with atoms dissociating from one another and combining with different atoms all the time at an extraordinarily small time-scale. That such a process can be represented as if each combination of atoms persists over an extended time-scale, yet still as nothing but a heuristic conceptualization consistent with all the measured results, has not been ruled out by the evidence. The shortcomings of trying to further represent the combinations as spheres or having spheres of influence underscore that the evidence has not yet managed to reveal sufficient specifics of the putative molecules to assure their existence as individuals.

This, however, is not to say that the step to *m-units* being a count of individual discrete items is unwarranted for purposes of ongoing research. Rather, it is to say that it places a great deal of stress on experimentally pursuing answers to questions about the integrity of molecules as individuals, especially on, in Nernst's words, "the nature of the forces which bind the atoms together in a molecule and which hinder them from flying apart in consequence of the heat motion."

On our construal, accordingly, the converging measures by no means established the reality of molecules once and for all. For, what the term "molecule" was being taken to designate remained, in spite of the added constraints, not only highly underspecified, but perhaps even failing to refer to definite persisting individuals at all. One might regard the extent to which the referent of "molecule" remained underspecified as a shortcoming of the evidence deriving from the agreeing measurements considered in this chapter. This, however, would be to miss the point. The evidence is so strong for the values of the molecular magnitudes in question precisely because it presupposes so little about molecules themselves, and hence is less in jeopardy of being overturned by future research than it would have been had it presupposed more. It thus contrasts sharply with the evidence achieved in the nineteenth-century efforts to obtain values of these very magnitudes, depending as it did on some form of assumption of sphericity. In this regard our construal seems closer to Alan Chalmers's response to van Fraassen than to Wesley Salmon's original invocation of common-cause reasoning for the case of the converging values for Avogadro's number; for Salmon's construal seems to require so much more of molecular-kinetic and chemical-molecular theory than either Chalmers's or ours does.[139]

All of this is in keeping with the force of Newton's same-effect-same-cause rule of reasoning, for it too, along with his other rules, licensed conclusions only to be *taken* to be true—his phrase—for purposes of ongoing research until experimental grounds should arise prompting their reconsideration.[140] Newton deployed these rules expressly with the goal of "arguing more securely"[141]—that is, of achieving comparatively secure grounds for some conclusions about gravity while leaving questions about its mechanism open to future research. Our construal of the force of the evidence in this chapter is accordingly not without precedent.[142] Nor, for that matter, is its presumption of persistence. For this presumption is fully akin to that of Clausius, Maxwell, and Boltzmann to the effect that all molecules of, say, oxygen are the same and hence interchangeable—a presumption that the discovery of isotopes showed to be not quite true. In both of these cases, the presumptions amounted to a research strategy, for not making them was tantamount to giving up hope that ongoing research was going to get much of anywhere. Here too Newton has given a warrant for the strategy, this time with his first Rule of Reasoning, which amounts to requiring that no complication be added to theory unless expressly called for by some specific empirical finding.

Notes

1. Salmon (1984), pp. 221ff.
2. Perrin (1990), p. 215f. The English translation, unlike the French original, does not set the final section of the book, called "Conclusions," aside as a separate section, thereby understating its intended force. Also, the 1913 and 1914 editions of *Les atomes*, and the 1916 English translation of the 1914 fourth edition, ended with a comparison of

13, not 16, values for Avogadro's number, all but two of them between 60×10^{22} and 75×10^{22}. See Perrin (1913), p. 289; (2014), p. 299; (1914), p. 293; and (1916), p. 206.

3. Mohr, Newell, and Taylor (2008), p. 86.
4. Huygens (1986). See also Smith (2007). In citing Huygens's pendulum measure, we should note that this was done first with cycloidal pendulums, but then subsequently with small-arc circular pendulums for which the length-to-period relationship matches that of the corresponding cycloidal pendulum in the asymptotic limit as the arc approaches 0. Correspondingly, his conical pendulum measure initially involved constant height pendulums, but then he added a measure implemented on a conical pendulum in which the bob is constrained to the surface of a paraboloid. For more details, see Yoder (1988).
5. Koyré (1968).
6. See Smith (2007) and (2001a), especially p. 334.
7. See Smith, "The Question of Mass in Newton's Law of Gravity," in preparation; also, Todhunter (1962), especially Volume 2; Sabine (1825); and Heiskanen and Vening Meinesz (1958).
8. Maxwell (1865), p. 499. Ironically, the "official" value of the speed of light for a time became the one obtained from Maxwell's equation—for the obvious reason that the quantities entering into it can be measured to higher precision than can be achieved by any measurement directly of light.
9. Thomson (1897). See also Smith (2001b).
10. Thomson (1899), p. 563. For more details, see Smith (2001b).
11. See Poynting (1894) and Zenneck (2007).
12. Perrin (1908d), p. 594.
13. Rutherford and Geiger (1908b), p. 171.
14. Ibid.
15. See Perrin (1910), p. 89, for a description of how Lorentz obtained his value.
16. Perrin (1990) and (1916), p. 105; in the original French, (1913), p. 150, or (2014), p. 201.
17. Ostwald (1912), p. 486.
18. Salmon (1984), p. 224.
19. Millikan (1911), p. 396.
20. Fletcher (1911), p. 106f.
21. Ibid., p. 95. Italics in the original.
22. Boltwood and Rutherford (1911); this paper had first appeared in German in March 1911. The actual value announced in the paper was 2.69×10^{19} for Loschmidt's number (p. 599); our value for Avogadro's number in the text was obtained from it.
23. Svedberg (1912), pp. 83 and 136.
24. Millikan (1913), p. 141.
25. Fletcher (1914), p. 453.
26. See, for example, Loeb (1934), p. 405; there he also cites Millikan's subsequent proposal that Perrin's reliance on Stokes's law contributed to the discrepancies as well.
27. To remind the reader, the product of Loschmidt's number and the volume occupied by a mole of a perfect gas at standard conditions—22,400 cm^3 to three significant figures, 22,413.96954 as of 2018—gives a value for Avogadro's number.

28. Birge (1929). Birge's values there (p. 59): for Faraday's constant, 2.8927×10^{14} abs es units; for the gas constant, 8.3136×10^{7} ergs per deg mole.
29. Ostwald, (1912), p. 510.
30. Einstein to Perrin, November 11, 1909, quoted from Nye (1972), p. 135f.
31. Bohr (1913), p. 9. Bohr's calculated value differed from the measured value of Rydberg's constant by 6 percent, largely because of his values for h and e both being too small and the calculation of Rydberg's constant depending on e to the fifth power. See Smith (2001b) for more details.
32. The Rydberg constant R is normally now stated as the inverse of a wave length. Bohr's formula thus corresponds to cR, that is, the product of R and the speed of light. Bohr's derivation assumed an infinite mass for the nucleus of the hydrogen atom. In subsequent years this led to a distinction among three Rydberg constants, R_{∞}, R_{H} for the hydrogen atom, and R_{He+} for the helium ion. That these three values calculated from the Bohr model are observationally distinct and agree to the extent that they did with the measured values for the latter two became significant evidence for it.
33. The Braggs had correspondingly adopted a value for the mass of a hydrogen atom 11 percent greater than Perrin's value (equivalent to a value of 61.5×10^{22} for Avogadro's number) in their landmark paper on x-ray diffraction crystallography published in the same summer as Bohr's paper; see Bragg and Bragg (1913), p. 437.
34. Perrin (1908d).
35. Rutherford and Geiger (1908a, 1908b). Rutherford, by the way, was awarded the Nobel Prize in Chemistry in 1908; perhaps Perrin cites his work only in a footnote because he had not appreciated the relevance of Rutherford's work to his until then.
36. Perrin (1910), p. 82.
37. Ibid., p. 83. The experiments in question followed the method of Harold Wilson, using water droplets in an electric field; see Millikan and Begemann (1908).
38. Perrin (1910), p. 85.
39. Perrin (1912a), p. 237. As the title of the Proceedings indicates, almost all of the other papers at the Conference concerned radiation, quanta, and the problem of specific heats. The one other paper tied to molecular-kinetic theory was by M. Knudsen, reviewing experimental results pertaining to ideal gases.
40. Perrin (1990), p. 185, (1916), p. 176; in the original French, (1913), p. 250, or (2014), p. 272.
41. Perrin (1912a), p. 249.
42. Perrin (1913), p. 289 or (2014), p. 299; (1914), p. 293; in English (1916), p. 206.
43. Perrin (1990), p. 215.
44. Ostwald (1912), p. 535f, p. 510, and p. 538.
45. Ibid., p. vi.
46. Ibid., pp. 533–548.
47. Ibid., p. 536.
48. Ibid.
49. Ibid., p. 538.
50. Fletcher (1911), p. 107.

51. See Birge (1942), pp. 116–119, where the source of systematic error in Millikan's value is discussed and reasons are given for abandoning his measure in favor of the value for *e* based on the grating space of crystals obtained from wave-lengths of x-rays determined with ruled gratings and the angles at which these x-rays are reflected by a crystal. This gave him a preferred value of Avogadro's number, $(6.0228 \pm 0.0011) \times 10^{23}$, from which he then derived his preferred value for *e*, $(4.8025 \pm 0.0010) \times 10^{-10}$ esu.
52. Rutherford and Geiger (1908a), p. 161.
53. Rutherford and Geiger (1908b), p. 168.
54. Ibid.
55. Ibid., p. 170.
56. Ibid., p. 172.
57. Millikan (1910) and (1911).
58. Perrin (1912a), p. 237.
59. Regener (1909).
60. Royds and Rutherford (1963), p. 135.
61. Dewar (1908); Boltwood and Rutherford (1909); Dewar (1910); and Boltwood and Rutherford (1911).
62. Perrin (1990), p. 210f, and (1916), p. 201ff; in the original French, (1913), p. 285ff, or (2014), p. 295f.
63. In the 1923 edition of *Atoms* but not in the French editions of (1913) and (1914) and not the English edition of (1916), the value quoted for granule displacements became 64×10^{22}; see section 4.8 and note 63 of Chapter 4 for the explanation.
64. See Kuhn (1978), p. 111.
65. Planck (1967), p. 82; (1972), p. 13.
66. See Rogers (2005) for details. Refining Debye's account still further was a paper by Born and von Kármán (1912).
67. In particular, see Perrin (1910), pp. 88–90, where he shows how little he had grasped Planck's theory with the remark, "Independently of Lorentz, by a more complicated theory, Max Planck had already arrived at the same formula."
68. Perrin (1912a), pp. 246–288.
69. Perrin (1990), p. 156. Perrin's brief discussion of Einstein's specific heat theory following this quotation ends up deferring the matter to Nernst. The citation of Bohr is a reference to the short final section of the "Light and Quanta" chapter, where Perrin notes Bohr's proposal of quantization in his theory of the hydrogen spectrum without mentioning the Bohr-Rutherford model of the atom. No such passage occurs in the 1913 or 1914 editions of *Les atomes* or the 1916 English translation.
70. Perrin (1990), p. 155 and (1916), p. 153; in the original French, (1913), p. 218, or (2014), p. 249.
71. Planck (1967), p. 89f.
72. In the Preface to the second edition of his *The Theory of Heat Radiation*, published in 1913, Planck remarked (1991), p. vii, "Meanwhile the experimental methods, improved in an admirable way by the labors of *E. Rutherford, E. Regener, J. Perrin, R. A. Millikan, The Svedberg* and others, have without exception decided in favor of

the value [of *e*] deduced from the theory of radiation which lies between the values of Perrin and Millikan." In 1900 he had deduced values for *k*, N_0, and *e* in support of his derivation, and he had compared his value for the last of these, 4.69×10^{-10}, with J. J. Thomson's value, 6.5×10^{-10}.

73. Perrin (1912a), p. 251. Three other topics were raised in the discussion: (1) whether Perrin had taken into consideration the "pressure of radiation" (by Langevin); (2) whether Ehrenhaft's results on charged particles in the air presented a challenge to the unit charge (by Einstein); and (3) the relationship between radiation from radium and from polonium and Regener's measurements (by Madame Curie).
74. The one place we have noted (in Chapter 4 and the preceding) in which Perrin retrospectively raised the possibility of systematic error in his measurements, specifically in those of ξ^2/τ, was in a later edition of *Les atomes*, following Constantin's suggestion that including granules very near the walls was introducing such an error (Perrin 1923, p. 124); this induced him to make some new measurements and then to change his value for Avogadro's number from this measure from 68.2×10^{22} to 64×10^{22}, but without any mention of the new discrepancy arising from this change when compared with his values obtained from his vertical-gradient and Brillouin's diffusion measurements. This revision dates from after 1914.
75. Warburg (1912), p. 84.
76. Perrin (1912a), pp. 248 and 249. Perrin's announced value of 64×10^{22} from blackbody radiation at Solvay appears to have come from averaging the two values for *e* that Warburg had listed, 4.69×10^{-10} and 4.32×10^{-10}, the validity of the second of which Warburg had expressly challenged. In his 1913 book Perrin retained this value for blackbody radiation, but then in the 1921 edition, as can be seen in Perrin (1990), p. 155, he switched to 62×10^{22}, that is, a rounded-off version of Planck's 1900 value, which of course corresponded to Warburg's preferred Solvay value for *e*.
77. Warburg (1912), p. 84f.
78. Perrin (1912a), pp. 228–233.
79. Ibid., p. 233.
80. On the basis of Weiss's more recent results showing that Ehrenhaft's dust particles actually carried a charge around 4.5×10^{-10} esu.
81. Millikan (1911), p. 385. Millikan revisited his approach to the frictional action of the air on oil droplets in his 1913 paper, resulting in a revised value for *e* of 4.774×10^{-10}.
82. Fletcher (1911). Fletcher's value was 2.88×10^{14} in electrostatic units, versus a then accepted value of 2.896×10^{14} for Faraday's constant. (Birge's value for Faraday's constant in 1929 was 2.892×10^{14}.)
83. Perrin (1912a), p. 237.
84. Ibid. Even in the eleventh edition of *Les atomes* in 1921, Perrin was still saying, "The precise value of this elementary charge, which we now know to exist, has yet to be determined" (Perrin, 1990, p. 184), offering the same line of reasoning that he had in 1911.
85. In all the editions of *Les atomes*, Perrin's preferred value is 67×10^{22} rather than the 64×10^{22} of the Solvay Conference paper; see Perrin (1913), p. 251, or (2014), p. 273, (1916), p. 177, and (1990), p. 186.

86. Fletcher (1911), p. 106.
87. Fletcher (1914), p. 453.
88. Sommerfeld (1923), p. 213.
89. Perrin (1912a), p. 249.
90. Madame Curie was, of course, present at the Conference, so perhaps Rutherford's silence was to avoid raising questions even indirectly about any of her work.
91. We know of only two sources that have given it appropriate recognition: Whittaker (1973), p. 8, where unfortunately a page later he attributes a value of 60.6×10^{22} to Perrin's determination from Brownian motion; and Post (1968), in a footnote on p. 226 with no mention of Boltwood or Rutherford by name.
92. Boltwood and Rutherford (1911), p. 599.
93. Svedberg (1912), p. 83.
94. Ibid., p. 136f.
95. Perrin (1990) and (1916), p. 128; in the original French, (1913), p. 182f, or (2014), p. 225f.
96. Nernst (1923), p. 510.
97. Ibid., p. 514.
98. Planck (1991), p. vii.
99. Poincaré (1963), p. 90.
100. Ibid.
101. According to Kuhn (1978, p. 209f), the question of the indispensability of quantizing something in the case of blackbody radiation was receiving increasing attention in the years preceding the Solvay Conference. In particular, "Ehrenfest produced a fuller proof than any previously given that no radiation law which agreed with Planck's in the high and low frequency limits could be derived without recourse to discontinuity" (p. 210). Kuhn specifically cites Ehrenfest (1911).
102. Namely, in his "The Dynamical Evidence of the Molecular Constitution of Bodies," Maxwell (1875).
103. See Rocke (1984).
104. Rutherford (1963); this abstract was followed, later in the year, by a full paper [Rutherford (1911)] in *Philosophical Magazine* that likewise drew limited favorable response until late 1913. So far as we have been able to determine, if Rutherford called attention to this paper while he was attending the Solvay Conference later that year, it was not during the recorded discussion of any of the papers.
105. Bohr (1913). This is the first of the three papers Bohr published in 1913 on the subject.
106. Moseley (1913). This again is the first of two papers Moseley published on the subject, with the second appearing in 1914.
107. van Fraassen (2009), p. 23. Italics in the original.
108. Perrin (1990), p. 215f; (1916), p. 207; in French, (1913), p. 289f, or (2014), p. 299.
109. van Fraassen (2009), p. 11.
110. Ibid., p. 23.
111. Ibid., p. 10.
112. Ibid., p. 23.
113. Warburg (1912), p. 84.

114. Lord Rayleigh's letter to the Conference and James Jeans's paper, which preceded Warburg's, Rubens's, and Planck's, were looking for a way to obtain Planck's results within the framework of classical physics, without any quantization.
115. Letter from Einstein to Heinrich Zangger, quoted in Mehra (1975), p. xiv.
116. The most ambitious efforts on the transfer properties of basically spherical molecules along Maxwell's lines, but this time in non-uniform gases, were carried out by Chapman and Enskog in the decade following the first Solvay Conference. For the details of the resulting theory, see Chapman and Cowling (1939); included, pp. 380–390, are a historical summary of the efforts and a bibliography of the principal publications.
117. The version of the letter sent to Poincaré can be found on the site of the Solvay Institutes. The translation in our text has been adapted to fit that version from Jagdish Mehra's translation of Nernst's draft proposed to Solvay, as given in Mehra (1975), p. 6.
118. In de Broglie and Langevin (1912), p. 436. Translation from Straumann (2011); this paper offers a thorough technical discussion of the points at issue on radiation and quanta at the Conference, particularly among Lorentz, Planck, and Einstein.
119. de Broglie and Langevin (1912), p. 439. See, as well, Einstein (1993), pp. 426–437.
120. The values Perrin lists in his table (1912a, p. 249), as well as his proposed preferred value of 67×10^{22}, seem to us biased toward supporting the values he had obtained from Brownian motion. One example of this is his citing in the text Millikan's recently published oil-drop value, 59×10^{22}, in the body of the paper, but then deciding a page later that the preferred value from measurements of e is 63×10^{22}, yet finally in the table giving for this value 64×10^{22}. Another is his choice of 64×10^{22} for the value obtained from the volume of helium obtained from α-particles versus the recently published Boltwood-Rutherford value of 60.25×10^{22}; Rutherford did not comment on this during the recorded discussion. A still more striking example is Perrin's value of 64×10^{22} from blackbody radiation, while earlier in the Conference, Warburg's review of the experimental evidence for Planck's equation had given a value for e that entailed a value of 61×10^{22} for N_0. Of course, the bias we are attributing to Perrin here may instead be our own, knowing as we do, and he did not, that his values from Brownian motion were consistently on the high side versus our current value.
121. Perrin (1912a), p. 249. Translation by Adwait Parker.
122. The exception we are allowing for is Poincaré, who attended the Conference, participated to some extent in the discussions of the individual papers, and even more so in the concluding discussion.
123. Perrin (1912a), p. 216.
124. Ibid., p. 161f.
125. Ibid., p. 222.
126. Ibid., p. 223.
127. By the time of the eleventh edition of *Les atomes*, improved measurements for the diffusion of light had yielded a value of 69×10^{22} for Avogadro's number from the blueness of the sky. See Perrin (1990), p. 139.
128. Nernst's (1912) paper at the Conference focused on departures of Einstein's equation for specific heats of solids at low temperatures from measured results and a

phenomenological equation proposed by Lindemann to eliminate the discrepancies. In the long discussion following his paper, Nernst (1912, p. 300f) proposed the dropping out of rotational degrees of freedom of molecules at low temperatures in response to a question from Poincaré, remarking on uncertainties in the dimensions and consequently the rotational energies of molecules. As we have noted earlier, the shortcomings in Einstein's formula disappeared in 1912 with Debye's refinement. Nernst's proposal on rotation did not come to full fruition until Dennison (1927).

129. Lorentz (1912), p. 34.
130. As often occurred at the Conference, Langevin's paper raised worries about how to achieve agreement with experiment when he acknowledged in the discussion period that he had adopted Sommerfeld's action principle for the motion of the electron rather than the *prima facie* more appropriate choice of Planck's because it yielded results in disagreement with experiment [de Broglie and Langevin, 1912, p. 405]. Sommerfeld's paper, which preceded Langevin's and had provoked a good deal of discussion, concerned the application of the theory of the quantum of action to non-periodic molecular phenomena, while Langevin's theory concerned a periodic phenomenon.
131. See Jeans (1916), Chapters 10 through 14. A comparison of these chapters with the parallel chapters of Meyer's 1899 edition of *The Kinetic Theory of Gases* reveals a great deal about both what had changed with the advent of a precise value for N_0, and what had not changed.
132. Perrin (1910), p. 91.
133. Ibid., p. 92.
134. Perrin (1912a), p. 249. Translation by Adwait Parker.
135. Ibid., p. 250.
136. Ibid. Perrin used the same word, "scaffolding"—"*échafandages*"—in a shorter version of the paper with the same title at the 1912 annual meeting of the Société française de Physique (1912b, p. 52). His 1909 monograph has no paragraph comparable to the one quoted, but in its place the single sentence, "The discovery of such relationships marks the point where the underlying reality of molecules becomes a part of our consciousness" (1910, p. 92). In his 1913 book he switches to a different metaphor, "In doing so we should not be removing the support from a thriving plant that no longer needed it; we should be cutting the roots that nourish it and make it grow" [(1913), p. 290f, or (2014), p. 300; (1916), p. 207; (1990), p. 216].
137. W. Ostwald (1907), p. 499f. Two pages earlier Ostwald had acknowledged his adopting of Mach's term, pseudo-problem (*Scheinprobleme*).
138. Nernst (1895), p. 237; (1911), p. 280; (1923), p. 327. The paragraph preceding this sentence that defended the further development of molecular theory in spite of the hypothetical status of molecules was dropped in the 1909 and 1921 editions.
139. Chalmers (2011); and Salmon (1984), pp. 211–227.
140. See Smith (2016).
141. Cohen and Koyré (1972), p. 298 (p. 188 in the original Latin, p. 589 in the Cohen and Whitman translation). The Latin phrase is "*tutius disputare licebit*"—"it will be possible to argue more securely."
142. See Harper (2012) and Smith (2014).

7
Our Initial Issues, Revisited

We began this monograph by posing two issues: (1) *What did the new standing of atomic-molecular theory at the end of the first decade of the twentieth century amount to?* (2) *What did the research by Perrin and others establish about Brownian motion* independently *of the molecular hypothesis and kinetic theory?* In posing these two, we chose not to challenge the "lore" that Perrin's Brownian motion results made a crucial difference to the change in standing of the theory. Our long discussion of the various determinations of Avogadro's number between 1908 and 1913 in Chapter 6 has raised a further question that we should answer as well: (3) *In what respects, if any, did Perrin contribute crucially to the change in the standing of atomic-molecular theory that occurred in those years?*

The second and third questions have been receiving partial answers along the way in Chapters 4 through 6. So, all we need do with them here is to pull together the points from those chapters that together comprise direct answers to the questions as stated—which we shall do in sections 7.1 and 7.2. The first question, by contrast, we have been sidestepping throughout Chapters 4 through 6, restricting our attention to the evidence bearing on chemical and kinetic molecular theory during the years in question. This evidence has turned out to be more complicated than it is usually portrayed. In particular, the question of a preferred value for Avogadro's number changed dramatically between Ostwald's remark in 1909, "The most probable value is $.71 \times 10^{24}$ (Perrin 1908)," and Millikan's statement in 1913 that the "estimated limits of uncertainty" for his newly refined value of 6.062×10^{23} were $\pm 0.012 \times 10^{23}$. These complications have led us to answer our question about the new standing in three steps. Section 7.3 will answer it from the perspective of Ostwald in 1909; section 7.4 will answer it, as best we can, from the perspective of someone writing a review article on the question in the aftermath of the publication of the Solvay Conference proceedings in 1912; and section 7.5 will then step back to comment on the different respects in which a hypothesis can legitimately be said to have become accepted within a scientific community.

Our answer to the question about the new standing thus forms the heart of the chapter. This change in standing, however, was inextricably linked to the emergence of a new way of using theory-mediated measurement as a source of evidence pertaining to the microphysical realm. This new approach to evidence had begun with J. J. Thomson's theory-mediated measurements of the charge-to-mass

Brownian Motion and Molecular Reality. George E. Smith and Raghav Seth, Oxford University Press (2020). © Oxford University Press. DOI: 10.1093/oso/9780190098025.001.0001.

ratio of the constituents of cathode rays in 1897 and Planck's appeal to complementary measurements of microphysical constants as evidence for his theory of blackbody radiation in 1900. Over the course of the twentieth century it became increasingly the most decisive form of evidence in microphysics, culminating in the formation of the Committee on Data for Science and Technology—that is, CODATA—and their Task Group on Fundamental Constants now publishing new preferred values for over two hundred constants every four years.

Perrin, and the 1905–1913 period on which we have been focusing, formed a key step in the transition between Thomson and Planck, on the one hand, and the full flowering of this new way of using theory-mediated measurement as a primary source of evidence, on the other. Just what that step was and how it fit into a twentieth-century revolution in theory-mediated measurement is the subject of section 7.6. This section thus completes the "study of theory-mediated measurement" promised in our subtitle by answering the question: (4) *How did the developments in theory-mediated measurement of Avogadro's number and constants linked to it during the period 1905–1913 contribute to the increasing reliance on such measurements for evidence that ensued thereafter?*

7.1. Conclusions Established about Brownian Motion

Throughout the previous three chapters we have focused almost exclusively on the results on Brownian motion coming out of Perrin's laboratory. In doing so, we have followed the lead of Perrin himself, who consistently dismissed the results on Brownian motion obtained by others, especially those by Svedberg.[1] Svedberg had been the first to publish experimental confirmation, faulty though it was, of Einstein's formula, and as we shall see in section 7.3, this confirmation seems to have been what had convinced Ostwald, before he had read Perrin's 1908 paper, that Brownian motion had established the kinetic theory of heat. Both at the Solvay Conference and in *Les atomes*, moreover, Perrin cited the value 2.88×10^{14} for *Ne* that Fletcher had published in 1911 from experiments on Brownian motion in air, noting its close agreement with the established value of Faraday's constant from electrolysis, 2.896×10^{14} esu, but without noting in either place that Fletcher's result implied a 57.5×10^{22} value for Avogadro's number, far below his own values; nor did he note any other respect in which Fletcher's announced results for Brownian motion might have a bearing on his own. (Fletcher's refined results, including the value 60.3×10^{22}, did not appear until 1914, outside the time frame we are considering.)

Perhaps largely because of what Perrin had said about research by others on Brownian motion, others at the time appear to have taken the results from his laboratory to be the definitive ones. To give just one example, Jeans's 1916 edition of *The Dynamical Theory of Gases* lists all four of Perrin's Solvay Conference values for Avogadro's number and no values published by any others on Brownian motion from their research.[2] We have thus not been idiosyncratic in taking the results by others at the time on Brownian motion as mostly just complementary to those of Perrin and his younger colleagues, Chaudesaigues and Léon Brillouin. If indeed we put aside issues about the preferred values of fundamental molecular constants, and focus instead on just what was established during the period about Brownian motion itself, the results of others were in fact almost entirely complementary. Arguably, then, we are not being unfair to them in confining our answer to the question of what was established about Brownian motion during the period to the four points established in Perrin's laboratory.

The basic result was the universality of the mean translational kinetic energy of (spherical) granules at any given temperature in dilute emulsions—that is, universality with respect to the size and mass of the granules, the density and viscosity of the liquid, and the range of temperatures examined. Even though the range of temperatures was limited, the most appropriate way to state this finding, on our view, is that Perrin and his colleagues established that the mean translational kinetic energy of granules varies as $W_{\mathrm{T}}T$, where W_{T}, they showed, is a constant. They showed this, of course, by determining its value, roughly 1.8×10^{-16} ergs per degree K, by means of three different theory-mediated measures. Moreover, this value did not differ all that much from the corresponding values for $3k/2$ that had been estimated for molecules, such as Planck's 2.02×10^{-16} from blackbody radiation.

As we emphasized in Chapter 4, Perrin does not himself ever quote a value for the mean translational kinetic energy of granules per degree Kelvin, but only what he takes to be the corresponding value (1.77×10^{-16} in 1909, 1.8×10^{-16} in 1911) for molecules. The 10 percent discrepancy between his value for W_T for granules and the values that others were obtaining for the corresponding value for molecules may be raising a question in some readers' minds about whether Perrin really did establish a value for W_T. One possibility is that he somehow managed to introduce a roughly 10 percent systematic error in all his results for Brownian motion, including those for the mean rotational energy of his granules. The four mediating equations he employed had only one parameter in common, the radius of the granules, and that occurs linearly in the measurements based on displacements and diffusion and cubed in the measurements based on vertical-gradients and rotation. We, therefore, are inclined not to attribute the discrepancy to an across-the-board systematic error in Perrin's four measures of W, for we do not see any plausible source for such an error. We are instead inclined to say that his values for W_{T} really do represent the mean translational kinetic energy per degree Kelvin of (spherical) granules in dilute emulsions, and that the discrepancy represents a real difference between that mean kinetic energy and the corresponding mean kinetic energy per degree Kelvin for molecules. Hence, we feel justified in saying that Perrin established the universality of W_T as a constant over the range of Brownian motions his experiments covered.

Perrin's value for W_T had a notable consequence: velocities of granules in Brownian motion obtained from the quotient of observed displacements and the times in which they occur were totally misleading. True root-mean-square granule velocities are orders of magnitude greater than those values. This in turn entailed that the granules were undergoing thousands, if not millions, of changes in motion over the course of the time in which observers could see only one change in motion. With this result, a great many prior findings on Brownian motion were overturned, once and for all.

Related to establishing the constancy of and a reasonable value for W_T was Perrin's result for the rotational kinetic energy of granules and its confirmation, at least for the case investigated, of the equipartition of energy across all degrees of freedom. While this result was important in Perrin's mind because of the long-standing controversy over equipartition of energy, we put less stress on it because it was confirmed only for one kind of granule; and of course, unlike Perrin, we know that the equipartition principle was about to be replaced by the partition function in statistical thermodynamics.

As important as establishing the constancy of W_T in Brownian motion was, we are inclined to agree with Deborah Mayo that still more important was Perrin's establishing that granule displacements versus time have a Gaussian statistical distribution. This, together with the finding that granules must be undergoing thousands, if not millions, of changes of motion per second, established the thoroughgoing randomness of Brownian motion at a scale far smaller than what can be observed. Needless to say, it was by virtue of this feature that "Brownian motion" came to denote a subject in mathematics, so-called random-walk stochastic processes. Brillouin's further confirmation of Einstein's formula for the diffusion of granules then showed that diffusion itself, at least in the case of granules in dilute emulsions, is a consequence purely of a random-walk process, not requiring any forces to drive it.

Finally, by virtue of the previously established indefinite persistence of the motion over time in combination with the preceding results, Perrin established important conclusions about the cause of Brownian motion, albeit conclusions that do not categorically require molecular theory. From a classic hydrodynamic perspective, what maintains Brownian motion while causing the granules to undergo so many changes of motion per second has to be extremely high-frequency pressure fluctuations in the liquid on a spatial scale far smaller than the smallest granule Perrin tested. These pressure fluctuations must themselves persist in the liquid indefinitely over time; and each pressure fluctuation must involve motion of liquid within it, and hence localized kinetic energy. But then, to maintain

Brownian motion indefinitely, the mean value of the kinetic energy of the liquid over the many pressure fluctuations must be at least as great as that of the mean translational kinetic energy of the granules, and hence it too must be a mean kinetic energy that varies linearly with the absolute temperature of the liquid. This result, we think, can legitimately be taken as decisive evidence for the kinetic theory of heat, at least so far as the liquid substrate.

Still further, as we argued in section 5.6 of Chapter 5, the thoroughgoing randomness of the granule motion on a time scale far smaller than can be observed requires the localized pressure fluctuations in the liquid to be correspondingly random on a no greater time scale. That, however, implies that the individual pressure fluctuations in the liquid must not be correlated with one another, and hence the highly localized fluctuating motions of the liquid must over time have some independence of one another. That in turn suggests—we choose the vague word scientists often use in this kind of situation—that the elements of liquid moving in each pressure fluctuation have at any one instant an individual integrity of their own, and hence in principle can be counted in the manner of discrete elements. At the very least, this is evidence that the liquid substrate is not a continuum at a scale below that of the granules, and hence (using Ostwald's word) can be taken to be granular.

To go further than this requires considerations beyond the results on Brownian motion that were established independently of molecular-kinetic theory. Perrin invoked the range of values for Avogadro's number that had been derived from the viscosity of gases in order to argue, in effect, that the mean kinetic energy of the granules is identical with that of molecules. With respect to the reasoning purely in terms of Brownian motion itself that we have just given, this amounts to arguing that the mean kinetic energy of the individual fluctuations in the liquid that drive Brownian motion matches $3R/2N_0$, and hence the local liquid elements in random motion are just the molecules posited in kinetic theory. After Rutherford and his colleagues had provided a robust value for N_0 under the assumption that α-particles become helium molecules when they lose their charge, a parallel line of reasoning could have been invoked to reach the same conclusion. As we said, however, such reasoning does involve considerations beyond the conclusions that were established about Brownian motion independently of molecular-kinetic theory.

We have derided the idea, expressed by Perrin and many others at the time and still today, that *Brownian motion is molecular motion writ large*—that is, that the proper way to think of it is as if it consisted of very large molecules in motion. According to Perrin's own estimates, molecules in the surrounding liquids are three orders of magnitude smaller in their principal dimension than the granules he tested. This means that the number of molecules required to fill the volume occupied by a single granule is of an order of magnitude greater than 10 million, and the number required to cover its surface is of an order greater than 100,000. Moreover, granules in Brownian motion in a dilute emulsion do not interact with one another, and their motion is resisted by forces on the surface arising from the viscosity of the fluid. Indeed, two of Perrin's three measures of the mean translational kinetic energy of the granules expressly presupposed the viscous action of the fluid on them, and the third presupposed their buoyancy in the fluid. In these respects, Brownian motion is a macrophysical, not a microphysical, phenomenon, even though a microscope is required to observe it.

There is, nevertheless, a version of the claim that we have derided that we regard as both warranted and informative: *Brownian motion is molecular motion made manifest.* We have two extraordinary aspects of Brownian motion in mind in granting this: (1) it persists indefinitely in time, a fact about it known long before Perrin began his research; and (2) it is thoroughly irregular to a degree that cannot begin to be seen even in an ultramicroscope, a fact about it that Perrin's laboratory has legitimate claim to having established experimentally. If "making molecular motion manifest" in these respects is what Nernst had in mind in using the phrase "*ocular* confirmations," we have no objections. Nor do we object in any way to the claim that Perrin's efforts left no candidate for explaining Brownian motion other than perpetual motion of invisible, discrete constituents of the underlying liquid substrata.

Perrin also claimed to have shown that the gas law holds for granules in dilute emulsions in full parallel with van't Hoff's result for the osmotic pressure of solutes in dilute solutions. Here, we think, Perrin was overstating what his experimental results had shown. They had indeed shown that the impact of granules on the wall of the container produces a pressure on it proportional to the product of the number-density of the granules and the absolute temperature of the liquid. Unlike van't Hoff, however, Perrin did not measure the osmotic pressure of the granules and therefore did not show that the constant of proportionality in this relationship matches the universal gas constant *R*. His sole grounds for concluding that it does was from his assumption that the mean kinetic energy of the granules matches that of molecules. Had he, independently of molecular theory, shown that the constant of proportionality amounts to *R*, somewhat stronger conclusions than those we granted earlier might have been drawn about the pressure fluctuations driving Brownian motion.[3]

7.2. Perrin's Contribution to the New Standing

"Lore" has it—and on this point van Fraassen seems to agree—that Perrin not only did the most of anyone at the time to change the standing of molecular-kinetic theory, but that he did so by obtaining from Brownian motion the first reasonably robust value for Avogadro's number. Most of the philosophic literature, however, including van Fraassen, has offered differing reasons for why the determination of a robust value for that constant had such an abrupt and decisive effect on the standing of the theory. On that question we side with van Fraassen's general answer: the value Perrin obtained for N_0 *grounded* molecular-kinetic theory by yielding values for several of its other parameters. Our primary reason for this conclusion is more historical than philosophical, for it comes partly from what Ostwald himself said before and then after 1908, most notably the remark, "The atomic hypothesis is thus raised to the position of a scientifically well-founded theory."[4] Our differences with van Fraassen, which we do regard as fundamental, concern the extent to which molecular-theory itself entered into the theory-mediated measurements that yielded Perrin's value for N_0, or for that matter the theory-mediated measurements of its value made at the time by others. We shall not belabor those differences further here.

Nevertheless, we are uncomfortable, looking at the history, with the conclusion that Perrin's singular contribution was his being first in obtaining a reasonably robust value for N_0. True, Ostwald specifically cited Perrin's value for the "molecular constant," along with the parallel between his verified formula for the vertical variation of the number density of granules and Maxwell's kinetic theory formula for the vertical variation of the density of different species in the atmosphere.[5] But Ostwald wrote this passage before Perrin had begun publishing his results testing Einstein's theory, and also, it would appear, before he read the Rutherford-Geiger paper in August 1908 giving their determinations of e and N_0. What Ostwald would have said about Perrin had he been writing his book a year later surely would have credited him for more with regard to the implications of Brownian motion and probably less with regard to determining the value of N_0. Moreover, those at the Solvay Conference in late 1911 would surely have concluded from Perrin's paper that his value for N_0, while robust, was but one of several independently determined robust values. And in fact, though somewhat suppressed in Perrin's presentation, the cluster of values from the two complementary α-particle determinations by Rutherford and his colleagues and Regener, from the various blackbody spectrum measurements supporting Planck, and from Millikan's oil-drop experiments agreed significantly more closely with each other than they did with Perrin's.

The shift in the standing of molecular-kinetic theory did not occur in a single instant. Any assessment of what each new result contributed to it therefore depends not only on which scientists' attitude one relies on in assessing it, but also on the date one chooses for considering it. Maybe because of our retrospective bias, we are inclined to single out the two determinations of N_0 made by Rutherford, respectively with Geiger (1908), supported by Regener's independent determination, and with Boltwood (1911), as the most decisive. For the "weak link" in them, especially in the latter, seems to us to have involved the least challengeable step from the primary quantity measured to a count of the number of molecules per mole. Granted that its conclusion applied directly only to helium and hence that other determinations of N_0 were needed to support its inductive extension to other molecules. Still, at the very least, the Rutherford determinations and Regener's, followed immediately by the support Millikan's result lent them, had lessened the importance of Perrin's determinations by the time of the Solvay Conference.

If anyone at the Conference challenged Perrin's insistence that his numbers were to be preferred to the others, it would appear to have been outside of the formal sessions, for we have found no recorded discussion of the gap between his numbers and those of the rest in the proceedings in French published in 1912. Arnold Eucken, however, raised the issue in the Appendix, "The Development of Quantum Theory from the Autumn of 1911 to the Summer of 1913," that he added as editor to the proceedings in German, published in 1914. In particular, he gave, in bold, two recent values for Avogadro's number from blackbody radiation, 59.7×10^{22} and 62.0×10^{22}, plus Millikan's 1913 value, 60.62×10^{22}. He went on then to note the earlier Rutherford-Geiger and Regener values, 62×10^{22} and 60×10^{22}, remarking "the gap between numbers found from Perrin remains."[6] Eucken was not an official participant in the Conference, though as a research assistant to Nernst he might have been present. Regardless, whether from Nernst or otherwise, he came to see the gap with Perrin's values as enough of an issue that he could single it out without needing to explain it further.

In point of fact, even though Perrin continued to stress his determinations of the molecular constants in the various editions of his *Les atomes*, others in need of values for any of these constants had already chosen not to use his in the year when its first edition appeared in print. To give two examples from the summer of 1913, Bohr adopted a value for *e* in his legendary paper on the structure of the hydrogen atom of 4.7×10^{-10} instead of Perrin's Solvay Conference value of 4.2×10^{-10};[7] and the Braggs had adopted a value for the mass of a hydrogen atom of 1.64×10^{-24} grams instead of Perrin's 1.47×10^{-24} in their seminal paper on the lattice structure of crystalline NaCl and related crystals.[8] Moreover, as we noted earlier, in the 1916 edition of his *The Dynamical Theory of Gases*, Jeans cites all four of Perrin's Solvay Conference values for Avogadro's number (68.3, 68.8, 65, and 69, all times 10^{22}) and then adopts for his book the value 61.2×10^{22}.[9] So, Perrin's values for the molecular constants had ceased being viewed as a still-standing reliable contribution scarcely before the ink of the first edition of *Les atomes* had dried.

This, however, does not mean that Perrin made no *singular* contribution to the change in the standing of the theory. While Einstein may have preceded him in championing the idea of settling molecular-kinetic theory once and for all, Perrin became the spokesperson for doing so after he began getting converging results on Brownian motion. Perrin was the first, in a short note in May 1908, to announce a value for Avogadro's number derived from Brownian motion ("supposing the deduction of the preceding equations rigorous"); he confirmed the value in the case of granules twice as large as in the first series in a September note; he followed this with a refined value and values implied by it for *e* and other constants in October in a note that cited the Rutherford-Geiger results as well as Planck's; and this was then followed in November by Chaudesaigues's note announcing the Gaussian distribution of the displacements.[10] Throughout this period, Perrin was notably more preoccupied than any of the others with using their measured values more to establish the reality of molecules than to establish a value for constants. In this regard, Perrin was unique in the emphasis he put on agreeing values of Avogadro's number and what they collectively imply about the existence of molecules.

Chaudesaigues's note we will consider in a moment. Perrin's October note, though it barely extended to a third page, nevertheless deserves further mention, and not just because it was the only publication of Perrin's that Ostwald cites. It opens (see Figure 7.1) by pointing out that, by virtue of Faraday's constant and the gas constant, a determination of any one of the three molecular magnitudes N_0, α, and e yields values for the other two. It then goes on to argue that his determination from Brownian motion is the "most direct and amenable to unlimited precision," and displays in bold his values for the three.[11] Because he singles out values obtained by other means—in particular, the 1903 Wilson-Thomson and 1908 Rutherford-Geiger values for *e*—he is calling attention to the possibility of playing off values obtained by different methods in pursuit of greater precision. Planck had cited the interconnection of the values for the magnitudes in 1900, but Perrin's 1908 note appears to be the first place where a comparison of values obtained by different approaches occurs side by side with a preoccupation with improving precision.

In addition to this, none of the other determinations of N_0 involved motion in any significant way, and hence none of the others had such direct bearing on kinetic theory as Perrin's. Ostwald credited Svedberg's testing of Einstein's theory with having given "a fairly satisfactory proof of the kinetic theory of heat," which theretofore had "remained an arbitrary though useful hypothesis."[12] Perrin's tests of Einstein's theory were far more definitive than Svedberg's, even without Brillouin's verification of Einstein's formula for diffusion; and with it included, Perrin's and his colleagues' results provided the most direct and decisive evidence for the kinetic theory of heat that had ever been put forward. So, this was part of Perrin's singular contribution. It may also be a proper spelling out of the substantive content lying behind the thought expressed by many at the time that,

594 ACADÉMIE DES SCIENCES.

En résumé, les gaz de l'atmosphère non liquéfiables dans l'air liquide sont constitués par du néon et de l'hélium, et notre procédé nous a permis de trouver des lignes nouvelles dans le spectre du néon pur.

PHYSIQUE. — *Grandeur des molécules et charge de l'électron*. Note de M. **Jean Perrin**, transmise par M. J. Violle.

Le nombre N de molécules que contient toute molécule-gramme, la charge e de l'électron et le quotient α de l'énergie moyenne d'une molécule par sa température absolue T sont des *constantes universelles* qui sont connues dès qu'une d'elles est connue. En effet, dans l'électrolyse d'un sel monovalent, 96550 coulombs sont transportés par N atomes, ce qui donne

$$Ne = 3.10^9.96\,550 = 29.10^{13} \text{ (unités électrostatiques C. G. S.)},$$

et, d'autre part, d'après la théorie cinétique, le produit 3RT mesure la force vive de translation que possèdent à un instant quelconque les N molécules d'une molécule-gramme, en sorte qu'on a

$$2N\alpha = 3R = 3.83{,}2.10^6.$$

La théorie cinétique donne au moins l'ordre de grandeur de N en s'aidant de considérations qu'on peut, je crois, résumer comme il suit : d'une part, dans un liquide, les molécules (qu'on suppose sphériques) ne peuvent être plus serrées que des boulets dans une pile de boulets, et de là résulte que N doit être plus grand que 4.10^{22}; d'autre part, le volume *vrai* des molécules ne peut qu'être supérieur à celui des sphères *parfaitement* conductrices qui, substituées à ces molécules, donneraient la même constante diélectrique au fluide, et de là résulte que N doit être plus petit que 204.10^{22}; enfin, d'après la théorie de Van der Waals, le volume *vrai* de N molécules est le douzième du volume critique de la molécule-gramme, et de là résulte pour N la valeur 60.10^{22}, déjà sans doute assez approchée, sans qu'on puisse dire au juste à combien près (un écart de 40 pour 100 en plus ou en moins n'étonnerait pas).

Les recherches du deuxième groupe utilisent le pouvoir qu'ont les ions d'un gaz de condenser la vapeur d'eau (C.-T.-R. Wilson). On peut alors obtenir e, soit en divisant la charge présente dans le gaz par le nombre de gouttelettes qui la portent (Townsend, J.-J. Thomson), soit en observant l'action d'un champ électrique sur ces gouttelettes chargées (Harold A. Wilson). Townsend a ainsi trouvé, en unités électrostatiques C. G. S., 3.10^{-10} (ions des gaz de l'électrolyse); J.-J. Thomson a trouvé $3{,}4.10^{-10}$ (ions des deux signes dus au radium) et $6{,}8.10^{-10}$ (ions négatifs produits par la lumière); Harold A. Wilson a trouvé $3{,}1.10^{-10}$ (ions négatifs produits par les rayons X). Les valeurs correspondantes de N sont comprises entre 43.10^{22} et 96.10^{22}.

Je discuterai ailleurs ces expériences qui, fixant un ordre de grandeur inconnu, ont

Figure 7.1a. Perrin on fundamental constants, 1908.

in Nernst's words, Brownian motion had provided "*ocular* confirmations of the kinetic theory of the molecular world."[13]

Still, singular that it was, it was only part of Perrin's unique contribution to the change in standing of the theory. No less important at the time, and probably more important in the years following the 1911 Solvay Conference, was

SÉANCE DU 5 OCTOBRE 1908. 595

imposé la notion capitale de *corpuscules* bien plus petits que les atomes (J.-J. Thomson), mais qui, je crois, n'ont pas pu être très précises (1).

En troisième lieu je dois citer les résultats de Max Planck et H.-A. Lorentz. Le premier, dans sa belle théorie électromagnétique du rayonnement noir, a trouvé pour α, à partir des mesures de Kurlbaum, la valeur $2,02.10^{-16}$, qui fait N égal à $\mathbf{61}.10^{22}$; mais, à partir des mêmes mesures, par une théorie différente, Lorentz a trouvé N égal à $\mathbf{77}.10^{22}$.

J'ai suivi, pour déterminer N, une méthode qui me semble *directe et susceptible d'une précision illimitée*. Elle se fonde sur ce que des grains égaux se répartissent dans une émulsion étendue comme feraient des molécules de même masse obéissant aux lois des gaz parfaits (2). Cela entraîne l'équation :

$$2,3\log_{10}\frac{n_0}{n} = \frac{N}{RT}\,\frac{4}{3}\pi a^3 g(d-\sigma)h$$

où n et n_0 désignent les concentrations des grains en des niveaux distants de h, a le rayon des grains et $(d-\sigma)$ l'excès de leur densité sur celle de l'eau.

Trois séries d'expériences m'ayant prouvé, comme je l'ai dit, que des grains de tailles très différentes conduisaient sensiblement à la même valeur de N, je me suis efforcé d'en faire une particulièrement soignée, dont je donnerai ailleurs le détail; T était égal à $273+20$, h à 30^{μ}, $(d-\sigma)$ à 0,2067, a à $0^{\mu},212$; enfin, pour avoir $\frac{n_0}{n}$, j'ai compté un à un 13000 grains en 16000 lectures. En conséquence de ces mesures, je crois pouvoir donner

$$N = \mathbf{71}.10^{22},$$

ce qui entraîne pour l'électron

$$e = \mathbf{4{,}1}.10^{-10}$$

et pour la constante d'énergie moléculaire

$$\alpha = \mathbf{1{,}7}.10^{-16}.$$

(1) Enfin Rutherford, en un beau Mémoire (août 1908) qui vient seulement aujourd'hui à ma connaissance, obtient pour e, par deux groupes de mesures de radioactivité, deux nombres qui placeraient N entre $\mathbf{62}.10^{22}$ (qu'il regarde comme le plus probable) et $\mathbf{72}.10^{22}$ (ceci en admettant égale à 2000 ans la constante de temps du radium).

(2) *Comptes rendus*, t. CXLVI, 1908, p. 967; t. CXLVII, 1908, p. 530.

C. R., 1908, 2e *Semestre*. (T. CXLVII, N° 14.) 78

Figure 7.1b. Perrin on fundamental constants, 1908, continued.

the evidence he supplied for the thoroughgoing randomness of Brownian motion and, with it, of the fluctuations in the liquid maintaining it as well. Joined with Brillouin's measurements of granule diffusion, this showed that diffusion itself involves nothing more than what we have now come to call a random-walk process. On the importance of this point we agree with Mayo. Earlier we noted both Einstein's (1911) and the Ehrenfests' (1912) appeal to Perrin's results when defending Boltzmann's probabilistic interpretation of entropy and with it statistical, in contrast to classical, thermodynamics. The statement by the latter bears repeating:

> A "visible" state S_A at time t_A completely determines the "visible" state for any subsequent time t_B.
>
> This statement is clearly not a direct expression of an experimental fact. Apart from a very small group of aerodynamical processes, we always deal with turbulent motions of the gas, where it is impossible to follow the "visible" state with our measuring instruments. Furthermore, even the best insulation against thermal conductivity and radiation is completely unsatisfactory unless we deal with very short periods of observation.
>
> The above statement is therefore a postulate, which goes considerably beyond the possibility of an experimental check. Our observations cannot tell us anything about what the sequence of state of an actual gas quantum would be over a very long period of time and whether it would satisfy the principle of determinacy.
>
> As soon as we include a microscope among the instruments of observation of the "visible" state, we discover the Brownian motion. For such a more precise "visible" state the principle of determinacy does not seem to hold any more. On the other hand, it has been shown that this phenomenon is surprisingly well suited to a statistical interpretation.[14]

This is undoubtedly the most justified respect in which those at the time could speak of Brownian motion as molecular motion made manifest.

The only other robust determination of N_0 at the time tied to Boltzmann's theory was Planck's, and the way he obtained it was part of what was behind others, including Perrin, questioning his derivation of the empirically well-founded blackbody equation. Consequently, for the very reason that Einstein cited it, Perrin's evidence for the thoroughgoing randomness of Brownian motion, which (as we saw) was independent of his determinations of N_0, made its own unique contribution at the Solvay Conference, where of course the question of how properly to derive Planck's formula was at the forefront.

As we acknowledged in Chapter 5, Perrin's emphasis on the visibility of Brownian motion—in words, diagrams, photographs, and films—had much the rhetorical effect on broader audiences that he intended. Of course, the visibility of the granules was indispensable to his theory-mediated measurements, which indeed were a major contribution. Equally, establishing the Gaussian distribution of their displacements required individual granules to be followed visually over time. What tends to get lost in this emphasis on visibility, however, is that the true motion of the granules is not visible at all, and as a consequence its full stochastic character is not within the reach of vision or film, but is instead a matter of inference.

7.3. The New Standing of Molecular-Kinetic Theory: Ostwald's "Conversion"

Part of the "lore" surrounding Perrin and his research on Brownian motion is its having been instrumental in Ostwald's reversing his stance on the reality of molecules in the 1909 edition of his *Grundriss der allgemeinen Chemie*. Even a text in applied mathematics on the subject labeled "Brownian motion" ends its introductory presentation with the quote from Ostwald's Preface, "this evidence now justifies even the most cautious scientist in speaking of the *experimental* proof of the atomic nature of space filling matter. What has up to now been called the atomistic hypothesis is thereby raised to the level of a well-founded theory"[15]—a statement we too have quoted more than once. The emphasis put on Ostwald's conversion is nevertheless misleading in two respects. First, in large part because of his prior view, Ostwald had long since ceased having influence within the physics community on the question of molecules, and hence his shift may have mattered more to himself than to anyone in it.[16] Second, Perrin and his research almost certainly was not a factor in Ostwald's shift in stance toward molecules in the new edition, for this shift had to have occurred before he was aware of Perrin's results. We will briefly explain both of these and then justify why Ostwald's conversion is nonetheless worthy of consideration before turning to the question of the new standing in his eyes.

While Ostwald was not the primary instigator of the "energetics movement," especially in comparison with Georg Helm, he did come to champion it in the first half of the 1890s. As a consequence he became a target for the strong criticisms of this movement, especially by such figures within German physics as Boltzmann and Planck.[17] Ostwald's continuing association with the movement over the next decade, and his continuing opposition to "mechanistic materialism," had led him to be regarded increasingly as a figure on the periphery of the physics community, if not outside it entirely.

Even so, his continuing insistence on the hypothetical status of molecules seems not to have been dismissed to the same extent. To give one significant example, Boltzmann responds to Ostwald's views in a single passage in his *Lectures on Gas Theory*, first rejecting in two paragraphs the claim that atoms lose their integrity and individuality when forming a chemical compound, and then adding a third paragraph:

> There is no question that for the calculation of natural processes the mere equations, without their foundations, are sufficient; likewise, empirically confirmed equations have a higher degree of certainty than the hypotheses used in deriving them. But on the other hand, it appears to me that the mechanical basis is necessary to illustrate the abstract equations, in the same way that geometric constructions illuminate algebraic relations. Just as the latter are not made superfluous by mere algebra, so I believe that one cannot completely dispense with the intuitive representation of the laws valid for the action of macroscopic masses provided by molecular dynamics, even if he doubts the possibility of knowledge of the latter, or indeed the existence of molecules. A clear understanding is just as important for knowledge as the establishment of results by laws and formulas.[18]

However askance Ostwald's energetics was viewed by his chief critics within the physics community, his insistence on the lack of experimental access to the molecular realm seems not to have been scorned to the same extent. Rather, it seems to have been widely acknowledged.

Among the four seminal figures within physical chemistry from 1890 on—Ostwald, van't Hoff, Arrhenius, and Nernst—only Ostwald championed energetics. Nevertheless, as we noted in quoting from van't Hoff's Nobel Prize lecture of 1901 in Chapter 2, as well as from many passages in Nernst's *Theoretical Chemistry* in that chapter and others, the other three continued to acknowledge the hypothetical status of molecules while remaining uncompromisingly focused on research into the molecular realm. So, we grant that Ostwald's conversion made little difference to their research, though it must have pleased them that this element of difference within their otherwise mutually supportive group had finally ceased.

Ostwald's conversion appeared in print in his *Grundriss* of 1909, the same year that he, perhaps not entirely coincidently, became the third of the four to be awarded the Nobel Prize. Perrin's first paper announcing results on Brownian motion (and a value for N_0 of 6.7×10^{23}) had appeared in May 1908, yet the only one of his papers from that year that Ostwald unmistakably cites in the book, as implied by its announced value of 7.1×10^{23}, appeared in October, the month preceding the date on Ostwald's Preface announcing his conversion. The departures of this new edition from the 1899 edition are far too extensive not to have been conceived and almost entirely put in place before Perrin's results had appeared in print. Perrin is cited in only two places in the book: a single sentence in the chapter on kinetic theory saying that 7.1×10^{23} is the most probable, immediately following Ostwald's derivation from van der Waal's formulation of the gas law of a value of 8.5×10^{23}, "or in round numbers 10^{24}";[19] and a single paragraph immediately following Ostwald's comparisons of Svedberg's 1906 Brownian motion results with calculations based on Einstein's and Smoluchowski's formulas[20]—comparisons that were providing evidence for the conclusion Ostwald had stated just before them:

> If the movements of a small particle suspended in a liquid are calculated on the basis of this hypothesis, the agreement with the movements actually observed is so close that we are compelled to regard this agreement as a fairly satisfactory proof of the kinetic nature of heat.

Though we are not in a position to insist on the point, both of these citations of Perrin's results appear to have been inserted into a finished manuscript, if not after the book was already in press. Added support for claiming this comes from the opening sentence of the second of the two: "Still more definite confirmation of the agreement of Brownian movements with the hypothesis of thermal agitation of the smallest particles has been afforded very recently by the measurements of Perrin (1908)."[21]

Our conclusion, accordingly, is that Ostwald's "conversion" and virtually all of the new edition reflecting it were already in place when he first learned of Perrin's results. He nevertheless would have had strong reasons for inserting the two citations of Perrin's October 1908 paper. Recall the demand Ostwald had made of molecular theory in the *Monist* translation of his 1907 article:

> When expressions or notations of magnitudes which can not be observed and measured and for which we can substitute *no definite and empirically determinable value*, occur in a formula by means of which some physical relations are to be represented, we have to deal with an hypothesis. For the task of the exact sciences is to establish the reciprocal relation of measurable and demonstrable quantities, or in other words to find the mathematical forms or functions by which these quantities are interrelated, so that one of them can be calculated when the others are given. In order to establish experimentally such a functional relation it is therefore necessary to measure singly all variable or constant magnitudes which appear in such an equation.[22] [emphasis added]

As we are about to explain, Ostwald had determined independent values for all three of the quantities in the equation eN_0 = Faraday's constant, before he had known of Perrin's October 1908 paper. This paper ends with values for N_0, e, and further molecular magnitudes, all inferred from the vertical gradient in Brownian motion; these values had two striking virtues from Ostwald's point of view: (1) they were independent of his determinations; and (2) they were not only consistent with his determinations, but had clear claim to being more precise than his.

Throughout our monograph we have been treating Ostwald's *Grundriss*, in all its editions, as a book-length review article, not as representing his view alone. This is the way those in chemistry at the time would have read not only the earlier editions on which we relied in Chapter 2, but also the 1909 edition. That Ostwald's "conversion" occurred before he knew of Perrin's results would not have been all that obvious to them. The conclusion we have reached now, however, makes the question of what led him to convert a question about Ostwald as an individual, not as a representative of the community. We shall answer it in this light, along with his view of the new standing of molecular theory, yet end by arguing that at least the latter can still be regarded as representative of the community.

Suppose, then, we assume that Ostwald had completed the manuscript for the new edition before he knew of Perrin's results and ask how the argument for the reality of molecules proceeded in it without the two passages citing him. Ostwald's new support for the kinetic theory of heat did not hinge on Svedberg's 1906 results supporting Einstein's and Smoluchowski's formulas. To begin with, the support was at best limited: "The paths calculated from Einstein's formula are roughly one-fourth the observed paths, from Smoluchowski's formula one-third," so that the agreement was only one of "approximate proportionality" as the solvent and hence its viscosity were varied.[23] More importantly, Ostwald reaches his key conclusion immediately prior to the section, "The Brownian Movements," in the chapter new to the 1909 edition, "Disperse Systems":

> It is highly significant that there is a continuous transition from suspensions and emulsions, which are readily recognisable as such, to true solutions (for which the laws of solutions hold), this fact indicating that the nature of both systems is the same. Or, in other words, the facts here discovered indicate that true solutions are really heterogeneous, and may therefore be regarded as mixtures of different kinds of particles in the sense of the molecular hypothesis. If we add to this the result of the investigations into the electrical discharges of gases (see later, p. 533), in which are recognisable discontinuities or a grained structure in the submicroscopic region, it appears that *the final proof of the grained or atomistic-molecular nature of matter has been obtained*, after a fruitless search during a whole century. Here we do not have relations which lead indirectly to atomic methods of representation, but which can also be explained in a more general, and therefore, more scientifically justifiable manner, and thus afford no decisive support to the hypothesis. They are experimentally demonstrated differences in the nature of matter, which exclude the possibility of a continuous occupancy of space. This discontinuity is below the limit directly reached by the ultramicroscope, but is not far removed from it. The "molecular dimensions" calculated from the surface tension lies roughly between 6×10^{-9} cm. and $6 \times$

> 10^{-8}, while the ultramicroscope goes down to 5×10^{-7}. The gap to be bridged is only a tenth power, which is no great distance.[24] [emphasis in original]

Ostwald's conclusion about the granular structure of matter, at least as he presents it, thus relies on observations made possible by the ultramicroscope, not on any specifics of Brownian motion.

Because of Ostwald's own use of emphasis in the paragraph, we have not italicized the two sentences that seem to us most important: "*Here we do not have relations which lead indirectly to atomic methods of representation, but which can also be explained in a more general, and therefore, more scientifically justifiable manner, and thus afford no decisive support to the hypothesis. They are experimentally demonstrated differences in the nature of matter, which exclude the possibility of a continuous occupancy of space.*" In other words, the grounds for Ostwald's conclusion is not its explanatory power, but "experimentally demonstrated differences" that allow only the possibility of discontinuous matter. The experimentally demonstrated differences in question are ones involving separation, filtration, and, most of all, optical phenomena presented on the preceding pages—*plus* electrical phenomena in gases, to be presented in the chapter, "Conduction in Gases and Radioactivity," that begins on p. 533. We leave it to readers to assess the strength of the evidence Ostwald gives on those pages. The key point is that he is appealing to experimental results that colloids and solutions have in common, on the one hand, and results on ions in gases and their relation to ions in solutions, on the other, to infer the discontinuity of matter below the level of visibility even in an ultramicroscope.

The new chapter on "disperse systems" ends with a section on "the size of disperse particles" that cites Zsigmondy's 1906 results in concluding that "the isolated existence of still smaller particles, which was to be expected considering the accidental nature of the limit [viz. that imposed by the intensity of the available light], has been demonstrated in other ways."[25] After repeating the point about the "unbroken continuity" down to the 10^{-8} cm called for by molecular theory, the chapter ends: "The limit of spatial discontinuity which is physically observable is still lower, for electrical phenomena take us at least a further power of ten."[26] We shall return to this and the reference to electrical phenomena in the preceding long quotation in a moment.

The section on "Brownian movements" immediately following the preceding long quotation addresses a topic never mentioned in the 1890 and 1899 editions. It concludes, "If the movements of a small particle suspended in a liquid are calculated on the basis of this [kinetic theory] hypothesis, *the agreement with the movements actually observed is so close that we are compelled to regard this agreement as a fairly satisfactory proof of the kinetic nature of heat.*"[27] Ostwald then gives the comparison of Svedberg's observations with the calculations based on

Einstein's and Smoluchowski's formulas. This is followed by a single paragraph (in German, two in the English translation) describing Perrin's 1908 vertical-gradient results. It concludes that Perrin "has developed the method to such an extent that it is probably the most accurate value for the molecular constant yet obtained."[28]

The chapter on the kinetic theory of heat, which had appeared very early in the book in 1890 and 1899, occurs some 20 pages after the chapter on disperse systems. It nevertheless remained virtually word-for-word the same as in the earlier editions until near the end, where two sections are added, the first entitled "Number of Molecules," and the second, "The Kinetic Theory of Liquids and Solids." The latter's opening sentence summarizes it: "While the kinetic theory of gases has led to a very remarkable series of conclusions, the later confirmation of which has given us some confidence in its expediency, the development of the corresponding theory of the two other states of aggregation is still in its infancy, though hopeful beginnings have been made."[29] The conclusion reached in the new section on the number of molecules, based on their size inferred from viscosity and van der Waals's formula, we quoted earlier: 8.5×10^{23} per mole, "or in round numbers 10^{24}." This is followed by a two-sentence paragraph, which we have concluded was inserted after the manuscript for the new edition was complete, noting first that other calculations "yield values which are a little higher or a little lower," and then giving Perrin's 7.1×10^{23} as the most probable value.[30]

We have gone into detail here in order to highlight the round-number value 10^{24} that we think Ostwald had settled on before he saw Perrin's October 1908 paper. Some 30 pages later (in both the German edition and the translation), Ostwald describes J. J. Thomson's isolating and counting ions and their charge in what we now would call a cloud chamber. He concludes, "*The electric charge on an electron is equal to the unit charge of an electrolytic ion, namely about 10^{-19}coulombs.*"[31] Ostwald does not cite a source, but this value corresponds so closely to the value Harold Wilson had announced from water-droplet experiments in 1903, 3.1×10^{10} esu,[32] that he might as well have cited him. More importantly, the product of the two round-numbers, 10^5, corresponds closely to a number Ostwald had emphasized as fundamental in his account of electrochemistry in all three editions, namely the value of Faraday's constant, 96,540 (coulombs per mole).[33] Ostwald had expressed the significance of this three pages before the sentence in italics we just quoted:

> If the method (p. 509) of calculating the number of molecules from the mean cross-section (which can be deduced from diffusion experiments as well as from viscosity) is applied to gas ions, and if the total electric charge is determined at the same time, values are obtained for the number of ions in a mol, and for the charge on a single ion. The former agrees with that given on p. 510,

> and the latter with the value given on p. 538. Or, gas ions behave from the point of view of the kinetic theory like molecules of ordinary gases, and each ion carries the same charge as a univalent electrolytic ion.[34]

So, at least in terms of round numbers, Ostwald had begun to meet his announced requirement with independently measured, mutually consistent values for the triad of N_0, e, and Faraday's constant before he saw Perrin's paper and had thereby tested at least to first order the claim eN_0 = Faraday's constant. Perrin's value for N_0 then became somewhat of a two-edged sword for him: it proffered greater precision, but at the same time threw into some question the value for *e* that Ostwald appears to have taken from Wilson and Thomson.

Ostwald nowhere cites Perrin in his new chapter (that begins on p. 533), "Gaseous Conduction and Radioactivity," from which the preceding quotations are taken. The sections on radioactivity recount what was known as of 1905, qualified by the remark, "the facts are still in course of evolution, and a fairly comprehensive review is not yet possible."[35] Nothing in the discussion indicates that the phenomena contributed to Ostwald's conversion, though of course he could scarcely have missed their consistency with atomism. Beyond the previous two quotations and the passages leading up to them, the sections of the chapter on J. J. Thomson's findings consist mostly of proposals about the implications of "electron theory." One of these is worth noting because it concerns the structure of atoms:

> The simplest application of this [electron] theory is to electrolytic ions, which, from this point of view are compounds of chemical atoms with as many electrons as its valency amounts to. These compounds are separated by electrolysis, the chemical atom remaining behind in the electrolyte or on the electrode, while the electrons travel further into the metal. Metallic conduction must be regarded as a migration of electrons in the metal.
>
> It must be remembered that positive electrons are not known, but only positive atoms, or gas ions and kations. We may either assume that positive electrons similar to negative electrons exist, but have not yet been isolated, or we may consider the positive state to be produced by the loss of a negative electron in the system. In view of the want of perfect symmetry between positive and negative electricity, as shown in certain electrical phenomena, the second assumption (which corresponds to Franklin's old theory of electricity, only the signs are reversed) is preferable.[36]

Ostwald goes on to remark, "it is more difficult to apply the theory to chemical compounds."[37] In other words, he leaves the question of molecular structure entirely open in the chapters new to his 1909 edition, much as Nernst continued to do in his 1909 edition.

Our conclusion, then, is that Ostwald's conversion came about from the combination of four sets of experimental findings primarily during the period from 1903 to 1907: (1) the counting of gas ions and, with it, the discovery of the electron, the fundamental negative charge, and its asymmetry with the positive charge—mostly under J. J. Thomson between 1897 and 1904; (2) the related determination of an order-of-magnitude value for the fundamental charge that, in combination with an order-of-magnitude value for Avogadro's number based on van der Waals's gas law, was fully consistent with the established value of Faraday's constant; (3) the observations of extremely small particles in constant rapid motion under the ultramicroscope reported by Zsigmondy in 1905 and 1906; and (4) Svedberg's 1906 tests of Einstein's and Smoluchowski's formulas for Brownian motion showing that "agreement with the movements actually observed is so close that we are compelled to regard this agreement as a fairly satisfactory proof of the kinetic nature of heat."[38] Insofar as Ostwald was still insisting on the mere hypothetical standing of molecules in an article published in German early in 1907, and his *Fundamental Principles of Chemistry: An Introduction to All Text-Books of Chemistry* totally eschewing atoms and molecules had been published later that year, the fourth of these appears to have been a tipping point.

Given the amount of attention that he had devoted to ions in solution over the preceding two decades, however, being able to determine the charge per ion in liquids and to count their number had to have been giving him reason to reconsider for a few years before this. (In his Faraday Lecture of 1904 he remarked, "I think there is no word that I have oftener spoken or written than the word 'ion' ";[39] yet Thomson's research is not mentioned in this lecture, nor in his 1907 *Fundamental Principles*, much less the 1899 edition of his *Grundriss*.) Subsequently encountering Perrin's October 1908 paper and its far more precise values for the molecular magnitudes either just before or while the new edition was in press must have markedly reassured him—so much so as to add the citations to it.

Finally, we turn to the question of the new standing of molecular and molecular-kinetic theory indicated by the 1909 edition of Ostwald's *Grundriss* and expressed in his Preface as its having been "raised to the position of a scientifically well-founded theory." The independent values, 10^{24} and 10^{-19} coulombs, for N_0 and e that agreed to their level of precision with the known value of Faraday's constant were not sufficiently precise to meet Ostwald's stated requirement for grounding the theory—namely, definite values that can be substituted "in a formula by means of which some physical relations are to be represented . . . in order to establish experimentally" the functional relations in question. Perrin's more precise value of 7.1×10^{23} for N_0 was a step in the right direction, but the value for *e* he inferred from it, 4.1×10^{-10} esu (i.e., 1.4×10^{-19}), was too far removed from the Wilson-Thomson value of 3.1×10^{10} esu to give Perrin's 7.1×10^{23} much of any claim to finality. *So far as grounding of molecular theory, then, the most that the 1909 edition indicated was sufficient progress toward independently measured values of the relevant molecular magnitudes to license the conclusions that definite values exist and hence too that physical relations are indeed represented by the formulas involving these magnitudes.* As such, having "the position of a scientifically well-founded theory" was tantamount to giving unqualified warrant to continuing experimental pursuit of increasingly precise values for the magnitudes.

Moreover, the two magnitudes in question both involved counts of discrete items, the number of molecules in a mole and the number of elementary charges comprising the amount of electricity needed to yield a mole of a monovalent substance in electrolysis. To give unqualified warrant to pursuit of increasingly precise values for them therefore amounted as well to unqualified warrant to *take* both molecules and discrete elementary charges to exist while engaged in this pursuit. Short of systematic failure to achieve more definite values, the question of whether molecules exist had therefore become closed. None of this was to deny Nernst's point that the question about what binds atoms together in molecules and hinders their flying apart remained entirely open, as did questions about the shapes and sizes of molecules of different kinds. These questions, however, had gained unqualified legitimacy, and the emerging experimental results on ions in gases and solutions and on radioactivity were showing promise of providing experimental means to begin addressing them. Here too, then, having "the position of a scientifically well-founded theory" was tantamount to giving unqualified warrant to questions that before had lacked it because of the failure to find experimental means for addressing them.

7.4. The New Standing of Molecular-Kinetic Theory: Post-Solvay

We shift perspective now to late 1912 or early 1913, following the publication in French in 1912 of the proceedings of the 1911 Solvay Conference.[40] The obvious advance since 1909 was the number of alternative independent determinations of each of N_0 and *e*. Perrin's values for the former ranged from 65×10^{22} to 69×10^{22}, implying a value for *e* from Faraday's law ranging from 4.45×10^{-10} to 4.19×10^{-10} esu, added to which was the Boltwood-Rutherford value of 60.3×10^{22}, implying a value for *e* of 4.80×10^{-10} esu. The various new determinations of *e* independent of those for N_0 ranged in values from 4.65×10^{-10} to 4.891×10^{-10}, implying a value for N_0 ranging from 62×10^{22} to 59.2×10^{22}. Within those bounds were as well the two post-Solvay values for N_0 cited by Eucken that had been derived from measurements of k in blackbody radiation, 59.7×10^{22} and 62.0×10^{22}.

Ignoring the discrepancy between Perrin's values and all the rest, the two sets combined provided strong evidence that definite values exist for N_0 somewhere between 59×10^{22} and 69×10^{22} and for e somewhere between 4.1×10^{-10} and 4.9×10^{-10} esu. Insofar as other molecular magnitudes were derivable from these—in particular, energy per molecule per degree Kelvin and absolute masses of atoms and molecules with known chemical formulas—molecular and molecular-kinetic theory had to a greater extent than a decade before become "grounded," in van Fraassen's sense.

Put another way, sufficiently precise values for these molecular magnitudes had become established to warrant not just their being taken for granted, but for their being employed *constitutively* in ongoing research. Anyone doing this, however, had to choose a specific value, with or without error-bars. Thus, for example, in the summer of 1913, Bohr, needing a value for e in order to obtain a theoretically derived value for Rydberg's constant to test against the known value, chose 4.7×10^{-10} esu;[41] the value for N_0 corresponding to this according to Faraday's law is 61.6×10^{22}. At almost the same time that summer, the Braggs, faced with having to derive the distance between sodium and chlorine atoms in the crystal lattice of salt, chose a value for the mass of a hydrogen atom of 1.64×10^{-24} grams;[42] we assume they obtained this from a value for N_0 of 61.5×10^{22}. Perrin himself, who was not at the time using his values for molecular magnitudes constitutively in this way in his ongoing research, continued to insist in his *Les atomes* of 1913 and subsequent editions that his notably larger values for N_0 were to be preferred. Nevertheless, by the summer of 1913, leading figures at the forefront of ongoing research on atoms and molecules were departing from him.

The link between Avogadro's number and the fundamental unit of charge mediated by Faraday's constant is associated with molecular theory, not molecular-kinetic theory. The other such link that we emphasized in Chapter 6—between Avogadro's number and Boltzmann's constant—is, by contrast, entirely associated with molecular-kinetic theory through the relationship $N_0 k = R$, where, according to that theory, the mean translational kinetic energy per molecule per degree K, designated α by Perrin, is $3k/2$. In this case, too, the independently measured values for k as of 1912 were not in full agreement. The value from blackbody radiation that Planck gave that year was 1.34×10^{-16} ergs per degree K, implying a value for α of 2.01×10^{-16} ergs per degree K.[43] Perrin's values for N_0 had come from assuming that his measured values for the mean translational kinetic energy of granules in Brownian motion correspond to the mean translational kinetic energy of molecules at the same temperature. Based on the values Perrin announced for N_0 at the Solvay Conference, his values for α ranged from

1.81×10^{-16} to 1.91×10^{-16} ergs per degree—that is, between 5 and 10 percent less than Planck's.

In this case, the question of a preferred value did not arise with the same force as it did with e and N_0, for α (and hence also k) had less of an immediate role to play in ongoing research. Nevertheless, here were two robust values, 2.01×10^{-16} from by then established blackbody measurements, and $(1.86 \pm 0.05) \times 10^{-16}$ from highly developed Brownian motion measurements that, according to the respective theories, were values for the same physical quantity, yet did not match to the level of precision thought to hold for them. Planck's derivation of the blackbody law was linked to kinetic-molecular theory only through Boltzmann's account of entropy, and even then his appeal to that account had been in some respects idiosyncratic. Perrin's various derivations, by contrast, were linked to the core of kinetic-molecular theory, and his assumption of the identity between granule and molecular energies seemed in no way idiosyncratic. Were it not for the fact that Planck's implied value for N_0, 62.0×10^{22}, was so much closer at the time to the values other than those of Perrin, the difference might have given grounds for questioning whether Planck's k really does amount to a molecular magnitude. Regardless, we have found no one at the time who instead proposed that the mean translational kinetic energies of granules in Brownian motion are in fact simply less than the mean translational kinetic energies of the molecules in the ambient liquid.

These concerns over the less than desired agreement among values are not why we included the qualifier "to a great extent" in our claim three paragraphs ago about molecular-kinetic theory having become grounded. The molecular magnitudes N_0, e, and α are all three generic, not ones that vary from one kind of molecule to another. The only molecular magnitudes determined by them that vary from one kind of molecule to another are the absolute masses, and they scarcely enter at all into kinetic-molecular theory. The prominent macroscopic phenomena that do vary from one kind of molecule to another were viscosity, diffusion, thermal conduction, and deviations from the ideal gas law; and the molecular magnitudes with which they vary, according to kinetic theory, are the sizes and shapes of different kinds of molecules. No less than before the grounding of it that we are here considering, kinetic theory continued to deal with viscosity, diffusion, thermal conduction, and deviations from the gas law by postulating that molecules are spherical, and then inferring sizes for different kinds of molecules from these phenomena. Yet, even more so after the values for N_0 between 59×10^{22} and 69×10^{22} had become available than before, it was clear that (in the words of Jeans from 1916) if "the molecules are *assumed* as a first approximation to be elastic spheres, experiment leads to discordant results for the diameters of these spheres, shewing that the original assumption is unjustifiable."[44] One must accordingly be careful not to overstate the extent to which molecular-kinetic theory had become grounded by the end of 1912.

In point of fact, some progress was made on the shape of molecules in 1911 and 1912, much of it in Nernst's laboratory by such figures as Arnold Eucken. They and others had produced strong experimental evidence supporting the Einstein-Debye theory of specific heats of solids at low temperatures. This amounted to evidence that degrees of freedom "freeze out" as the temperature drops, or in other words, that energy thresholds have to be crossed in order for some degrees of freedom to absorb energy.

Eucken had also shown that the ratio of the specific heats of hydrogen does not remain at its room temperature value of 1.4 as the temperature is lowered, but increases, appearing to approach the 1.67 value of monoatomic gases as the temperature approaches absolute zero. This then was evidence that diatomic molecules like hydrogen, oxygen, and nitrogen have five degrees of freedom at room temperature, but as the temperature is lowered, two of these degrees of freedom "freeze out"; diatomic molecules like chlorine for which the ratio of specific heats is less than 1.4 at room temperature, on the other hand, have a vibrational degree of freedom already absorbing energy at room temperature. All of this not only began to address the long-standing quandary over the specific heats of gases, but also provided evidence that diatomic molecules have a dumbbell shape, with three translational degrees of freedom, plus two rotational once the requisite energy threshold is exceeded, and various vibrational at still higher energy levels.[45]

Suppose then that diatomic molecules are taken to have a dumbbell shape, and consider the problem of incorporating this into the kinetic theories of viscosity, diffusion, thermal conduction, and deviations from the ideal gas law for just the gases in question. The pivotal parameter in the molecular-kinetic theories of these phenomena is the *mean free path* of a molecule between its encounters with other molecules. Consider now the complexity of the problem of defining mathematically such a mean free path in the case of rotating dumbbell-shaped molecules, taking their number per unit volume and the statistical distribution of their translational velocities as given; and then using measured values of the viscosity, diffusivity, thermal conductivity, and deviations from the ideal gas law for the different gases in question to infer the sizes of their respective molecules, or even their respective effective "spheres of influence." One should scarcely be surprised that Jeans, after denying that molecules are spherical and subsequently noting the recent results on specific heats, nevertheless stayed with spherical molecules in the 60 pages of the 1916 edition of his *Dynamical Theory of Gases* devoted to viscosity, thermal conduction, and diffusion; nor that the 1916–1917 mathematical theories of viscosity, heat conduction, and diffusion in non-uniform gases developed independently by Enskog and Chapman concerned gases composed of spherical molecules. Simply put, notwithstanding the grounding provided by the values of N_0, e, α, and the masses of a huge range of molecules as of 1913, the molecular-kinetic theories of the phenomena of viscosity, diffusion, thermal conductivity, and deviations from the ideal gas law remained theories about hypothetical—or, more accurately, fictional—spherical molecules.

These considerations are reminiscent of what we regard are the most provocative remarks van Fraassen makes in "The Perils of Perrin, in the Hands of Philosophers":

> It is still possible, of course, to also read these results as providing evidence for the reality of molecules. But it is in retrospect rather a strange reading—however, much encouraged by Perrin's own prose and by the commentaries on his work in the scientific and philosophical communities. For Perrin's research was entirely in the framework of the classical kinetic theory in which atoms and molecules were mainly represented as hard but elastic spheres of definite diameter, position, and velocity.[46]

Leaving aside our claims about the nuanced relationship between Perrin's research on Brownian motion and molecular-kinetic theory, the considerations prompting our here quoting van Fraassen once again are reminiscent as well of part of the argument we made in section 2.8 of Chapter 2 for why molecular theory still had the standing of merely a hypothesis in 1900:

> Similarly, the statistical mechanical formulation of the molecular-kinetic theory gave generic qualitative explanations of the phenomena of viscosity and its independence of density in gases, yet absent any determination of the sizes and shapes of molecules of different kinds, it offered only a hand-waving explanation of the specific differences in viscosity from one gas to another or the specific systematic variation of viscosity with temperature in each gas. The inability to determine the sizes and shapes of molecules of different kinds resulted in the theory of viscosity and diffusion being a theory about hypothetical spherical molecules, and even then, efforts to determine their sizes had failed to yield determinate values. In short, the molecular-kinetic theory of heat, even with everything that could be said in its favor, remained a theory about hypothetical discrete entities.

We have now conceded that the grounding molecular-kinetic theory gained from the comparatively precise values determined for N_0, e, α, and the masses of a large range of molecules still had not met the requirement called for in this paragraph. Are we, then, to conclude, in agreement with van Fraassen, that the reality of molecules still had merely a hypothetical status in 1913?

We claim no, yet at the same time concede that the reality of molecules had not yet become established once and for all. The new standing of the molecular hypothesis, we claim, derived from adequate grounds having been given for *taking* molecules to exist for purposes of ongoing research, thereby closing the question of their existence—or, perhaps better put, eliminating the question from the scope of that research. The word "take" we are lifting from Newton's fourth rule of reasoning:

> In experimental philosophy, propositions gathered from phenomena by induction should be taken to be true, either exactly [*accuratè*] or very nearly [*quamproximè*], notwithstanding any contrary hypotheses, until yet other phenomena should render such propositions either more exact [*accuratiores*] or liable to exceptions.[47]

The propositions to be taken to be thus true as of 1913 were: (1) All moles of any substance contain the same definite number of chemically demarcated molecules, which to current approximation is around 64×10^{22}; (2) the molecules in all gases have the same definite mean translational kinetic energy per degree of absolute temperature, which to current approximation is around 1.95×10^{-16} ergs per degree Kelvin; and (3) there exists a definite fundamental, universal unit of negative charge, which to current approximation is around 4.5×10^{-10} esu, and it is always joined with a definite rest mass, the combination constituting the *electron*, with the number of them required to liberate a mole of molecules of any monovalent substance in electrolysis always the same. These propositions are to be presupposed as fully established in ongoing research—that is, they are licensed to enter *constitutively* into the design of experiments and rules for measurement—until yet other phenomena should render them either more exact or liable to exceptions. Keep these three propositions in mind, for we shall be returning to them repeatedly in the next section.

Our first chapter included the following quotation from Stephen Brush that, in effect, was rejecting the claim we have just made:

> For the next few years, he [Perrin] seems to have devoted much of his time to popularizing the significance of his work on Brownian motion, in particular the idea that atoms have now been proved to exist. In this he was surprisingly successful. In fact, the willingness of scientists to believe in the "reality" of atoms after 1908, in contrast to previous insistence on their "hypothetical" character, is quite amazing.

> The evidence provided by the Brownian-movement experiments of Perrin and others seems rather flimsy, compared to what was already available from other sources. The fact that one could determine Avogadro's number and the charge on the electron by one more method seems hardly sufficient to justify such profound metaphysical conclusions. Several independent methods of determining these parameters had been known since 1870 or before, to say nothing of the many successes of kinetic theory in predicting the properties of gases.[48]

The obvious question, therefore, is how, according to us, the situation had changed between 1900 and 1913. The answer is simple. *All* of the determinations of Avogadro's number and the mean translational kinetic energy of molecules per degree Kelvin before 1908 (save for Planck's) had presupposed that molecules are spheres of a definite diameter, or at least have a definite sphere of influence. *None* of the determinations that emerged between 1908 and 1913 presupposed anything about their shape or size. In other words, before 1908 all the determinations of Avogadro's number were giving the number of hypothetical spherical molecules per mole, while the series of new determinations between 1908 and 1913 were not of the number of hypothetical spherical molecules, but rather simply of the number of molecules per mole. Questions about their sizes and shapes remained entirely open, to be pursued in further research taking full advantage of the newly determined molecular magnitudes.

We take this point to be grounds for dismissing van Fraassen's contention quoted earlier, as well as Brush's. There was nothing at all strange in reading the results of the new mutually complementary theory-mediated measurements of the quantities in question as providing evidence for the reality of molecules of a sort that had never been provided by their efficacy in conceptualizing and explaining phenomena. This efficacy had never yielded anything specific that could enter constitutively into ongoing research without begging questions and hence without the results of any such research having a merely hypothetical standing.

Because the distinction we are invoking with the notion of *take to be true* is generic, and not specific to molecules, we are saving until the next section of the chapter a more developed account of it. We shall end this section with a few brief points in response to questions that are likely to have arisen about the case at hand.

First, what difference did it make whether the molecular and molecular-kinetic hypotheses were taken to be true at the time versus simply continuing to grant that they were offering the only viable explanation of the chemical and thermodynamic phenomena? Part of the answer we gave in our discussion of Ostwald's conversion: questions about the specifics of the many kinds of molecules and what binds atoms together in them came to have not just a legitimacy, but beyond this an immediacy that they had not had before. Such questions had come to occupy the cutting edge of research precisely because further progress hinged on finding experimental means for answering them. Still more important than this, moreover, was that taking them to be true carried with it license to take as given in pursuing such experimental means the three claims we listed earlier concerning the number of molecules per mole, their mean translational kinetic energy per degree Kelvin, and the fundamental negative charge. Theory is needed to gain experimental access to regimes we cannot observe. Ampère's law and the galvanometer it licensed as a means of measuring it provided a degree of experimental access to electric current that had not existed before them. Taking these three claims as now given for purposes of research amounted to taking them to be providing means for gaining a degree of experimental access to the microphysical realm that had not existed before.

This, however, invites the question Brush raised: What stood in the way of the values for N_0, α, and e that had been determined by 1900 serving this same purpose? The answer is twofold. First, those values for each of these three had not converged remotely to the extent that the values in 1912 had. Tables 2.1 and 2.2 in Chapter 2 display this lack of convergence in the case of the supposedly universal number of molecules per mole or per cubic centimeter at standard conditions; the only way, then, to obtain values for α were via values for N_0, and hence these values too were not remotely convergent; and the values for e consisted of Townsend's 3.1×10^{-10} esu for the negatively charged and 2.8×10^{-10} esu for the positively charged hydrogen ion and Thomson's 6.5×10^{-10} esu for ions formed by x-ray in air and roughly the same for ions in hydrogen.

We noted earlier that an issue in the case of the values as of 1913 was which specific ones to adopt to incorporate in ongoing research. The risk involved in this choice was one of degree of precision, in particular at the time between a second and a third significant figure. By contrast, to select from the ranges available any specific values for purposes of ongoing research in 1900 was to render any results from that research at best precarious, just as Gibbs had said in 1901.[49]

Moreover, all the values for N_0 and α as of 1900 had been obtained from phenomena covered by kinetic theory under the assumption that molecules are spherical, or at least have definite spheres of influence in their interactions with one another. Had the values for N_0 in 1900 converged to the extent that they had by 1913, there would have been grounds for concluding that this assumption, despite the many misgivings that had been expressed in adopting it, had some merit. The failure of convergence, however, made it still more suspect. By contrast, the only assumptions that the several 1913 values for N_0 and α had in common were the reality of molecules and Avogadro's hypothesis concerning the number of molecules per unit volume at standard conditions or, as recast in the last decades of the nineteenth century, the number per mole. The combination of the convergence of the different 1913 values plus the diversity of phenomena from which those values had been obtained constituted strong evidence that both of these assumptions, even if not true, were safe to adopt at least for purposes of attaining robust theory-mediated measurements.

Still, granting all of this, the point we stressed in section 2.8 of Chapter 2 about the failure as of 1900 to gain experimental access to specific contrasting features of molecules of different kinds that could be correlated with contrasting features of macroscopic phenomena still seemed true at the end of 1912. Why wasn't that still giving reason to question whether experimental access had really been gained to molecules?

The concern as of early 1913 was not whether sufficient experimental access to the microphysical realm had been gained to determine the characteristic sizes and shapes of molecules of different kinds and the forces that bind the atoms forming them together. Clearly it had not. The concern was rather whether the experimental access that had been gained, as reflected in the values of N_0, α, and e that had then been determined, provided a basis for further experimental developments that would ultimately enable questions about the specifics of molecules beyond just their individual masses to be resolved. This concern hinged on whether the promise of such progress adequately outweighed the risk that research constitutively presupposing the values in question would ultimately have to all be thrown out should those values turn out not to have been well-founded after all. We leave to the reader whether the evidence we reviewed in Chapter 6 supporting the values adequately addressed this risk.

Needless to say, we have to eliminate from consideration the subsequent history of the research predicated on those measured values if we are to assess the potential they offered at the end of 1912—that is, before Bohr and the Braggs had published their papers and while von Laue's 1912 breakthrough in x-ray diffraction was still being assimilated. So, let us offer an example entirely of our own devising to indicate the sort of promise they were offering.

In section 2.8 we appealed to the phenomenon of NO_2 gas progressively transforming into a mixture of it and nitrogen peroxide, N_2O_4, as the temperature is lowered as an illustration of something about the specifics of molecules that molecular theory had not begun to address. Consider this phenomenon in the light of the measured values of N_0 and α and Perrin's verification of the Gaussian character of Brownian motion. As the temperature is diminished, the rate of change of mole density can be measured. But then so too can the number of NO_2 molecules per unit volume that have joined to form N_2O_4 molecules. The mean translational kinetic energy of both kinds of molecules is the same, and its variation with temperature can be calculated from α. From this and the masses of the two kinds of molecules, their mean translational velocities can be calculated as a function of temperature, with those for N_2O_4 roughly 70 percent of those for NO_2. Finally, taking the distribution of velocities among the NO_2 to be Gaussian, at least estimates can be made of the number of NO_2 molecules per unit volume with velocities close to the mean velocities of N_2O_4. All of this points to the possibility of finding some systematic correlation between the number of NO_2 molecules per unit volume with velocities in various ranges and the rate versus temperature at which they join to form N_2O_4 molecules. As conjectural as this is, its promise for revealing something instructive was notably greater at the end of 1912 than it had been before 1900.

Even granting the virtue of having comparatively specific values for these molecular magnitudes, however, if so many questions about the specifics of molecules remained open at the end of 1912, how was research predicated on taking the claims associated with these values to be true any different from research predicated on them in which they continued to be taken as mere hypotheses? Not any difference to speak of, so long as the research in question is making sustained progress. There are three ways in which it can fail to be making sustained progress. First, central questions can remain recalcitrant, beyond the reach of all experiments. This was true throughout the nineteenth century in the case of questions about the forces binding atoms together in molecules, and it continued to be so for more than a decade after 1913. Second, the results of experiments can fail to provide sufficiently unequivocal evidence to resolve questions. The experimental attempts predicated on kinetic theory reported in Meyer's 1899 book to determine the sizes of molecules and, with them, a constant value for Avogadro's number, provide an example, one that again continued, in the case of the sizes, for more than a decade, even after the value of Avogadro's number was available. Third, clear results can be obtained, but only by adopting further hypotheses that seem irreconcilable with established physics. The violation of classical physics by the Rutherford-Bohr model of the atom is an example.

In each of these cases, when the research is predicated on a claim that still has the status of a hypothesis, its continuing failure puts the hypothesis itself into dispute. When, by contrast, it is predicated on a claim for which evidence has provided adequate grounds for it to be taken to be true, it does not. It then has a status like the legal presumption of innocence until proven guilty beyond reasonable doubt. Decisive evidence is required even to raise doubts about its truth. This, of course, was the whole point of Newton's introducing the phrasing, *take to be true, either accuraté or quamproximé, notwithstanding any contrary hypotheses,* in his fourth rule. It was also the point Euler was making when, in 1752, he announced, in the introduction of his *Recherches sur les irrégularités du mouvement de Jupiter and de Saturne*:

> since M. CLAIRAUT has made the important discovery that the movement of the apogee of the Moon is perfectly in accord with the Newtonian hypothesis . . . , there can no longer remain the least doubt about this proportion. . . . One can now maintain boldly that the two planets Jupiter and Saturn attract each other mutually in the inverse ratio of the squares of their distance, and that all irregularities that can be discovered in their movement are infallibly caused by this mutual action. . . . *And if the calculations that one claims to have drawn from this theory are not found to be in good agreement with observations, one will be always justified in doubting the correctness of the calculations, rather than the truth of the theory.*[50] [emphasis added]

Our answer to the question of the standing of the molecular and molecular-kinetic hypotheses as of early 1913 is that a remark to the same effect held for them. The reception of Bohr's model a few months later—when it clearly did not work for atoms with two or more electrons, and the only hard evidence for it was the agreement to within 6 percent between the calculation drawn from it of Rydberg's constant and the observed value—is one of several illustrations of these hypotheses then having this standing.

7.5. On the Standing of Hypotheses

The principal distinction Newton was making with his fourth rule was between "hypotheses," for which he had long had a low regard (to say the least), and "propositions gathered from phenomena." As any remotely knowledgeable reader at the time realized, the hypothesis he had in mind was the view that planets are carried around in their orbits by fluid vortices, hypothesized to provide a contact mechanism to counteract their centrifugal tendency;[51] and, needless to say, the propositions gathered from phenomena with which he was contrasting this hypothesis were his law of gravity and its account of planetary motions according to which they are perturbed departures from Keplerian motion. He had come to use the phrase, "take to be true," between the first and second editions of the *Principia*, initially in licensing the inductive leap from the evidence that his law of gravity holds for the planets to taking it to hold universally for all matter. Two considerations appear to have led him to do so: the realization that in any such leap to a sweeping generalization, one is always simply *taking* the general claim to be true; and his coming to appreciate the potential for much stronger evidence for his theory of gravity than he could muster in the *Principia* if others in the future were to use it, instead of relying on observation, to pin down the enormous systematic complexities of planetary motion that result from their mutual interactions.[52]

Smith has invoked Newton's phrasing elsewhere to argue for a narrower than customary conception of what it is for a scientific community to accept a proposal.[53] His main point was to distinguish between the evidence that warrants its initially taken to be true for purposes of predicating ongoing research on it and the evidence that can accrue to it from the success of such research over extended periods of time. His aim there was to respond to historians and sociologists of science who had challenged the epistemic standing of the advanced sciences by emphasizing the extent to which many of their most fundamental tenets had become an integral part of those sciences on the basis of quite limited and sometimes even misdirected evidence. In response, Smith argued that this epistemic standing does not derive from canons of initial acceptance, but from the history of evidence that accrues to propositions from their continuing to be tested, often tacitly and *en passant*, by research constitutively predicated on them. Correlatively, he argued that philosophers assessing the epistemic standing of the sciences need to look carefully at the history of evidence that accrues to fundamental claims in this way over decades and, in some cases, centuries. The extraordinary evidence supporting Newtonian gravity—that is, the evidence that Einstein felt he somehow had to accommodate—was not to be found in the *Principia*, but in the subsequent two centuries of research in celestial mechanics.

Our aim in invoking Newton's phrasing here, by contrast, is to elucidate a distinction between, on the one hand, evidence taking broadly the form of inference to the best explanation that leads at least substantial fractions of a scientific community to believe a hypothesis and, on the other hand, evidence that leads a community to grant it a presumption of inviolability as a constitutive element in ongoing research. We say "elucidate" because we claim that the distinction by no means originates with us, but rather has been part of the practice of scientific communities since at least as long ago as the late seventeenth century. The quote from Euler in 1752 at the end of the preceding section is evidence of this. So too is our quote in Chapter 1 from the 1909 edition of Nernst's *Theoretische Chemie*, which immediately follows a list of values for Loschmidt's number: "the atomistic conception begins to lose its hypothetical character."[54]

An especially clear statement of the distinction can be found on the first page of the 1904 and 1916 editions of Jeans's *The Dynamical Theory of Gases*. In the first edition he says,

> It need hardly be said that this identification of heat and motion is only a hypothesis; it never has been, and from the nature of things never can be, proved. At the same time this hypothesis shows an ability to explain and even to predict natural phenomena, such that there can be little doubt that it rests upon a foundation of truth.

As we have quoted earlier, in the second and all subsequent editions he says,

> This hypothesis was for long regarded as pure conjecture, incapable of direct proof, and probable just in proportion to the number of phenomena which could be explained by its help. In recent years, however, the study of the Brownian movements has provided brilliant visual demonstration of the truth of this conjecture, and the actual heat-motion of molecules—or at least of particles which play a role exactly similar to that of molecules—may now be seen by anyone who can use a microscope.[55]

(Jeans's statement notwithstanding, we do not mean to be suggesting here that visual access is in any way indispensable to the distinction we are elucidating.)

The distinction is in the same spirit as Poincaré's distinction between "indifferent hypotheses" and "real generalizations," where, in the case of the former, conclusions reached would be the same with or without them.[56] The sole reason for conducting a series of experiments may be to test whether the results are consistent with a hypothesis, and hence that it can explain them, yet the results of the experiment would be the same whether they are consistent with it or not. We saw an example of this with Perrin's Brownian motion experiments in Chapter 4. His reason for conducting the experiments, and Einstein's reason for developing his account of such motion, was to support the molecular-kinetic theory of heat as the sole viable explanation of the phenomenon. Perrin's measured values of the mean kinetic energies of his granules nevertheless did not logically presuppose either that molecules exist or that he was measuring their mean kinetic energies. Our analysis in Chapter 4 showed this as a matter of logic. The subsequent history showed it as well when the discrepancy between Perrin's values for α and those obtained through other approaches raised questions about whether his value for granules really does match the molecular value.

We are not belittling the importance of inference to the best explanation in the history of the physical sciences. We are only calling attention to the difference between evidence of that form and evidence that warrants granting hypotheses the status of a presumption of inviolability as a constitutive element in ongoing research. As we have noted more than once, "inference to the best explanation" initially entered the philosophic literature in the late nineteenth century as the name of a fallacy, specifically the fallacy of taking a conclusion to be established in spite of not having eliminated unconceived alternatives.[57] It is, of course, this very possibility that lies behind Jeans's distinction in the preceding quote between "direct proof" and "probable just in proportion to the number of phenomena which could be explained." Equally, we are not impugning the legitimacy of the widespread acceptance of the molecular and molecular-kinetic hypotheses among physicists in the years before the period we have examined.

It did indeed have enormous heuristic value, just as Ostwald himself acknowledged in so many places in the 1889 edition of his *Grundriss*.

We do, however, remind readers that a material ether was similarly widely accepted as the only viable explanation of the oscillating transfer between kinetic and potential energy required for sustained wave motion. In his *Encyclopedia Britannica* on the ether, Maxwell reviews the failures of attempts to measure any specific aspect of the ether, including the pre-Michelson-Morley attempts to measure the velocity of the earth through it. He still, however, concludes,

> Whatever difficulties we may have in forming a consistent idea of the constitution of the aether, there can be no doubt that the interplanetary and interstellar spaces are not empty, but are occupied by a material substance or body, which is certainly the largest, and probably the most uniform body of which we have any knowledge.[58]

Again, we are not impugning the acceptance of the existence-of-the-ether hypothesis by Maxwell, Lorentz, and a great many others during the nineteenth century. We are only calling attention to the difference between his grounds for concluding that the ether exists and the grounds he might have had if any of the efforts to determine specific properties of it that he considers in the article had succeeded.

So much for one side of the distinction in evidence that we are elucidating; what about the other? As a step toward giving more content to it, consider the experimental efforts presented by Meyer in his 1899 *Kinetic Theory of Gases* to leverage measurements of the different sizes of molecules from phenomena for which molecular-kinetic theory offered an explanation, arguably the only viable explanation at the time. The molecular hypothesis *did* enter constitutively into those efforts, specifically the spherical molecule hypothesis, but scarcely with a presumption of inviolability. The failure of the different approaches to yield convergent values for the sizes, or even for the individual approaches to meet the cross-check of yielding a constant value for Loschmidt's number,[59] gave more reason to question the spherical molecule hypothesis than to accept it. In the end, Meyer nevertheless takes 2×10^{-8} cm to be "the average diameter of a gaseous molecule," prefacing his choice with the remark, "although we may not pretend to see [in it] an exact evaluation of the size."[60] Granted that this was at best only an order-of-magnitude value, but the question still arises why it was not grounds for taking the existence of molecules to be a settled question, yet the 1912 values for N_0, α, and e were.

One difference is that Meyer's 2×10^{-8} cm has virtually no claim to being a physically specific value. Consequently, were it to have been taken to be true and presupposed constitutively in further research, the most it could have contributed to measurement or evidence would have been an order-of-magnitude cross-check on results reached without it. Hence it would have added little more, if anything, toward advancing research than just the bare claim that molecules exist.

More generally, not much is to be gained from taking a claim to be true and presupposing it constitutively in further research unless it has some form of specificity. Specific claims have far more potential for being leveraged into something more. We hazard the guess that all that would have been needed for the existence of the ether to have been taken to be true in our sense was a reasonably specific value of the speed of the earth through it—that is, a well-behaved positive result from the Michelson-Morley experiment or its predecessors—even though all the other questions about it that Maxwell reviewed in his *Britannica* article would have remained totally open. Regardless, the claims about N_0, α, and e as of 1913 were not only comparatively specific, but beyond this, their claim to specificity was strengthened by the fact that a specific value for each of the three quantities had been determined that was independent of all of the convergent determinations of the other two. This, and the fact that there were the three of them, gave them a potential for being leveraged into something more, even though virtually all the questions about the specifics of molecules themselves remained open. What is the point of taking a claim to be true and adopting it constitutively in further research if it lacks such potential?

A second difference is that whatever claim to specificity that Meyer's 2×10^{-8} cm might have been taken to have was undercut by its presupposing something specific about molecules themselves, namely that they are, at least to high approximation, spherical in shape, or if not, that they have well-defined "spheres of influence" in their interactions with one another. So, any research predicated constitutively on this value would at the same time have been predicated on a claim about molecules that had nothing to recommend it except mathematical tractability. The whole point of taking a claim to be true and adopting it constitutively in further research is to pursue experimental answers to open questions. To presuppose an answer to one of the principal open questions in doing so puts all the answers that might emerge from the further research at risk of begging the question. By contrast, none of the assumptions we listed in section 6.5 of Chapter 6 that bridged the gap between macrophysically measured quantities and N_0, α, and e presupposed anything specific about molecules, and hence answers to further questions achieved by research predicated on them did not risk begging the question in this way.

The two preceding paragraphs have taken for granted that the step from a claim's having a merely hypothetical standing to its being granted a presumption of inviolability has to turn on risk versus potential gain. The risk is not that the research will not yield robust results, but that those results will all have to be discarded at a later point when the claim on which they are predicated turns out not to be true. Its later turning out not to be true, however, does not automatically entail that the results predicated on it will have to be discarded. Newton's law of gravity has now given way to Einstein's field equations, yet the two centuries of results concerning which details in our solar system make a difference and what differences they make that presupposed it were not discarded, thanks to Einstein's showing that it holds asymptotically in the static, weak field limit of his equations. Sheldon Glashow has generalized this point in a trenchant sentence: "*A physical principle cannot be said to be true until it has been shown to be false and its envelope of applicability delineated*."[61]

The question, then, is what requirements a claim presumed to be inviolable in ongoing research has to satisfy in order for the results that presuppose it to survive its later turning out not to be true. The answer we propose, in the spirit of Glashow's remark, is that it has to satisfy some combination of three requirements: (1) it has to hold at least to an appropriately high level of approximation; (2) it has to hold, if not universally, then at least over a domain that reasonably approximates the one the research on which it is predicated is trying to cover; and (3) the terms in which it is stated, if not entirely adequate in their own right, will at least be effective proxies over the domain in question, especially for purposes of measurement, for terms that allow it to better meet the first two requirements. In the preceding section of this chapter we proposed that three claims, interrelated by the convergent measurements reviewed in Chapter 6, had by the end of 1912 come to have a standing of presumed inviolability for purposes of ongoing research: two, stated in terms of molecules, centered on the measured values of N_0 and α, and a third, stated in terms of electrons, centered on the measured values of e. We here propose that the evidence supporting these three should be assessed with respect to the three requirements we have just listed.

In Chapter 6 we concentrated on the convergence of the measured values for N_0, α, and e as evidence. In doing so, we somewhat understated the strength of the evidence that in each case a definite physical quantity exists for which approximate values were being determined in the measurements. Values for e were being determined from two separate phenomena, ionization in gases and α-particles. Values for α were also being determined from two separate phenomena entirely independent of those from which values for e were being determined, namely blackbody radiation and Brownian motion. Even in the case of N_0, values had been determined from one phenomenon entirely independent of those from which values for e and α were being determined, namely the collection of a countable number of α-particles to form a given fraction of a mole of helium gas. Moreover, as Perrin had stressed, thanks to the established values of the gas constant and Faraday's constant, each determination of a value for any one of e, α, or N_0 implied a value for each of the other two as well. So, every determination of a value for any one of the three was subject to cross-checks not only from other values determined for it, but from all the values determined for the other two.

All this, together with the degree of convergence of the measured values for all three between 1908 and 1913, gave strong grounds for concluding that definite physical values exist for (1) the number of molecules in a mole, (2) the mean translational kinetic energy per degree Kelvin of the molecules, and (3) the fundamental unit of charge—that is, definite values at least over the ranges of the various phenomena for which determinations had been made; and these definite values lie within the ranges of the announced measured values from the most reliable experimental determinations of each of the three. Moreover, the strong grounds that a definite number of molecules exist per mole were grounds too that molecules exist, or (using Jeans's phrasing) something akin to them that within this evidence "plays the same role"; and correspondingly for the existence of a fundamental unit of charge and with it the electron. Beyond this, the range of the measured values for N_0 implied a corresponding range of values for the absolute masses of all atoms for which atomic weights had been established, and hence too a range of values for the masses of all molecules for which chemical formulas had been established in terms of these atoms. Finally, insofar as a comparatively precise value had been determined for the charge-to-mass ratio of the electron, the range of the measured values for e gave a range for its mass as well.

In short, notwithstanding how many questions about molecules and, for that matter, electrons remained entirely open, the evidence coming out of the theory-mediated measurements considered in Chapter 6 had licensed quite a range of interrelated propositions to be taken to be true and presupposed constitutively in continuing research. We are not denying that the evidence for the existence of molecules was limited, for so many questions about them remained open. The evidence for the claims listed in the preceding nevertheless gave good reason (1) that taking them to be true would enhance the pursuit of answers to those questions and (2) that experimental results constitutively presupposing them would not themselves be undercut should those answers require the claims to be reconsidered.

One way to avoid an extended garden path of research predicated on a claim that has been taken to be true on limited evidence is to demand a great deal of that research right away. In some respects, the 1913 Bohr model did this in the case of the charge and mass of the electron, and since the charge of the electron had become interlinked with the number of molecules in a mole and their mean kinetic energy, for them too. The Bohr model offers a lesson of a different sort as well. After the Bohr model was shown to hold for the helium ion and Sommerfeld had accounted for a doublet in the Balmer series by adding elliptical to Bohr's circular orbits, the tenet that electrons describe definite orbits around nuclei was accorded just the sort of privileged status in the case of ongoing quantum research that we have been highlighting.[62] Yet it was just this tenet that Heisenberg, a decade later, rejected as having been a continuing impediment in his watershed paper of 1925—that is, the paper in which he insisted on restricting quantum theory to what he called "observable quantities": "it is necessary to bear in mind that in quantum theory it has not been possible to associate the electron with a point in space, considered as a function of time, by means of observable quantities."[63] We cite this example to emphasize again that the presumption of inviolability of a tenet that has been taken to be true in our sense is provisional. Notice, too, that the distinction marked by Bohr orbits—between stable energy states for which they turned out to have been a surrogate—survived.

The point of this section has been to clarify our distinction between a hypothesis having been generally accepted within a community, yet still having the standing of a hypothesis, and the standing we claim the molecular and molecular-kinetic hypotheses had come to have at the end of 1912 and the beginning of 1913. This moment corresponded to the time immediately after the results of the Solvay Conference became available to the community at large with the 1912 publication of the proceedings in French. Our choice of this moment is nevertheless rather arbitrary insofar as so many groundbreaking results of research at least bearing on, and usually constitutively presupposing, these hypotheses had already appeared or were about to. Experimental results supporting Debye's version of Einstein's theory of specific heats of solids had been published in 1912, as had both results on the specific heats of diatomic gases at low temperatures and von Laue's breakthrough with x-ray diffraction. The Braggs' paper on the crystal structure of halides appeared in July 1913, followed by further papers, including one on the structure of diamond, in September. The first of Bohr's three papers also appeared in July 1913, and the first of Moseley's two papers tying x-ray spectra to atomic number appeared in the same journal five months later. By the end of 1913, therefore, significant new evidence had already accrued to the claims associated with the values for N_0, α, and e from groundbreaking results presupposing them. The time at the end of 1912 and beginning of 1913 for which the molecular and molecular-kinetic hypotheses had just the standing we are attributing to them, and nothing beyond it, was a fleeting moment.

However brief the moment was, we nonetheless contend that the best way to understand the new standing the molecular and molecular-kinetic hypotheses had acquired as a consequence of the measurements of N_0, α, and e between 1908 and 1913 is ours—namely, a middle stage between their being widely accepted yet still having the standing of a hypothesis, and a subsequent stage in which they became increasingly entrenched by virtue of the results of research constitutively presupposing them.

An added virtue of viewing the new standing in this way is the incisive rejoinder it provides to van Fraassen's remark, that to "read these results as providing evidence for the reality of molecules . . . is in retrospect a rather strange reading."[64] For it justifies this reading without implying that the evidence for the reality of molecules in 1913 was in any respect definitive. From the first edition of his *Theoretische Chemie* in 1893 to the "eighth-tenth" edition in 1921, Nernst had said, "At present scarcely anything definite is known about the nature of the forces which bind atoms together and which hinder them from flying apart."[65] Nothing in the results for N_0, α, and e amounted to any advance on this. Worse, so far as the molecular-kinetic hypothesis is concerned, the only specific variation from one kind of molecule to another that had been determined was their masses. No advance had been made on the specific variations in their sizes and shapes needed to link different kinds of molecules to the macroscopic differences they display in phenomena of viscosity, diffusion, heat conduction, and deviations from the ideal gas law, not to mention chemical kinetics. The referent of the term "molecule" remained experimentally thoroughly underdetermined.

We noted at the beginning of this section that Smith has elsewhere invoked Newton's "take to be true" phrasing to contrast the standing that the law of gravity had on the basis of the evidence presented in the *Principia* with the increasingly entrenched standing that it came to have over the course of the next two centuries as a consequence of evidence accruing to it from the research that presupposed it. This contrast is no less apropos in regard to the reality of molecules. Anyone asking now what the evidence is that molecules exist is misconceiving the way evidence works in physics if they focus on the time before 1913. This question can be meaningfully answered only through a critical review of *all* the evidence bearing on molecular reality that has been, and is still being, generated since 1913.[66]

7.6. 1905–1913 within the History of Theory-Mediated Measurement

We have noted in earlier chapters that theory-mediated measurement in what became known as the physical sciences goes back at least as far as Ptolemy's *Almagest*. The developments during the years 1905 to 1913, with Perrin at the center, were a central part of a major transition, initiated by J. J. Thomson in 1897 and Planck in 1900, in how theory-mediated measurement can contribute to evidence in microphysics. The primary contribution to evidence before then was to confirm that various quantities that according to theory should be constant actually do have determinate single values. We have more than once cited the effort that went into determining such values in the case of Faraday's constant from electrolysis and the ideal gas constant. A further example from the second half of the nineteenth century was the series of measurements of the gravitational constant G from a variety of variations of the Cavendish experiment.[67] A review article of 1901 cites the agreement among the values obtained from Cavendish forward as the only unqualified evidence that gravity varies linearly with the mass of the attracting body—evidence that took the form of the constant of proportionality in the law of gravity being verified, within the limits of experimental precision, to be constant.[68]

A new era began with J. J. Thomson's measurements of the mass-to-charge ratio, m/e, of the constituents of cathode rays in 1897.[69] This value turning out to be more than three orders of magnitude smaller than the smallest known value of m/e, of the hydrogen ion in electrolysis, led him to conclude, "Thus on this view we have in cathode rays matter in a new state, a state in which the subdivision of matter is carried out much further than in the ordinary gaseous state."[70] Two years later, after finding the same order of magnitude for m/e for the discharges in both the photoelectric effect and thermionic emission, Thomson concluded, "From what we have seen, this negative ion must be a quantity of fundamental importance in any theory of electrical action; indeed, it seems not improbable that it is the fundamental quantity in terms of which all electrical processes can be expressed."[71] Thomson's "discovery of the electron" has properly been termed the birth of microphysics. It also has claim to being the first successful measurement of any microphysical quantity. And, as such, it opened a new era in theory-mediated measurement by virtue of such measurement of various microphysical quantities being the primary way of gaining experimental access to this realm.

In a paper of 1898, midway between the two papers cited in the preceding paragraph, Thomson published his first measured values for the charge of ions in air, 6.5×10^{-10}. He then inferred from this value and the known value of the total quantity of electricity per cubic centimeter of hydrogen released in electrolysis—tantamount to Faraday's constant—a value for Loschmidt's constant, 20×10^{18}, which corresponds to a value of 45×10^{22} for Avogadro's number.[72] Thomson then cited a value of Loschmidt's number, 20×10^{18}, that he said was from experiments on the viscosity of air, as evidence supporting his 6.5×10^{-10} value for the charge. In a related paper the next year, John Townsend provided evidence from experiments on the diffusion of gaseous ions to conclude, independently of the specific value for either the charge or Loschmidt's number, that the charge per ion generated by x-rays is the same as the charge on the hydrogen ion in electrolysis.[73] With these steps in evidential reasoning, the putative link provided by Faraday's measurements in electrolysis between the charge on ions and the number of molecules in a mole became a pivotal source of evidence for claims in microphysics.

Thomson's point in deriving a value for Loschmidt's number was to confirm the value for the charge of gaseous ions he had obtained, under the assumption that this charge is the same as the charge on hydrogen ions in electrolysis. Townsend's point was to confirm this assumption. As we have seen, the existing value for Loschmidt's number with which Thomson compared the value he had obtained was somewhat arbitrarily chosen among the many available, resulting in not much more than order-of-magnitude confirmation. This became clear five years later when Harold Wilson and he obtained a value for the charge from water droplets in an electric field less than half of the value from 1898. Our main point here, however, is that Thomson's goal in invoking Faraday's constant was simply to confirm his theory-mediated measurement of the microphysical charge.

Planck's derivation in 1900 of the blackbody radiation formula added a further step in the endeavor to extract evidence from theory-mediated measurements of microphysical quantities. This derivation had yielded a coefficient of temperature k in the formula that he claimed to be Boltzmann's constant, a measure of the mean kinetic energy per molecule per degree Kelvin of a gas. In tandem with a value for what became known as Planck's constant, blackbody radiation determines a comparatively precise value for k, but there was no existing value for it with which this value could be compared. So, from the measured value of k Planck derived values for Avogadro's number, $N_0 = R/k$, Loschmidt's number, and "the elementary quantum of electricity e," the last of these through eN_0 = Faraday's constant. His paper then ends with a paragraph that we quote once again, for it marks a landmark in the history of evidence deriving from theory-mediated measurement:

> If the theory is at all correct, all these relations should not be approximately, but absolutely, valid. The accuracy of the calculated numbers is thus essentially the same as that of the relatively worst known, the radiation constant k, and is thus much better than all determinations up to now. To test it by more direct methods should be both an important and a necessary task for further research.[74]

As we have noted earlier, Planck was invoking these other constants not to confirm his value for k, but to lend support to his derivation. In other words, Planck's goal in invoking the gas constant and Faraday's constant was not to confirm his value for k, but for other independent theory-mediated measurements of N_0, Loschmidt's number, and e to provide evidence for his theory of the blackbody formula.

The trouble, of course, was the absence at the time of values for these other constants of remotely the precision as that of the value for k, or for that matter for h, obtained from blackbody radiation. We can do no better than Planck's summary in his Nobel Prize lecture of 1920 of the situation in 1900 and what then ensued over the course of the next few years.

The first constant is of a somewhat formal nature; it is connected with the definition of temperature. If temperature were defined as the mean kinetic energy of a molecule of a perfect gas, which is a minute energy indeed, this constant would have the value 2/3. But in the conventional scale of temperature the constant assumes (instead of 2/3) an extremely small value, which naturally is intimately connected with the energy of a single molecule, so that its accurate determination would lead to the calculation of the mass of a molecule and associated magnitudes. This constant is frequently termed Boltzmann's constant, although to the best of my knowledge Boltzmann himself never introduced it (an odd circumstance, which no doubt can be explained by the fact that he, as appears from certain of his statements, never believed it would be possible to determine this constant accurately). Nothing can better illustrate the rapid progress of experimental physics within the last twenty years than the fact that during this period not only one, but a host of methods have been discovered by means of which the mass of a single molecule can be measured with almost the same accuracy as that of a planet.

While at the time when I carried out this calculation on the basis of the radiation law an exact test of the value thus obtained was quite impossible, and one can scarcely do more than test the admissibility of its order of magnitude, it was not long before E. Rutherford and H. Geiger succeeded, by means of a direct count of α-particles, in determining the value of the electrical elementary charge as 4.65×10^{-10}, the agreement of which with my value 4.69×10^{-10} could be regarded as a decisive confirmation of my theory. Since then further methods have been developed by E. Regener, R. A. Millikan, and others, which have led to a but slightly higher value.[75]

Although Planck did not mention Perrin, and Perrin's values had by then been superseded, what he was referring to was the 1905–1913—or, perhaps better, 1908–1913—period in which Perrin was in many respects the central figure. The list of measured values of N_0 (and e) in our Table 6.1 in Chapter 6 is intended to encapsulate the advances to which Planck was referring that were made in this period. During these years, there was a quantum jump, so to speak, in both the kind and the strength of the evidence that was coming out of theory-mediated measurements of microphysical quantities. Hopefully we have said more than enough about this already. Here we need only remind readers that the progress made in the measurements themselves during this period amounted to an increase in reliability from order-of-magnitude to Poincaré's first significant figure—or perhaps better, paraphrasing Planck's 1913 assessment, to values for N_0 between 58 and 70 times 10^{22}, if not between 59 and 69 times 10^{22}, and correspondingly for e at least between 4.1 and 4.9 times 10^{-10} esu. Even at that level of reliability, we have seen how much impressive evidence was marshaled out of the various measurements of the quantities in question.

We chose to bound our study by the publication of Bohr's first paper of his 1913 trilogy not merely because it opened the way for the first time to theory-mediated measurements of quantities pertaining to the structure of atoms and hence changed the entire character of molecular research, but no less so because it happens also to have been a landmark event in the history of theory-mediated measurements of microphysical quantities generally. Bohr's proposal that electrons in a hydrogen atom emit radiation only when they jump from one quantized energy state to another was, needless to say, a radical departure from classical physics. That it then allowed the energy differences between such states to be specified by the energy $h\nu$ corresponding to lines in the spectra allowed the spectra, for the first time ever, to provide more empirical information than just a fingerprint, so to speak, of the atom or molecule emitting or absorbing radiation. Beautiful and promising though that was, it did not in and of itself provide any empirical evidence for the Bohr model, only evidence of its promise. The sole empirical evidence to support the model in Bohr's first paper was his derivation of the value of the Rydberg frequency from values of e, h, and m/e of the electron. The agreement Bohr's calculation achieved, 3.1×10^{15} versus the known 3.29×10^{15} value, was, he said, "inside the uncertainty due to experimental errors in the constants entering into the expression for the theoretical value."[76] What more could anyone have asked for?

The answer, of course, was for the value Bohr used for e, 4.7×10^{-10}, actually to have been "inside the uncertainty due to experimental errors." As we noted at the end of section 6.2 in Chapter 6, e entered Bohr's expression for the theoretical value of the Rydberg frequency, as calculated, to the fifth power. Had he used the value for e that Perrin was defending in 1913, 4.21×10^{-10}, the discrepancy in the calculation would have been 55 percent instead of merely 6 percent, and the evidence supporting Bohr's radical departure from classical physics would have been far less compelling than it was. Consequently, Bohr's appeal to Rydberg's constant as his source of evidence required far tighter error bounds on the value of e than Perrin had demanded of the various values of N_0 which he had cited as evidence for the existence of molecules at the end of his 1909 monograph, his 1911 paper, and his *Les atomes*. This is one respect in which Bohr's paper amounted as well to a landmark event in the history of theory-mediated measurements of microphysical constants: his evidence needed an error-band on the value of e not much greater than ±1 percent—or, in other words, a value in which the *second* significant figure, not just the first, was reasonably reliable. The demands on the precision of the theory-mediated measurements for purposes of evidence in ongoing research thus increased with Bohr's paper by virtually an order of magnitude.

A second respect is that the test against Rydberg's constant linked h and e to one another, and, because the measurement of this constant was of a wave length, to the speed of light c as well. Theretofore, only Planck's derivation of the black-body radiation equation and Einstein's yet-to-be-confirmed proposal on the photoelectric effect[77] had linked h and e to one another and to quantities outside of microphysics. Over the course of the next decade and a half, as quantum theory was developing out of the Bohr model, several other such combinations of microphysical constants yielded calculated values of quantities that could be measured independently, and hence provided the same kind of test of theory as Bohr's appeal to Rydberg's constant. To give just a couple of examples, Sommerfeld's fine-structure constant, $\alpha = 2\pi e^2/ch$, gave the fine structure in the spectral lines of hydrogen that Bohr had ignored; and the so-called Bohr magneton, $eh/4\pi m$—that is, the elementary quantum of magnetic moment—entered into the calculation of magnetic properties of gases and, at a later stage, into electron spin. The force of the evidence provided by the calculations of these quantities depended on the level of agreement with independently measured quantities, and hence depended as well on the precision of the theory-mediated values for e, h, and m. Equally, however, having several different combinations of these three linked to independently measurable quantities opened the way to an increasing number of theory-mediated measures of the three, and hence to being able to play these off against one another in pursuit of preferred values.[78]

A milestone in the development of quantum theory, and with it a transition to a new stage—in effect, from quantum mechanics to the beginnings of quantum electrodynamics—occurred in 1928 with the publication of Dirac's relativistic extension of quantum mechanics in his two-part paper "The Quantum Theory of the Electron," and Darwin's "The Wave Equations of the Electron" confirming that the full spectrum of the hydrogen atom, fine-structure and all, plus the Zeeman effect, could be derived from Dirac's theory.[79]

Though only peripherally related to it, a transition to a new stage in the history of theory-mediated measurement occurred a few months later with the publication of Birge's "Probable Values of the General Physical Constants (as of January 1, 1929)."[80] Preferred values of the fundamental constants in orbital astronomy had been adopted in the last decades of the nineteenth century in order to eliminate differences in calculated planet locations arising purely from different ephemerides using slightly different values for the constants.[81] As preoccupation with precise agreement between theoretically calculated and observed values in microphysics increased, a corresponding need arose there as well. Thus Birge notes in his opening paragraph,

> Some of these constants can be evaluated by various methods. Each has been investigated by various persons, at various times, and each investigation

normally produces a numerical result more or less different from that of any other investigation. Under such conditions there arises a general and continuous need for a searching examination of the *most probable* value of each important constant. The need is general since every physical scientist uses such constants. The need is continuous since the most probable value of to-day is not that of to-morrow, because of the never ending progress of scientific research. These remarks appear to the writer so self-evident that the mere statement of them may be deemed superfluous. However, in spite of these facts, an investigation of the values of general constants in current use in the literature reveals a surprising lack of consistency, both in regard to the actually adopted values and to the origin of such values.[82]

Over the course of the next 40 years, Birge and his protégés took it upon themselves to issue updates to the values of the "fundamental" constants, the number of which kept increasing as a consequence of continuing research in microphysics. One of these updates, Birge's "The 1944 Values of Certain Atomic Constants with Particular Reference to the Electronic Charge," is of special interest in the context of the present monograph.[83] As we noted earlier, in the wake of Bohr's 1913 paper, measured values for e became primary, with preferred values for N_0 inferred from them via Faraday's constant. Birge's paper revisits these three constants, examining not only the precision with which each was being measured independently of the others, but also possible sources of systematic error in them in response to their imperfect consistency with one another. He concluded that x-ray crystallographic measurements of N_0 from five different crystals were most reliable, yielding a value of $(60.2338 \pm 0.00043) \times 10^{22}$; and the measurement of Faraday's constant was sufficiently precise and reliable[84] to give a value for e from it, $(4.8021 \pm 0.0006) \times 10^{-10}$, that was to be preferred over the most tenable value from measures entirely independent of the Avogadro and Faraday constants, $(4.803 \times 0.002) \times 10^{-10}$. In other words, by 1945 the situation had become the reverse of what it had been at the end of the period covered in our study: the measured value for N_0 had become primary, with the preferred value for e inferred from it.

The further development of quantum theory took a dramatic turn two years later with the announcement at the Shelter Island Conference of discrepancies between Dirac's theory and highly precise measurements first in the fine structure of hydrogen—the Lamb shift—and then in the hyperfine structure of hydrogen and deuterium—the anomalous magnetic moment of the electron.[85] Although nearly two decades elapsed, the new quantum electrodynamics that emerged after Shelter Island led to a new era in the history of theory-mediated measurement with the 1969 publication by three of Birge's protégés, Taylor, Parker, and Langenberg, of the landmark article, "Determination of e/h, Using

Macroscopic Quantum Phase Coherence in Superconductors: Implications for Quantum Electrodynamics and the Fundamental Constants"[86]—the same year as CODATA, which had been created the year before by the International Committee for Science, formed its Task Group on Fundamental Constants. This Task Group of CODATA has thereafter assumed the responsibility that Birge and his protégés had been shouldering for the previous 40 years. What Taylor et al. announced in 1969 that helped to spur this change was a determination of a value for the fine-structure constant "that did not require the use of QED theory" and that thereby provided "a test of QED in which a priori information from QED itself is not essential."[87] With this, an approach to testing QED emerged involving comparison of measured values of fundamental constants at most weakly presupposing QED with ones strongly presupposing QED.[88]

In other words, the role in evidence of the theory-mediated determinations of microphysical constants expanded from one of supplying values needed for comparisons of theory and experiment to one of themselves comprising the immediate content of the evidence. The 2006 CODATA report lists six distinct theory-mediated values of the fine-structure constant, depending to different degrees on different aspects of QED, that agreed with one another to seven significant figures, that is, to better than 1 part in 10 million. The two with the smallest standard uncertainty, one strongly dependent on QED, the other weakly, diverge in the tenth significant figure, and even there by less than half of the standard uncertainty. To quote the report once again, "This is a truly impressive confirmation of QED theory."[89]

As significant as such increasingly stringent testing of QED has been, focusing on it takes us away from the history of measurement of the three constants Perrin singled out in 1908. From 1969 until 1998, CODATA published only two sets of recommended values, in 1973 and 1986. Starting in 1998, however, they decided that the pace of advances toward greater precision had increased to a point that adjustments to the values of the fundamental constants should be made every four years. To get a sense of the extent to which the precision was increasing, consider the estimated relative uncertainty in the values for N_0: 6.6 parts per million in 1969, 7.9 parts per 100 million in 1998, and 1.2 parts per 100 million in 2014; correspondingly, the uncertainties in values for e were 4.4 parts per million in 1969, 3.9 parts per 100 million in 1998, and 6.1 parts per billion in 2014.[90] In 2014 sufficient precision had been achieved with the constants singled out by Planck in 1900 and Perrin in 1908 to warrant their assigned values being the basis of a 2018 revision to the International System of Units (the SI) in which "the physical concepts underlying the definition of the kilogram, the ampere, the kelvin, and the mole have been changed."[91]

On our account, then, the history of theory-mediated measurement is to be thought of as having had five stages: (1) before 1897; (2) from 1897 to 1929;

(3) from 1929 to 1969; (4) from 1969 to 1998; and (5) from 1998 to 2018. The second of these saw the successful development of convergent, stable theory-mediated measurements of microphysical quantities, N_0, α, and e in particular, from at best order-of-magnitude to values that had claim to being within ±1 to 2 percent. The present monograph has focused on the middle portion of this stage, from 1905 to 1913, during which convergent, stable measurements of these constants came to have claim to being within ±5 percent, that is, claim to Poincaré's one significant figure. Insofar as that step was so immediately superseded by a demand for values within ±1 to 2 percent, one might think it as not all that important to the history of theory-mediated measurement. Any such thought is wrong. Before the 1905–1913 period, the question of whether meaningful measurements of microphysical quantities could be achieved at all was still open. Achieving convergent, stable measurements of such quantities to within ±5 percent settled that question: ***it was clearly possible to gain sufficient experimental access to the microphysical realm by means of theory-mediated, mutually corroborating measurements to attain meaningful values for at least some highly important quantities, in particular*** N_0, α, **and** ***e***—that is, the three "universal constants" that Perrin had singled out in the opening paragraph of his October 1908 paper.

We list here the CODATA 2018 recommended values for these three constants, together with the two macrophysical constants that linked them to one another throughout our study and, for that matter, continuing to the present:[92]

Avogadro's number:	$60.2214076 \times 10^{22}$
e:	$1.602176634 \times 10^{-19}$ C (= $4.80320471 \times 10^{-10}$ esu)
α:	$2.0709735 \times 10^{-23}$ J/degK
Faraday's constant:	96485.33212 C/mol (= $2.892557488 \times 10^{14}$ esu)
R:	8.31446261 J/mol degK

For completeness, we include as well the 2018 values of the two constants of Planck's blackbody formula: $k = 2\alpha/3 = 1.380649 \times 10^{-23}$ J/degK and $h = 6.62607015 \times 10^{-34}$ J/Hz. As of 2018, all of these values have come to be stipulated as *exact* for purposes of defining the new SI units, specifically for the mole, the ampere, the kelvin, and the kilogram. The research into the values of these constants that Planck set in motion in 1900 thus came to a fulfillment of sorts in 2018, though doubtless not in a form that he had anticipated.

The preceding sections gave our answer to the question of how the standing of molecular-kinetic theory changed during the 1905 to 1913 period. Perhaps we should instead have been answering the question of how the standing of research into microphysics changed during this period. We have just given our answer in italics in the last sentence of the paragraph before last. On its face, that answer seems not to be that far removed from van Fraassen's claim that what changed was molecular-kinetic theory becoming "grounded." But thinking of it in that way is missing the point. For among the microphysical quantities that had come to be shown to be measurable were the number of molecules in a mole of gas, and by virtue of that their masses and hence their numbers in liquids and solids as well; the mean translational kinetic energy per degree Kelvin of the individual molecules in at least gases and liquids, if not solids as well; and the number of elementary charges in any macroscopic quantity of electricity, whether static or in electric current, and hence too the number of electrons in any aggregate of negative charge. On our view, this was the most important contribution Perrin and others during this period made to science. It is why we wrote this monograph.

Notes

1. See, most notably, the footnotes on pages 120 and 128 of Perrin (1990) and (1916) [(1913), pp. 171 and 182f, or (2014), pp. 218 and 225f], the latter of which even dismisses Svedberg's results on diffusion while granting that they support Brillouin's. Perrin does not cite Svedberg (1912) in which his diffusion results are elaborated in some detail (pp. 61–83), along with the value for Avogadro's number drawn from them of 58×10^{22}.
2. Jeans (1916), p. 8.
3. Perrin's high values for Avogadro's number of course imply that, had he derived a value for R from his results, he would not have confirmed its agreement with the established value of the gas constant. In his Nobel Lecture (Perrin, 1926, p. 6f) he offered an indirect argument, derivative from one by van't Hoff, that any osmotic pressure in his emulsions must be the same as the partial pressure of the granules on the wall of the container. As we noted in Chapter 4, however, even granting him this argument, he still would not have done what van't Hoff did, namely confirm that the constant of proportionality for the osmotic pressure is equal to the gas constant.

   We ignored Perrin's Nobel lecture in Chapters 4 through 6 because it fell outside the 1905–1913 period on which we were focusing. One other oddity of the lecture is nevertheless worth noting here. Oseen's presentation speech lists Perrin's values for Avogadro's number: 68.2×10^{22} from vertical gradients, 68.8×10^{22} from mean square displacements, and 65.0×10^{22} from rotation. In his lecture, by contrast, Perrin lists 68×10^{22} for the first, 64×10^{22} for the second on the basis of Constantin's preliminary results (which he did not identify as preliminary), and ends by stating "a crude average" of 64×10^{22} after citing some results on concentrated emulsions not involving Brownian motion. We can only wonder whether he ever acknowledged in print that his values for Avogadro's number from Brownian motion were notably higher than the 60.6×10^{22} Millikan value that had long since become standard.
4. W. Ostwald (1912), p. vi.
5. Ibid., p. 485f.

6. Eucken (1914), p. 376.
7. Bohr (1913), p. 5.
8. Bragg and Bragg (1913), p. 437. The Braggs needed a value for the mass of hydrogen in order to infer the spacing between atoms in the crystal lattice, and with it the wave length of the x-ray. Their value for the mass had to have been derived from a value for Avogadro's number, specifically 61.5×10^{22}.
9. Jeans (1916). In the subsequent 1921 and 1926 editions, Jeans adopts a value of $(60.62 \pm 0.06) \times 10^{22}$, obtained from Millikan's 1917 value for e, 4.774×10^{-10}.
10. Perrin (1908a), (1908c), (1908d); Chaudesaigues (1908). At the time his new edition went to press, Ostwald appears to have read only the last of these notes by Perrin, which initially lists 7.1×10^{23} as the "approximate" value of Avogadro's number, the value Ostwald cited in his book.
11. Perrin's (1908d) ends with two short paragraphs on the top of the next page, the first pointing out that the masses of atoms and molecules can be inferred from his value for Avogadro's number "with the same precision," giving 0.45×10^{-22} grams for the mass of the oxygen molecule, 1.4×10^{-24} for the hydrogen atom, and 0.75×10^{-27} for the mass of the electron; and the second infers from the density of liquid oxygen a diameter for a molecule of oxygen of 2.6×10^{-8} cm and for a molecule of helium, 1.7×10^{-8}.
12. W. Ostwald (1912), p. 485.
13. Nernst (1911), p. 207.
14. Ehrenfest and Ehrenfest (1959), p. 37.
15. Mazo (2002), p. 9. The translation is Mazo's, as is the italicization of the word *experimental*.
16. We thank Paul Forman for repeatedly calling our attention to the low standing of Ostwald in the physics community in the first decade of the twentieth century.
17. The literature on the controversy and the critiques of Helm, Ostwald, and energetics is too large to cover here. A classic source is Hiebert (1971). More recently Robert Deltete has published a number of works on energetics, of which we found the following especially informative for our purposes: (2007a), (2007b), (2007c), (2010), and (2012).
18. Boltzmann (1995), p. 407. The original German edition appeared in 1896 and 1898.
19. W. Ostwald (1909), p. 571; in translation (1912), p. 510. The chapter on the kinetic theory of gases, by the way, reads more or less word-for-word as in the earlier editions up to the point that Ostwald derives a value for Avogadro's number near its end; it has, however, been moved to a much later point in the book, inserted within the new chapters on disperse systems.
20. W. Ostwald (1909), p. 544; (1912), p. 485, where the translator has chosen to split the German paragraph into two paragraphs.
21. W. Ostwald (1909), p. 544. The English translator chose to shorten this sentence: "Still more definite confirmation has been afforded very recently by the measurements of Perrin (1908)" (Ostwald, 1912, p. 485).
22. W. Ostwald (1907), p. 500.
23. W. Ostwald (1912), p. 485.

24. W. Ostwald (1912), p. 483f; (1909), p. 541f in the German.
25. W. Ostwald (1912), p. 491; (1909), p. 550f in the German.
26. Ibid.
27. W. Ostwald (1912), p. 485; the emphasis is not included in the translation, but is present in the German (1909), p. 543.
28. W. Ostwald (1912), p. 485f; (1909), p. 544; as we noted earlier, the constant Ostwald cites in the paragraph is *R*, the gas constant, not Perrin's featured molecular value, Avogadro's number. This is one more sign that the paragraph was inserted in proof, allowing a fairly egregious mistake to escape unnoticed. In the subsequent citation of Perrin, in the chapter on kinetic theory, Ostwald correctly identifies the constant as Avogadro's number, specifically "$.71 \times 10^{24}$."
29. W. Ostwald (1912), p. 510; (1909), p. 571. The conclusion Ostwald offers in the case of solids is worth quoting:

    > While the molecular motion in liquids must be assumed to be uniform in all directions, since they are isotropic in every property, crystalline solids exhibit ordered anisotropism, i.e. suitable properties are functions of the direction in accordance with the general laws of symmetry, which find their highest expression in the crystal form. Hence the assumption that the molecules have different properties in different directions, and that in crystals they are arranged either parallel to each other or at least in congruent relative positions (e.g. alternately perpendicular to each other), so giving rise to anisotropism.
    >
    > The immediate conclusion from this, that the form of the molecule, indicated by the chemical constitution, must find expression in the crystal form, has not yet led to any simple general results, in spite of the attention which has recently been devoted to the idea; in certain special cases, however, some relations of this kind have been recognized.

    The 1909 edition includes a chapter on crystals much earlier in the book, which remained largely word-for-word the same as in the 1899 edition; the latter, however, departs significantly from the corresponding chapter of the 1890 edition.
30. W. Ostwald (1912), p. 510.
31. W. Ostwald (1912), p. 538; (1909, p. 602). Emphasis in both.
32. H. Wilson (1903), p. 439.
33. W. Ostwald (1912), p. 383.
34. Ibid., p. 535f. A footnote attached to the end of the quote says, "Isolated instances have been noticed in which ions occurred with double charges; still the simple ions were apparently in marked abundance." This sentence must have been written before Ostwald had read the two Rutherford-Geiger papers of 1908 and the clear case they make that α-particles are doubly positively charged, and that a more reliable value for the basic charge is 4.65×10^{-10} esu (= 1.55×10^{-19} coulombs). Insofar as Perrin cites the Rutherford-Geiger results in the 1908 paper cited in two places by Ostwald, this is still further evidence that the citations in these two places were added either while the book was in press or after the manuscript had been completed.
35. Ibid., p. 545.
36. Ibid., p. 539. Here too Ostwald might as well have cited Thomson (1900), for the emphasis put on the asymmetry of electric charge and electric current consisting

of the flow of negative, not positive, charges is little more than a summary of that landmark paper.

37. Ibid., p. 540.
38. Ibid., p. 485.
39. W. Ostwald (1904, p. 506).
40. The proceedings in German were not published until 1914, well after publication of the Bohr model.
41. Bohr (1913), p. 9. We shall discuss Bohr's reasons for this choice in section 7.6.
42. Bragg and Bragg (1913), p. 437. That it was atoms of sodium and chlorine occupying the lattice points in the crystal, and not salt molecules as generally thought before this publication, was notable. It must have been especially gratifying to the four seminal figures of physical chemistry.
43. Planck (1991), p. 172f. The original, *Wärmestrahlung*, was published in 1913, with a Preface dated November 1912.
44. Jeans (1916), p. 9. The passage continues, "The divergences arise not only from the fact that the shape of the molecules is not spherical, but also from the fact that the molecules are surrounded by fields of force, and in most experiments it is the extension of this field of force, rather than that of the molecules themselves, with which we are concerned."
45. These results were summarized by Eucken (1914) in the appendix he added, as editor, to the 1914 publication in German of the proceedings of the 1911 Solvay Conference. A summary of the progress that had been made by 1913 in confirming the Einstein-Debye theory of specific heats of solids can also be found in Jeans (1914), pp. 66–78.
46. van Fraassen (2009), p. 22f.
47. Cohen and Koyré (1972), p. 555 (p. 389 in the original Latin, p. 796 in the Cohen and Whitman translation [Newton, 1999]). Newton himself, by the way, appears to have maintained a distinction between the words *exacté* and *accuraté*— hence our including the Latin. The original Latin is, "*In philosophia experimentalis, propositiones ex phaenominis per inductionem collectae, non obstantibus contrariis hypothesibus, pro veris aut accurate aut quamproxime haberi debent, donec alia occurrerint phaenomena, per quae aut accuratiores reddantur aut exceptionibus obnoxiae.*"
48. Brush (1976b), p. 697f; originally published as Brush (1968).
49. Gibbs (1948), p. ixf.
50. Euler, from the Preface of his *Recherches sur les irrégularités du mouvement de Jupiter and de Saturne*, 1752; translation from C. Wilson (1980), p. 144.
51. The vortex theory was originally put forward by Descartes in his *Principia*, but then was formulated in a more mathematical version by Leibniz in his "Tentamen de Motuum Coelestium Causis," published in *Acta Eruditorum* in 1689, two years after the first edition of Newton's *Principia*. For a history of vortex theories, see Aiton (1972).
52. See Smith (2014); and "25 Years of Second Thoughts: Substantive Changes in the 1713 Edition of Newton's *Principia*," in preparation.
53. Smith (2010).

54. Nernst (1909), p. 438. The sentence occurs in the 1911 English translation as well (Nernst, 1911, p. 437), but Nernst had meanwhile authorized the translator to add a paragraph on Perrin's results from Brownian motion that had not been included in the list in German of recent results for Loschmidt's number that was providing the basis for the quote. The list in the original was: 4.5×10^{19} from kinetic theory; 3.7×10^{19} from values for the charge per ion from the method "of cloud formation"; 2.78×10^{19} from the charge of α-particles (as per Rutherford and Geiger); and 2.76×10^{19} from blackbody radiation.
55. Jeans (1904), p. 1; (1916), p. 1.
56. Poincaré (1952), p. 152f.
57. Sidgwick (1884), pp. 250 and 270ff.
58. Maxwell (1965), p. 775.
59. See Table 2.2 in Chapter 2.
60. Meyer (1899b), p. 332; (1899a), p. 334 in the German original.
61. Glashow (1997), p. 276, italics added. Smith thanks the late Sylvan (Sam) Schweber for having called his attention to this remark.
62. See Kragh (2010) and (2003).
63. Heisenberg (1967), p. 263.
64. van Fraassen (2009), p. 22.
65. Nernst (1911), p. 280.
66. For a limited example of such a review that reveals how far from simple and straightforward that evidence can be (at least in the case of diatomic molecules), see Miyake and Smith (2020).
67. See Poynting (1894) for a thorough history of such measurements plus his own contribution to them.
68. Zenneck (2007). For those who find the claim curious, keep in mind that there is no way to test the claim in question celestially because all bodies at any point respond to the gravitational attraction toward any other body in the same way, regardless of their mass. So, as Zenneck notes, the claim that the attraction toward, for instance, the Sun varies as its mass presupposed that the third law of motion applies to gravity. The experiments Zenneck cites do not presuppose the third law of motion. In fact, the masses that appear in the equations of celestial mechanics are all derived from the strength of the gravitational attraction toward the bodies in question, and they are then deployed always in the form *GM*, that is, as an acceleration; hence those masses serve as nothing but placeholders for the field strengths, expressed as accelerations.
69. Thomson (1897).
70. Thomson (1897), p. 312.
71. Thomson (1899), p. 565. The paragraph that opens with the quoted passage ends with, "These considerations have led me to take as a working hypothesis the following method of regarding the electrification of a gas, or indeed of matter in any state"; what follows is the twentieth-century conception of electrical action. For a review of these developments, see Smith (2001b).
72. Thomson (1898).
73. Townsend (1900).

74. Planck (1967), p. 89f.
75. Planck (1922), p.11.
76. Bohr (1913), p. 9. Bohr's formula, technically for Rydberg's constant times the speed of light, presupposes that the nucleus is of infinite mass. As research developed following this paper, the distinction among the Rydberg constant for this case and Rydberg constants for hydrogen and the helium ion became an important source of evidence historically. See Bohr (1914).
77. This relation was confirmed in 1916 by Millikan in his (1916).
78. Sommerfeld in one place went so far as to propose that the measured values for the Rydberg frequencies for the hydrogen atom and helium ion together with that of the fine-structure constant "provide us *with three determining equations for the three unknowns, e, m, and h,* that is, for the three most important universal constants of physical nature" (1923, p. 526).
79. Dirac (1928a), (1928b); and Darwin (1928).
80. Birge (1929). As others have noted, this was the first article in the journal that was subsequently retitled *Reviews of Modern Physics.*
81. See Seidelmann (1992), p. 616.
82. Birge (1929), p. 1f.
83. Birge (1945).
84. Birge's 1945 value for Faraday's constant, in abs.-coulomb units, was 96,487.7 ± 10. The CODATA preferred value, as of 2018, is 96,485.33212, taken to be exact in conjunction with the new SI units of 2019.
85. See Schweber (1994), Chapter 5, and (2015), pp. vii–lvi, especially pp. xvff.
86. Taylor et al. (1969a). This article appeared almost immediately in book form (1969b).
87. Ibid., p. 375.
88. For a 900-page discussion of such evidence, see T. Kinoshita (1990).
89. Mohr et al. (2008), p. 721.
90. See Taylor et al. (1969a), p. 477; Mohr and Taylor (2000), p. 447; and Mohr et al. (2016), p. 56.
91. Stock et al. (2019), p. 1. For details on the CODATA 2017 adjustment that formed the basis for the new SI, see Mohr et al. (2018).
92. https://physics.nist.gov/constants.

Postscript on the Realism-Instrumentalism Debate

Our monograph, from its original form as an undergraduate senior essay by Seth to its present, has been in response to van Fraassen's "Perils of Perrin" paper. Hence we could not avoid having it linked to some degree to the realism-instrumentalism debate within philosophy of science. We nevertheless have endeavored throughout to avoid directly addressing, and thereby risk becoming enmeshed in, the vast philosophic literature on this topic. Ever since Part One of Duhem's *The Aim and Structure of Physical Theory*, if not before, that literature has centered on the question, *What conclusions can be drawn about posited underlying entities and processes from the explanatory successes of the theories positing them*? That question we have done all we can to avoid. Our question, again from Seth's beginning, has been far more narrow: *To what extent, if any, did the converging theory-mediated measurements heralded by Perrin of Avogadro's number and related constants gain experimental access to molecules themselves*? Van Fraassen had argued, *none at all*. We have concluded, *arguably some, yet not so much as many, including Perrin himself, have claimed.*

In particular, the experimental access to atoms and molecules gained through those measurements paled in comparison to the potential for access being gained from spectra following the Bohr model and from x-ray diffraction following the groundbreaking results of von Laue and the Braggs. From our point of view, therefore, our monograph is meant to be just what its subtitle says, a contribution to the literature on the evidential significance of converging theory-mediated measurements, not a contribution to the realism-instrumentalism literature. Hopefully, that serves as an excuse, or at least an apology, for why we have engaged that literature so little.

We shall still not be engaging that literature in this Postscript, or the question we claimed above that it has centered on. Our efforts, however, have led us to think that we can offer some useful suggestions to those engaged in the realism-instrumentalism debate. These suggestions fall into three categories covered in the three sections that follow: on the scientific literature that we found most instructive, on some distinctions we found indispensable, and, at greatest length, on the question of what kinds of scientific results have, on their face, the strongest claim to being permanent.

Brownian Motion and Molecular Reality. George E. Smith and Raghav Seth, Oxford University Press (2020). © Oxford University Press. DOI: 10.1093/oso/9780190098025.001.0001.

P.1. On the Scientific Literature

The writings by physical scientists on the reality of molecules often tended toward polemics after the dispute over "energetics" came to a head following Georg Helm's address at the meeting in Lübeck in 1895. This tendency shows up to some extent in Perrin's *Brownian Movement and Molecular Reality* of 1909, though much more so in the various editions of Perrin's *Les atomes*, which of course was written for the general public. Another reason caution is needed in relying on Perrin's book is that the scientific community had moved beyond his values for Avogadro's number, if not already by the time its first edition appeared, by the time of its first English translation in 1916. That said, in all of his writings that we have emphasized—the October 1908 note, the 1909 monograph, the 1911 paper at the Solvay Conference, and the book—Perrin gives detailed, authoritative accounts of the experimental work done under him.

The problem with relying on Perrin is not with his accounts of his own work, but with his accounts of the measured results for molecular magnitudes published by others. We explained this in section 6.4 of Chapter 6, and hence need not repeat it here. One should always turn to the original publications to assess anything Perrin says about measurements from outside his laboratory. One point from Chapter 6 merits noting again here. Nothing surprised us more, from the time Seth first realized it, than the comparative obscurity of the Boltwood-Rutherford 1911 determination of Avogadro's number. The assumption it required to infer that value from what was actually measured, the number of α-particles that had been collected, seems to us the least adventurous of any of the assumptions required at the time to infer a value of Avogadro's number from what was actually measured—namely, that those α-particles retain their individuality when they lose their positive charge to yield helium. Moreover, of all the determinations of Avogadro's number published between 1900 and 1913, it came closest to our post-1941 value. Of course, since the experiment was not repeated, this may have been merely a fluke. It nevertheless remains an interesting question why it never attracted remotely the attention that Perrin's and Millikan's values have, both then and ever since.

The two Rutherford-Geiger papers of 1908 also merit more attention than they tend to receive for the breakthroughs they made toward the converging measurements achieved over the next few years. First, they determined a rate of α-particle emission that made possible the Boltwood-Rutherford measurement of Avogadro's number. Second, again from this rate of emission, they determined a value for the positive charge per α-particle that, far more than the determinations of e, had strong claim to truly being a count of the number of such charges. Third, as they expressly claimed, their implied value for e, the negative counterpart of half their charge per α-particle, was more to be trusted than any of the direct determinations of *e* preceding it. Their value for *e* from $2e = 9.3 \times 10^{-10}$, especially after it was supported by Regener the next year, became a standard against which to compare other directly determined values, even after Millikan began refining his oil-drop measurements.

Most of the attention Ostwald's *Grundriss der allgemeinen Chemie* has received within the philosophic literature has been confined to the 1909 edition and its 1912 English translation. And even then, most of the attention seems to have been centered on the three citations of Perrin, the most often quoted one from the Preface and the two scattered in the newly added Book V on Microchemistry. Assuming that we are correct that these were all inserted after the manuscript of the new edition was complete, they give a quite misleading impression of the book and the reasons Ostwald offers in it for abandoning his former position toward molecules and the kinetic theory. At least in part because so much of the manuscript had to date from before mid-1908, it precedes virtually all of the progress in microphysical measurement that was made so rapidly after it. As a result, the 1909 edition is of interest only as personal testimony about Ostwald himself and as summarizing the state of molecular science just prior to Perrin's initial publications on Brownian motion and the two Rutherford-Geiger papers.

By contrast, we regard the 1889–1890 editions of *Grundriss der allgemeinen Chemie*, both of which are available in English translation, as far more deserving of philosophers' attention than the 1909 edition. To begin with, even though Ostwald insists on the lack of evidence for the existence of atoms and molecules, his exposition is in terms of them throughout a large fraction of the book. His exposition thus displays their value in conceptualizing a large range of chemical and physical chemical phenomena, or said from another point of view, their "explanatory power." More important than this are Ostwald's frequent excursions in fine print explaining precisely why all the evidence pertaining to a set of phenomena just explicated in terms of atoms and molecules in no way presupposed them—that is, *precisely why the evidence would be the same even if the exposition had eschewed them*. We quoted some passages like this in Chapter 2, but that represented a mere sampling. As we said there, Ostwald was first and foremost an experimentalist. As such, he was able to spell out the specific respects in which nineteenth-century research in chemistry and physical chemistry had fallen short of gaining experimental access to atoms and molecules themselves, however close it sometimes appeared that it had.

The one area of research involving molecules which the 1889–1890 editions of Ostwald's *Grundriss* do not cover in any detail is the kinetic theory of gases; they simply present the theory and its account of phenomena in gases in a brief seven pages early in the book. The second, 1899 edition of Meyer's *Die kinetische Theorie der Gase*, translated that same year into English, is remarkable for the extent to which Meyer—in his own words—"collected together, as completely as I could, and summarized, the observations by which the admissibility of the theory might be tested and its correctness proved."[1] It is a compendium of the diverse efforts during the second half of the nineteenth century to leverage molecular-kinetic theory into means for gaining experimental access to molecules and their motions. Our citations of it in Chapter 2 do not begin to show the full range of the efforts to do this covered in the book. It was unique among all the expositions of kinetic theory well into the twentieth century in this regard—which is why others at the time, including Ostwald, Nernst, and Jeans, repeatedly cited it. Philosophers arguing about the extent to which phenomena covered by kinetic theory had provided support for the reality of molecules before 1905 have been derelict in how little attention they have given to it.

Because Meyer's book considers only phenomena covered by kinetic theory, it includes nothing on chemical phenomena and the evidence they were providing for the reality of molecules. Nernst's *Theoretische Chemie vom Standpunkte der Avogadroschen Regel und der Thermodynamik*, provides a more summary account of the evidence associated with kinetic theory in its 1893 edition, though the results it cites are in Meyer's 1899 book as well; and in the editions after 1900, Nernst cites Meyer's book as a source of more details. The special virtue of Nernst's book is that it is the one comprehensive exposition of all the evidence bearing on molecules, from chemistry and physical chemistry as well as from physics. Its first edition ran to almost 700 pages, the 1909 edition, to just short of 800, and the 1921 edition, to more than 900. Nernst as a trained physicist was always more favorably disposed to the molecular hypothesis than most chemists, yet from 1887 on he was at the cutting edge of experimental work in physical chemistry, and after Einstein's 1907 paper on specific heats, no less at the cutting edge of low-temperature experiments on specific heats—not to mention, of course, his contribution to thermodynamics, his "new heat theorem" of 1906, sometimes called "the third law" of thermodynamics.[2] Moreover, perhaps in part because his first edition was published as a complement to Ostwald's *Grundriss*, Nernst appears to us to have at all times been meticulous about when the evidence bearing on molecular theory was merely supportive of it versus when it had legitimate claim to having gained experimental access to the molecular realm.

As noted in the Appendix, Nernst's *Theoretische Chemie* appeared in 15 editions, the first in 1893, the second in 1898, and the remainder over the course of the first 26 years of the twentieth century. Five of these were translated into English, the first in 1895, followed by editions in 1904, 1911, 1916, and 1923. We consulted the first, third, and fifth of these in writing this monograph, and in the case of the 1911 and 1923 editions, we limited our attention mostly to material related to the measurements of molecular magnitudes. As far as we can tell, all editions have gone virtually unnoted in the recent philosophic literature on molecular reality. Yet, even the English translations alone provide the most thorough picture of how the evidence bearing on molecular reality developed from the time of van't Hoff's and Arrhenius's breakthrough results on solutions in the 1880s to the 1919 edition of Sommerfeld's *Atombau und Spektrallinien*. Moreover, no one was sociologically more in the middle of all these developments than Nernst, both before and after he instigated the first Solvay Conference.[3] Surely his *Theoretische Chemie* merits more attention than it has received from philosophers arguing about whether, when, and how the reality of molecules was established.

P.2. On Some Key Distinctions

Our monograph has relied heavily on two distinctions that are largely absent from the realism-instrumentalism debates, while trying to skirt around a third distinction (involving "approximate truth") that has been widely invoked in post-Laudan defenses of realism.[4] Here we shall explain why we think our two distinctions can make those debates more focused, even if not more readily resolvable, and why we think the third distinction is best replaced, to the same end. This will prepare the way for our thoughts on how to respond to Larry Laudan's challenge in the section that follows this one.

P.2.1. On Entering Constitutively into Research

The first distinction is between a proposition entering constitutively into further research—or, more specifically, into specific results of that research—and its not doing so. As noted in section 7.5 of Chapter 7, our distinction is at least akin to Poincaré's distinction between indifferent hypotheses and real generalizations, where, in the case of the former, results of calculations remain the same whether they are adopted or not. For us, a proposition enters constitutively into a research result just in case either the statement of or the evidential reasoning supporting the result indispensably presupposes it. Put another way, the result ceases to be warranted by the putative evidence without the proposition in question. Correlatively, a proposition does not enter constitutively into a result if it can be rejected or eliminated while leaving the result and the evidence for it intact.

Insofar as propositions can also motivate an experiment or serve heuristically in its design, careful logical analysis is often needed to determine whether any one proposition has or has not been constitutively presupposed by it. Chapter 4 has offered an example of such reasoning by showing how all of Perrin's results on Brownian motion itself did not constitutively presuppose the reality of molecules. To give just one example of why analysis was needed to conclude this, a proposition common to Einstein's derivation of his Brownian motion equation for ξ^2/τ and Perrin's experimental design testing it is that the motion of the granules is Gaussian insofar as Gaussian motion of molecules is producing it. Chaudesaigues, however, working under Perrin in the development of the experiment in question, took the trouble to confirm that the granule displacements are Gaussian, thereby eliminating any need to invoke molecular theory in the design of the experiment. A review of Chapter 4 will show that Perrin and his cohort similarly took the trouble to develop direct evidence, independent of molecular theory, for each proposition presupposed in the experiments yielding his Brownian motion results.

Readers might ask how we concluded in Chapter 5 that those results provided strong evidence for the kinetic theory of heat, yet not for the reality of molecules. Recall that what we concluded is that the evidence in question confirmed the existence of individual localized pressure fluctuations, on a scale notably smaller than the granules, that are in no way correlated with one another, and thus independent of one another and hence discrete. We had to invoke principles of the continuum mechanics of fluids to reach this conclusion, but not the existence of molecules. Granted that such pressure fluctuations are entirely consistent with molecular-kinetic theory and that it can therefore explain them; for that matter, it is not clear how else they may be explained. The conclusion nevertheless stands without requiring any recourse to the existence of molecules.

A handful of other examples shows how widely applicable the distinction is, beyond anything discussed in our monograph. The ether never entered constitutively into any research in optics or electromagnetism, however instrumental it was heuristically. Neither Descartes's nor Leibniz's vortices ever entered constitutively into orbital research. Leibniz's vortex theory was expressly designed to explain Keplerian orbits, but it offered no help in addressing either comet trajectories or the deviations from Keplerian motion exhibited by our Moon, Jupiter, and Saturn. It disappeared historically when research in which Newton's law of gravity was a central constitutive element handled not only all of these, but over the next century and a half so many others, in case after case revealing just what factors in the world, often quite subtle, are responsible for specific deviations from the Keplerian ideal. It entered constitutively even into the 43 arc-seconds per century residual in the precession of the perihelion of Mercury for which it failed to reveal a source, for the 43 arc-seconds are the result of subtracting from the observed 574 arc-seconds precession the 531 arc-seconds associated with the Newtonian actions on Mercury of all the other planets. In other words, Newton's law of gravity entered constitutively even into the only evidence Einstein initially had for general relativity, which helps explain why it was so important to him for Newton's law to hold in an asymptotic limit in our planetary system.

This last example underscores the need for careful logical analysis in assessing whether a proposition is entering constitutively into further research. A good example of such analysis is Duhem's producing an entire book, expanding on his dissertation, detailing how none of the evidence for Maxwell's equations constitutively presupposed the ether.[5] Whether a proposition, as stated, enters constitutively into an experiment may become clear only at a later time. Anyone at the time would have said that the proposition, *the constituents of cathode rays are charged particles*, entered constitutively into J. J. Thomson's 1897 measurements of their mass-to-charge ratio; and anyone at the time would have taken this proposition to entail that these constituents do not exhibit wave-like properties. Yet when Thomson's son and Davisson demonstrated the diffraction of electrons three decades later, the earlier measurements were not invalidated, for what they had presupposed, in effect, was that the constituents respond in the manner of charged particles to electric and magnetic fields and upon impact with a solid. The distinction between heuristic and ineliminable constitutive elements presupposed in an experiment can be quite subtle.

The example of Newton's law entering constitutively into Einstein's evidence brings out the most important reason for insisting on the distinction. When a proposition enters constitutively into ongoing research, it will in general continue to be tested, if only tacitly and *en passant*, by that research. Moreover, as the history of research in celestial mechanics shows, this can result in increasingly stringent tests of it—as, for example, the 8 percent discrepancy *per century* in the precession of the perihelion of Mercury. We saw an example of such a stringent test in this monograph with Bohr's needing to presuppose a value for e in order to test his calculated value for the Rydberg constant against the known value. Precisely because e entered to the fifth power in the calculation, this test of Bohr's model *en passant* provided the first stringent test of the value for e, in the process historically giving the first clear evidence that Perrin's value for it, and hence for Avogadro's number as well, was not to be relied on. How stringently any experimental result tacitly, or *en passant*, tests any proposition constitutively presupposed by it requires close analysis of the design of the experiment and of the evidential reasoning associated with it. Some such indirect test is nevertheless always occurring.

Of course, ongoing research can also yield evidence for propositions not constitutively presupposed in it. The chemical phenomena that led to the introduction of structured formulas and the optical results that led to van't Hoff's stereochemical proposals added significantly to the explanatory possibilities of atoms and molecules. Those new results, however, did not test atomic or molecular theory precisely because, as we had Ostwald explain in his own words in Chapter 2, they did not require it. Van't Hoff's subsequent discovery that the constant of proportionality associated with osmotic pressure in non-electrolytic solutions matches the gas constant provided clear support for molecular-kinetic theory by showing that its capacity to explain phenomena of pressure in gases almost certainly extends to liquids as well. We acknowledged as much by titling the section of Chapter 2 presenting van't Hoff's results for non-electrolytic solutions "A Major Development in Support of the Molecular Hypothesis." Nevertheless, here again his research on osmotic pressure did not constitutively presuppose the molecular-kinetic theory of pressure, and hence it did not test it. At least in regard to whether these advances in chemistry and physical chemistry actually tested the atomic-molecular or molecular-kinetic hypotheses, therefore, they can legitimately be said to fall in the category of Poincaré's "indifferent hypotheses." Again, however, we stress that the logical role of any proposition in any ongoing research calls for careful analysis.

One of the central, if not *the* central, issue in the realism-instrumentalism debates has concerned what conclusions can be drawn from the explanatory successes of a theory. We do not think of ourselves in any way engaging this issue in arguing for our distinction between propositions that do and do not enter constitutively into ongoing research. The most that we are suggesting is that this distinction might add a further, hopefully clarifying dimension to some of the disputes on the issue. It brings with it, as a corollary, a distinction as well between a claim being believed without its truth or falsity making any substantive difference to the empirical content of a science and a claim, whether believed or not, that does make a substantive difference to the empirical content by virtue of research in that science constitutively presupposing it.

That molecules exist *did* enter constitutively into the research reviewed in Meyer's 1899 *Die kinetische Theorie der Gase* that was trying to use kinetic theory to obtain measured values of their sizes. The specific proposition which that research constitutively presupposed was that molecules have a characteristic dimension s, at least corresponding to a diameter of a sphere, that varies from one kind of molecule to another and that is a dominant factor governing the variations from one kind of gas to another of their respective deviations from the ideal gas law, their viscosities and thermal conductivities, the rates of diffusion within them, and their dielectric capacities. (We have here tried to state the proposition in the minimal form in which it was indispensable to the evidential reasoning employed in the efforts to derive values for s from the phenomena in question.) As Meyer acknowledged, and as we summarized in Chapter 2, the values for s from the five different approaches did not converge to a sufficient extent to allow anything more than an order-of-magnitude estimate for s, 10^{-8} cm, for molecules generally. Worse, within any one of the approaches, the values obtained for s for different kinds of molecules failed to meet the cross-check of yielding a single common value for Loschmidt's number—that is, the number of molecules per cubic centimeter at standard conditions—and hence too for Avogadro's number.

The upshot of the research, repeatedly acknowledged by Meyer, was doubt, if not outright rejection, of the claim that any such *s* is physically meaningful.

Suppose, however, that the research by 1900 had yielded convergent values for *s* from at least most of the alternative approaches and that the cross-checks against the constancy of the implied values for Loschmidt's and Avogadro's numbers had been comparable to the 1912 range of ±8 percent. The conclusion would surely have been that molecules do have a characteristic dimension *s* corresponding to the diameter of a sphere—whether a dimension of the molecules themselves or, following van der Waals, of their "spheres of influence." Moreover, the values of *s* in question would have been taken as a fixed constraint in continuing research on such questions as (in Nernst's words) what "binds atoms together in molecules and hinders their flying apart." In other words, the values of *s* in question would have been seen as sufficiently well authenticated to be adopted constitutively in ongoing research on such questions. And, with that, the existence of molecules also would have acquired a standing of a very different sort than it had had before—or that it, as a matter fact, had in 1900.

P.2.2. On Taking to Be True

The standing in question is at the center of the second distinction that we think may be of some benefit in the debate between realism and instrumentalism. It is a threefold distinction among (1) a proposition, whether constitutively presupposed in ongoing research or not, continuing to have the status of a hypothesis; (2) a proposition being taken to be true and granted a presumption of inviolability in ongoing research predicated on it; and (3) a proposition that is accruing evidence and thereby becoming increasingly entrenched as a result of continuing to be tested, *en passant* yet often stringently, in conjunction with research predicated on it. The latter half of the distinction provides a response to historians and sociologists who challenge the epistemic authority of the sciences by stressing how often their most fundamental principles have been accepted on the basis of little or no evidence. The principle of inertia, for example, seems to have become a fixture well before Newton's *Principia* on the basis of nothing but Descartes's sling; but it became indirectly tested, first by Huygens's various measurements involving conical pendulums, and then by all the Newtonian determinations of celestial forces. The response generally to these historians and sociologists of science is to turn the issue of epistemic authority into one concerning the extent and stringency of the *continuing* tests, usually over decades, and sometimes centuries.

The main benefit of the latter half of the distinction for the realist-instrumentalist debate is also, we think, to shift the issue to the long-term evidence accruing to a "realist" assumption by virtue of the research constitutively presupposing it. We fully grant that this shift will scarcely end the disputes; it can instead focus the debate on just how indispensable "realist" claims have been to the results achieved in the continuing research.

The potential benefit for the realist-instrumentalist debate we see from the first half of the distinction is also one of refocusing. This debate often centers on what it takes to close questions like, *Do atoms and molecules exist?* once and for all—where by "once and for all" we mean the way in which perception can generally close questions like *Is there a tree in that backyard?* Sometimes the issue is whether specific evidence, or whether for that matter any evidence, can close such a question. After Worrall's "structural realism" paper in 1989[6] the issue has instead often been how an answer to such a question needs to be qualified in order for evidence to close it once and for all. The middle alternative in our threefold distinction offers a further option in these disputes: *taking* a question to be closed and predicating further research on its answer "until yet other phenomena make such propositions either more accurate or liable to exceptions." When the research predicated on it proves to be highly successful and no phenomena have yet to emerge countermanding the results of that research, a retrospective impression can be created to the effect that the question had actually been taken to be closed once and for all at the time of initial community-wide acceptance of an answer, when in truth it had then been taken to be closed only provisionally—specifically for the purpose of pursuing research along a path not available without it.

Our quoting from Newton's *Principia* is meant to remind readers that we have taken our threefold distinction from him. Our reason for emphasizing it in the body of our monograph was not because it offers a new option to those engaged in the realism-instrumentalism debate, but because it gives the correct account of the new standing the atomic-molecular and molecular-kinetic hypotheses had come to have within the scientific research community by the end of 1912. Specifically, when values for the number of molecules in a mole, their mean translational kinetic energy per degree Kelvin, and the number of unit charges required to produce a mole of monovalent gas in electrolysis had been established to sufficient precision that they were being taken for granted in ongoing research, the questions of whether molecules and unit charges exist had come to be taken as closed. While our monograph has argued for such a standing only in this specific case, we submit that its like has occurred repeatedly in the history of science.

The phrase "take to be true" alone is enough to indicate that the evidence required for a proposition to acquire this intermediate standing is something short of evidence that it is true. The point in granting it is to promote further research, yet presumably in a way that will to some extent continue to test it in the process. One requirement on the evidence, therefore, is that constitutively presupposing it in ongoing research will enable new evidence to be brought to bear on open questions to a greater extent than without it. The more important requirement, however, is evidence to the effect that results established in the new research which presuppose the proposition in question not be nullified should the research turn out to give reasons to revise it. The evidence must therefore give grounds at least for there being something correct about those respects in which the proposition is most indispensable to any new results predicated on it. Newton appears to have expressly rejected the viability—even the apparently unique viability—of a hypothesis's capacity to explain phenomena as adequate. The qualifying phrase he used was "propositions gathered from phenomena by induction," which on its face requires that the propositions have been shown to hold at least over a range of specific cases and hence not be purely hypothetical.

Newton's further phrasing—literally "until yet other phenomena are to have been come upon by means of which they would be rendered either more accurate or liable to exceptions"—suggests that a requirement for propositions to be taken to be true is that they be at least approximately true. The distinction invoked here, between being true and being approximately true, is the one we think is best replaced in the realism-instrumentalism debate. "Approximate truth" is an oxymoron and hence at best a metaphor. Propositions are either true or false. The conclusion Planck stated in 1913 that the fundamental unit charge is $(4.5 \pm 0.4) \times 10^{-10}$ esu was not merely approximately true.[7] Worse, as used in the realism-instrumentalism literature, "approximate truth" is much too vague to help clarify what requirements the evidence for a proposition taken to be true should meet in order to obviate the risk of results predicated on it being nullified should it subsequently turn out not to be true. What is needed in place of "approximate truth" is consideration first of the ways in which a proposition with sufficient support for it to be taken to be true can turn out not to be so, and then of provisions under which results predicated on it can nevertheless remain in place.

(Newton himself had initially adopted a threefold adverbial distinction—*exacté, accuraté*, and *quamproximé*—only to abandon *exacté* in connection with measurable quantities in the *Principia* itself.[8] He uses these adverbs to modify not the word "truth," but the entire assertion made in clauses, as in "If the areas are proportional to the times *quamproximé*, then the remaining force will tend toward the body T *quamproximé*."[9] The clause, so qualified, is itself either true or false. His main concern, in other words, was to assure that valid—that is, truth preserving—inferences could still be drawn from the antecedent to the consequent of such conditionals, even when he had decisive grounds for thinking that the empirical world deviates from it in some systematic, yet limited way.[10] Our talk of "provisions under which results predicated on it can nevertheless remain in place" is in this same spirit, even though the fact that the risks involved are in connection with inferences is not so apparent.)

How then can such a supported proposition turn out not to be true? One way, just as Newton said, is when subsequent results indicate that it holds to less precision than the original evidence for it revealed. Famous examples of such imprecisions are the not-yet-called "ideal" gas law, as revealed by Regnault's meticulous measurements in the 1840s, and Millikan's universally accepted 1917 value for e, $4.774 \pm 0.005 \times 10^{-10}$ esu, as revealed a decade and a half later by x-ray diffraction determinations of Avogadro's number.[11] Experimental results that presuppose any such proposition are at risk only if they depend on its having a level of precision tighter than it turns out to have. Any result not requiring it to have so much precision will remain intact, and even results requiring this can often be adjusted instead of having to discard them. An example of the latter is Huygens's measured results into the fourth significant figure for the strength of surface gravity. These presupposed uniform gravity acting along parallel lines, required as well in order to guarantee isochronism of the pendulums by means of which the measurements were made; Newton then validated the measured results by showing that they can still hold in the case of his universal gravity in an asymptotic limit as the ratio of the length of the pendulum used to the radius of the earth approaches zero.[12]

The risk of having to discard results because of limited precision can be obviated to some extent by requiring that the evidence leading to the proposition being taken to be true give clear grounds for its holding to reasonably high precision, or even better, that it provide specific bounds on precision. The marked reluctance to predicate any research on the values of the molecular magnitudes obtained prior to the 1908–1913 measurements is a sign that this requirement was being respected in practice. Keep in mind that very few in the physics community at the time had been showing any such reluctance to believe in the atomic-molecular and molecular-kinetic hypotheses.

A second way in which a proposition taken to be true can turn out not to be so is when subsequent results show that the domain over which it holds, to whatever precision, has to be markedly circumscribed. The most famous example of this is the principle Newton was insisting should be taken to be true when he enunciated his rule of reasoning, namely his law of gravity. This example, along with Maxwell's laws, was surely what Glashow had in mind when he wrote, "A physical principle cannot be said to be true until it has been shown to be false and its envelope of applicability precisely delineated."[13] As long, of course, as what new results presuppose are instances of the proposition that lie within the domain in which it holds to an appropriate level of precision, they will remain intact when that domain subsequently has to be circumscribed. The problem lies in the lack of any way to characterize, except in retrospect, the boundaries of the domain over which the evidence licenses the proposition to be taken to be true, and hence as well the boundaries of the domain within which results from the research predicated on it are least at risk. One way to obviate the risk is to require the evidence licensing the proposition to be taken to be true to have a wide range itself. Another is to look for complementary evidence for the results—that is, evidence for them that is less dependent on the evidence for the presupposed proposition—and ad hoc ways of cross-checking the respects in which it is dependent.[14]

Perrin's Brownian motion research included two instances of initially putting forward a conclusion, with no restriction on its "envelope of applicability," on the basis of extremely limited evidence. One was his conclusion that the mean translational kinetic energy of granules varies linearly with temperature. His evidence for this consisted of a single vertical-gradient test involving a change of temperature of less than 50 degrees Kelvin.[15] This alone gave little license to conclude, after taking the granule kinetic energy to match that of the molecules in the liquid, that α, the mean translational kinetic energy per degree Kelvin of molecules, is a constant. This was of no importance historically, for blackbody radiation measurements confirmed that $k = (2/3)\alpha$ is constant over a wide range of temperatures.

The other instance was Perrin's proposing, on the basis of the match between his measured mean translational and rotational kinetic energies of granules, "*In this, I think, will be found the most extended verification up to this day of the equipartition of the kinetic energy of translation*."[16] The quotation, in italics in the original, is from his 1909 monograph. He presents the rotational results and their match with the translational in his 1911 Solvay Conference paper, but nothing akin to the proposal of their verifying equipartition of energy. This is scarcely surprising insofar as one of the two central concerns of the Conference was the "freezing out" of degrees of freedom at low temperature and its implications for equipartition of kinetic energy.

A third way in which results predicated on a proposition taken to be true can turn out to require reconsideration is less widely discussed in the literature and hence requires a little more in the way of explanation. The proposition can come to be regarded as having been all along merely a surrogate for another proposition, usually stated in different terms, that for one reason or another is to be preferred over it—higher precision, wider range of applicability, greater compatibility with general theory, or combinations of these. To illustrate what we mean by "surrogate," consider the second law of thermodynamics.[17] It was initially formulated simply as "*Heat never flows from cold to hot*." Following Clausius's introduction of the term, this was replaced by "*The change in entropy is always greater than or equal to 0*," where *change in entropy* was taken to be an integral over a reversible process. Both of these amounted to surrogates, as we are using the term, for the twentieth-century statement of the law in terms of the probabilistic conception of entropy laid out by Boltzmann and augmented by Nernst's "new heat theorem."[18]

While this example illustrates what we mean by "surrogate," it does not display the risk attached to the possibility of a proposition taken to be true turning out to be merely a surrogate for a further proposition. For this we turn to the proposition that the constituent common to cathode rays, photoelectric and thermionic emission, and beta radiation—that is, the electron—is a charged particle. All the experimental determinations of the charge-to-mass ratio of these constituents, from Thomson's 7.7×10^6 abs emu in 1897 to Neumann's 1.765×10^7 in 1914,[19] presupposed them to be charged particles. So too did Bohr's model of the hydrogen atom presuppose that the electron is a charged particle, specifically a well-demarcated particle in orbit around the nucleus. None of the charge-to-mass ratio measurements had to be discarded, or even adjusted, following the conclusion that the electron is, as they say, a wave-particle, but some of the results that presupposed electrons are well-demarcated particles orbiting the nucleus did.

The original evidence on the basis of which the charged-particle hypothesis had come to be taken to be true was the response of cathode rays in magnetic and electric fields, the accumulation of charge and the increase in temperature when they impact a collector, as well as their having a measurable mass-to-charge ratio. Further evidence accrued to the proposition with each successful, increasingly precise measurement of this ratio for cathode rays and other discharges as well, especially after correction to the raw measurements introduced with the shift from *mass* to *rest mass* shortly after 1900.[20] None of this evidence, however, entailed anything about electrons in stable orbits bound to a nucleus, or for that matter, electrons anywhere except in discharges.

We intend this example to illustrate that there is a form of risk in taking a proposition to be true and predicating research on it, in addition to the risks from limitations in precision and range of applicability it may turn out to have. When the proposition turns out to be merely a surrogate for a proposition stated in different terms, some of the logical implications it had in its original statement may have to be abandoned. Any experimental results that presuppose not just the proposition taken to be true, but elements implied by the specifics of its statement, may not survive its replacement. A way of obviating this risk is to be meticulous, when predicating research on a proposition taken to be true, about which aspects of it the evidence for it specifically supports and which aspects reach beyond that evidence. Another way to express this caution is, be careful about respects in which the proposition is serving merely heuristically in contrast to its constitutively enabling a result. As we remarked in section 7.5 of Chapter 7, this distinction is sometimes more readily drawn in retrospect than while the research is unfolding.

These three forms of risk are not mutually exclusive. One can argue that Newton's law of gravity, for example, fell short not only in its holding to more limited precision, and that over a more highly circumscribed domain, than had been thought for two centuries, but also in its turning out, by virtue of its statement in terms of *force* akin to all other kinds of force, to be merely a surrogate for a more discriminating characterization of gravity. Needless to say, it is commonplace in the realism-instrumentalism literature to say that Newton's law is nevertheless "approximately true," and this is why so much of the research predicated on it remained in place after it gave way to general relativity.

Even if sense can be made of "approximate truth"—which we doubt—invoking it in this way is uninformative. Which results predicated on a proposition taken to be true will likely survive its replacement and which will not depends on (1) the specifics of how it enters into the different results that presuppose it *in relation to* (2) the specifics of the evidence supporting it, including both the evidence on the basis of which it was taken to be true in the first place and any further evidence that has accrued to it from its having continued to be tested, *en passant*, in the research predicated on it. *No* blanket statement of the requirements a proposition should meet before it is taken to be true for purposes of ongoing research can capture the nuances of the different ways in which such research is at risk, much less statements like "it has to be at least approximately true" or "it has to be approximately true in the right ways."

We have so far been focusing on the risks in predicating research on a proposition taken to be true when the evidence for it is limited. The most compelling reason why no blanket statement can be given of requirements a proposition should meet to obviate the risks in taking it to be true is that the risks always have to be considered in relation to the potential gains from doing so. The potential gains always concern research that would not be feasible without it. This can consist of little more than enabling a test of some novel hypothesis. An example is Bohr's taking the values of the charge and charge-to-mass ratio of the electron for granted in order to compare the value of the Rydberg constant implied by his model for hydrogen with the known value. The potential gain can also consist of being able to determine values for various quantities that otherwise were resisting measurement. Examples are Planck's contention in 1900 that taking the coefficient of temperature in his blackbody formula to be what came to be known as Boltzmann's constant allowed the values of a whole host of molecular magnitudes to be determined to an accuracy "much better than all determinations up to now," and Perrin's 1908 contention to the same effect made possible by taking the mean translational kinetic energy of granules in Brownian motion to be equal to that of molecules in the liquid.

A still broader potential gain can consist of enabling already accessible data to be turned into much richer evidence than had been possible before. A striking example is Bohr's proposal that the lines in spectra correspond to quantum transitions from one stable energy state of an atom or molecule to another. Taking this proposal to be true enabled 80 years of spectra, which theretofore were being used at most as "fingerprints" for identifying their chemical sources, to be turned into evidence for atomic and molecular structure. The Bohr model is an especially striking example of the extent to which potential gains can outweigh the risks in taking a proposition to be true and granting it a presumption of inviolability. For, even though the experimental evidence for Bohr's revolutionary claims was initially so limited, the risks seem to have been historically outweighed by their potential for bringing an immense amount of evidence to bear on questions about atomic and molecular structure that had, until then, been purely matters for conjecture.

The Bohr model illustrates as well how research can indirectly address the risks arising from its presupposing a proposition supported by limited evidence. As we have noted, Bohr's testing his model against the Rydberg constant provided, *en passant*, a far more stringent test of the claimed values for the charge of the electron than had been possible before 1913, yielding clear evidence that Perrin's value could not be relied on, and hence neither could any of Perrin's values for the other molecular magnitudes. Finding ways for ongoing research to provide a stringent *en passant* test of propositions it is presupposing may be the most effective way to limit the range of results that might have to be discarded at some later time.

The main point of this entire section of our Postscript has concerned the virtues of recognizing an intermediate standing that a claim can have in scientific practice between that of a mere hypothesis that can explain so much and that of an entrenched principle supported by a large amount of accumulated evidence. Specifically, it can be taken to be true and employed constitutively in ongoing research on the basis of limited evidence that nonetheless shows that doing so (1) has promise of major advances in being able to develop more telling evidence and (2) has means of obviating the risks of all that research subsequently having to be discarded. We contend that a failure to recognize this intermediate standing has led to much confusion not just in the realism–anti-realism debates, but in the philosophic analysis of acceptance of claims within scientific communities generally.

Our immediate intent in this section, however, has been just to suggest a way for narrowing the gap in realism–anti-realism disputes. Realists, for example, can offer the proposed intermediate standing in response to anti-realist complaints that the only evidence for the existence of molecules as of 1913 required inference to the best explanation; and anti-realists can offer it in response to realist appeals to the converging values for Avogadro's number by reminding all of Nernst's words in 1921, "At present scarcely anything definite is known either of the forces which bind the atoms together in the molecule and which hinder them from flying apart in consequence of heat motion, or of their laws of action."[21]

P.3. On What Has Claim to Being Permanent

We noted in the preceding section that the value for the charge-to-mass ratio of the electron, once Lorentz's velocity correction was included, became comparatively well-defined, ranging from Kaufmann's 1.77×10^7 abs emu early in the century to Neumann's 1.765×10^7 in 1914.[22] This value changed less than 1 percent from Kaufmann's value through the advent of first the old quantum theory and then quantum mechanics and finally quantum field theory; Birge's recommended value in 1941 was $1.7592 \pm 0.0005 \times 10^7$, while the 2014 CODATA recommended value is 1.758820024×10^7.[23] One might say that, at least within tight bands of precision, this value has a clear claim to permanence. Much the same can be said for the measured values of the acceleration of gravity in Paris, from Huygens's 1673 value (in modern form) of 980.7 cm/sec^2, obtained through converging measurements from small-circular-arc and conical pendulums, to the value listed in 1958, 980.970, a change of less than 0.03 percent.[24] Huygens's value, of course, presupposed uniform gravity acting along parallel lines, not Newtonian gravity. When he put it forward he considered it a constant for the entire earth, but within a decade and a half he had concluded that the acceleration of gravity varies from one latitude to another; and we have now determined that it varies a little with time as well, in part because of seasonal and long-term redistributions of matter near the surface of the earth. Our point, however, is that the value measured at Paris has remained the same, within tight bands of precision, even though the theory of gravity has changed.

Even more notable in this regard are Ptolemy's measured values for the eccentricities and radius-ratios—that is, the ratio of the radius of his deferent to that of his epicycle—for the outer planets. Ptolemy presupposed his epicyclic model to infer values for these orbital elements from patterns in their irregular motions. Put in modern form, his values for the radius ratios inverted were 1.52, 5.22, and 9.23 for Mars, Jupiter, and Saturn, respectively, and his values for the corresponding eccentricities were 0.1, 0.046, and 0.057.[25] Kepler's 1627 values for these were 1.523, 5.20, and 9.51, taken to be the mean distances of the three in units in which the mean distance between the sun and the earth is 1.0; and 0.093, 0.047, and 0.057, taken to be the eccentricities of ellipses instead of Ptolemy's eccentric circles.[26] Our current values are 1.5236, 5.2026, and 9.5219, and 0.0934, 0.0485, and 0.0532.[27] Needless to say, both the meanings of these numbers and the theory presupposed in deriving them from observations changed radically from Ptolemy to Kepler, and to a lesser extent from Kepler to now. Yet the specific values obtained by Ptolemy, though more limited in precision, have more claim to correspond to current values than J. J. Thomson's original 1897 values for the mass-to-charge ratio of the electron, 0.77×10^7 and 1.17×10^7 abs emu, have.

We could give still more examples from the history of theory-mediated measurements in which the values have remained the same once allowance is made for increasing precision, yet both their interpretation and aspects of the mediating theory have changed. Obvious questions are why a theory-mediated value should ever persist in spite of theory change, what conditions have to be met for this to occur, and what significance it should be given. These questions, however, seem far removed from the realism-instrumentalism debate. So, first we should review why a claim any scientific result has to permanence has become important in this debate.

Larry Laudan's "Confutation of Convergent Realism" made this issue central to the recent debate through a series of historical examples challenging whether *any* evidence for a claim at the time can ever assure that it will not be rejected in the future.[28] This line of argument goes back at least as far as Chapter 3 of Part I of Duhem's *The Aim and Structure of Physical Theory*, where he argued that no form of experimental evidence can ever assure that claims about posited entities and processes underlying phenomena will not be discarded in the future unless the existence of the entities and processes can be confirmed by our senses.[29] Much has been written, both before and after Laudan, about the arbitrariness of Duhem's demand for confirmation by the senses. We agree with the literature on this point—indeed, throughout this monograph we have resorted to the phrase "gaining experimental access to" as providing a principled alternative for what Duhem was getting at, namely the widespread insistence within the scientific community at the time he was writing that the existence of the ether, and of atoms and molecules as well, amounted to nothing more than a hypothesis put forward in an effort to provide various phenomena with mechanical explanations. We nevertheless think that the general rejection of demands for sensible access has obscured an important element of legitimacy in the appeals that anti-realists, especially van Fraassen, continue to make to the primacy of perception. We suspect, further, that it is this element that makes questions about permanence unavoidable in the realist-instrumentalist debate.

Whatever else is to be said about perception, it does have *prima facie* claim to being able to establish some conclusions once and for all. Stated more precisely, perception *can* eliminate virtually all possibility of future information overthrowing a conclusion based on it. (The word "virtually" is included here in part to remove skeptical arguments that challenge the possibility of *any* empirical knowledge from consideration.) This is not to deny that perception can be misleading, or that perceptual judgments can turn out later to be false. We have so many means at our disposal for addressing such possibilities, especially collectively, that perceptual judgments nevertheless provide an exemplar of permanence in the case of empirical claims.[30]

The challenge posed by anti-realists centers on whether any claims in science can match perceptual judgments in this regard. Again put more precisely, *the challenge is whether any evidence for propositions in science can ever eliminate virtually all possibility of future information refuting them.* We submit that the charitable way to construe anti-realists' insistence on perceptual access to underlying entities and processes is to take them as invoking perceptual judgments as *the* exemplar of permanence in the empirical realm and posing the challenge of whether any propositions in science can match them. If we are right about this, arguments that perceptual access is an unfounded arbitrary requirement, however legitimate they may be in their own right, fail to come to grips with what gives lifeblood to anti-realism.

P.3.1. On Levels of Representation

With this as context, let us return to the questions we posed two paragraphs ago about theory-mediated values persisting through theory change. Central to any theory-mediated measurement is the formula relating the less accessible target quantity to the more accessible quantities in terms of which the measurement is being effected. We are calling such formulas *measures*. For purposes of illustration here we shall use two measures from Chapter 4 of the mean translational kinetic energy W of granules in Brownian motion at a given temperature:

$$\ln\left(\frac{n_0}{n}\right) = \frac{2\pi a^3\left(\rho_{gran} - \rho_{fluid}\right)g}{W}h$$

where the numerator involves the buoyant weight of the individual granules and h is the increment in the vertical distance from the plane for which the number of granules is n_0; and

$$\xi^2 = \tau\frac{2W}{9\pi a\zeta}$$

where ξ is the root-mean-square of the displacements of the granules over the time increment τ, ζ is the viscosity of the liquid, and a the radius of the granules. Notice that the only variable quantity these two measures have in common is the radius of the granules, which Perrin made every effort to assure was uniform across all granules in any one experiment. (We have chosen measures of W at a given temperature for purposes of illustration instead of W_T, the mean translational kinetic energy per degree of absolute temperature, because Perrin's efforts to verify that W varies linearly with temperature were so limited, and the range of temperatures feasible to test this in the case of solutions is so small.)

In an effort to avoid begging questions, we are going to analyze what enters into theory-mediated measurements in terms taken from van Fraassen's *Scientific Representation*. He would of course say that what we call a *measure* is itself a *representation*, where his point in using this term is that it resembles what it represents in some respects, but does not resemble it in others. To obtain values for the target quantity in any measurement requires observations, but he would be quick to point out that the values of the accessible quantities employed in any one determination of a value for the target quantity are rarely the observations themselves. We follow him in saying that at least two usually distinct levels of representation of the observations need to be distinguished in order to characterize the values of the accessible quantities used in a measure. The first consists of what scientists call *data*, values obtained from individual observations, but generally expressed in terms not accessible to observation. The magnitude of electric current in a given conductor, for example, is never accessible to simple observation, yet is taken as data all the time. The same is true of viscosity. In the case of the two measures we have highlighted, Perrin inferred the radii of granules in different indirect ways because he could not simply put them against a scale;[31] nor could he simply count the number of granules at any height owing to their motion, but had to resort to a series of photographs.[32]

Van Fraassen invokes Patrick Suppes's much neglected concept of a *data model* as a further level of representation of observations, or, perhaps more accurately said, a representation of the data.[33] In the case of our second measure, for instance, the data for displacements ξ have to be squared and then averaged since the measure calls for a mean value over each time interval. Similarly, both measures require a single value for the radius of the granules to be extracted from the values Perrin obtained indirectly for them in more than one way, and even with all his efforts, the uniformity of his granules is not a mere datum, but part of his data model. Suppes's original portrayal of data models was quite broad, allowing for a range of different ways "to incorporate all the information about the experiment which can be used in statistical tests of the adequacy of the theory."[34] Our appeal to the concept here is narrower: data generally have to be distilled into a form appropriate to the use they are to serve, in our case to provide values for more readily accessible quantities in a theory-mediated measurement of less accessible quantities. Data, without some form of distillation, are rarely suited for this purpose.

A specific example of the contrast between data and a corresponding data model can clarify why the distinction is important for our purposes. The data in the one-second small-arc circular pendulum measurements of surface gravity initiated by Huygens consisted of the length of the pendulum, measured to the center of a spherical bob, and the duration over which the pendulum beat synchronously with a well-tuned pendulum clock, typically required to be of the order of a half-hour or more. Huygens called for two corrections to these observations: a calculated adjustment of the length to the center of oscillation of the pendulum, resulting in a length slightly longer than the measured length to the center of the bob; and a further correction to the length to compensate for any imprecision in the measurement of time by the clock versus sidereal time, checked over the duration of a sidereal day.[35] Newton subsequently introduced a small correction for the reduction in effective gravity owing to buoyancy of the bob in air. The data model yielding the measurement of acceleration at the surface of gravity included these corrections of the raw data. Over the course of subsequent centuries other corrections came to be included, such as one for the resistance of the air to the motion of the bob, another for the deviations from isochronism as a function of arc-length of any circular-arc pendulum.[36] This example illustrates why the distinction is so important in precision measurement: the distillation of data into a data model, among other things, can compensate for sources of systematic error that would otherwise limit the amenability to increasing precision of a measure.

While much more can be said about data and data models as levels of representation of observations, it does not have much bearing on our concern here—why results of theory-mediated measurements can persist through theory change. Our point so far is that two levels or layers of representation precede the use of a measure in obtaining values of a target quantity. The measure itself is a still further level or layer of representation in van Fraassen's sense insofar as it usually represents a relation among quantities in some, but not in other, respects. More important than this, however, is that the measure does not come out of nowhere. It requires justification first for the specific mathematical form it asserts to hold among the various quantities entering into it, and second for what it claims its target quantity amounts to physically. This is true even when the target quantity is, historically, being invented with the measure. As Hasok Chang stressed in his *Inventing Temperature*, efforts were required first to show that two specific heights of mercury in a cylindrical tube consistently correspond to the freezing and boiling points of water at atmospheric conditions, and second that the relationship between the height of mercury intermediate between these two heights had claim to varying linearly with intermediate stages between the freezing and boiling points.[37]

Putting cases in which a quantity is being invented with a measure to one side, the justification for a measure usually consists of a derivation of it from broader mathematically expressed principles. Thus, the measurement of electric current by the magnetic torque on a coil carrying it serves to define the measure of it incorporated into galvanometers and ammeters. The specific expression relating current in the coil to the torque is derived from the more general principle specifying the force on a current carrying conductor in a magnetic field, and that principle in turn can be derived from the general mathematical theory defined by Maxwell's laws. As we summarized in Chapter 4, the two measures for *W* cited earlier were both derived, by applying the *mathematics* of kinetic theory to solutions, under assumptions about the motions of granules, several of which were independently tested. Insofar as the justification for a measure involves considerations reaching beyond it, it too must either itself be or be part of a further level or layer of representation, namely one that links variables to one another in some more general fashion. For want of a fully descriptive name here, let us call this level of representation *mathematical theory*, leaving open the extent of theory invoked in justifying a measure. We repeat that what is required of this fourth level is to justify first the specific mathematical form of the relationship among the quantities in the measure and second what the measure takes its target quantity to amount to physically.

These four levels of representation invariably enter into any theory-mediated measurement, and hence the question of why values from some of these measurements persist through theory change needs to focus on them. Duhem explicitly (and van Fraassen at least by implication) identifies a still further level of representation, the "explanatory part" of physical theories that he then goes on to license in the form of heuristic "as-if" models.[38] What he had in mind were the ether models of electromagnetic phenomena put forward by Maxwell and his followers, but surely too Ostwald's 1890 concessions to the conceptual virtues of atoms and molecules in physical chemistry.[39] Here we prefer to draw the distinction between this level of representation and the level we have called "mathematical theory" in the manner of an avowed realist, the physicist Steven Weinberg. He distinguishes between a "part of modern physical theories . . . that usually consists of the equations themselves, together with some understandings about what the symbols mean operationally and about the sorts of phenomena to which they apply" and a part that "is the vision of reality that we use to explain why the equations work."[40] Taking this last phrase as the key, let us call the fifth level of representation *explanatory theory*.

Again, a lot more can be said about the distinction between what we are calling *mathematical* and *explanatory* theory, much of it contentious because of the ways in which it ties into so much of the realism-instrumentalism debate. For our purposes here, fortunately, we need not say much more, for the main function of the five levels of representation that we have distinguished is to situate the middle of the five, what we have called the *measure* itself, in its broader logical context. We can now concentrate on it, asking why it can, at least in some cases, yield values of its target quantity that persist through theory change, where the latter allows for changes in both our "mathematical" and our "explanatory" theory.

P.3.2. "Stability" as a Factor in Claims to Permanence

In Chapter 7 we concluded that Perrin had provided evidence that each of his two measures for *W* that we are using for illustration is *stable*—that is, evidence that the value for *W* obtained by means of each measure remains the same to reasonably high precision over a reasonably wide range of their respective manipulable quantities. To quote Perrin in regard to the first measure, "the values of the granular energy deduced from the preceding series are concordant within the limits of accuracy of the experiments, although the mass of the granules has varied as 1:40, the difference of density between the granules and the medium has varied as 1:4.7, and the rapidity of rarefaction as a function of height has varied as 1:30."[41] A similar statement holds for the other measure, for which Perrin varied the radii of the granules from 0.212 to an extreme of 5.5 microns. Evidence for the stability of each measure is evidence as well that a specific linear relationship holds in each of the two cases: between $\ln(r/r_0)$ and h in the first case, with its slope varying with the radius of the granules and their buoyant density; and between ξ^2 and τ in the second case, with its slope varying with the radius of the granules and the viscosity of the liquid.

Now for the important point. The values of each of the manipulable quantities just listed were determined completely independently of the mathematical justification of the respective measures! Therefore, the *value* of *W* obtained through the measure, in and of itself, amounts to nothing more than a factor in the value of the coefficient of proportionality of each of two empirically determined linear relationships between two variables whose values in no way depend on the mathematical justification of the measure. In other words, *the empirically determined value for W and its empirically turning out, at least to reasonable precision, to be a constant over a range of the manipulable quantities are, in each case, independent of the mathematical justification of the respective measures.*

This is not to say that the value determined for *W* and its constancy are independent of the respective data and data models employed in the two determinations. And, of course, that *W* represents the mean translational kinetic energy of the granules at the given temperature in each case is entirely dependent on the mathematical justification of the respective measures. Our point is that the numerical value itself obtained for *W*, putting to one side what it amounts to physically, is in each of the two cases simply an empirically determined numerical factor in the coefficient of proportionality in an empirical relationship among variables; and the values of all those variables, and hence too the *value* of *W*, are determined independently of the mathematical justification of the measure. As a consequence, evidence for the stability of each of the measures of *W* is evidence that a linear relationship holds empirically among certain specific quantities over a certain range of values of the manipulable variables, and this evidence is entirely independent of the mathematical justification of the measure.

This point can be put in another way. Evidence for the stability of each of the measures of *W* is evidence for the existence of a *regularity* in the world corresponding to each of the measures, a regularity that is being *represented* as a relation among comparatively accessible quantities: in the first case, among the ratio of the number of granules in a narrow slice at different heights, those heights, the radii of the granules, and their buoyant densities; in the second case, among the mean of the square of the displacements over different time intervals, the magnitudes of those time intervals, the radii of the granules, and the viscosity of the liquid. To complete the specification of the relationship in each case, a value for *W* is needed. Evidence for the stability of the measure is evidence that this value of *W* is a numerical constant in the relationship, and hence a constant whose value can be determined from values of the other quantities in the relationship, themselves represented in the data model employed in the measurements. The *value* of *W* accordingly has an empirical integrity in its own right, independent of the mathematical theory underlying the measure, and hence independent too of what it physically amounts to.

The same point can be made for Huygens's theory-mediated measures of the strength of surface gravity and Ptolemy's theory-mediated measures of the radius ratios and orbital eccentricities of the outer planets. The value of Huygens's distance of fall in the first second in the absence of air resistance, half our *g*, taken unto itself, was a factor in a coefficient of proportionality of a linear relationship between the length of pendulums and the square of their periods. The regularity represented in the case of his simple pendulum measure, which was restricted to cycloidal and small-arc circular pendulums, was between two quantities that could be determined independently of any assumptions about gravity, and hence so too could the value itself for the strength of surface gravity. (The same is true of Huygens's conical pendulum measures, of which he had two, the first restricted to constant-height pendulums and the second to ones in which the center of the bob is restricted to the surface of a paraboloid; both involved a regularity represented as a relationship between the square of the period and, in the case of the former, the height of the pendulum and, in the case of the latter, the latus rectum of the paraboloid.[42])

In saying that the measures represent a regularity, we do not mean that first there was the regularity and then came its representation by the measure. That such a regularity exists, in both the Perrin and the Huygens cases, was initially derived from theoretical considerations. These derivations, however, could not guarantee that the regularity actually exists. That it does exist gets confirmed by what we are calling the stability of the measure—a conclusion that is reached by means of inductive generalization from cases. One can legitimately take its empirically established stability as evidence supporting its derivation and hence the theoretical claim that it exists. Much of our Chapter 4 was devoted to spelling out just what the evidence of this sort consisted of in the case of Perrin's two measures. Here we are making a different point. Once the stability of the measure confirms the existence of a regularity among its accessible variables, *the existence of this regularity no longer depends on the theoretical derivation of the measure*. What that derivation then provides is a physical interpretation of the less accessible quantity that is being measured. This explains how the value of that quantity itself, though of course not its physical interpretation, can gain an integrity in its own right, independently of the theoretical considerations underlying the measure.

The regularities we are singling out here are instances of the "gross regularities" that Jody Azzouni has stressed in various places, most notably in his *Knowledge and Reference in Empirical Science*.[43] Two distinguishing features of such regularities, according to him, are their highly local character and their existence being amenable to confirmation independently of theory and hence their withstanding theory change. He has argued that they are not only ubiquitous, but that they are far more crucial to understanding science than has generally been recognized.

Our point here is more modest, yet nonetheless again more crucial to understanding measurement in science than has generally been recognized. Most theory-mediated measurements involve a comparatively local regularity, the existence of which and hence the utility of which for purposes of measurement are independent of overarching theory, thereby tending to insulate them from theory change. This is true not just for the examples we are considering here, but for ones establishing quantities in science in the first place, like the regularity between the boiling point and freezing point of water, on the one hand, and the length of a column of mercury, on the other, or the correlation Ampère discovered between the number of voltaic cells to which a wire is connected and the response of a magnetic compass adjacent to that wire. Our main point here is narrow, namely that the stability of a measure typically provides evidence for both the existence of the regularity at its heart and its independence of theory.

Because Ptolemy's measures are expressed in a geometric form, the corresponding points are more hidden in their case. Nevertheless, his values for the radius ratios and eccentricities of Mars, Jupiter, and Saturn were part of a representation of regularities in their respective observed departures from uniform angular motion; and the values themselves were determined from these departures, and thus had an integrity of their own, independent of his epicycle model.[44]

In none of these cases are we denying that the target quantities of the measures are being taken to represent physical quantities. Our claim is that their physical interpretation does not come from the data or data model used in determining their values. The physical interpretation in every case lies in the mathematical justification of the measure—which, in our terminology, is part of the mathematical theory underlying it. This terminology is misleading in one respect, however. The measures are not simply being deduced from a mathematical theory, established or otherwise. Their mathematical justification invariably involves assumptions. The key assumption in the case of Huygens was that gravity acts along parallel lines to yield uniform vertical acceleration. In the case of our second illustrative example from Brownian motion, the assumption was more complex: (1) the granule displacements over time are random, with a Gaussian distribution over time; (2) the resistance force from the viscosity of the liquid on the spherical granules is in accord with Stokes's law; and (3) the duration τ covers a very large number of changes in velocity. Perrin took steps to confirm the first two of these independently of the measure, in the process yielding a historically landmark conclusion in the case of the first. Nevertheless, the claim that W represents the mean translational kinetic energy of the granules depended on these assumptions, and thus had a hypothetical status that the value obtained for W, in and of itself, did not have.

This stands out even more clearly with our other illustrative measure from Brownian motion, for it required an assumption about the relationship between the pressure exerted on the walls by impact of the granules and the fraction of the vertical pressure increase in the fluid from the weight of the granules, neither of which could be independently measured.[45]

To summarize, while the value determined for the target quantity in each of these cases depended only on its data and data model, not on the mathematical justification of the measure, the physical interpretation of the target quantity depended indispensably on the latter, and hence had a hypothetical status owing to the status of the assumptions made as part of the justification. The numerical value Perrin obtained for W in each of our two cases was accordingly not open to challenge in the same way as his claim that this value is the mean translational kinetic energy of the granules in Brownian motion. Among the challenges at the time was whether Stokes's law holds for granules so small.[46]

P.3.3. "Convergence" as a Factor in Claims to Permanence

We have so far been considering our two illustrative measures of W separately from one another. We now consider them together. At the time of the Solvay Conference in 1911, Perrin's values for them at the same temperature agreed with one another to within 0.6 percent—far closer than the roughly 10 percent and greater gap between his values for W and the values for the mean translational kinetic energy of molecules at the corresponding temperatures implied by the measurements of e in the laboratories of Rutherford and Millikan. To put the point in the way we have in the main body of this monograph, Perrin's two stable measures of W had far stronger claim in 1911 to having yielded *convergent* values than the various stable measures of N_0, e, and α had. The question now is, what further conclusions, beyond those we have drawn from the stability of the two measures, are to be drawn from their convergence?

The first point to notice is that the convergence was of no consequence at all for the specific *values* obtained from the two measures. The values themselves were obtained as factors in the coefficient of proportionality in the linear relationships between, respectively, $\ln(n/n_0)$ and h, on the one hand, and ξ^2 and τ, on the other. Each value for W depended only on the data and data model from which the values of all the other quantities in the measure were drawn. That the two values agree with one another, however, does provide evidence that neither one of them, nor the stability of the respective measures, was merely an artifact of the data, data model, or assumptions underlying the measure. Absent the convergence, the claim that neither value is merely such an artifact relied only on the mathematical justification of the measure. Evidence that the values are not artifacts is evidence that they are values representing a physical magnitude. Moreover, insofar as a lack of convergence would have been evidence that W does not stand for the same quantity in the two cases, the convergence was evidence that the symbol does stand for the same physical quantity in both measures.

Nothing in the data, data models, or measures, considered in isolation from the theoretical justifications of the latter, assured that W stands for the same quantity in both. Indeed, the only other term common to both is the radius of the granules, a, and it enters as a cube in one of them and linearly in the other. Why these two measures should have a factor in their respective coefficients of proportionality in common can therefore be answered only by turning to their respective mathematical justifications. The mean translational kinetic energy of the granules enters into the two justifications in very different ways—in the first case in connection with the pressure gradient needed to support the weight of the granules; and in the second in connection with their motion resisted by viscous forces.[47] The convergence of the two measures was accordingly evidence that two distinct regularities in Brownian motion, represented by the respective measures, have a governing physical magnitude in common, represented in each measure as the mean translational kinetic energy of the granules at the given temperature. But then the combination of the stability of each measure and their convergence was evidence in support of the assumptions made in the theoretical derivation of each of them—not so much that these assumptions are true, but that they were safe to adopt for purposes of measurement.

We have expressed this point in such a cautious fashion in an effort not to beg any questions in the realism-instrumentalism debate. The point can be made in a less roundabout way. Before turning to their convergence, we acknowledged that the claim of each to be a measure of the mean translational kinetic energy of the granules at the given temperature was open to challenge because of assumptions made in its mathematical justification. The stability of each measure provided evidence that its W was a constant at any given temperature over a range of globule sizes and liquids—a remarkable finding in itself—but not grounds for dismissing challenges to the claim of each that what it measures physically is the mean translational kinetic energy of the granules. The convergence of the two measures, by contrast, did give grounds for dismissing such challenges, for it provided evidence that these two measures are measures of the same physical quantity, with each measure holding to a certain level of precision over a certain range of values of its manipulable variables. In other words, their convergence provided evidence that any shortcomings in the assumptions entering into the justification of each measure—or, for that matter, in the respective data models—were at worst imposing limits on its precision, on the range over which it holds, or on some combination of these two.

In the preceding section of this Postscript we spoke of one proposition coming to be regarded as all along having been merely a surrogate for another proposition that for one reason or another replaces it—higher precision, wider range of applicability, greater compatibility with general theory, or combinations of these. Insofar as the factors listed at the end of the last sentence of the preceding paragraph parallel these, we can restate its point: *the combination of the stability of the two measures and their convergence provided evidence that each of them would turn out to be no worse than a surrogate for some other measure that has higher precision, a wider range of applicability, greater compatibility with advances made in mathematical and explanatory theory, or some combination of these.*[48]

To conclude that the evidence showed each measure to be at least a surrogate for some more preferable one is equally to conclude that this evidence is imposing a constraint on replacing either measure with any alternative to it: any replacement must either retain the measure as holding to a certain level of precision over the range of the manipulable variables for which it was stable and converged with the other one; or it must explain why the evidence arising from their stability and convergence was somehow illusory. Putting aside the second of these alternatives, no new measure can meet the first demand unless it preserves, at least to high approximation, the *values* for *W* that had been obtained by means of the original measures. We are thus beginning to see why the values themselves obtained through theory-mediated measurements *can*, at least in some cases, persist through theory change. The values, when supported by sufficient evidence, tend themselves to become constraints on any change in theory in which the original measures are replaced.

The examples we have been using as illustrations are historically of no interest when it comes to theory change, for the values obtained for *W* ceased being of much consequence to physics once it became clear—surely no later than 1915—that they are 10 percent (or more) less than the mean translational kinetic energy of molecules at the same temperatures. A more historically illustrative example involving theory change is Huygens's converging measures of the distance of fall in the first second in the absence of air resistance. Both his simple and his conical pendulum measures presupposed, in modern terms, that the acceleration of gravity does not vary with any varying heights of the pendulum during the measurements, and that it is vertical, along strictly parallel lines. Both measures proved to be highly stable, and in his own statement of the results in 1673, the values from the two agreed to better than three significant figures. Both measures therefore amounted to mere surrogates in the context of Newtonian gravity—including the replacement of *g* taken to be a constant vertically with *g* varying with height. To use Glashow's phrasing, in showing Huygens's measures to be false, Newton also delineated their envelope of applicability, thereby validating the value Huygens's measures had obtained for *g* to the better than three significant figures claimed for them. Newton then invoked this value in his evidence that terrestrial gravity extends to our moon.[49] In this case, therefore, the value was more than just a constraint on theory change; it was crucial to the evidence for extending terrestrial gravity to the celestial realm.

P.3.4. "Amenability to Increasing Precision" as a Factor in Claims to Permanence

In Chapter 4 we noted a further demand that is customarily made of theory-mediated measures: their amenability to increasing precision. A classic example was Huygens's redefining the length of the pendulum once he had solved the center of oscillation problem.[50] Perrin himself, in one of his first papers on Brownian motion, had claimed that the first of his measures for *W* that we have been considering "was susceptible to an unlimited precision."[51] We also noted in Chapter 4 that Perrin, in a later edition of *Les atomes*, had said that René

Constantin had convinced him that the motion of granules near the walls was too constrained to be representative and had then added:

> Working at a sufficient distance from the walls with the grains I had used, he obtained the value $N = 64 \times 10^{22}$; unfortunately the number of observations (about 100) was too small. These measurements will be repeated.[52]

This change concerned the second of our two measures, amounting to a 7.5 percent increase in the value of *W*. Notice that the pursuit of higher precision here concerned the data and data model employed, not the measure itself, though what prompted Constantin to propose it was the conflict between the assumption of unbounded fluid in Einstein's derivation of his relationship between ξ^2 and τ, on the one hand, and the contained emulsion, on the other. What difference might have emerged in the value for *W* from the second measure had Perrin followed up this proposal we have no way of knowing. We should nevertheless ask, while we are at it, what contribution the amenability to increasing precision can make to the persistence of values from theory-mediated measures across theory change.

To begin with, evidence that a measure is amenable to being made more precise is evidence that it is measuring a definite physical quantity. Planck illustrates this point in the 1913 edition when he reformulates his derivation of the molecular magnitudes from phenomena of radiation in terms of two more highly measurable constants, one, *a*, from the Stefan-Boltzmann relation and the other, c_2, from the "specific intensity of a monochromatic plane polarized ray of the wave length λ which is emitted from a black body at the temperature *T* into a vacuum perpendicular to the surface."[53] He then derives values for his *h* and the various molecular magnitudes to three significant figures, ending with the claim: "Since absolute accuracy is claimed for the formulae here employed, the degree of approximation to which these numbers represent the corresponding physical constants depends only on the accuracy of the measurements of the two radiation constants *a* and c_2."[54] In other words, if the formulas are exact, the amenability to increasing precision of the various measures depends only on the data and data model for the two radiation constants. Equally, however, if increasing the precision of the measurements of the two radiation constants were to result in a more precise value for any of the molecular magnitudes in conflict with a value of it from any other measure, the question would arise whether Planck's formulas are "absolutely accurate."

Amenability to increasing precision of a measure thus opens the way to assessing the limits of precision of alternative convergent measures of the same quantity against one another and deciding which one is to be preferred. To put this in a more customary manner, amenability to increasing precision in the case of stable, convergent measures becomes a research tool for exposing sources of systematic error in the measures. This is precisely what happened in the case of Millikan's 1917 value for *e*, $4.774 \pm 0.005 \times 10^{-10}$ esu. It was being taken as the preferred basis for all other molecular magnitudes until the 1930s, when comparison with its implied value for Avogadro's number with a value obtained from x-ray crystallography exposed a roughly 0.7 percent systematic error in it, introduced by a mistaken value for viscosity of air. This error in the data model was corrected to yield an oil-drop value for $e = 4.8130 \pm 0.0006 \times 10^{-10}$ esu; in the process, however, it was decided that limits of precision in the measurement of viscosity entailed limits to the amenability of increasing precision in the oil-drop measure, and this measure then ceased being taken as the preferred basis for all other molecular magnitudes.[55]

The upshot, then, is that amenability to increasing precision amounts to a way of cross-checking values obtained from stable, convergent measures, a way of addressing the risk of being misled by a value that appears to have greater precision from its determination than it should be taken to have. The greater the precision that a value can come to claim, the greater the reason to assure its persistence, with suitable bounds on its claim to precision, through any subsequent changes in theory. Indeed, what actually persists historically, as the Millikan example shows, is not the original value, nor even one subsequent value, but the increasingly precise variants of them. That still leaves the original value as an approximation to subsequent values, and the fact that precision unfolded in a sequence from the original amounts to evidence not just that the original value itself held to a certain limited level of precision, but that the sequence of more precise values is at least approximating, to an increasing level of precision, the value of a definite physical quantity.

The history of our two illustrative measures for *W* brings out another way in which the demands for convergence and amenability to increasing precision can lead to reconsideration of a measure. We have presented them as measures of the mean translational kinetic energy of granules. Both Einstein and Perrin assumed that *W* is at the same time a measure of the mean translational kinetic energy of molecules in the solution at the same temperature, licensing the two of them to transform both of these measures into measures of Avogadro's number. And such they were viewed throughout the period covered in this monograph. Their values for this constant, however, remained 10 percent (and more) greater than any of the values that were being obtained from the measures based on gaseous ionization, α-particles, and blackbody radiation, all of which were within 3 percent of one another. As Millikan's value for the first became increasingly precise, the Brownian motion measures of Avogadro's number came to be considered as, at best, merely approximate.[56]

Insofar as the Brownian motion measures converged with one another, this implied either that they contained a systematic error or that *W* should be taken to be a merely rough, approximate measure of the mean translational kinetic energy of molecules. But the two convergent measures of *W* we have been considering had no parameter in the data from which they were obtained in common. For their deviation from other measures of Avogadro's number to be coming from systematic error therefore required two separate systematic errors of nearly matching magnitude. Far more likely was that the mean translational kinetic energy of Perrin's granules was at least 10 percent less than that of the molecules in his emulsions and therefore was a quantity that only correlated with that of the molecules, not a measure of it amenable to increasing precision. We do not know of anyone who expressly made this point at the time, but as interest in the Brownian motion measures of Avogadro's number diminished, no one may have had occasion to ask just what Perrin's measures were in fact measuring. Notice, nonetheless, that this in no way undercut the evidence from their stability and convergence that they were yielding reliable values for the mean translational kinetic energy of his granules.

P.3.5. Claims to Permanence

The point of both this and the Millikan example is that the demands for stability, convergence, and amenability to increasing precision can provide very strong cross-checks on claims made for the value yielded by a measure. At the beginning of this section we proposed that in the case of perception, one *can* eliminate virtually all possibility of future information overthrowing a conclusion based on it, and in that regard anti-realists have some grounds for insisting on it as a standard against which claims to permanence in science should be compared. Our examination of why certain values from theory-mediated measurement have persisted through theory change suggests an at least analogous point for them: *the demands for stability of a measure, its convergence with other measures of nominally the same quantity, and its amenability to increasing precision* **can** *eliminate virtually all possibility of future information showing that the* **value** *itself obtained from it was all along just an artifact of the data or data model from which the value was extracted, and therefore all along not something corresponding, at least to a degree of approximation, to a bona fide physical quantity.*

The emphasis on the word "can" is to acknowledge that this does not happen automatically; it depends on the strength of the evidence for stability, convergence, and amenability to increasing precision, and perhaps as well to specifics about the measure. The emphasis on the word "value" is to distinguish it from its physical interpretation, for we do not want to be suggesting that the physical interpretation of the value can have a comparable claim to permanence. We have seen earlier with Perrin, Huygens, and Ptolemy that a value obtained from theory-mediated measurement can have a claim to permanence, and with it a claim to being at least an approximate value for a physical quantity, without its physical interpretation becoming closed to reconsideration.

How the value can have such a claim to permanence stems from two aspects of theory-mediated measurement. First, the whole point of such measurement is to determine the values of less accessible quantities from those of more accessible ones. It does this by finding a regularity among the more accessible quantities—or, expressed more in keeping with van Fraassen, a representation of a regularity expressed in terms of the more accessible quantities—in which the target quantity is a parameter. Examples of such regularities are the period of a small-arc circular pendulum and its length and the rate of rarefaction with height of granules in a dilute emulsion and their buoyant weight. The target quantity is required to have a specific value in any one instance of the regularity, but it need not be a constant: *g* turned out to vary from one location to another, *W* was all along expected to vary with temperature, and Ptolemy's radius ratio and eccentricity varied from one outer planet to another. The regularity, moreover, need not be some general law of nature; it need hold only under a quite restricted, localized set of conditions. For determining values for *W*, Perrin's rarefaction regularity was required to hold only over a limited range of heights for dilute emulsions formed with uniform spherical granules of a limited range of sizes—scarcely a fundamental law of nature, notwithstanding the fact that Perrin's Nobel Prize citation included the phrase, "especially for his discovery of sedimentation equilibrium."

Second, insofar as the measures require only relatively localized regularities among comparatively accessible quantities, they are subject to extensive testing and hence numerous cross-checks. Evidence showing that a measure meets the demands for stability and for amenability to increasing precision requires just such testing. The aim with both demands is evidence that the measure represents, at least to a certain level of approximation over a certain range of its accessible quantities, a physical regularity, and hence that the value it yields is not an illusory artifact of the data and how they are modeled.

The key point here is that what is being tested is a regularity within the accessible quantities, and whether this regularity exists in fact is independent of the theoretical justification of the measure, and hence of the physical interpretation of its target quantity. The demand for convergence with another measure of nominally the same quantity seems primarily to be a test of its physical interpretation. Even so, an agreement between two numerical values obtained from different regularities holding among different accessible quantities is evidence that in neither case is the value itself somehow merely an artifact of how values are being determined for the respective accessible quantities. At least in principle, therefore, these three demands can address and eliminate a great many ways in which future developments might lead to the conclusion that the *value* obtained for the target quantity all along failed to correspond to any physical magnitude.

The parallel we are suggesting with perception should now be clear. The eliminate-virtually-all-possibility claim in the case of perception is based first on the many resources we have for addressing different sources of perceptual illusion, second on the extent to which these resources depend on perceptual regularities, and third on the extent to which these regularities allow such sources, save for ones involving radical skepticism, to be addressed locally. Verifying that a perception is not illusory, moreover, can be done independently of the words used to describe its content. The process of verifying it may even lead to a reverbalization, yet with no change beyond refinement of detail in the perceptual content itself.[57]

The parallel with perception extends beyond the ways we have to eliminate sources of illusion. The theoretical justification of any measure amounts to a prediction that a certain regularity is going to be observed among the accessible quantities and that, by virtue of this regularity, the measure will yield a stable value for its target quantity. This is a prediction, however, not a guarantee. Whether the regularity exists depends on the data that assign values to the accessible quantities, and the evidence that it exists depends on whether the measure determines a stable value for its target quantity from those values. Whether the measure is stable is a question put to the empirical world that only it can answer.

Still further, the theoretical justification cannot predict what the value is going to turn out to be. Any such prediction can only come from having obtained a value for nominally the same quantity from some other measure. Here again, therefore, only the empirical world can answer the question of what the value of the target quantity is. As a consequence, for a measure to yield a stable value—or, even better, a stable value amenable to increasing precision that is convergent with a value from another measure—in and of itself gives that value a *prima facie* claim to at least represent a physical magnitude in much the way that one's running into a table in the dark gives the perception of the contact a *prima facie* claim to involve an encounter with a physical object.

That the theoretical justification of a measure cannot dictate that the value of its target quantity will turn out to be stable, much less the value itself, is reminiscent of a requirement for measurements to "ground" a theory that van Fraassen adapted from Clark Glymour: "there must be an alternative possible outcome for the same measurements that would have refuted the hypothesis *on the basis of the same theoretically posited connections*."[58] The requirements of stability, convergence, and amenability to increasing precision in theory-mediated measurement parallel van Fraassen's requirements of "determinability" and "concordance" for grounding that he adapted from Herman Weyl. A measurement cannot fully ground a theory in van Fraassen's sense unless its measure is stable, convergent, and amenable to increasing precision.

He imposes a further requirement, however: "this determination can, may, and generally must be made *on the basis of theoretically posited connections*." Perrin's values for *W*, for example, could not have grounded molecular-kinetic theory without the assumption that the mean translational kinetic energies of the granules and the molecules of the liquid are the same. We do not disagree with van Fraassen about this. The contrast between him and us lies in our setting apart the measures in theory-mediated measurements and their relation to the data and data models from which they extract values in and of themselves, independently of their theoretical justification. And our point is that the measures and the values they yield *can* acquire special standing, a claim to permanence, in their own right that is independent of their theoretical justification.

Van Fraassen's term "grounding" is of course a metaphor, but one that we agree is appropriate for describing, for example, what the determinations of Avogadro's number examined in the body of this monograph did at the time for the molecular-kinetic theory of heat and the atomic-molecular theory of chemical composition. The question we have been considering in this section of our Postscript, by contrast, concerns why the values of theory-mediated measurement have sometimes persisted through change in theory. The corresponding metaphor for this, we submit, is that values which have acquired the sort of claim to permanence that we have proposed *anchor* theory change. They, and the measures yielding them (once suitably qualified), become constraints on theory change.

P.3.6. So What?

One might well ask why this should matter to anti-realists and realists disputing the scope and limits of theoretical knowledge. The reason does not lie in the values themselves, but in a further implication of evidence being able to eliminate virtually all possibility of their not corresponding to a physical magnitude. Evidence that does this is evidence that the measures yielding those values represent, at least to some precision over a circumscribed domain, physical regularities. Granted, these regularities tend to be highly local, and the measures representing them may turn out to be surrogates for representations that in some respect or other are preferable to them. Nevertheless, evidence supporting the permanence of a value obtained through a measure is evidence that the measure supports counterfactual claims. For example, the rate of rarefaction with height of granules in a dilute emulsion would be different, indeed in a specifiable way, if the value of *W* were different, say from a change in temperature, or if the granules of the emulsion were of a different buoyant weight. The relationship between the period of a small-arc circular pendulum and the length of its string would be different, again in a specifiable way, if the value for g were different, say from making the measurement at a different latitude, or if the size of the bob and hence the center of oscillation of the pendulum were different. And the pattern of departures from uniform angular motion, as viewed from the Earth, would be different in a specifiable way if the value of the eccentricity for Mars determined by Ptolemy's measure were more like that for Jupiter.

Several points need to be made about such counterfactuals. To begin with, they concern only quantities that enter into a measure, its so-called accessible quantities, manipulable or otherwise, and the value of its target quantity. Like the measures, they too therefore tend to be highly local. What supports them is the evidence that the measure represents a physical regularity at least to some precision over a circumscribed domain. We have argued earlier that evidence that the measure does this need focus only on the value of the target quantity, without having to take into consideration the physical interpretation assigned to it by the theoretical justification for the measure. But then, to the extent that the evidence for the stability and amenability to increasing precision of the measure is independent of its theoretical justification, and the evidence for its convergence with other measures is restricted to being so, the counterfactuals supported by the measure are also independent of its theoretical justification. The counterfactual about the rate of rarefaction of granules and their buoyant weight stated earlier, for example, holds independently of what W represents physically. Therefore, evidence that the value of the target quantity is going to persist through theory change is evidence as well for the counterfactuals supported by the measure. In other words, to the extent that this value constrains future theory change, so too do these counterfactuals.

This, however, entails that the counterfactuals be highly local in another respect. The counterfactuals concern alternative values of the quantities entering into the measures. The evidence supporting the measure, taken in isolation from its theoretical measure, involved variations of those values over a limited range. Consequently, the alternative values entering into the counterfactuals supported by the measure need to be restricted to the same range. If the alternative values are not so restricted, the counterfactual may well be invalidated by future theory change, for only the theoretical justification of a measure can license the range of its projectability. To give a historical example of this, not so far removed from theory-mediated measurement as it may at first appear, consider the counterfactuals pertaining to our solar system licensed by Newton's law of gravity. Before general relativity, they included counterfactuals in which the mass of the sun could be varied by orders of magnitude. The weak-field limit imposed on Newton's law by general relativity restricted the validity of the counterfactuals supported by it to a narrow range—or, in Glashow's phrasing, a narrow "envelope of applicability."

Narrow though they may be, the counterfactuals nevertheless identify which details in the world make a difference and, at least to some level of precision, the differences they make. The buoyant weight of the granules, for example, makes a difference to the rate of their rarefaction in Perrin's sedimentation equilibrium, but the viscosity of the fluid does not; and the size of the granules, but not their mass, makes a difference in their rate of diffusion in a dilute emulsion. We showed in Chapter 4 that the evidence Perrin developed for both of these claims was independent of molecular-kinetic theory, and hence that they hold independently of whether the mean translational kinetic energy of the granules is or is not the same as the mean translational kinetic energy of molecules in the fluid. We are now arguing that at least much of the evidence for them was also independent of whether *W* is the mean translational kinetic energy of the granules. But for just this reason, the details singled out in the counterfactuals as making a difference and the differences they make acquire a claim to permanence through theory change—whether as specified or under a re-description or a limitation of scope—when the value of the target quantity has such a claim.

The instrumentalism-realism debate has focused on overarching, usually fundamental theories. Laudan's once-accepted-subsequently-rejected examples involved mostly such theories too, challenging whether any evidence for them can ever assure that they will not be rejected in the future. If one were to consider with any care the question of what knowledge the sciences have given us over the last four centuries that Galileo did not have, surely the answer would not center on overarching fundamental theories. The best candidate for knowledge that the sciences have given us that Galileo could scarcely have imagined consists of literally millions of details in the world that make a difference and, at least to some level of precision, the differences they make. Correspondingly, the sense that we understand the world around us so much more thoroughly than those in Galileo's time derives far more from what the sciences have revealed about which details in the world make various differences, and which do not, than it does from their overarching, fundamental theories.

We are by no means denying that these theories have entered indispensably into establishing many of these details. Our principal point in this section of the Postscript is that whatever claim to permanence at least some such details have has often been far less dependent on overarching theory than is generally recognized. We submit, as a corollary to this point, that the extent to which theory has entered into claims to permanence, and precisely how it has, requires the sort of detailed analysis of the evidence illustrated in this monograph.

Nothing in this Postscript should be read as taking sides, one way or another, in the instrumentalist-realist debate. We have merely been putting forward some suggestions that might give the sides of the debate a more extensive common ground with one another and that might accordingly make it more tractable than it has tended to be. In this regard, we intend the Postscript to be an addendum to Howard Stein's more cited than heeded paper of 30 years ago, "Yes, but . . . Some Skeptical Remarks on Realism and Anti-Realism."[59] At the risk of oversimplifying what he said, we take his main claim to have been that close analysis of the history of the development of different areas of science will leave the two sides of the debate less at irreconcilable loggerheads with one another. We are suggesting that one promising topic for such analysis is the role of theory-mediated measurement in identifying which details in the world make a difference and what differences they make.

The twentieth century abounds with examples for which, at first glance, realists and anti-realists would seem to be diametrically opposed, but on further inspection might have trouble remaining so. We end with two more closely complementary examples than one might at first think of details revealed by theory-mediated measurements: (1) the angle between the two oxygen atoms in carbon dioxide is 180 degrees and hence the shape of the molecule is linear, in distinct contrast to other dioxides such as NO_2 (134.1 degrees) and SO_2 (119.5 degrees);[60] and (2) roughly 2,890 km below the surface of the earth, its interior changes from solid to liquid, and some further roughly 1,270 km below that, it changes from liquid to solid.[61] These both involve measured quantities to which we will almost surely never gain direct access. The listed values of the quantities were all obtained through representations that, in one way or another, simplified or idealized a more complicated state of affairs than was being taken into account. Nevertheless, all these quantities have claim to being permanent—the geophysical ones only once they are referenced to our present epoch insofar as the earth is gradually solidifying. We submit that examples like these can do more to clarify what, if anything, is ultimately under dispute between realists and anti-realists than the ether and molecule examples that have dominated the literature ever since Duhem.

Notes

1. Meyer (1899b), p. vi.
2. See Nernst (1969).
3. See Barkan (1999) for support for this claim.
4. That is, defenses of realism postdating Laudan (1981).
5. Duhem (2015); the French original was published in 1902.
6. Worrall (1989).

7. We repeat here what Planck specifically said in the Preface to the second edition of his *The Theory of Heat Radiation*: "Meanwhile the experimental methods, improved in an admirable way by the labors of *E. Rutherford, E. Regener, J. Perrin, R. A. Millikan, The Svedberg*, and others, have without exception decided in favor of the value [of the elementary electric charge] deduced from the theory of radiation which lies between the values of *Perrin* and *Millikan*." Later in the book he derives a value of 4.67×10^{-10} esu from the value 1.34×10^{-18} ergs per degree Kelvin for k derived from blackbody radiation (along with values of 6.20×10^{23} for N_0 and 2.01×10^{-16} for α), and then adds: "Since absolute accuracy is claimed for the formulae here employed, the degree of approximation to which these numbers represent the corresponding physical constants depends only on the accuracy of the measurements of the two radiation constants, a and c_2" (Planck, 1991, p. 172f).
8. The adverb *exacté* occurs once in the first edition of the *Principia*: "I suspended the same weight of gold as exactly as I could in the center of oscillation of the other [pendulum]" (Book 3, Prop. 6). By contrast, *accuraté* has more than 80 occurrences (generally translated as "exactly"), and *quamproximé* (literally, *as nearest as possible*) more than 130.
9. Book 1, Proposition 3, Corollary 2.
10. For more on Newton's concern with the validity of inferences drawn from such claims, see Smith (2002).
11. Millikan's error, of course, was using a wrong value for viscosity. For details, see Birge (1945). Birge's revised value for e derived from the value for Avogadro's number and the Faraday constant was $4.8021 \pm 0.0006 \times 10^{-10}$ esu. The 2018 value, $4.80320471 \times 10^{-10}$ esu, readers should note, lies outside Birge's 1944 bands, though not so far outside as his 1944 value fell outside Millikan's bands. Estimates of error-bands often fall prey to undetected sources of systematic error, as illustrated by Millikan's 1917 bands.
12. See Newton's *Principia*, Book 1, Propositions 50–52 and the two corollaries of the latter. The formula Newton gives there has the imprecision arising from Huygens's assumption appearing in the seventh significant figure in the case of a uniformly dense earth.
13. Glashow (1997), p. 276.
14. In his doctoral dissertation, Dustin King (2018) has examined examples of both cross-checks and complementary evidence in his analysis of Cecilia Payne's groundbreaking determination of the extraordinarily high abundances of hydrogen and helium in stars. To calculate abundances at all, she had to take a substantial number of propositions to be true, some of which she was quite sure at the time were not true, at least as stated. Her dissertation, Payne (1925), provides a model for how to obviate the risks of leveraging results from propositions taken to be true for which the available evidence is at best limited and at worst, little different from wishful thinking.
15. Perrin (1990), p. 106f; Perrin (1912), p. 185f.
16. Perrin (1910), p. 73. Italics in the original.
17. See section 2.6 of Chapter 2 for a sketch of the struggle with the second law during the second half of the nineteenth century.

18. To see why we have included Nernst's new heat theorem, consider the following remark by Planck in the Preface to the second edition of his *The Theory of Heat Radiation* (1991), p. vii:

    For the hypothesis of quanta as well as the heat theorem of *Nernst* may be reduced to the simple proposition that the thermodynamics probability (Sec. 120) of a physical state is a definite integral number, or, what amounts to the same thing, that the entropy of a state has a quite definite positive value, which, as a minimum, becomes zero, while in contrast therewith the entropy may, according to the classical thermodynamics, decrease without limit to minus infinity. For the present, I would consider this proposition as the very quintessence of the hypothesis of quanta.

19. See Thomson and Thomson (1928), p. 264.
20. Ibid. Lorentz had recommended the correction to Kaufmann well before Einstein published his special relativity paper.
21. Nernst (1923), p. 327; (1921), p. 285 in the original German edition.
22. Thomson and Thomson (1928), p. 264.
23. Birge (1942), p. 127; and Mohr et al. (2016), p. 35009–35057.
24. The value Huygens listed for the distance of fall in the first second in his *Horologium Oscillatorium* of 1673 (Part 4, Prop. 26) was 15 Paris feet 1 inch, based on the length of 3 Paris feet, 8.5 lines (where a line is a twelfth of an inch) for a one-second pendulum (Huygens, 1986, p. 171); we have converted the latter value to a value for g in metric units. The 1958 value we have quoted is from Heiskanen and Vening Meinesz (1958), p. 85.
25. These values are taken from Neugebauer (1975), pp. 177ff. We have inverted Ptolemy's radius ratios, halved his double-eccentricities, and converted his hexadecimal values to decimal. Ptolemy's value for the eccentricity of Mars came from a third trial that has received some commentary; the value, in modern form, for the second trial was 0.098. G. W. Hill has shown why Ptolemy's value for the eccentricity of Mars was less accurate than his values for Jupiter and Saturn, namely an assumption that the eccentricities are small enough for him to have, in effect, dropped a term. See Hill (1900).
26. C. Wilson (1989), p. 180.
27. Fegley and Lodders (1998), p. 87f.
28. Laudan (1981).
29. Duhem (1991), pp. 7–9 and Part I, Chapter 3. Duhem concedes in these pages that the vibrational theory of sound explains, in his sense, acoustic phenomena, citing our being able to confirm the existence of the vibrations with our senses as distinguishing this case of the explanation of optical phenomena in terms of vibrations in the ether, the existence of which we are unable to confirm with our senses. In saying the latter, by the way, Duhem does not reject the electromagnetic wave theory of light, taken to consist of measurable oscillations of the strength of the local electric and magnetic potentials in the electromagnetic field. See his (2015).
30. This point and its implications with respect to the realism-instrumentalism debate is stressed by Jody Azzouni in his (2004) and in his paper coauthored with Otávio Bueno (Azzouni and Bueno, 2016). While these papers do not focus on theory-mediated

measurement in developing the point, our doing so has been significantly influenced by them and by remarks Azzouni made, a propos of it, in a seminar on realism in which our monograph was a central element.

31. See Perrin (1910), pp. 34–40.
32. Ibid., p. 32f.
33. van Fraassen (2008), pp. 166ff; Suppes (1962). As we acknowledged in Chapter 1, we have elected to ignore the philosophic issues arising from the gaps that almost always occur between what are usually called "raw data" and the statements of the numerical results of experiments and measurements in the ensuing publications, instead granting the authority of the latter. This was in keeping first with our endeavor to view the topics of this book as much as possible as someone writing a review article at the time would have, and second with our focus on the theoretical relations—that is, the "measures"—bridging the gap between parameters that are physically measured and target parameters. For more recent investigations of the representation of data than those of Suppes, we recommend Tal (2012).
34. Suppes (1962), p. 258.
35. For more details on the seconds-pendulum measure and why its precision came to be so important, see Schliesser and Smith (1996).
36. See Heiskanen and Meinesz (1958), pp. 84–101.
37. See Chang (2004), especially pp. 49–56 on the first claim and pp. 62–64 on the second. For original sources, see Cavendish et al. (1777), on the first claim; and de Luc (1772), on the second.
38. Duhem (1991), Part I, Chapter 4.
39. See section 2.1 in Chapter 2 for a summary of Ostwald's use of atoms and molecules in his expositions of phenomena in physical chemistry, in spite of his constant reminders that they are purely hypothetical.
40. Weinberg (2001), p. 198. Weinberg terms the first part "hard"—in the sense of durable—and the second "soft," adding of the latter that "it does change; we no longer believe in Maxwell's ether, and we know that there is more to nature than Newton's particles and forces." He then adds,

    > The changes in the soft part of scientific theories also produce changes in our understanding of the conditions under which the hard part is a good approximation. But after our theories reach their mature forms, their hard parts represent permanent accomplishments. If you have bought one of those T-shirts with Maxwell's equations on the front, you may have to worry about its going out of style, but not about its becoming false. We will go on teaching Maxwellian electrodynamics as long as there are scientists. I can't see any sense in which the increase in scope and accuracy of the hard parts of our theories is *not* a cumulative approach to truth.

    In a subsequent response to a challenge to these remarks, he added (p. 208),

    > But approximate theories are not merely approximately true. They can make a statement that, though it refers to an approximation, is nevertheless precisely true. For instance, although Maxwell's equations give only an approximate account of electric and magnetic fields, it is precisely true that the error introduced by using Maxwell's equations to calculate these fields can be made as

small as one likes by considering fields that are sufficiently weak and slowly varying. This is part of the reason that Maxwell's equations are a permanent part of physical science.

We acknowledge that the points made here by Weinberg and conversations Smith has had with him have influenced the way we put a number of things in this monograph, especially in this Postscript.

41. Perrin (1910), p. 46.
42. See Huygens (1986), especially pp. 69ff and 170–175; and Yoder (1988), pp. 16–33.
43. Azzouni (2002), especially p. 2 and pp. 63–70; and (2010), pp. 187–201.
44. For the case of the radius ratios, see Swerdlow (2004).
45. See section 4.3 of Chapter 4.
46. See, for example, Perrin's response to such a challenge in his (1908b).
47. See section 4.2 of Chapter 4 for how the mean translational kinetic energy enters into the first measure, and section 4.4 for how it enters into the second.
48. We include the greater-compatibility possibility to allow for advances in theory to give grounds for replacing the physical interpretation of *W* with one for which the mean translational kinetic energy of the granules at a given temperature comes to be viewed as a surrogate under certain qualifications.
49. Newton's "Moon test," readers should note, involved a convergence between Huygens's value for the acceleration of gravity at the surface of the Earth and the inverse-square acceleration of the moon toward the Earth inferred from its orbital motion.
50. See Huygens (1986), p. 168ff, where he discusses not only the need to measure the length to the center of oscillation to achieve greater precision, but some other moves as well.
51. Perrin (1908d), p. 595. Perrin, of course, made this claim about the measure for Avogadro's number obtained from the one we are considering by assuming that his value for *W* is the same as the value for the mean translational kinetic energy of molecules at the same temperature; that this assumption turned out to be false does not entail that the corresponding claim about *W* was false.
52. Perrin (1990), p. 124.
53. Planck (1991), p. 168.
54. Ibid., p. 173.
55. Birge (1945).
56. See, for example, Jeans (1916), p. 8.
57. Here we are very much echoing the corresponding parallel Azzouni put forward in his (2004), though the point we are making with it is much narrower than his, since here too ours is confined to the results of measurement.
58. van Fraassen (2009), p. 12. Italics in the original.
59. Stein (1989).
60. See Bernath (2016), especially pp. 262–274. For some early history of this discovery, see Dennison (1931).
61. See Masters and Shearer (1995). For some early history of this discovery, see Brush (1996), pp. 184–202.

APPENDIX

On Ostwald (1889–1890), Nernst (1893), and Meyer (1899)

This Appendix provides historical context for the three books on which Chapter 2 relies for its assessment of the state of the evidence on molecules and kinetic theory as of 1900. As such, it adds some explanation, beyond that given in the chapter, for why we regard these three as uniquely appropriate for this purpose.

The first edition of Wilhelm Ostwald's *Grundriss der allgemeinen Chemie*[1] appeared in 1889, to be followed almost immediately by a second edition in 1890. The second edition involves only corrections to the first, leaving it with the same 402-page pagination and most of those pages remaining word-for-word the same. This explains why we have chosen 1889–1890 to designate the work. Both editions were translated into English by James Walker of the University of Edinburgh, the first published in 1890 and the second in 1895, under the literal translation of the title, *Outlines of General Chemistry*. A third edition, published in 1899 and never translated into English, was an expansion of the second, with notably more material on electrochemistry and chemical "dynamics." The fourth edition, revised extensively in keeping with Ostwald's conversion to atomic-molecular theory, was published in 1909 and translated in 1912 into English by W. W. Taylor, again of the University of Edinburgh. A fifth German edition appeared in 1917. Insofar as the second edition served mostly to correct the first, all our quotations in English from these editions are taken from the 1895 edition, having checked them against the first.

Ostwald's use of "general" in his title meant that the book covers developments in what was coming to be known as "physical chemistry," thereby reaching beyond standard textbooks at the time in chemistry. He had completed a two-volume textbook, *Lehrbuch der allgemeinen Chemie*, in 1887, the same year he had initiated, with van't Hoff as co-editor, the first journal devoted to physical chemistry, *Zeitschrift für physikalische Chemie*. His purpose in publishing the single volume *Grundriss* two years later was "to meet the requirements of the student who, while not intending to devote himself to the detailed study of General Chemistry, still wishes to follow intelligently the progress made recently in this important branch of science."[2] Later in the 1889 Preface he adds,

> Within the last three or four years an enormous advance has been made in chemistry by the theories of solutions and electrolytic dissociation, due respectively to van't Hoff and Arrhenius. I hope not only to have rendered a service to students by the elementary exposition of these epoch-making theories and the experiments on which they are based, but also to have contributed towards their general recognition amongst my fellow teachers—a recognition which can scarce be longer delayed.[3]

The translator's Preface reinforces this feature of the book with the remark, "The singular disregard of the discoveries of van't Hoff and Arrhenius amongst the English-speaking scientific public must be in great measure attributed to the want of a connected account of them from one uniform point of view,—a want which this volume amply supplies."[4]

We, of course, originally turned to the initial editions of Ostwald's *Grundriss* because his 1909 "conversion" to atomic-molecular theory has been so widely heralded in the literature on Perrin. It would have been appropriate for us to do so, however, entirely independently of this. In the years after 1890, Ostwald engaged increasingly in conjecture, both scientific and philosophic, in many of his writings. At the time of the initial editions of his *Grundriss*, however, he was still first and foremost an experimental chemist who had helped spearhead the introduction of physics-based measurements into chemistry. In the remark quoted in the preceding, Ostwald referred to "epoch-making theories and the experiments on which they are based." Not just in its chapters on solutions and electrochemistry, but throughout, *Grundriss* presents not only the state of "general" chemistry at the time, but also the experiments on which that state was based. His constantly insisting in the book on atoms and molecules as offering nothing more than a helpful way to picture substantive results did not appeal to any sort of "metaphysical' qualms, but simply to the absence of experimental results yielding any substantive specifics about atoms and molecules themselves. As such, the book reflected a view widespread among chemists, and by virtue of this was uniquely qualified to serve the purpose to which we put it in Chapter 2.

The first edition of Nernst's *Theoretische Chemie vom Standpunkte der Avogadroschen Regel und der Thermodynamik* was published in 1893, with a translation into English by Charles Skeele Palmer of the University of Colorado published in 1895 by Macmillan[5]—the publisher as well of the 1890 and 1895 translations of Ostwald's *Grundriss*. The book was updated in new editions continually for the next 33 years, starting with the second in 1898 and ending with the fifteenth in 1926. Subsequent translations into English by different translators appeared in 1904, 1911, 1916, and 1923. We call these subsequent editions "updates" because so many passages remained word-for-word the same, and none of the editions we have consulted involved radical revision in the manner of Ostwald's fourth edition. Throughout our monograph we have stressed one such passage that appeared in every edition, though with words surrounding it varying; we repeat it here, including the final clause that appeared in the 1921 edition here italicized by us: "At present scarcely anything definite is known either of the nature of the forces which bind the atoms together in the molecule and which hinder them from flying apart in consequence of the heat motion, or of their laws of action, *though in this case also the application of the quantum theory is helping to throw light on the problem*."[6]

While an assistant to Ostwald in 1887–1889—that is, during the time Ostwald was writing his *Grundriss*—Nernst had worked closely with him on the experiments confirming van't Hoff's and Arrhenius's theories. The first version of his *Theoretische Chemie* had appeared in a handbook of organic chemistry; Nernst appears to have hesitated in putting it out as a book, given its overlap with Ostwald's *Grundriss*, until Ostwald encouraged him to do so. "Overlap" is slightly misleading, for Nernst's first edition is 75 percent longer than Ostwald's 1889 and 1890 editions. As quoted earlier, Ostwald's book was for a student "not intending to devote himself to the detailed study of" the subject; Nernst's is for students intending to devote themselves to physical chemistry. A paragraph from the Preface specifies the need the book was to fill:

> Of course in a treatise on theoretical chemistry, the different chapters of physics and chemistry must find their place. And the essential contents of this will invariably imply so much of chemistry as shall be indispensable for the physical investigator, and so much of physics as shall be indispensable for the chemical investigator; and in all this the physicist must conduct himself as a specialist in physics, and the chemist

> as a specialist in chemistry. Thus the development of physical chemistry as a special branch of natural science means—and I would lay particular emphasis on this—not so much the shaping of a new science, but rather the co-operation of two sciences which hitherto have been, on the whole, quite independent of one another.[7]

Trained as a physicist, Nernst was better qualified than any chemist, including Ostwald, to write the chapters on physics for chemists, and the experimental research in chemistry with which he had been engaged from 1887 made him uniquely qualified to write the chapters on chemistry for physicists.

Much of our reason for originally turning to Nernst's book lay in his being trained as a physicist and his consequently reflecting that discipline's attitude toward both molecular and kinetic theory more than Ostwald could have. From the outset Nernst was more favorably disposed toward the promise of these theories than Ostwald was. That said, as the quote in the paragraph before last about "chemical forces" indicates, he was no less insistent than Ostwald on the distinction between conjecture and experimental success in testing and establishing claims about atoms and molecules themselves. From our point of view, accordingly, Nernst, while complementing Ostwald, had credentials no less unparalleled for assessing the state of the molecular and kinetic hypotheses from the perspective of cutting-edge research in chemistry and physical chemistry—credentials that no physicist at the time could have had.

Oskar Emil Meyer, by contrast, was never involved in research in chemistry or physical chemistry. The full title of his 1899 book on which we have relied is *Die kinetische Theorie der Gase: In elementarer Darstellung mit mathematischen Zusätzen*—in English, *The Kinetic Theory of Gases: Elementary Treatise with Mathematical Appendices*.[8] The English translation, by Robert Baynes of Oxford, came out in the same year. It is the "second revised" edition, the previous edition of which was published in 1877, with no English translation. Unlike the other two books on which we have relied, this one has no subsequent 1909 or later edition incorporating the new results on Brownian motion, Meyer having died in the spring of that year. (Meyer was decades older than the other two, having been born in 1834; Ostwald's dates are 1853–1932, and Nernst's, 1864–1941.)

Three other treatises on kinetic theory came out in the decade between 1895 and 1905 by authors far more renowned now than Meyer: Boltzmann's *Lectures on Gas Theory* in 1896–1898, Gibbs's *Elementary Principles in Statistical Mechanics* in 1902, and the first edition of Jeans's *The Dynamical Theory of Gases* in 1904. Our reason for relying on Meyer's book rather than any of these can be found in his Preface:

> I undertook therefore to exhibit the kinetic theory of gases in such a way as to be more easily intelligible to wider circles, and especially to chemists and other natural philosophers to whom mathematics are not congenial. To this end I endeavoured, much more than was otherwise usual, not only to develop the theory by calculation, *but rather to support it by observation and found it on experiment. I therefore collected together, as completely as I could, and summarised, the observations by which the admissibility of the theory might be tested and its correctness proved.*[9] [italics added]

An indication of how unique at the time the book was in the respects we have highlighted in the quotation is Jeans's singling it out in the Preface of his 1904 book as the work in English his reader should turn to who "wishes a fuller account of experimental work."[10] To put the point in van Fraassen's terms, a central aim of Meyer's book was to summarize the extent to which molecular-kinetic theory had become experimentally "grounded" by the end of the nineteenth century.

The body of Meyer's book consists of three Parts: "Molecular Motion and Its Energy," "The Molecular Free Paths and the Phenomena Conditioned by Them," and "On the Direct Properties of Molecules." As these titles make clear, the book presupposes the existence of chemical molecules throughout. Also throughout it is forced to assume, with stated misgivings, that molecules are spherical in shape, or at least have a definite sphere of influence. This is especially significant in the Part devoted to "direct properties," the cornerstone of which is to derive values for the diameters of these spheres from various phenomena. Unlike the books by Ostwald and Nernst, accordingly, Meyer's does not deal with any of the developments in chemistry between the two editions of his book, most notably structured formulas and stereochemistry, nor developments in physical chemistry that occupy so much of the other two. But then neither of those two goes into remotely so much detail on the experimental basis of kinetic theory. In this respect, therefore, Meyer's book uniquely complements the other two for our purposes.

Notes

1. All consulted editions and translations listed in Bibliography.
2. W. Ostwald (1895), p. v.
3. Ibid. p. vi.
4. Ibid., p. vii.
5. All consulted editions and translations listed in Bibliography.
6. Nernst (1923), p. 327.
7. Nernst (1895), p. xii.
8. All consulted editions and translations listed in Bibliography.
9. Meyer (1899b), p. vi.
10. Jeans (1904), p. vi.

Glossary

The range and number of technical scientific terms in this monograph are so large that a glossary covering them all would begin to look like a small dictionary. Fortunately, the internet offers access to a large fraction of them even more quickly, perhaps, than turning to a glossary at the back of a book. Accordingly, we are restricting ourselves here to scientific terms that are either at the center of our account or less prominent yet tied closely to it; and for them we are indicating only how they were generally being used during the period of our study, not giving definitions that accord with present-day textbooks. By contrast, the technical terms pertaining to theory-mediated measurement that we have introduced are not accessible over the internet. Even though we define them in the text when initially introducing them, their novelty can put a burden on readers that we have tried to relieve by including a separate section here on our technical terminology pertaining to measurement. For these too our goal is to help readers with our terminology, not to supply philosophically rigorous definitions.

Scientific Terminology

Avogadro's number In accord with Avogadro's hypothesis, the universal number of units in a mole of any substance, where the mole is defined in terms of the molecular weight of the substance in grams; the latter in the nineteenth century was often called a "gramme-molecule" of the substance. During the period covered in this book, the units in question were taken to be molecules.

Boltzmann's constant In statistical formulations of thermodynamics a universal (that is, fundamental) constant, in units of energy per degree temperature, relating the mean kinetic energy of units forming a substance to its absolute temperature.

Brownian motion The phenomenon of indefinitely continuing, complex motion of particles in a fluid, first announced by Robert Brown in 1828 for grains of pollen in water.

Chemical-molecular theory Here taken to designate the atomic theory of chemical compounds according to which they are formed by specific numbers of atoms of different elements bound to one another to comprise a molecule.

Colloid solution A solution in which small, undissolved particles of one material remain distributed in a liquid in spite of the action of gravity.

Dielectric capacity A characteristic of non-conducting—that is, dielectric materials—pertinent here because it provided an indirect way for determining values for mean free paths in gases; Maxwell's electromagnetic theory linked the dielectric capacity of gases to their refractive index, and an extension of kinetic theory by Stefan linked the latter to mean free path.

Diffusion The movement of a constituent in a fluid from a region of high concentration toward regions of lower concentration.

Elementary charge As a consequence of J. J. Thomson's research from 1897 to 1900, the universal negative charge of what was coming to be known as the electron, or the equal-in-magnitude positive charge.

Equipartition of energy In statistical mechanics, the thesis that the kinetic energy of the molecules forming a gas is statistically distributed equally across all their degrees of freedom of motion, be these translational, rotational, or internal vibrational.

Faraday's constant (or the Faraday constant) As used here, as per Faraday's laws of electrolysis, the quantity of electricity required to yield a mole of hydrogen or any other univalent substance; in the context of chemical-molecular theory, it is the product of the elementary charge and Avogadro's number.

Gamboge The material—specifically, tree resin—of one kind of particles employed in Brownian motion experiments; when rubbed into water, it tends to form spherules.

Gramme-molecule The amount of material whose mass in grams is that of its molecular weight.

Granules The generic term, adopted by Perrin, for the particles engaged in Brownian motion.

Isomerism Two or more compounds with the same chemical formula, but different chemical properties—in the context of chemical-molecular theory, with different spatial arrangements of atoms within respective molecules.

Kinetic theory As used here, the theory that heat takes the form of motion of constituents of a gas, liquid, or solid.

Loschmidt's number In accord with Avogadro's hypothesis, the number of units—here, generally molecules—in a cubic centimeter of an ideal gas at standard pressure and temperature; Avogadro's number is the product of Loschmidt's number and the universal molar volume of an ideal gas, taken at the time to be roughly 22,400 cubic centimeters.

Mastic The material—specifically, resin obtained from a mastic tree—of one kind of particles employed in Brownian motion experiments.

Maxwell-Boltzmann distribution Within statistical mechanics, the statistical distribution of the velocities (in three degrees of freedom) of the particles forming a gas at conditions of thermodynamic equilibrium.

Mean free path Within molecular-kinetic theory, the average distance traveled by a particle in a gas between successive collisions, or interactions, with other particles that change its speed or direction.

Molar gas constant Also known as the universal or ideal gas constant, in units of energy per degree temperature per mole, relating the energy of a quantity of an ideal gas to its absolute temperature; within molecular-kinetic theory, the product of Boltzmann's constant and Avogadro's number.

Mole As defined by Ostwald (1912, p. 41), "the quantity of a gas which has the volume of 22412 cubic centimeters" at standard pressure and temperature; here, more generally, the quantity of any substance the mass of which in grams matches its molecular weight.

Molecular-kinetic theory Here taken to designate kinetic theory in which the constituents in motion are molecules—usually chemical molecules, though with allowance in some places to be Maxwell's "small portion of the substance which moves as one lump during the motion of agitation" (1875, p. 430).

Osmotic pressure Following van't Hoff, the pressure exerted by the solute on a semi-permeable membrane or partition that allows free passage of the solvent.

Rydberg constant Before Bohr's 1913 paper on the atomic structure of hydrogen, the constant in the formula devised by Johannes Rydberg giving the distribution of wavelengths in the spectral series of hydrogen.

Sedimentation equilibrium An equilibrium state in which the tendency of sediment to settle to the bottom of a liquid as a consequence of gravity is offset by its tendency to diffuse from higher to lower concentrations.

Solution A mixture, typically homogenous, of two (or more) substances in which the two do not combine chemically. Excepting "colloid solutions," solutions consist of a solute, like sugar or salt, dissolved in a solvent, like water. An electrolytic solution is one that conducts electricity.

Specific heat The quantity of heat required to increase the temperature of a unit mass of a material—in gas, liquid, or solid form—a given amount. In the case of gases, the specific heat with pressure held constant is distinguished from the specific heat with volume held constant, with the ratio of the former to the latter comprising a characteristic feature called the ratio of the specific heats, $\gamma = C_p/C_V$.

Stokes's law Proposed by Stokes, it gives the value of the drag force on a sphere in the limit as the Reynolds number approaches zero—that is, as the product of the density of the fluid, the velocity of the sphere, and its diameter divided by its viscosity approach zero.

Structural formula A refined form of chemical formula displaying the constituent structure of a molecule—that is, the distinct constituents forming it, especially when these include radicals as well as atoms, which experiment had revealed occurs commonly in organic compounds.

Thermionic emission The discharge of electrically charged particles from a hot filament or surface; in 1899, J. J. Thomson confirmed that the mass-to-charge ratio of these discharges closely matches that of cathode rays and photoelectric discharges.

Ultramicroscope An optical microscope invented in 1902 by Zsigmondy and Siedentopf working at Carl Zeiss AG that uses light scattering from high illumination at an angle to gain enhanced resolution in viewing small objects against a dark background.

Van der Waals's formula The most familiar of many alternative "real gas" modifications of the ideal gas law intended to capture measured deviations from the latter; it introduces two constants to represent the effects of, respectively, the volume taken up by the constituents of a gas and the forces among them—"constants" that generally turned out not to be constant under sufficiently large variations in temperature, thereby showing that it amounts to a curve-fit, not a physically well-constituted lawlike relation.

Theory-Mediated Measurement Terminology

Theory-mediated measurement Sometimes called "indirect measurement," the practice of obtaining values for a less accessible quantity by inferring them from measured values of more accessible quantities under the license of a theoretical relation among them.

Measure The specific equation licensing inference of values of a less accessible quantity from measured values of more accessible quantities. A classic example is the use of the following form of Huygens's equation for a small-arc circular pendulum to measure the acceleration of gravity, *g*, at the surface of the earth:

$$P = 2\pi\sqrt{\frac{l}{g}}$$

where P is the period of the (isochronous) oscillations of the pendulum and l its length. What we are calling the measure in theory-mediated measurement is not to be confused with theoretical presuppositions from which it is derived—this, to allow for the same measure to be derived from different theoretical presuppositions.

Stable Measure A measure that yields, to appropriately high precision, the same value for a constant—or the same functional relationship between a less accessible and a more accessible quantity—as the manipulatable factors entering into it are varied. In the case of the pendulum measure, it proved to be stable when it continued to yield the same value for *g* to at least three significant figures as the length of the pendulum and such further factors as the size, shape, and material of the bob were varied. The stability of a measure generally needs to be specified in terms of a degree of precision with respect to a range of variation of its manipulatable factors.

Complementary measures Two mathematically distinct measures of the same quantity or of the same functional relationship between a less accessible and a more accessible quantity. The requirement is for the equations comprising the two measures to be theoretically, as well as mathematically, independent of one another, leaving open the extent to which the immediate presuppositions from which each is derived differ from one another. Continuing with the Huygens example, his small-arc circular pendulum measure and his constant-height conical pendulum measure of surface gravity were complementary—a case in which both the measures and some of the immediate theoretical presuppositions from which each is derived differ.

Converging measures Two or more complementary measures that yield to appropriately high precision the same value for the less accessible quantity—or the same functional relationship among values of the less accessible and a more accessible quantity. An historical example of four converging measures are Huygens's small-arc circular pendulum measure, his cycloidal pendulum measure, his constant-height conical pendulum measure, and his paraboloidal conical pendulum measure; they all yielded the same value for the strength of local surface gravity to at least three significant figures over a range of variation of their manipulatable factors.

Amenability to increasing precision of a measure The potential of a measure to yield more precise values for its less accessible quantity—or a more precise functional relationship between a less accessible and more accessible quantity—either through introducing corrections to compensate for systematic errors within it and in its physical implementation, or by increasing the precision of the measurement of its more accessible quantities. Two measures that were historically not comparably amenable to increasing precision were Huygens's small-arc circular pendulum and his cycloidal pendulum measure, for he was able to increase the precision of the former by taking the length of the pendulum to be to the center of oscillation of the bob instead of its center of gravity, a correction that he could not correspondingly make in the case of the latter because its center of oscillation varies with the arc subtended by the pendulum.

Target quantity of a measure In theory-mediated measurement, the less accessible quantity, the values of which are to be inferred from those of more accessible quantities.

Proxy A quantity from which values for a less accessible quantity are to be obtained in theory-mediated measurement or, more generally, a quantity taken to stand for a less accessible quantity. The length of a one-second pendulum was historically taken as a proxy for the strength of surface gravity, and the strength of the magnetic field around a conductor was taken as a proxy for the electric current flowing through it (as implemented in Ampère's galvanometer).

Bibliography

Achinstein, P. (2001). *The Book of Evidence*. New York: Oxford University Press.

Aiton, E. J. (1972). *The Vortex Theory of Planetary Motions*. London: MacDonald.

Anderson, J. B., B. M. Boghosian, and C. A. Traynor. (1993). "An Exact Quantum Monte Carlo Calculation of the Helium-Helium Intermolecular Potential." *Journal of Chemical Physics* 99: 345–351.

Arrhenius, S. (1903). "Development of the Theory of Electrolytic Dissociation." Nobel lecture delivered in Stockholm, December 11, 1903. Available at https://www.nobelprize.org/nobel_prizes/chemistry/laureates/1903/arrhenius-lecture.html

Atkins, P., and J. de Paula. (2010). *Atkins' Physical Chemistry*. 9th edition. Oxford: Oxford University Press.

Azzouni, J. (2002). *Knowledge and Reference in Empirical Science*. London; New York: Routledge.

Azzouni, J. (2004). "Theory, Observation, and Scientific Realism." *The British Journal for the Philosophy of Science* 55: 371–392.

Azzouni, J. (2010). *Talking about Nothing*. Oxford; New York: Oxford University Press.

Azzouni, J., and O. Bueno. (2016). "True Nominalism: Referring versus Coding." *The British Journal for the Philosophy of Science* 67: 781–816.

Barkan, D. K. (1999). *Walther Nernst and the Transition to Modern Physical Science*. New York: Cambridge University Press.

Begeman. (1910). "An Experimental Determination of the Charge of an Electron by the Cloud Method." *The Physical Review* 31: 41–54.

Bernath, P. F. (2016). *Spectra of Atoms and Molecules*. 3rd edition. New York: Oxford University Press.

Bigg, C. (2008). "Evident Atoms: Visuality in Jean Perrin's Brownian Motion Research." *Studies in the History and Philosophy of Science* 39: 312–322.

Bigg, C. (2011). "A Visual History of Jean Perrin's Brownian Motion Curves." In L. Daston and E. Lunbek, eds., *Histories of Scientific Observation*. Chicago: University of Chicago Press, 156–179.

Bigg, C. (2014). "Representing the Experimental Atom." In U. Klein and C. Reinhardt, eds., *Objects of Chemical Inquiry*. Sagamore Beach, MA: Science History Publications, 171–201.

Birge, R. T. (1929). "Probable Values of the General Physical Constants (as of January 1, 1929)." *The Physical Review Supplement* 1: 1–73.

Birge, R. T. (1942). "The General Physical Constants: As of August 1941, with Details on the Velocity of Light Only." *Reports on Progress in Physics* 8: 90–134.

Birge, R. T. (1945). "The 1944 Values of Certain Atomic Constants with Particular Reference to the Electronic Charge." *American Journal of Physics* 13: 63–73.

Bohr, N. (1913). "On the Constitution of Atoms and Molecules." *Philosophical Magazine* 26: 1–25.

Bohr, N. (1914). "The Spectra of Helium and Hydrogen." *Nature* 92: 231–232.

Boltwood, B. B., and E. Rutherford. (1909). "The Production of Helium by Radium." *Memoirs and Proceedings of the Manchester Literary and Philosophical Society* 54: 1–2.

Boltwood, B. B., and E. Rutherford. (1911). "The Production of Helium by Radium." *Philosophical Magazine* 22: 586–609.

Boltzmann, L. (1995). *Lectures on Gas Theory*. Translated by S. G. Brush. New York: Dover.

Born, M., and T. von Kármán. (1912). "Uber Schwingungen in Raumgittern." *Physikalische Zeitschrift* 13: 297–309.

Bragg, W. H., and W. L. Bragg. (1913). "The Reflection of X-Rays by Crystals." *Proceedings of the Royal Society of London, Series A* 88: 428–438.

Bragg, W. H., and W. L. Bragg. (1915). *X Rays and Crystal Structure*. London: G. Bell.

Braun, E., L. Hoddeson, J. Teichmann, and S. Weart. (1992). *Out of the Crystal Maze: Chapters from the History of Solid State Physics*. New York: Oxford University Press.

Brewster, D. (1829). "Observations Relative to the Motions of the Molecules of Bodies." *Edinburgh Journal of Science* 10: 215–220.

Brown, R. (1828). "A Brief Account of Microscopical Observations Made in the Months of June, July and August 1827, on the Particles Contained in the Pollen of Plants; and on the General Existence of Active Molecules in Organic and Inorganic Bodies." *Philosophical Magazine* 4: 161–173. Also published 1828 in *Edinburgh New Philosophical Journal* 5: 358–371. Republished 1866 in *The Miscellaneous Botanical Works of Robert Brown*. London: Robert Hardwick, 463–479.

Brush, S. G. (1968). "A History of Random Processes: Brownian Movement from Brown to Perrin." *Archive for History of Exact Sciences* 5: 1–36.

Brush, S. G. (1976a). *The Kind of Motion We Call Heat: A History of the Kinetic Theory of Gases in the 19th Century; Book 1, Physics and the Atomists*. Amsterdam; New York: North-Holland.

Brush, S. G. (1976b). *The Kind of Motion We Call Heat: A History of the Kinetic Theory of Gases in the 19th Century; Book 2, Statistical Physics and Irreversible Processes*. Amsterdam; New York: North-Holland.

Brush, S. G. (1996). *Nebulous Earth: The Origin of the Solar System and the Core of the Earth from Laplace to Jeffreys*. Cambridge, UK: Cambridge University Press.

Brush, S. G., C. W. F. Everitt, and E. Garber, eds. (1986). *Maxwell on Molecules and Gases*. Cambridge, MA: MIT Press.

Bub, J., W. Demopulos, and M. Frappier. (2012). "Poincaré's 'Les conceptions Nouvelles de la matière.'" *Studies in History and Philosophy of Science Part B: Studies in History and Philosophy of Modern Physics* 43: 221–225.

Cahan, D. (1996). "The Zeiss Werke and the Ultramicroscope: The Creation of a Scientific Instrument in Context." In J. Buchwald, ed., *Scientific Credibility and Technical Standards in 19th and Early 20th Century Germany and Britain*. Boston: Kluwer Academic, 67–115.

Cartwright, N. (1983). *How the Laws of Physics Lie*. New York: Oxford University Press.

Cavendish, H., et al. (1777). "The Report of the Committee Appointed by the Royal Society to Consider of the Best Method of Adjusting the Fixed Points of Thermometers; and of the Precautions Necessary to Be Used in Making Experiments with Those Instruments." *Philosophical Transactions of the Royal Society* 67: 816–857.

Chalmers, A. (2009). *The Scientist's Atom and the Philosopher's Stone: How Science Succeeded and Philosophy Failed to Gain Knowledge of Atoms*. Dordrecht: Springer.

Chalmers, A. (2011). "Drawing Philosophical Lessons from Perrin's Experiments on Brownian Motion: A Response to van Fraassen." *British Journal for the Philosophy of Science* 62: 711–732.

Chandrasekhar, S. (1943). "Stochastic Problems in Physics and Astronomy." *Reviews of Modern Physics* 15: 1–89.

Chang, H. (2004). *Inventing Temperature: Measurement and Scientific Progress*. Oxford; New York: Oxford University Press.

Chapman, S., and T. G. Cowling. (1939). *The Mathematical Theory of Non-Uniform Gases*. Cambridge, UK: Cambridge University Press. Republished 1952.

Chaudesaigues, M. (1908). "Le mouvement brownien et la formule d'Einstein." *Comptes Rendus* 147: 1044–1046.

Chung, K. L. (2002). *Green, Brown, and Probability & Brownian Motion on the Line*. Singapore: World Scientific.

Clark, H. (1976). "Atomism versus Thermodynamics." In C. Howson, ed., *Method and Appraisal in the Physical Sciences*. Cambridge, UK: Cambridge University Press, 41–105.

Clausius, R. (2003). "The Nature of the Motion Which We Call Heat." In S. G. Brush and N. S. Hall, eds., *The Kinetic Theory of Gases: An Anthology of Classic Papers with Historical Commentary*. London: Imperial College Press, 111–134.

Cohen, E. R., K. M. Crowe, and J. W. M. Dumond. (1957). *Fundamental Constants of Physics*. New York: Interscience.

Cohen, I. B., and A. Koyré. (1972). *Isaac Newton's Philosophiae Naturalis Principia Mathematica: The Third Edition (1726) with Variant Readings*. With the assistance of A. Whitman. Cambridge, MA; Cambridge, UK: Harvard University Press and Cambridge University Press. For an English translation of the third edition, see Newton (1999).

Cohen, I. B., and R. E. Schofield. (1978). *Isaac Newton's Letters and Papers on Natural Philosophy*. Revised edition. Cambridge, MA: Harvard University Press.

Conybeare, J. H. (1890). Letter addressed to Reverend Dr. Buckland. *Journal of the Royal Microscopical Society* 10: 120.

Darwin, C. G. (1928). "The Wave Equations of the Electron." *Proceedings of the Royal Society A* 118: 654–680.

Daub, E. D. (1971). "Waterston's Influence on Krönig's Kinetic Theory of Gases." *Isis* 62: 512–515.

de Broglie, M., and P. Langevin, eds. (1912). *La théorie du rayonnement et les quanta, rapports et discussions de la réunion tenue à Bruxelles, du 30 octobre au 3 novembre 1911, sous les auspices de M. E. Solvay*. Paris: Gauthier-Villars.

de Haas-Lorentz, G. L. (1913). *Die Brownsche Bewegung und einige verwandte Erscheinungen*. Braunschweig: Friedr. Vieweg & Sohn.

Delsaulx, J. (1877). "Thermo-dynamic Origin of the Brownian Motions." *The Monthly Microscopic Journal* 18: 4.

Deltete, R. J. (2007a). "Wilhelm Ostwald's Energetics 1: Origins and Motivations." *Foundations of Chemistry* 9: 3–57.

Deltete, R. J. (2007b). "Wilhelm Ostwald's Energetics 2: Energetic Theory and Applications, Part I." *Foundations of Chemistry* 9: 265–316.

Deltete, R. J. (2007c). "Wilhelm Ostwald's Energetics 3: Energetic Theory and Applications, Part II." *Foundations of Chemistry* 10: 187–221.

Deltete, R. J. (2010). "Thermodynamics in Wilhelm Ostwald's Physical Chemistry." *Philosophy of Science* 77: 888–899.

Deltete, R. J. (2012). "Planck, Ostwald, and the Second Law of Thermodynamics." *HOPOS: The Journal of the International Society for the History of Philosophy of Science* 2: 1–33.

de Luc, J. (1772). *Recherches sur les modifications de l'atmosphere*. 2 vols. Geneva: n.p.

Dennison, D. (1927). "A Note on the Specific Heat of the Hydrogen Molecule." *Proceedings of the Royal Society A* 115: 483–486.

Dennison, D. (1931). "The Infrared Spectrum of Polyatomic Molecules." *Reviews of Modern Physics* 3: 280–345.

de Regt, H. W. (1996). "Philosophy and the Kinetic Theory of Gases." *The British Journal for the Philosophy of Science* 47: 31–62.

Dewar, J. (1908). "The Rate of Production of Helium from Radium." *Proceedings of the Royal Society of London A* 81: 280–286.

Dewar, J. (1910). "Long-Period Determination of the Rate of Production of Helium from Radium." *Proceedings of the Royal Society of London A* 83: 404–408.

Dirac, P. A. M. (1928a). "The Quantum Theory of the Electron." *Proceedings of the Royal Society A* 117: 610–624.

Dirac, P. A. M. (1928b). "The Quantum Theory of the Electron, Part II." *Proceedings of the Royal Society A* 118: 351–361.

Duclaux, J. (1908). "Pression osmotique et mouvement brownien." *Comptes Rendus* 147: 131–134.

Duhem, P. (1991). *The Aim and Structure of Physical Theory*. Translated by P. P. Wiener. Princeton, NJ: Princeton University Press.

Duhem, P. (2015). *The Electric Theories of J. Clerk Maxwell: A Historical and Critical Study*. Translated by A. Aversa. New York: Springer.

Ehrenfest, P. (1911). "Welche Züge der Lichtquantenhypothese spielen in der Theorie der Wärmestrahlung eine wesentliche Rolle?" *Annalen der Physik* 36: 91–118.

Ehrenfest, P., and T. Ehrenfest. (1959). *The Conceptual Foundations of the Statistical Approach in Mechanics*. Translated by M. J. Moravcsik. Ithaca, NY: Cornell University Press.

Einstein, A. (1901). "Folgerungen aus den Kapillaritätserscheinungen." *Annalen der Physik* 4: 513–523. Reprinted in Einstein (1989a), 10–21; (1989b), 1–11.

Einstein, A. (1902). "Ueber die thermodynamische Theorie die Potentialdifferenz zwischen Metallen und vollständig dissociirten Lösungen ihrer Salze, und über eine elektrische Methode zur Erforschung der Molekularkräfte." *Annalen der Physik* 8: 798–814. Reprinted in Einstein (1989a), 23–40; (1989b), 12–29.

Einstein, A. (1904). "Allgemeinen molekularen Theorie der Wärme." *Annalen der Physik* 14: 361. Reprinted in Einstein (1989a), 99–108; (1989b), 68–77.

Einstein, A. (1905). "Über die von der molekularkinetischen Theorie der Wärme geforderte Bewegung von in ruhenden Flüssigkeiten suspendierten Teilchen." *Annalen der Physik* 17: 549–560. Reprinted in Einstein (1989a), 224–236; (1989b), 123–134; (1956), 1–18.

Einstein, A. (1906a). "Eine neue Bestimmung der Moleküldimensionen." *Annalen der Physik* 19: 289–306. Corrections published 1911, *Annalen der Physik* 34: 591–592. Reprinted in Einstein (1989a), 184–205, 347f; (1989b), 104–122, 191; (1956), 36–62.

Einstein, A. (1906b). "Zur Theorie der Brownschen Bewegung." *Annalen der Physik* 19: 371–381. Reprinted in Einstein (1989a), 334–345; (1989b), 180–190; (1956), 19–35.

Einstein, A. (1907). "Theoretische Bemerkungen über die Brownsche Bewegung." *Zeitschrifft für Elektrochemie* 13: 41–42. Reprinted in Einstein (1989a), 399f; (1989b), 229–231; (1956), 63–67.

Einstein, A. (1908). "Elementare Theorie der Brownschen Bewegung." *Zeitschrift für Elektrochemie* 14: 235–239. Reprinted in Einstein (1989a), 497–502; (1989b), 318–328; (1956), 68–85.

Einstein, A. (1956). *Investigations on the Theory of the Brownian Movement.* Edited by R. Fürth. Translated by A. D. Cowper. New York: Dover.

Einstein, A. (1979). *Autobiographical Notes: A Centennial Edition.* Translated and edited by P. A. Schlipp. La Salle, IL: Open Court.

Einstein, A. (1987). Letter to Mileva Marić. In *The Collected Papers of Albert Einstein.* Vol. 1, *The Early Years, 1879–1902 (English Translation Supplement).* Translated by A. Beck. Princeton, NJ: Princeton University Press, 167.

Einstein, A. (1989a). *The Collected Papers of Albert Einstein.* Vol. 2, *The Swiss Years: Writings, 1900–1909.* Edited by J. Stachel: Princeton, NJ: Princeton University Press.

Einstein, A. (1989b). *The Collected Papers of Albert Einstein.* Vol. 2, *The Swiss Years: Writings, 1900–1909 (English Translation Supplement).* Translated by A. Beck. Princeton, NJ: Princeton University Press.

Einstein, A. (1993). *The Collected Papers of Albert Einstein.* Vol. 3, *The Swiss Years: Writings, 1909–1911 (English Translation Supplement).* Translated by A. Beck. Princeton, NJ: Princeton University Press.

Einstein, A. (1997). "On the Quantum Theory of Radiation." In *The Collected Papers of Albert Einstein.* Vol. 6, *The Berlin Years: Writings, 1914–1917 (English Translation Supplement).* Translated by A. Engel. Princeton, NJ: Princeton University Press, 220–233.

Eucken, A. (1914). "Die Entwicklung der Quantentheorie vom Herbst 1911 bis Sommer 1913." In A. Eucken, ed., *Die Theorie der Strahlung und der Quanten. Verhandlungen auf einer E. Solvay einberufenen Zusammenkunft (30. Oktober bis 3. November 1911), mit Anhang über die Entwicklung der Quantentheorie vom Herbst 1911 bis Sommer 1913.* Halle a s.: Druck und Verlag Wilhelm Knapp, 371–405.

Exner, F. (1900). "Notiz zu Brown's Molekularbewegung." *Annalen der Physik* 2: 843–847.

Fegley, B., and Lodders, K. (1998). *The Planetary Scientist's Companion.* New York: Oxford University Press.

Feynman, R. P., R. B. Leighton, and M. Sands. (1964). *Feynman Lectures on Physics.* Vol. 1. Reading, MA: Addison-Wesley.

Fitzgerald, G. F. (1987). "The Dissociation of Atoms." *The Electrician* 39: 103–104.

Fletcher, H. (1911). "A Verification of the Theory of Brownian Movements and a Direct Determination of the Value of *NE* for Gaseous Ionization." *The Physical Review* 33: 81–110.

Fletcher, H. (1914). "A Determination of Avogadro's Constant *N* from Measurements of the Brownian movements of Small Oil Drops Suspended in Air." *The Physical Review* 4: 440–453.

Freundlich, H. (1926). *Colloid and Capillary Chemistry.* Translated by H. S. Hatfield. New York: E. P. Dutton.

Gardner, M. R. (1979). "Realism and Instrumentalism in 19th Century Atomism." *Philosophy of Science* 46: 1–34.

Giancoli, D. C. (2008). *Physics for Scientists and Engineers*. Vol. 1. Boston: Pearson Education, 477–478.

Gibbs, J. W. (1948). *Elementary Principles in Statistical Mechanics*. In *The Collected Works of J. Willard Gibbs*. Vol. 2. New Haven, CT: Yale University Press.

Glashow, S. L. (1997). "On Being Almost Lorentz Invariant." In A. Zichichi, ed., *Highlights of Subnuclear Physics, 50 Years Later*. Singapore: World Scientific, 276–286.

Glymour, C. (1975). "Relevant Evidence." *Journal of Philosophy* 72: 403–426.

Gouy, L. (1888). "Note sur le mouvement brownien." *Journal de Physique Théorique et Appliquée*. 7: 561–564.

Gouy, L. (1889). "Sur le mouvement brownien." *Comptes Rendus* 109: 102–105.

Gouy, L. (1895). "Le mouvement brownien et les mouvements moléculaires." *Revue Générale des Sciences* 6: 1–7.

Halley, E. (1686). "A Discourse of the Rule of the Decrease of the Height of the Mercury in the Barometer, According as Places Are Elevated above the Surface of the Earth, with an Attempt to Discover the True Reason of the Rising and Falling of the Mercury, upon Change of of Weather" [*sic*]. *Philosophical Transactions of the Royal Society* 16: 104–116.

Harman, G. (1965). "Inference to the Best Explanation." *Philosophical Review* 74: 88–95.

Harper, W. (2012). *Isaac Newton's Scientific Method: Turning Data into Evidence about Gravity and Cosmology*. Oxford: Oxford University Press.

Hecht, C. E. (1990). *Statistical Thermodynamics and Kinetic Theory*. Mineola, NY: Dover.

Heisenberg, W. (1967). "Quantum-Theoretical Re-Interpretation of Kinematic and Mechanical Relations." In B. L. van der Warden, ed., *Sources of Quantum Mechanics*. New York: Dover, 261–276.

Heiskanen, W. A., and F. A. Vening Meinesz. (1958). *The Earth and Its Gravity Field*. New York: McGraw Hill.

Hiebert, E. (1971). "The Energetics Controversy and the New Thermodynamics." In D. H. D. Roller, ed., *Perspectives on the History of Science and Technology*. University of Oklahoma Press, 67–86.

Hiebert, E. (1978). "Nernst, Hermann Walther." *Dictionary of Scientific Biography*, Vol. XV, Supplement I. New York: Scribners, 432–453.

Hiebert, E. (1982). "Developments in Physical Chemistry at the Turn of the Century." In C. G. Bernard, E. Crawford, and P. Sörbom, eds., *Science, Technology and Society in the Time of Alfred Nobel*. Oxford: Pergamon Press for the Nobel Foundation, 97–115.

Hiebert, E., and H.-G. Körber. (1978). "Ostwald, Friedrich Wilhelm." In *Dictionary of Scientific Biography*, Vol. XV, Supplement I. New York: Scribners, 455–469.

Hill, G. W. (1900). "Ptolemy's Problem." *The Astronomical Journal* 21: 33–35.

Hund, F. (1974). *The History of Quantum Theory*. Translated by G. Reece. New York: Harper & Row.

Huygens, C. (1673). *Horologium Oscillatorium, sive De motu Pendularum as Horologia Aptato Demonstrationes Geometricae*. Paris: F. Muguet. Reprinted in *Oeuvres Complètes de Christiaan Huygens*. Vol. 18. The Hague: Publiées par la société hollandaise des sciences, 1934, 86–368. In English, Huygens (1986).

Huygens, C. (1690). *Discours de la Cause de la Pesanteur*. Leiden: Pierre Vander. Reprinted in *Oeuvres Complètes de Christiaan Huygens*. Vol. 21. The Hague: Publiées par la société hollandaise des sciences, 1944, 443–488. Translation by K. Bailey, available from G.E.S.

Huygens, C. (1986). *The Pendulum Clock, or Geometrical Demonstrations Concerning the Motion of Pendula as Applied to Clocks*. Translated by R. J. Blackwell. Ames: Iowa State University Press.

Jeans, J. H. (1901). "The Distribution of Molecular Energy." *Philosophical Transactions of the Royal Society* 196: 397–430.

Jeans, J. H. (1904). *The Dynamical Theory of Gases*. Cambridge, UK: Cambridge University Press.

Jeans, J. H. (1914). *Report on Radiation and the Quantum Theory*. London: "The Electrician."

Jeans, J. H. (1916). *The Dynamical Theory of Gases*. 2nd edition. Cambridge, UK: Cambridge University Press.

Jeans, J. H. (1921). *The Dynamical Theory of Gases*. 3rd edition. Cambridge, UK: Cambridge University Press.

Jeans, J. H. (1954). *The Dynamical Theory of Gases*. 4th edition. New York: Dover. Originally published in 1925.

Kerker, M. (1974). "Brownian Movement and Molecular Reality Prior to 1900." *Journal of Chemical Education* 51: 764–768.

Kerker, M. (1976). "The Svedberg and Molecular Reality." *Isis* 67: 190–216.

King, D. (2018). *Flawed Assumptions and Robust Evidence: On the Role of Theoretical Presuppositions in Grounding Empirical Science*. PhD dissertation, Department of Philosophy, Stanford University.

Kinoshita, T. (1990) (ed.). *Quantum Electrodynamics*. Singapore: World Scientific.

Kinoshita, T. (1996). "The Fine Structure Constant." *Reports on Progress in Physics*, 59: 1459–1492.

Kipnis, A. Y., J. S. Rowlinson, and B. E. Yavelov. (1996). *Van der Waals and Molecular Science*. Oxford: Oxford University Press.

Klein, U. (2003). *Experiments, Models, Paper Tools: Cultures of Organic Chemistry in the Nineteenth Century*. Palo Alto, CA: Stanford University Press.

Koyré, A. (1968). "An Experiment in Measurement." In *Metaphysics and Measurement: Essays in Scientific Revolution*. Cambridge, MA: Harvard University Press, 89–117.

Kragh, H. (2003). "Magic Number: A Partial History of the Fine-Structure Constant." *Archive for History of Exact Sciences* 57: 395–431.

Kragh, H. (2010). *The Early Reception of Bohr's Atomic Theory (1913–1915): A Preliminary Investigation*. RePoSS: Research Publications on Science Studies 9. Aarhus: Center for Science Studies, University of Aarhus. Available at http://www.css.au.dk/reposs.

Kragh, H., and S. J. Weininger. (1996). "Sooner Silence than Confusion: The Tortuous Entry of Entropy into Chemistry." *Historical Studies in the Physical and Biological Sciences* 27: 91–130.

Kuhn, T. (1978). *Black-Body Theory and the Quantum Discontinuity, 1894–1912*. New York: Oxford University Press.

Laidler, K. (1985). "Chemical Kinetics and the Origins of Physical Chemistry." *Archive for History of Exact Sciences* 32: 43–75.

Langevin, P. (1908). "Sur la théorie du mouvement brownien." *Comptes Rendus* 146: 530–533.

Laudan, L. (1981). "A Confutation of Convergent Realism." *Philosophy of Science* 48: 19–49.

Leibniz, G. W. (1689). "Tentamen de Motuum Coelestium Causis." In *Acta Eruditorum*, 1689; English translation in D. B. Meli, *Equivalence and Priority: Newton versus Leibniz*, Oxford: Oxford University Press, 1993, 126–142.

Loeb, L. B. (1934). *The Kinetic Theory of Gases: Being a Text and Reference Book Whose Purpose Is to Combine the Classical Deductions with Recent Experimental Advances in a Convenient Form for Student and Investigator*. New York: McGraw-Hill.

Lorentz, H. A. (1912). "Sur l'application au rayonnement du théoreme de l'équipartition de l'énergie." In M. de Broglie and P. Langevin, eds., *La théorie du rayonnement et les quanta, rapports et discussions de la réunion tenue à Bruxelles du 30 octobre au 3 novembre, 1911, sous les auspices de M. E. Solvay*. Paris: Gauthier Villars, 12–39.

Loschmidt, J. (1865). "On the Size of the Air Molecule." *Proceedings of the Academy of Science of Vienna* 52: 395–413.

MacDonald, D. K. C. (1982). *Noise and Fluctuations: An Introduction*. New York: John Wiley & Sons.

Maddy, P. (2001). "Naturalism: Friends and Foes." *Philosophical Perspectives* 15: 37–67.

Maddy, P. (2007). *Second Philosophy: A Naturalistic Method*. Oxford University Press.

Maiocchi, R. (1990). "The Case of Brownian Motion." *The British Journal for the History of Science* 23: 257–283.

Masters, T. G., and P. M. Shearer. (1995). "Seismic Models of the Earth, Elastic and Anelastic." In T. J. Ahrens, ed., *Global Earth Physics: A Handbook of Physical Constants*. Washington, DC: American Geophysical Union, 88–103.

Maxwell, J. C. (1860). "Illustrations of the Dynamical Theory of Gases—Part I. On the Motions and Collisions of Perfectly Elastic Spheres." *The London, Edinburgh, and Dublin Philosophical Magazine and Journal of Science* 19: 19–32.

Maxwell, J. C. (1865). "A Dynamical Theory of the Electromagnetic Field." *Philosophical Transactions of the Royal Society of London* 155: 459–512.

Maxwell, J. C. (1873). "On Loschmidt's Experiments on Diffusion in Relation to the Kinetic Theory of Gases." *Nature* 8: 298–300.

Maxwell, J. C. (1875). "On the Dynamical Evidence of the Molecular Constitution of Bodies," a lecture delivered at the Chemical Society and then published in *Nature*, 11: 357–359, 374–377; reprinted in Maxwell (1965), 418–438, and in Garber et al. (1986), pp. 216–237; page citations are to Maxwell (1965).

Maxwell, J. C. (1878). "Ether," originally published in the *Encyclopedia Britannica*, 9th edition, reprinted in Maxwell (1965), 763–775.

Maxwell, J. C. (1965). *The Scientific Papers of James Clerk Maxwell*. Edited by W. D. Niven. Vol. 2. New York: Dover.

Maxwell, J. C. (2003). "Illustrations of the Dynamical Theory of Gases." In S. G. Brush and N. S. Hall, eds., *The Kinetic Theory of Gases: An Anthology of Classic Papers with Historical Commentary*. London: Imperial College Press, 148–171.

Mayo, D. (1986). "Cartwright, Causality, and Coincidence." *PSA: Proceedings of the Biennial Meeting of the Philosophy of Science Association*: 42–58.

Mayo, D. (1988). "Brownian Motion and the Appraisal of Theories." In A. Donovan, L. Laudan, and R. Laudan, eds., *Scrutinizing Science: Empirical Studies of Scientific Change*. Hingham, MA: Kluwer, 219–243.

Mayo, D. (1996). *Error and the Growth of Experimental Knowledge*. Chicago: University of Chicago Press.

Mazo, R. M. (2002). *Brownian Motion: Fluctuations, Dynamics and Applications*. Oxford University Press.

Mehra, J. (1975). *The Solvay Conferences on Physics: Aspects of the Development of Physics since 1911*. Dordrecht: D. Reidel.

Meyer, O. E. (1877). *Die kinetische Theorie der Gase*. 1st edition. Breslau: Maruschke & Berendt.

Meyer, O. E. (1899a). *Die kinetische Theorie der Gase: In elementarer Darstellung mit mathematischen Zusätzen*. 2nd revised edition. Breslau: Maruschke & Berendt.

Meyer, O. E. (1899b). *The Kinetic Theory of Gases: Elementary Treatise with Mathematical Appendices*. 2nd revised edition. Translated by R. E. Baynes. New York: Longmans, Green.

Millikan, R. A. (1910). "A New Modification of the Cloud Method of Determining the Elementary Charge and the most Probable Value of that Charge." *Philosophical Magazine* 19: 209–228.

Millikan, R. A. (1911). "The Isolation of an Ion, a Precision Measurement of Its Charge, and the Correction of Stokes's Law." *The Physical Review* 32: 349–397.

Millikan, R. A. (1913). "On the Elementary Electrical Charge and the Avogadro Constant." *The Physical Review* 2: 109–143.

Millikan, R. A. (1916). "A Direct Photoelectric Determination of Planck's '*h*.'" *The Physical Review* 7: 355–388.

Millikan, R. A. (1917). "A New Determination of e, N, and Related Constants." *Philosophical Magazine* 34: 1–30.

Millikan, R. A., and L. Begeman. (1908). "On the Charge Carried by the Negative Ion of an Ionized Gas." *The Physical Review* 26: 197–198.

Miyake, T., and G. E. Smith. (2020). "Realism, Physical Meaningfulness, and Molecular Spectroscopy." In P. Vickers and T. Lyons (eds.), *Contemporary Scientific Realism and the Challenge from the History of Science*. Oxford: Oxford University Press.

Mohr, Peter J., and Barry N. Taylor. (2000). "CODATA Recommended Values of the Fundamental Physical Constants: 1998." *Reviews of Modern Physics* 72: 351–495.

Mohr, P. J., D. B. Newell, and B. N. Taylor. (2008). "CODATA Recommended Values of the Fundamental Physical Constants: 2006." *Reviews of Modern Physics* 80: 633–730.

Mohr, P. J., D. B. Newell, and B. N. Taylor. (2016). "CODATA Recommended Values of the Fundamental Physical Constants: 2014." *Reviews of Modern Physics* 88: 35009-1–35009-73.

Mohr, Peter J., David B. Newell, Barry N. Taylor, and Eite Tiesing. (2018). "Data and Analysis for the CODATA 2017 Special Fundamental Constants Adjustment." *Metrologia* 55: 125–146.

Moseley, H. (1913). "The High Frequency Spectra of the Elements." *Philosophical Magazine* 26: 1024–1034.

Nägeli, K. (1879). "Ueber die Bewegungen kleinster Körperchen." *Sitzungsberichte der mathematisch-physikalischen Classe der K. Bayerischen Akademie der Wissenschaften zu München* 9: 389–453.

Nernst, W. (1893). *Theoretische Chemie vom Standpunkte der Avogadroschen Regel und der Thermodynamik*. 1st edition. Stuttgart: Ferdinand Enke.

Nernst, W. (1895). *Theoretical Chemistry from the Standpoint of Avogadro's Rule and Thermodynamics*. Translated by C. S. Palmer from the 1st German edition. London: Macmillan.

Nernst, W. (1904). *Theoretical Chemistry from the Standpoint of Avogadro's Rule and Thermodynamics*. Translated by R. H. Lehfeldt from the 4th German edition. London: Macmillan.

Nernst, W. (1909). *Theoretische Chemie vom Standpunkte der Avogadroschen Regel und der Thermodynamik*. 6th edition. Stuttgart: Ferdinand Enke.

Nernst, W. (1911). *Theoretical Chemistry from the Standpoint of Avogadro's Rule and Thermodynamics*. Translated by H. T. Tizard from the 6th German edition. London: Macmillan.

Nernst, W. (1912). "Application de la théorie des quanta a divers problèmes physico-chimiques." In M. de Broglie and P. Langevin, eds., *La théorie du rayonnement et les quanta, rapports et discussions de la réunion tenue à Bruxelles, du 30 octubre au 3 novembre 1911, sous les auspices de M. E. Solvay*. Paris: Gauthier-Villars, 254–303.

Nernst, W. (1916). *Theoretical Chemistry from the Standpoint of Avogadro's Rule and Thermodynamics*. Translated by H. T. Tizard from the 7th German edition. London: Macmillan.

Nernst, W. (1921). *Theoretische Chemie vom Standpunkte der Avogadroschen Regel und der Thermodynamik*. 8th–10th edition. Stuttgart: Ferdinand Enke.

Nernst, W. (1923). *Theoretical Chemistry from the Standpoint of Avogadro's Rule and Thermodynamics*. Translated by L. W. Codd from the 8th–10th German edition. London: Macmillan.

Nernst, W. (1969). *The New Heat Theorem*. Translated by G. Barr. New York: Dover. Originally published in 1926.

Neugebauer, O. (1975). *A History of Ancient Mathematical Astronomy*. New York: Springer-Verlag.

Newton, I. (1999). *The Principia: Mathematical Principles of Natural Philosophy*. Translated by I. B. Cohen and A. Whitman, with the assistance of J. Budenz. Berkeley: University of California Press.

Nye, M. J. (1972). *Molecular Reality: A Perspective on the Scientific Work of Jean Perrin*. London: Macdonald; New York: American Elsevier.

Nyhof, J. (1988). "Philosophical Objections to the Kinetic Theory." *The British Journal for the Philosophy of Science* 39: 81–109.

Odhner, C. T. (1901). "Nobel Prize Presentation Speech Delivered in Stockholm, 10 December 1901." Available at https://www.nobelprize.org/nobel_prizes/chemistry/laureates/1901/press.html.

Oreskes, N. (1999). *The Rejection of Continental Drift: Theory and Method in American Earth Science*. New York: Oxford University Press.

Oreskes, N., ed. (2001). *Plate Tectonics: An Insider's History of the Modern Theory of the Earth*. Boulder, CO: Westview Press.

Oseen, C. W. (1926). "Nobel Prize Presentation Speech Delivered in Stockholm, 10 December 1926." Available at https://www.nobelprize.org/nobel_prizes/physics/laureates/1926/press.html.

Ostwald, C. W. W. (1912). *Grundriss der Kolloidchemie*. 3rd edition. Dresden: Steinkopff.

Ostwald, C. W. W. (1915). *A Handbook of Colloid-Chemistry: The Recognition of Colloids, the Theory of Colloids, and Their General Physico-chemical Properties*. Translated by M. H. Fischer. Philadelphia: P. Blackiston's Son.

Ostwald, C. W. W. (1917b). *An Introduction to Theoretical and Applied Colloid Chemistry: The World of Neglected Dimensions*. Translated by M. H. Fischer. New York: John Wiley & Sons.

Ostwald, Wilhelm. (1889). *Grundriss der allgemeinen Chemie*. 1st edition. Leipzig: Wilhelm Engelmann.

Ostwald, Wilhelm. (1890a). *Grundriss der allgemeinen Chemie*. 2nd edition. Leipzig: Wilhelm Engelmann.

Ostwald, Wilhelm. (1890b.) *Outlines of General Chemistry*. 1st edition. Translated by J. T. Walker. London: Macmillan.

Ostwald, Wilhelm. (1891). *Solutions: Being the Fourth Book, With Some Additions, of the Second Edition of Ostwald's "Lehrbuch der Allgemeinen Chemie."* Translated by M. M. P. Muir. London: Longmans, Green.

Ostwald, Wilhelm. (1894). *Manual of Physico-Chemical Measurements*. Translated by J. T. Walker. London: Macmillan.

Ostwald, Wilhelm. (1895). *Outlines of General Chemistry*. 2nd edition. Translated by J. T. Walker. London: Macmillan.

Ostwald, Wilhelm. (1899). *Grundriss der allgemeinen Chemie*. 3rd edition. Leipzig: Wilhelm Engelmann.

Ostwald, Wilhelm. (1904). "Faraday Lecture: Elements and Compounds." *Journal of the Chemical Society, Transactions* 85: 506–522.

Ostwald, Wilhelm. (1907). "The Modern Theory of Energetics." *The Monist* 17: 481–515. Originally published in German in *Rivista di Scienza*, 1907.

Ostwald, Wilhelm. (1907–1909). *The Fundamental Principles of Chemistry: An Introduction to All Text-Books of Chemistry*. Translated by Harry W. Morse. New York: Longmans and Green. This is the 1909 translation of *Prinzipien der Chemie: Eine Einleitung in alle chemischen Lehrbücher*. Leipzig: Akademische Verlagsgesellschaft, 1907.

Ostwald, Wilhelm. (1909). *Grundriss der allgemeinen Chemie*. 4th edition. Leipzig: Akad. Verl.-Ges.

Ostwald, Wilhelm. (1912). *Outlines of General Chemistry*. 3rd edition. Translated by W. W. Taylor. London: Macmillan.

Ostwald, Wilhelm. (1917a). *Grundriss der allgemeinen Chemie*. 5th edition. Dresden; Leipzig: Th. Steinkopff.

Ostwald, Wilhelm. (1980). "The Theory of Electrolytic Dissociation." In *Electrochemistry: History and Theory*. Translated by N. P. Date. New Delhi: Published for the Smithsonian Institution and the National Science Foundation, Washington, DC, by Amerind, 1067–1124.

Pais, A. (1982). *'Subtle is the Lord . . . ': The Science and the Life of Albert Einstein*. New York: Oxford University Press.

Partington, J. R., and W. G. Shilling. (1924). *The Specific Heats of Gases*. New York: Van Nostrand.

Pauling, L. (1970). *General Chemistry*. New York: Dover.

Payne, C. H. (1925). *Stellar Atmospheres: A Contribution to the Observational Study of High Temperature in the Reversing Layers of Stars*. Cambridge, MA: Harvard University Press.

Perrin, J. (1903). *Traité de Chimie Physique. Les Principes*. Paris: Gauthier-Villars.

Perrin, J. (1905). "Mécanisme de l'électrisation de contact et solutions colloïdales." *Journal de chimie physique* 3: 50–110.

Perrin, J. (1908a). "L'agitation moléculaire et le mouvement brownien." *Comptes Rendus* 146: 967–970.

Perrin, J. (1908b). "La loi de Stokes et le mouvement brownien." *Comptes Rendus* 147: 475–476.

Perrin, J. (1908c). "L'origine du mouvement brownien." *Comptes Rendus* 147: 530–532.

Perrin, J. (1908d). "Grandeur des molécules et charge de l'électron." *Comptes Rendus* 147: 594–596.

Perrin, J. (1910). *Brownian Movement and Molecular Reality*. Translated by F. Soddy. London: Taylor and Francis. Originally appeared in 1909, *Annales de Chimie et de Physique* 18: 5–14.

Perrin, J. (1912a). "Les preuves de la réalité moléculaire (Étude spéciale des émulsions)". In M. de Broglie and P. Langevin, eds., *La théorie du rayonnement et les quanta: Rapports et discussions de la réunion tenue à Bruxelles, du 30 octubre au 3 novembre 1911, sous les auspices de M. E. Solvay*. Paris: Gauthier-Villars, 153–253.

Perrin, J. (1912b). "Les preuves de la réalité moléculaire." In *Les idées modernes constitution de la matière, conférences faites en 1912*; proceedings, Paris: Gauthier-Villars, 1913, pp. 1–53.

Perrin, J. (1913). *Les atomes*. Paris: Alcan.

Perrin, J. (1914). *Les atomes*, 5th edition. Paris: Alcan.

Perrin, J. (1916). *Atoms*. Translated by D. L. Hammick. New York: Van Nostrand; translation of the "4th revised edition" of *Les atomes* of 1914.

Perrin, J. (1926). "Discontinuous Structure of Matter." Nobel lecture delivered in Stockholm, December 11, 1926. Available at https://www.nobelprize.org/nobel_prizes/physics/laureates/1926/perrin-lecture.html.

Perrin, J. (1990). *Atoms*. Translated by D. L. Hammick. Revised 2nd English edition. Woodbridge, CT: Ox Bow Press. Originally published in 1923, from the 11th French edition of 1921.

Perrin, J. (2014). *Les atomes*. Paris: CNRS. A centennial reprint of the original 1913 edition, preceded by historical commentary.

Piasecki, J. (2007). "Centenary of Marian Smoluchowski's Theory of Brownian Motion." *Acta Physica Polonica B* 38: 1623–1629.

Planck, M. (1882). "Verdampfen, Schmelzen und Sublimiren." *Annalen der Physik*: 446–475.

Planck, M. (1913). *Vorlesungen über die Theorie der Wärmestrahlung*. 2nd edition revised. Leipzig: Barth; for an English translation, see Planck (1991).

Planck, M. (1922). *The Origin and Development of the Quantum Theory: Being the Nobel Prize Address Delivered before the Royal Swedish Academy of Sciences at Stockholm, 2 June, 1920*. Translated by H. T. Clarke and L. Silberstein. Oxford: Clarendon Press.

Planck, M. (1949). "A Scientific Autobiography." In *Scientific Autobiography and Other Papers*. Translated by F. Gaynor. New York: Philosophical Library.

Planck, M. (1967). "On the Theory of the Energy Distribution Law of the Normal Spectrum." Translated by D. ter Haar. In D. ter Haar, ed., *The Old Quantum Theory*. Oxford: Pergamon Press, 82–90.

Planck, M. (1972). *Zur Theorie des Gesetzes der Energieverteilung in Normalspectrum*. In H. Kangro, ed., *Planck's Original Papers in Quantum Physics*. London: Taylor & Francis, 13.

Planck, M. (1991). *The Theory of Heat Radiation*. 2nd edition. Translated by M. Masius. Reprint. New York: Dover.

Poincaré, H. (1952). "Hypotheses in Physics." In *Science and Hypothesis*. New York: Dover, 140–159.

Poincaré, H. (1963). "The Relations between Matter and Ether." In *Mathematics and Science, Last Essays (1913)*. Translated by J. W. Bolduc. New York: Dover, 89–101.

Post, H. R. (1968). "Atomism 1900." *Physics Education* 3: 225–232 and 307–312.

Poynting, J. H. (1894). *The Mean Density of the Earth*. London: Charles Griffen.

Psillos, S. (2011a). "Making Contact with Molecules: On Perrin and Achinstein." In G. J. Morgan, ed., *Philosophy of Science Matters: The Philosophy of Peter Achinstein*. New York: Oxford University Press, 177–190.

Psillos, S. (2011b). "Moving Molecules above the Scientific Horizon: On Perrin's Case for Realism." *Journal for General Philosophy of Science* 42: 339–363.

Psillos, S. (2014). "The View from Within and the View from Above: Looking at van Fraassen's Perrin." In W. C. Gonzalez, ed., *Bas van Fraassen's Approach to Representation and Models in Science*. Dordrecht: Springer, 143–166.

Ramsay, W. (1882). "On Brownian or Pedetic Motion." *Proceedings of the Bristol Naturalists' Society* 3: 299–302.

Ramsay, W. (1892). "Pedetic Motion in Relation to Colloidal Solutions." *Chemical News and Journal of Industrial Science* 65: 90.

Regener, E. (1909). "On Counting the Alpha Particles by Scintillation and on the Size of the Elementary Electrical Particle," i.e. "Über Zählung der α-Teilchen durch die Szintillation und über die Grösse der elektrischen Elementarquantums." *Sitzungsberichte der Königlich Preussischen Akademie der Wissenschaften zu Berlin* 38: 948–965.

Regnauld, M. J. (1858). "Études relatives au phénomène désigné sous le nom de mouvement Brownien." *Journal de Pharmacie et de Chimie* 34: 141–142.

Renn, J. (1997). "Einstein's Controversy with Drude and the Origin of Statistical Mechanics: A New Glimpse from the 'Love Letters.'" *Archive for History of Exact Sciences* 51: 315–354.

Renn, J. (2005). "Einstein's Invention of Brownian Motion." *Annalen der Physik* 14: 23–37.

Riedi, R. H. (2013). "Brownian Motion." In A. H. El-Shaarawi, W. W. Piegorsch, eds., *Encyclopedia of Environmetrics*. Vol. 1. Wiley Online Library. Available at https://doi.org/10.1002/9780470057339.vnn072.

Rocke, A. J. (1984). *Chemical Atomism in the Nineteenth Century: From Dalton to Cannizzaro*. Columbus: Ohio State University Press.

Rogers, D. W. (2005). *Einstein's Other Theory: The Planck-Bose-Einstein Theory of Heat Capacity*. Princeton, NJ: Princeton University Press.

Royds, T., and E. Rutherford. (1963). "The Nature of the α-Particle." In J. Chadwick, ed., *The Collected Papers of Lord Rutherford of Nelson*. Vol. 2. London: Interscience, 134–135.

Rücker, A. W. (1901). "Address of the President of the British Association for the Advancement of Science." *Science* 14: 425–443.

Rutherford, E. (1911). "The Scattering of α and β Particles by Matter and the Structure of the Atom." *Philosophical Magazine* 21: 669–688.

Rutherford, E. (1963). "The Scattering of α and β Rays and the Structure of the Atom." In J. Chadwick, ed., *The Collected Papers of Lord Rutherford of Nelson*. Vol. 2. London: Interscience, 212.

Rutherford, E., and H. Geiger. (1908a). "An Electrical Method of Counting the Number of α-Particles from Radioactive Substances." *Proceedings of the Royal Society of London A* 81: 141–161.

Rutherford, E., and H. Geiger. (1908b). "The Charge and Nature of the α-Particle." *Proceedings of the Royal Society of London A* 81: 162–173.

Sabine, E. (1825). *An Account of Experiments to Determine the Figure of the Earth: By Means of the Pendulum Vibrating Seconds in Different Latitudes, as Well as on Various Other Subjects of Philosophical Inquiry*. London: John Murray.

Salmon, W. C. (1984). *Scientific Explanation and the Causal Structure of the World*. Princeton, NJ: Princeton University Press.

Schliesser, Eric, and George E. Smith. (1996). "Huygens's 1688 Rapport to the Directors of the Dutch East India Company on the Measurement of Longitude at Sea and the

Evidence It Offered against Universal Gravity," available online at the Centre for Digital Philosophy, Western University, Canada. uwo.ca/philosophy/research/Centre%20 for%20Digital%20Philosophy.html.

Schweber, S. S. (1994). *QED and the Men Who Made It: Dyson, Feynman, Schwinger, and Tomonaga*. Princeton, NJ: Princeton University Press.

Schweber, S. S. (2015). "Introduction." In *Shelter Island II: Proceedings of the 1983 Shelter Island Conference on Quantum Field Theory and the Fundamental Problems of Physics*. Edited by R. Jackiw, N. N. Khuri, S. Weinberg, and E. Witten. Mineola: Dover, vii–lvi.

Seddig, M. (1908). "Über die Messung der Temperaturabhägigkeit der Brownschen Molekularbewegung." *Physikalische Zeitschrift* 9: 465–468.

Seidelmann, P. K., ed. (1992). *Explanatory Supplement to the Astronomical Almanac*. Sausalito, CA: University Science Books.

Servos, J. W. (1990). *Physical Chemistry from Ostwald to Pauling: The Making of a Science in America*. Princeton, NJ: Princeton University Press.

Sidgwick, A. (1884). *Fallacies: A View of Logic from the Practical Side*. New York: D. Appleton.

Siedentopf, H., and R. Zsigmondy. (1903). "Über Sichtbarmachung und Grössenbestimmung ultramikroskopischer Teilchen mit besonderer Anwendung auf Goldrubingläser." *Annalen der Physik* 10: 1–39.

Smith, G. E. (forthcoming). "25 Years of Second Thoughts: Substantive Changes in the 1713 Edition of Newton's *Principia*."

Smith, G. E. (forthcoming). "The Question of Mass in Newton's Law of Gravity."

Smith, G. E. (2001a). "Comments on Ernan McMullin's 'The Impact of Newton's *Principia* on the Philosophy of Science.'" *Philosophy of Science* 38: 327–338.

Smith, G. E. (2001b). "J. J. Thomson and the Electron, 1897–1899." In J. Z. Buchwald and A. Warwick, eds., *Histories of the Electron: The Birth of Microphysics*. Cambridge, MA: MIT Press, 21–76.

Smith, G. E. (2002). "From the Phenomenon of the Ellipse to an Inverse-Square Force: Why Not?" In D. B. Malament, ed., *Reading Natural Philosophy: Essays in the History and Philosophy of Science and Mathematics*. Chicago: Open Court, 31–70.

Smith, G. E. (2007). "Getting Started: Building Theories from Working Hypotheses." Delivered at Stanford University, March 1, 2007. Available at https://web.stanford.edu/dept/cisst/events0506.html.

Smith, G. E. (2010). "Revisiting Accepted Science: The Indispensability of the History of Science." *The Monist* 93: 545–579.

Smith, G. E. (2014). "Closing the Loop: Testing Newtonian Gravity, Then and Now." In Z. Biener and E. Schliesser, eds., *Newton and Empiricism*. New York: Oxford University Press, 262–351.

Smith, G. E. (2016). "The Methodology of the *Principia*." In R. Iliffe and G. E. Smith, eds., *The Cambridge Companion to Newton*. 2nd edition. Cambridge: Cambridge University Press, 187–228; also in 1st edition, eds. I. B. Cohen and G. E. Smith, 2002, 138–173.

Smoluchowski, M. (1906a). "Zur kinetischen Theorie der Brownschen Molekularbewegung und der Suspensionen." *Annalen der Physik* 21: 756–780.

Smoluchowski, M. (1906b). "Essai d'une théorie cinétique du movement brownien et des milieux troubles." *Krakau Anzeiger* 7: 577.

Söderbaum, H. G. (1926). "Nobel Prize Presentation Speech Delivered in Stockholm, 10 December 1926." Available at https://www.nobelprize.org/nobel_prizes/chemistry/laureates/1926/press.html.

Somerville, M. (1840). *On the Connexion of the Physical Sciences*. 5th edition. Reissued in replica by Elibron Classics, 2005.

Sommerfeld, A. (1917). "Zum Andenken an Marian von Smoluchowski." *Physikalische Zeitschrift*. 18: 533.

Sommerfeld, A. (1923). *Atomic Structure and Spectral Lines*. Translated by H. L. Brose. New York: Dutton.

Stein, H. (1989). "Yes, but . . . Some Skeptical Remarks on Realism and Anti-Realism." *Dialectica* 43: 47–65.

Stock, M., R. Davis, E. de Mirandés, and M. J. T. Milton. (2019). "The Revision of the SI: The Result of Three Decades of Progress in Metrology." *Metrologia* 56: 1–14.

Stokes, G. (1880). "On the Theories of the Internal Friction of Fluids in Motion and of the Equilibrium and Motion of Elastic Solids." *Mathematical and Physical Papers*. Vol. 1. Cambridge, UK: Cambridge University Press, 75–129.

Straumann, N. (2011). "On the First Solvay Congress in 1911." Cornell University Library. Available at https://arxiv.org/abs/1109.3785.

Strutt, J. W., Lord Rayleigh. (1900). "The Law of Partition of Kinetic Energy." *Philosophical Magazine* 49: 98–118.

Suppes, P. (1962). "Models of Data." In E. Nagel, P. Suppes, and A. Tarski, eds., *Logic, Methodology, and the Philosophy of Science: Proceedings of the 1960 International Conference*. Palo Alto, CA: Stanford University Press, 252–261.

Sutherland, W. (1899). "Cathode, Lenard, and Röntgen Rays." *Philosophical Magazine* 47: 269–284.

Sutherland, W. (1904). "The Measurement of Large Molecular Masses." *Report of the 10th Meeting of the Australasian Association for the Advancement of Science*. Dunedin: Australasian Association for the Advancement of Science, 117–121.

Sutherland, W. (1905). "A Dynamical Theory of Diffusion for Non-Electrolytes and the Molecular Mass of Albumin." *Philosophical Magazine* 9: 781–785.

Svedberg, T. (1906). "Über die Eigenbewegung der Teilchen in Kolloidalen Lösungen." *Zeitschrift für Elektrochemie* 12: 853–860.

Svedberg, T. (1907). *Studien zur Lehre von den kolloiden Lösungen*. Upsala.

Svedberg, T. (1912). *Die Existenz der Moleküle: Experimentelle Studien*. Leipzig: Akademische Verlagsgesellschaft.

Swerdlow, N. (2004). "The Empirical Foundations of Ptolemy's Planetary Theory." *Journal for the History of Astronomy* 35: 249–271.

Tal, E. (2012). *The Epistemology of Measurement*. PhD dissertation, Department of Philosophy, University of Toronto.

Taylor, B. N., W. H. Parker, and D. N. Langenberg. (1969a). "Determination of *e/h*, Using Macroscopic Quantum Phase Coherence in Superconductors: Implications for Quantum Electrodynamics and the Fundamental Constants." *Reviews of Modern Physics* 41: 375–496.

Taylor, B. N., W. H. Parker, and D. N. Langenberg. (1969b). *The Fundamental Constants and Quantum Electrodynamics*. A Reviews of Modern Physics Monograph. New York: Academic Press.

Thomson, J. J. (1897). "Cathode Rays." *Philosophical Magazine* 44: 293–316.

Thomson, J. J. (1898). "On the Charge of Electricity Carried by the Ions Produced by Röntgen Rays." *Philosophical Magazine* 46: 528–545.

Thomson, J. J. (1899). "On the Masses of the Ions of Gases at Low Pressures." *Philosophical Magazine* 48: 547–567.

Thomson, J. J. (1900). "Indications relatives à la constitution de la matière par les recherches récentes sur le passage de l'électricité à travers les gas." *Congrès International de Physique 3*. Paris: Gauthier-Villars.

Thomson, J. J. (1906). "Carriers of Negative Electricity." Nobel lecture delivered in Stockholm, December 11, 1906. Available at https://www.nobelprize.org/nobel_prizes/physics/laureates/1906/thomson-lecture.html.

Thomson, J. J., and G. P. Thomson. (1928). *Conduction of Electricity through Gases*. Vol. 1. 3rd edition. Cambridge, UK: Cambridge University Press; reissued in 1969 by Dover; page citations to Dover edition.

Thomson, W., Lord Kelvin. (1870). "The Size of Atoms." *Nature* 1: 551–553.

Thomson, W., Lord Kelvin. (1901). "Nineteenth Century Clouds over the Dynamical Theory of Heat and Light." *Philosophical Magazine* 2: 1–40.

Törnebladh, H. R. (1903). "Nobel Prize Presentation Speech Delivered in Stockholm, 10 December 1903." Available at https://www.nobelprize.org/nobel_prizes/chemistry/laureates/1903/press.html.

Todhunter, I. (1962). *A History of the Mathematical Theories of Attraction and the Figure of the Earth*. New York: Dover.

Townsend, J. (1900). "The Diffusion of Ions into Gases." *Philosophical Transactions of the Royal Society A* 193: 129–158.

van der Waals, J. D. (1873). *Over de Continuiteit van den Gas-en Vloeistoftoestand*. Doctoral thesis. Leiden.

van Fraassen, B. C. (2008). *Scientific Representation*. New York: Oxford University Press.

van Fraassen, B. C. (2009). "The Perils of Perrin, in the Hands of the Philosophers." *Philosophical Studies* 143: 5–24.

van Fraassen, B. C. (2012). "Modeling and Measurement: The Criterion of Empirical Grounding." *Philosophy of Science* 79: 773–784.

van't Hoff, J. H. (1901). "Osmotic Pressure and Chemical Equilibrium." Nobel lecture delivered in Stockholm, December 13, 1901. Available at https://www.nobelprize.org/nobel_prizes/chemistry/laureates/1901/hoff-lecture.html.

Warburg, E. (1912). "Vérification expérimentale de la formule de Planck pour le rayonnement du corps noir." In M. de Broglie and P. Langevin, eds., *La théorie du rayonnement et les quanta, rapports et discussions de la réunion tenue à Bruxelles du 30 octobre au 3 novembre, 1911, sous les auspices de M. E. Solvay*. Paris: Gauthier-Villars, 78–86.

Weinberg, S. (2001). "The Non-Revolution of Thomas Kuhn." Reprinted in Weinberg, *Facing Up: Science and Its Cultural Adversaries*. Cambridge, MA: Harvard University Press, 187–209.

Weinberg, S. (2008). *Cosmology*. New York: Oxford University Press.

Weyl, H. (1949). *Philosophy of Mathematics and Natural Science*. Translated by O. Helmer. Princeton, NJ: Princeton University Press.

Whittaker, E. (1973). *A History of the Theories of Aether and Electricity*. Vol. 2. New York: Humanities Press.

Wichmann, E. H. (1967). *Quantum Physics: Berkeley Physics Course*. Vol. 4. New York: McGraw-Hill.

Wigmore, J. H. (1913). *The Principles of Judicial Proof*. Boston: Little, Brown.

Wilson, C. (1980). "Perturbations and Solar Tables from Lacaille to Delambre: The Rapprochement of Observation and Theory." Part I. *Archive for History of Exact Sciences* 22: 53–188.

Wilson, C. (1989). "Predictive Astronomy in the Century after Kepler." In R. Taton and C. Wilson, eds., *Planetary Astronomy from the Renaissance to the Rise of Astrophysics, Part A: Tycho Brahe to Newton*. Cambridge, UK: Cambridge University Press, 161–206.

Wilson, H. A. (1903). "A Determination of the Charge on the Ions produced in Air by Röntgen Rays." *Philosophical Magazine* 5(28): 429–441.

Worrall, J. (1989). "Structural Realism: The Best of Both Worlds?" *Dialectica* 43: 99–124.

Yoder, J. (1988). *Unrolling Time: Christiaan Huygens and the mathematization of nature*. Cambridge, UK: Cambridge University Press.

Zenneck, J. (2007). "Gravitation." In J. Renn, ed., *The Genesis of General Relativity*. Vol. 3, *Gravitation in the Twilight of Classical Physics*. Edited by J. Renn and M. Schemmel. Dordrecht: Springer: 77–112.

Zsigmondy, R. (1905). *Zur Erkenntnis der Kolloide: Über irreversible Hydrosole und Ultramikroskopie*. Jena: Gustav Fischer.

Zsigmondy, R. (1909). *Colloids and the Ultramicroscope*. Translated by J. Alexander. New York: John Wiley & Sons.

Index

Tables and figures are indicated by *t* and *f* following the page number

For the benefit of digital users, indexed terms that span two pages (e.g., 52–53) may, on occasion, appear on only one of those pages.

www.ingramcontent.com/pod-product-compliance
Ingram Content Group UK Ltd.
Pitfield, Milton Keynes, MK11 3LW, UK
UKHW020850150926
475595UK00003B/108

9 780190 098025